New Trends
in Electrical Engineering
Automatic Control, Computing
and Communication Sciences

Edited by

Carlos A. Coello Coello

Alexander Poznyak

José Antonio Moreno Cadenas

Vadim Azhmyakov

LOGOS Verlag, Berlin, Germany

Bibliographic information published by the Deutsche Nationalbibliothek

The Deutsche Nationalbibliothek lists this publication in the Deutsche
Nationalbibliografie; detailed bibliographic data are available
in the Internet at http://dnb.d-nb.de.

ISBN 978-3-8325-2429-6

Logos Verlag Berlin GmbH
Comeniushof, Gubener Str. 47,
10243 Berlin
Tel.: +49 (0)30 42 85 10 90
Fax: +49 (0)30 42 85 10 92
INTERNET: http://www.logos-verlag.de

Editors:

Dr. Carlos A. Coello Coello
CINVESTAV-IPN
Departamento de Computación
Av. IPN No. 2508
Mexico, D.F. 07360
Mexico.
e-mail: ccoellocs.cinvestav.mx

Dr. Alexander Poznyak and Dr. Vadim Azhmyakov
CINVESTAV-IPN
Departamento de Control Automático
Av. IPN No. 2508
Mexico, D.F. 07360
Mexico.
e-mail: apoznyak (vazhmyakov)@ctrl.cinvestav.mx

Dr. José Antonio Moreno Cadenas
CINVESTAV-IPN
Departamento de Ingeniería Eléctrica
Av. IPN No. 2508
Mexico, D.F. 07360
Mexico.
e-mail: jmoreno@cinvestav.mx

Preface

This book contains extended versions of selected papers presented at the *2008 5th International Conference on Electrical Engineering, Computing Science and Automatic Control (CCE'2008)*, which took place in Mexico City, México during 12–14 November, 2008.

This volume contains 28 chapters organized in four parts, each of them corresponding to one of the major topics covered in CCE'2008. Next, we provide a short description of each of these 28 chapters.

Part I contains the following 10 chapters related to automatic control and mechatronics:

In chapter 1, Lu investigates mixed finite element methods for semilinear optimal control problems. The author derives *a priori* error estimates for the coupled state and control approximation, and presents a numerical example that confirms the author's theoretical results.

Robles-Aguirre et al. present in chapter 2, the design of first and second order sliding mode controllers for the synchronous generator speed, Unified Power Flow Controller (UPFC) series power flows, UPFC voltage magnitude and UPFC internal DC link voltage. The authors use combinations of sliding mode and block control techniques to design these controllers, which are implemented in a small power system. Digital simulations are adopted by the authors to illustrate the effectiveness of the proposed approach.

Chapter 3, by Garrido and Miranda, presents a method for DC servomotors working in a closed loop. They use a proportional integral controller to stabilize the servomotor, and another one to close the loop around a linear model of the servomotor. A laboratory prototype is used to validate the proposed method.

In chapter 4, Villafuerte and Mondié propose a new approach for the analysis of the case of neutral and retarded type time delay systems. Some examples and a case study are adopted to illustrate the proposed approach.

Galvan-Guerra and Azhmyakov consider, in chapter 5, the linear quadratic impulsive hybrid optimal control problem and apply to it the corresponding Pontryagin-type maximum principle. The aim of the authors was to investigate the relationship between such a maximum principle and the Bellman dynamic programming approach. They also formulate the necessary optimality conditions for the problem of their interest and derive the associated Riccati-type equation, which allows an implementable numerical algorithm.

Chapter 6, by Garrido et al., presents a task-space controller for robot manipulators, which uses two proportional actions at joint and task levels. Such a topology exploits the fact that, in general, measurements at the task level are noisier than the measurements at the joint level. A planar 2-link revolute joint robot under visual feedback is used by the author to assess the performance of the proposed approach.

Sánchez and Collado propose in chapter 7, a modified version of the standard truncated Carleman linearization, which reduces the error in the truncation process. This modified approach is applied to two Van der Pol oscillators with slightly different frequencies.

Arias-Montiel and Silva-Navarro present in chapter 8 a LQR control scheme for unbalance compensation using an active magnetic bearing in a rotor-bearing system. Simulation and experimental results are included in the chapter to show the transient and steady-state behavior of the proposed closed-loop system.

In chapter 9, Cabrera-Amado and Silva-Navarro address the problem of unbalance compensation in a rotor-bearing system. The authors apply a semiactive balancing control scheme based on two radial MR dampers mounted in one of the supports. Some numerical simulations and experimental results on a physical platform are presented to validate the dynamic and robust performance of the proposed control system.

Finally, in chapter 10, Silva-Navarro et al. present the design of a passive/active autoparametric pendulum-type absorber to control the resonant vibrations in damped Duffing systems. Some simulation results are included in the chapter in order to illustrate the dynamic performance of the proposed system.

Part II contains 3 chapters related to solid-state materials and electron devices:

Chapter 11, by Juárez-Díaz et al., studies the effect of postgrowth thermal annealing processes on the optical characteristics of the zinc oxide (ZnO) films grown on (001) silicon substrates by DC reactive magnetron sputtering. The main motivation for this work was to enhance the PL response and to identify the origin of deep-level luminescence bands. Results indicate that the ZnO films annealing have potential applications in optoelectronic devices.

Albarrán et al. present in chapter 12 a description of the spinodal decomposition of the $GaSb_xN_yAs_{1-x-y}$ quaternary alloys lattice-matched to the GaAs as the result of the internal deformation and coherency strain energies. The authors show ranges of spinodal decomposition of the $GaSb_xN_yAs_{1-x-y}$ alloys up to $y \leq 0.035$ with and without coherency strain energy.

Finally, in chapter 13, Morales-Sánchez et al. investigate the properties of amorphous and fcc polycrystalline $Ge_2Sb_2Te_5$ phases as well as the crystallization kinetics. The authors indicate that the various resistance states reached by annealing at different temperatures, suggest the possibility of using the $Ge_2Sb_2Te_5$ material as the active layer in a multi-state memory device.

Part III contains the following 7 chapters related to biomedical engineering, circuits and communication systems:

Vera et al., present in chapter 14, a study that compares four time-delay estimation methods, which relate temperature changes inside tissue replica (phantom) with time shifts in ultrasound echo-signals. The simulated signals were obtained from a numerical phantom which is proposed by the authors, and experimental signals were acquired from a specially developed agar phantom with uniformly distributed scatterers.

Chapter 15, by Ramírez-Mireles and Almada, presents a method for assessing the statistical nature of the multiple-access interference in wireless sensor networks. The authors propose a low-complexity method to establish "Gaussianity regions" in a two dimensional plane and determine the boundary of such regions for three different conditions: an ideal propagation channel with perfect and imperfect power control, as well as a multipath channel with "perfect average" power control.

In chapter 16, Cortez et al. present a family of hybrid codes, known as *LD STBC-VBLAST* codes, along with a receiver architecture suitable for low-complexity hardware implementation. The authors also present a correlated MIMO channel model and explore the impact of correlation on code performance. They show that *LD STBC-VBLAST* codes exhibit higher performance than other (recently-proposed) hybrid codes.

Lozano et al. present in chapter 17 a numerical integration considering shadowed and illuminated currents in physical optics in order to calculate the radiation pattern of antennas on complex structures modeled by NURBS. This approach is adopted for the case in which the antenna is placed at a distance less than one wavelength of the structure. The results are compared to measurements obtained with other physical optics techniques, such as the stationary phase method or numerical integration considering only illuminated currents.

Medina-Vázquez et al. show in chapter 18, the differences and advantages that the multi-input floating gate MOS (MIFGMOS) transistor has versus the conventional CMOS transistor. In order to illustrate such differences, the authors design and implement both a voltage to current converter cell and a memory current cell using MIFGMOS transistors. The development is based in mathematical and simulation analysis as well as in experimental results.

In chapter 19, Aguirre-Hernandez and Linares-Aranda present an 8×8-bits CMOS pipelined multiplier built using a full-adder cell with a new internal logic structure and a pass-transistor logic style that allow to get reduced delay and power consumption. A test chip containing the multiplier was fabricated using a $0.35 \mu m$ CMOS technology. The authors could confirm, through experimental measurements, its operation at 1.2GHz with a power consumption of 180mW, for a supply voltage of 3.3V.

Finally, chapter 20, by Medina Hernández et al., presents a new model for the production of autowaves which is used for navigation control of a mobile robot. Necessary analytic conditions based on linear theory are established in such a way that a reaction-diffusion system has an oscillating behavior in the initial phase.

Part IV contains the following 8 chapters related to computer science and computer engineering:

Chapter 21, by Cazarez-Castro et al., introduces a type-2 fuzzy logic controller whose intended task is to achieve the output regulation of a servomechanism with backlash. The design of this controller is optimized by a genetic algorithm aiming to obtain the closed-loop system in which the load of the driver is regulated to a desired position.

In chapter 22, Ferretti et al. presents what aims to be a general approach to combine autonomous robot navigation with high-level reasoning. In their proposal, they integrate vision-based motion planning with defeasible decision making (using logic programming) for differential-wheeled robots.

Galán Hernández and León Chávez present, in chapter 23, an object-oriented model of an open source software e-learning platform called Moodle. The model was developed using the Unified Model Language, and includes an analysis of its security services and vulnerabilities.

In chapter 24, Algredo-Badillo et al., provide a study in which they show how the software radio paradigm can be used to implement in hardware several standard security architectures within a single flexible platform.

Chapter 25, by de la Fraga, presents a method to efficiently build a mesh of triangles from a contour map of a terrain. The proposed approach includes intermediate points among contour lines to avoid the problem of flat triangles. The intermediate points are selected from the skeleton of the contour lines, and their heights are calculated automatically.

Barilla and Spann present in chapter 26 an experimental analysis of color-based texture image classification in order to evaluate whether or not the color and texture information should be used jointly or separately. Their results show that, indeed, color and texture information should be treated separately.

In chapter 27, López-García et al. report a software performance comparison of ten blind signature schemes that have been proposed between 1983 until 2008. The chapter provides a brief introduction to the basic concepts of blind digital signatures, and then provides an algorithmic description of the ten schemes studied along with the corresponding main arithmetic building blocks.

Finally, in chapter 28, Hernández León et al. propose a new algorithm for mining frequent itemsets. The proposed approach compresses the data while maintaining the necessary semantics for the frequent itemsets. This algorithm was found to be more efficient than other approaches that use traditional compression algorithms.

We we wish to thank all the authors for their high-quality contributions to this volume.

Finally, we wish to express our appreciation to LOGOS Verlag for their accomplished handling of the manuscript, for their understanding and for their patience.

Carlos A. Coello Coello
Alexander Poznyak
José Antonio Moreno Cadenas
Vadim Azhmyakov
Editors

CINVESTAV-IPN, Mexico City, Mexico, March 2010

Contents

Part I

Automatic Control and Mechatronics

1

Error estimates of mixed finite element methods for semilinear optimal control problems

Zuliang Lu

Hunan Key Laboratory for Computation and Simulation in Science and Engineering, Department of Mathematics, Xiangtan University, Xiangtan 411105, Hunan, P.R.China

Summary. In this chapter, we investigate mixed finite element methods for semilinear optimal control problems. The state and the co-state are discretized by the lowest order Raviart-Thomas mixed finite element spaces and the control is discretized by piecewise constant elements. We derive a priori error estimates for the coupled state and control approximation. Finally, we present a numerical example which confirm our theoretical results.

1.1 Introduction

The finite element methods of optimal control problems has been extensively investigated in many early literature. It is crucial to many aspects of the science and engineering numerical simulation. There are two early papers on the numerical approximation of linear quadratic elliptic optimal control problems by Falk [11] and Geveci [12]. In [1], the authors studied the numerical approximation of distributed optimal control problems governed semilinear partial differential equations with pointwise constraints on the control. The uniform convergence of discretized controls to optimal controls are stated and the error estimates are established. A posteriori error estimators are derived for a class of distributed elliptic optimal control problems in [15]. These error estimators are shown to be useful in adaptive finite element approximation for the optimal control problems and are implemented in the adaptive approach. Systematic introduction of the finite element method for optimal control problems can be found in [16], [17], and [18]. Most of this research has been, however, for the standard finite element methods for optimal control problems.

We have done some primary study on a priori error estimates and superconvergence for linear optimal control problems by mixed finite element methods in [5], [6] and [7]. Some realistic regularity assumptions are presented and applied to error estimation by using an operator interpolation. We derive L^2-superconvergence properties for the flux functions along the Gauss lines and for the scalar functions at the Gauss points via mixed projections. Also, L^∞-priori error estimates for general optimal control problem using mixed finite element methods are considered in [19] and

[25]. In [8] and [20], a posteriori error estimates of mixed finite element methods for general convex optimal control problems were addressed.

The model problem that we shall investigate is the following semilinear elliptic optimal control problems:

$$\min_{u \in K \subset U} \{g_1(p) + g_2(y) + j(u)\} \tag{1.1}$$

subject to the state equation

$$-\text{div}(A\nabla y) + \Phi(y) = f + u, \quad x \in \Omega, \tag{1.2}$$

with the boundary condition

$$y = 0, \quad x \in \partial\Omega, \tag{1.3}$$

which can be written in the form of the first order system

$$\text{div}p + \Phi(y) = f + u, \quad x \in \Omega, \tag{1.4}$$
$$p = -A\nabla y, \quad x \in \Omega, \tag{1.5}$$
$$y = 0, \quad x \in \partial\Omega, \tag{1.6}$$

where $\Omega \subset \mathbb{R}^2$ is a regular bounded and convex open set with the boundary $\partial\Omega$, $f \in L^2(\Omega)$, Ω_U is a bounded open set in $\mathbb{R}^2$ with the Lipschitz boundary $\partial\Omega_U$, g_1, g_2, and j arc convex functionals and K is a closed convex set in $U = L^2(\Omega_U)$.

We adopt the standard notation $W^{m,p}(\Omega)$ for Sobolev spaces on Ω with a norm $\|\cdot\|_{m,p}$ given by $\|v\|_{m,p}^p = \sum_{|\alpha| \leq m} \|D^\alpha v\|_{L^p(\Omega)}^p$, a semi-norm $|\cdot|_{m,p}$ given by $|v|_{m,p}^p = \sum_{|\alpha|=m} \|D^\alpha v\|_{L^p(\Omega)}^p$. We set $W_0^{m,p}(\Omega) = \{v \in W^{m,p}(\Omega) : v|_{\partial\Omega} = 0\}$. For $p=2$, we denote $H^m(\Omega) = W^{m,2}(\Omega)$, $H_0^m(\Omega) = W_0^{m,2}(\Omega)$, and $\|\cdot\|_m = \|\cdot\|_{m,2}$, $\|\cdot\| = \|\cdot\|_{0,2}$. In addition C or c denotes a general positive constant independent of h.

Let us state the assumptions on the operator A and the functional Φ.

(A1) The coefficient matrix $A(x) = (a_{i,j}(x))_{2 \times 2} \in L^\infty(\Omega; \mathbb{R}^{2 \times 2})$ is a symmetric 2×2-matrix and there is a constant $c > 0$ satisfying for any vector $\mathbf{X} \in \mathbb{R}^2$, $\mathbf{X}^t A \mathbf{X} \geq c \|\mathbf{X}\|_{\mathbb{R}^2}^2$.

(A2) Φ is of class C^2 with respect to the variable y, for any $R > 0$ the function $\Phi(\cdot) \in W^{2,\infty}(-R,R)$, $\Phi'(y) \in L^2(\Omega)$ for any $y \in H^1(\Omega)$, and $\Phi'(y) \geq 0$.

Now, we recall a result from Bonnans [3].

Lemma 1. *For every function $G \in L^p(\Omega)$, the solution y of*

$$-\text{div}(A\nabla y) + \Phi(y) = G \quad \text{in } \Omega, \quad y|_{\partial\Omega} = 0, \tag{1.7}$$

belongs to $H_0^1(\Omega) \cap W^{2,p}(\Omega)$. Moreover, there exists a positive constants C such that

$$\|y\|_{W^{2,p}(\Omega)} \leq C \|G\|_{L^p(\Omega)}. \tag{1.8}$$

Due to Lemma 1, the state equation (1.2) admits a unique solution in $H_0^1(\Omega) \cap H^2(\Omega)$.

Next, we introduce the co-state elliptic equation

$$-\text{div}(A(\nabla z + g_1'(p))) + \Phi'(y)z = g_2'(y), \quad x \in \Omega, \tag{1.9}$$

with the boundary condition

$$z = 0, \quad x \in \partial\Omega. \tag{1.10}$$

The existence of a unique solution of (1.9) is justified by Lemma 1. Moreover, we make the following realistic assumption [5] on the regularity of the solution of the optimal control problems (1.1)-(1.3) and the co-state problems (1.9)-(1.10):

$$u \in W^{1,\infty}(\Omega), \quad y, z \in H^{2+s}(\Omega), \quad \text{for} \quad 0 < s \leq 1.$$

The outline of this chapter is as follows. In section 1.2, we construct the mixed finite element discretization for the elliptic optimal control problems (1.1)-(1.3). Then, we briefly state the properties of some interpolation operators in section 1.3. In section 1.4, we derive a priori error estimates for the control and state approximations. A numerical example is presented in section 1.5. Finally, we analyze the conclusion and the future work in section 1.6.

1.2 Mixed finite element discretization

In this section we briefly discuss the mixed finite element discretization of semilinear convex elliptic optimal control problems (1.1)-(1.3). Let

$$V = H(\text{div}; \Omega) = \{v \in (L^2(\Omega))^2, \text{div}\, v \in L^2(\Omega)\}, \quad W = L^2(\Omega).$$

The Hilbert space V is equipped with the following norm:

$$\| v \|_{\text{div}} = (\| v \|_{0,\Omega}^2 + \| \text{div}\, v \|_{0,\Omega}^2)^{1/2}.$$

Then, the weak formulation of the optimal control problems (1.1)-(1.3) is to find $(p, y, u) \in V \times W \times U$ such that

$$\min_{u \in K \subset U} \{g_1(p) + g_2(y) + j(u)\} \tag{1.11}$$

$$(A^{-1}p, v) - (y, \text{div}\, v) = 0, \qquad \forall v \in V, \tag{1.12}$$

$$(\text{div}\, p, w) + (\Phi(y), w) = (f + u, w), \qquad \forall w \in W, \tag{1.13}$$

where the inner product in $L^2(\Omega)$ or $(L^2(\Omega))^2$ is denoted by $(\cdot, \cdot)$. It is well known (see, e.g., [10]) that the optimal control problem (1.11)-(1.13) has a solution (p, y, u), and that a triplet (p, y, u) is the solution of (1.11)-(1.13) if and only if there is a co-state $(q, z) \in V \times W$ such that (p, y, q, z, u) satisfies the following optimality conditions:

$$(A^{-1}p,v) - (y,\text{div}v) = 0, \qquad\qquad \forall v \in V, \qquad\qquad (1.14)$$

$$(\text{div}p,w) + (\Phi(y),w) = (f+u,w), \qquad \forall w \in W, \qquad\qquad (1.15)$$

$$(A^{-1}q,v) - (z,\text{div}v) = -(g_1'(p),v), \qquad \forall v \in V, \qquad\qquad (1.16)$$

$$(\text{div}q,w) + (\Phi'(y)z,w) = (g_2'(y),w), \qquad \forall w \in W, \qquad\qquad (1.17)$$

$$(j'(u)+z,\tilde{u}-u)_U \geq 0, \qquad\qquad \forall \tilde{u} \in K, \qquad\qquad (1.18)$$

where g_1', g_2', and j' are the derivatives of g_1, g_2, and j, and $(\cdot,\cdot)_U$ is the inner product of U. In the rest of the paper, we shall simply write the product as $(\cdot,\cdot)$ whenever no confusion should be caused.

Moreover, we suppose that the following assumptions are satisfied.

(A3) g_1', g_2', and j' are locally Lipschitz continuous, that is

$$|g_1'(p_1) - g_1'(p_2)| \leq C|p_1 - p_2|, \qquad \forall p_1,p_2 \in H(\text{div};\Omega),$$
$$|g_1'(y_1) - g_1'(y_2)| \leq C|y_1 - y_2|, \qquad \forall y_1,y_2 \in L^2(\Omega).$$

(A4) There exists a constant $c > 0$ such that

$$(j'(u_1) - j'(u_2),u_1 - u_2) \geq c \| u_1 - u_2 \|^2, \qquad \forall u_1,u_2 \in K.$$

We are now able to introduce the discretized problem. To this aim, we consider a family of triangulations or rectangulations $\mathbb{T}_h$ of Ω. With each element $T \in \mathbb{T}_h$, we associate two parameters $\rho(T)$ and $\sigma(T)$, where $\rho(T)$ denotes the diameter of the set T and $\sigma(T)$ is the diameter of the largest ball contained in T. The mesh size of the grid is defined by $h = \max_{T \in \mathbb{T}_h} \rho(T)$. We suppose that the regularity assumptions are satisfied.

(A5) There exist two positive constants ρ_1 and ρ_2 such that

$$\frac{\rho(T)}{\sigma(T)} \leq \rho_1, \qquad \frac{h}{\rho(T)} \leq \rho_2$$

hold for all $T \in \mathbb{T}_h$ and all $h > 0$.

(A6) Let us define $\bar{\Omega}_h = \bigcup_{T \in \mathbb{T}_h} T$, and let Ω_h and Γ_h denote its interior and its boundary, respectively. We assume that $\bar{\Omega}_h$ is convex and that the vertices of $\mathbb{T}_h$ placed on the boundary of Γ_h are points of $\partial\Omega$. We also assume that $|\Omega \setminus \Omega_h| \leq Ch^2$.

Similarly, we assume that $\mathbb{T}_h(\Omega_U)$ be triangulations or rectangulations of Ω_U. With each element $T_U \in \mathbb{T}_h(\Omega_U)$, the two parameters $\rho(T_U)$ and $\sigma(T_U)$ are assumed to satisfy the assumptions (A5)-(A6). Next, to every boundary triangle or rectangle T (T_U) of $\mathbb{T}_h$ $(\mathbb{T}_h(\Omega_U))$ we associate another triangle or rectangle $\hat{T}$ $(\hat{T}_U)$ with curved boundary. We denote by $\hat{\mathbb{T}}_h$ $(\hat{\mathbb{T}}_h(\Omega_U))$ the union of these curved boundary triangle with interior triangles of $\mathbb{T}_h$ $(\mathbb{T}_h(\Omega_U))$, such that

$$\bar{\Omega} = \bigcup_{\hat{T} \in \hat{\mathbb{T}}_h} \hat{T}, \qquad \bar{\Omega}_U = \bigcup_{\hat{T}_U \in \hat{\mathbb{T}}_h(\Omega_U)} \hat{T}_U.$$

Let $V_h \times W_h \subset V \times W$ denote the Raviart-Thomas space [24] of the lowest order associated with the triangulations or rectangulations $\mathbb{T}_h$ of Ω. P_k denotes the space of polynomials of total degree at most k, $Q_{m,n}$ indicates the space of polynomials of degree no more than m and n in x and y, respectively. If T is a triangle, $V(T) = \{v \in P_0^2(T) + x \cdot P_0(T)\}$ and if T is a rectangle, $V(T) = \{v \in Q_{1,0}(T) \times Q_{0,1}(T)\}$. We define

$$V_h := \{v_h \in V : \forall T \in \mathbb{T}_h, v_h|_T \in V(T); \ v_h = 0, \ \text{on } \overline{\Omega} \setminus \Omega_h\},$$
$$W_h := \{w_h \in W : \forall T \in \mathbb{T}_h, w_h|_T = \text{constant}; \ w_h = 0, \ \text{on } \overline{\Omega} \setminus \Omega_h\}.$$

Associated with $\hat{\mathbb{T}}_h(\Omega_U)$ is another finite dimensional subspace U_h of U:

$$U_h := \{\tilde{u}_h \in U : \forall \hat{T}_U \in \hat{\mathbb{T}}_h(\Omega_U), \tilde{u}_h|_{\hat{T}_U} = \text{constant}\}.$$

The mixed finite element discretization of (1.11)-(1.13) is as follows: compute $(p_h, y_h, u_h) \in V_h \times W_h \times U_h$ such that

$$\min_{u_h \in K_h \subset U_h} \{g_1(p_h) + g_2(y_h) + j(u_h)\} \tag{1.19}$$

$$(A^{-1}p_h, v_h) - (y_h, \text{div} v_h) = 0, \qquad \forall v_h \in V_h, \tag{1.20}$$
$$(\text{div} p_h, w_h) + (\Phi(y_h), w_h) = (f + u_h, w_h), \qquad \forall w_h \in W_h, \tag{1.21}$$

where $K_h = U_h \cap K$.

The optimal control problem (1.19)-(1.21) again has a solution (p_h, y_h, u_h), and that a triplet (p_h, y_h, u_h) is the solution of (1.19)-(1.21) if and only if there is a co-state $(q_h, z_h) \in V_h \times W_h$ such that $(p_h, y_h, q_h, z_h, u_h)$ satisfies the following optimality conditions:

$$(A^{-1}p_h, v_h) - (y_h, \text{div} v_h) = 0, \qquad \forall v_h \in V_h, \tag{1.22}$$
$$(\text{div} p_h, w_h) + (\Phi(y_h), w_h) = (f + u_h, w_h), \qquad \forall w_h \in W_h, \tag{1.23}$$
$$(A^{-1}q_h, v_h) - (z_h, \text{div} v_h) = -(g_1'(p_h), v_h), \qquad \forall v_h \in V_h, \tag{1.24}$$
$$(\text{div} q_h, w_h) + (\Phi'(y_h)z_h, w_h) = (g_2'(y_h), w_h), \qquad \forall w_h \in W_h, \tag{1.25}$$
$$(j'(u_h) + z_h, \tilde{u}_h - u_h)_U \geq 0, \qquad \forall \tilde{u}_h \in K_h. \tag{1.26}$$

1.3 Some preliminaries

Let $P_h : W \to W_h$ be the orthogonal $L^2(\Omega)$-projection into W_h [2], which satisfies:

$$(P_h w - w, \chi) = 0, \qquad w \in W, \ \chi \in W_h, \tag{1.27}$$
$$\| P_h w - w \|_{0,q} \leq C \| w \|_{t,q} h^t, \qquad 0 \leq t \leq k+1, \text{ if } w \in W \cap W^{t,q}(\Omega), \tag{1.28}$$
$$\| P_h w - w \|_{-r} \leq C \| w \|_t h^{r+t}, \qquad 0 \leq r, \ t \leq k+1, \text{ if } w \in H^t(\Omega). \tag{1.29}$$

Let $\pi_h : V \to V_h$ be the Raviart-Thomas projection [22], which satisfies

$$(\text{div}(\pi_h v - v), w) = 0, \qquad v \in V, \ w \in W_h, \tag{1.30}$$
$$\| \pi_h v - v \|_{0,q} \leq C \| v \|_{t,q} h^t, \ 1/q < t \leq k+1, v \in V \cap W^{t,q}(\Omega)^2, \tag{1.31}$$
$$\| \text{div}(\pi_h v - v) \|_0 \leq C \| \text{div} v \|_t h^t, \ v \in V \cap H^t(\text{div}; \Omega), t \leq k+1. \tag{1.32}$$

We have the commuting diagram property

$$\text{div} \circ \pi_h = P_h \circ \text{div} : V \to W_h \quad \text{and} \quad \text{div}(I - \pi_h)V \perp W_h, \qquad (1.33)$$

where and after, I denote identity matrix.

Furthermore, we also define the standard $L^2(\Omega_U)$-orthogonal projection $Q_h : U \to U_h$, which satisfies: for any $\tilde{u} \in U$

$$(\tilde{u} - Q_h\tilde{u}, \tilde{u}_h)_U = 0, \qquad\qquad \forall \tilde{u}_h \in U_h, \qquad\qquad (1.34)$$

$$\| \tilde{u} - Q_h\tilde{u} \|_{-t,r,U} \leq C|\tilde{u}|_{1,r,U} h_U^{1+t}, \quad t = 0,1 \text{ for } \tilde{u} \in W^{1,r}(\Omega_U). \qquad (1.35)$$

For $\varphi \in W_h$, we shall write

$$\phi(\varphi) - \phi(\rho) = -\tilde{\phi}'(\varphi)(\rho - \varphi) = -\phi'(\rho)(\rho - \varphi) + \tilde{\phi}''(\varphi)(\rho - \varphi)^2, \quad (1.36)$$

where

$$\tilde{\phi}'(\varphi) = \int_0^1 \phi'(\varphi + s(\rho - \varphi)) ds,$$

$$\tilde{\phi}''(\varphi) = \int_0^1 (1 - s)\phi''(\rho + s(\varphi - \rho)) ds$$

are bounded functions in $\overline{\Omega}$ (see, e.g., [13]).

1.4 A priori error estimates

In the rest of the paper, we shall use some intermediate variables. For any control function $\tilde{u} \in K$, we first define the state solution $(p(\tilde{u}), y(\tilde{u}), q(\tilde{u}), z(\tilde{u}))$ associated with $\tilde{u}$ that satisfies

$$(A^{-1}p(\tilde{u}), v) - (y(\tilde{u}), \text{div} v) = 0, \qquad\qquad \forall v \in V, \qquad (1.37)$$

$$(\text{div} p(\tilde{u}), w) + (\Phi(y(\tilde{u})), w) = (f + \tilde{u}, w), \qquad\qquad \forall w \in W, \qquad (1.38)$$

$$(A^{-1}q(\tilde{u}), v) - (z(\tilde{u}), \text{div} v) = -(g_1'(p(\tilde{u})), v), \qquad \forall v \in V, \qquad (1.39)$$

$$(\text{div} q(\tilde{u}), w) + (\Phi'(y(\tilde{u}))z(\tilde{u}), w) = (g_2'(y(\tilde{u})), w), \qquad \forall w \in W. \qquad (1.40)$$

Correspondingly, we define the discrete state solution $(p_h(\tilde{u}), y_h(\tilde{u}), q_h(\tilde{u}), z_h(\tilde{u}))$ associated with $\tilde{u} \in K$ that satisfies

$$(A^{-1}p_h(\tilde{u}), v_h) - (y_h(\tilde{u}), \text{div} v_h) = 0, \qquad\qquad \forall v_h \in V_h, (1.41)$$

$$(\text{div} p_h(\tilde{u}), w_h) + (\Phi(y_h(\tilde{u})), w_h) = (f + \tilde{u}, w_h), \qquad\qquad \forall w_h \in W_h, (1.42)$$

$$(A^{-1}q_h(\tilde{u}), v_h) - (z_h(\tilde{u}), \text{div} v_h) = -(g_1'(p_h(\tilde{u})), v_h), \qquad \forall v_h \in V_h, \quad (1.43)$$

$$(\text{div} q_h(\tilde{u}), w_h) + (\Phi'(y_h(\tilde{u}))z_h(\tilde{u}), w_h) = (g_2'(y_h(\tilde{u})), w_h), \quad \forall w_h \in W_h. \quad (1.44)$$

Thus, as we defined, the exact solution and its approximation can be written in the following way:

$$(p, y, q, z) = (p(u), y(u), q(u), z(u)),$$
$$(\mathbf{p}_h, y_h, q_h, z_h) = (\mathbf{p}_h(u_h), y_h(u_h), q_h(u_h), z_h(u_h)).$$

By Lemma 2.1 in [22], we can obtain the following technical results:

Lemma 2. *Suppose* (**A1**) *hold. Let* $\gamma \in C^1(\Omega)$, $\omega \in V$, $\varphi \in L^2(\Omega)^2$, *and* $\psi \in L^2(\Omega)$. *If* $\tau \in W_h$ *satisfies*

$$(A^{-1}\omega, v_h) - (\tau, \mathrm{div} v_h) = (\varphi, v_h), \qquad \forall v_h \in V_h,$$
$$(\mathrm{div}\omega, w_h) + (\gamma\tau, w_h) = (\psi, w_h), \qquad \forall w_h \in W_h,$$

then, there exists a constant C such that

$$\| \tau \|_0 \le C \left(h \| \omega \|_0 + h \| \mathrm{div}\omega \|_0 + \| \varphi \|_0 + \| \psi \|_0 \right), \tag{1.45}$$

for h sufficiently small.

Set some intermediate errors:

$$\varepsilon_1 := p - p_h(u) \quad \text{and} \quad e_1 := y - y_h(u). \tag{1.46}$$

To analyze the intermediate errors, let us first note the following error equations from (1.14)-(1.15) and (1.41)-(1.42) with the choice $\tilde{u} = u$:

$$(A^{-1}\varepsilon_1, v_h) - (e_1, \mathrm{div} v_h) = 0, \qquad \forall v_h \in V_h, \tag{1.47}$$
$$(\mathrm{div}\varepsilon_1, w_h) + (\tilde{\Phi}'(y)e_1, w_h) = 0, \qquad \forall w_h \in W_h. \tag{1.48}$$

By (1.27)-(1.35) and Lemma 2, we can establish the following error estimates:

Lemma 3. *Suppose that* (**A1**)-(**A6**) *are fulfilled. There is a positive constant C independent of h such that*

$$\| y - y_h(u) \|_0 \le Ch, \tag{1.49}$$
$$\| p - p_h(u) \|_0 \le Ch, \tag{1.50}$$
$$\| \mathrm{div}(p - p_h(u)) \|_0 \le Ch. \tag{1.51}$$

Proof. Let $\tau = P_h y - y_h(u)$ and $\sigma = \pi_h p - p_h(u)$. Rewrite (1.47)-(1.48) in the form

$$(A^{-1}\varepsilon_1, v_h) - (\tau, \mathrm{div} v_h) = 0, \qquad \forall v_h \in V_h, \tag{1.52}$$
$$(\mathrm{div}\varepsilon_1, w_h) + (\tilde{\Phi}'(y)\tau, w_h) = (\tilde{\Phi}'(y)(P_h y - y), w_h), \qquad \forall w_h \in W_h. \tag{1.53}$$

It follows from (1.28) and Lemma 2 that

$$\| \tau \|_0 \le C \left(h \| \varepsilon_1 \|_0 + h \| \mathrm{div}\varepsilon_1 \|_0 + \| \tilde{\Phi}'(y)(P_h y - y) \|_0 \right)$$
$$\le C \left(h \| \varepsilon_1 \|_0 + h \| \mathrm{div}\varepsilon_1 \|_0 + h \| y \|_1 \right). \tag{1.54}$$

Using again (1.28) that

$$\| e_1 \|_0 = \| y - y_h(u) \|_0 = \| P_h y - y \|_0 + \| \tau \|_0$$
$$\leq C\big(h \| \varepsilon_1 \|_0 + h \| \operatorname{div}\varepsilon_1 \|_0 + h \| y \|_1 \big). \tag{1.55}$$

If we now again rewrite (1.47)-(1.48) as

$$(A^{-1}\sigma, v_h) - (\tau, \operatorname{div}v_h) = (A^{-1}(\pi_h p - p), v_h), \qquad \forall v_h \in V_h, \tag{1.56}$$
$$(\operatorname{div}\sigma, w_h) = -(\tilde{\Phi}'(y)e_1, w_h), \qquad \forall w_h \in W_h. \tag{1.57}$$

Using the standard stability results of mixed finite element in [4] and the assumption (**A2**), we can establish the following results:

$$\| \sigma \|_{\operatorname{div}} \leq C(\| \pi_h p - p \|_0 + \| e_1 \|_0)$$
$$\leq C(h \| y \|_2 + \| e_1 \|_0). \tag{1.58}$$

From (1.58), (1.31), and the commuting diagram property (1.33) we now deduce the bounds

$$\| \varepsilon_1 \|_0 \leq C(\| \pi_h p - p \|_0 + \| \sigma \|_0)$$
$$\leq C(h \| y \|_2 + \| e_1 \|_0), \tag{1.59}$$

and

$$\| \operatorname{div}\varepsilon_1 \|_0 \leq C(\| \operatorname{div}\circ\pi_h p - \operatorname{div}p \|_0 + \| \operatorname{div}\sigma \|_0)$$
$$= C(\| P_h \circ \operatorname{div}p - \operatorname{div}p \|_0 + \| \operatorname{div}\sigma \|_0)$$
$$\leq C(h \| y \|_3 + \| e_1 \|_0), \tag{1.60}$$

which, when substituted into (1.55), yields the estimates

$$\| e_1 \|_0 \leq Ch(\| e_1 \|_0 + h \| y \|_3). \tag{1.61}$$

Then (1.61) implies (1.49) holds if h is small enough. Applying (1.61) to (1.59) and (1.60) shows that (1.50) and (1.51) also hold. □

Now, we set some other intermediate errors:

$$\varepsilon_2 := q - q_h(u) \quad \text{and} \quad e_2 := z - z_h(u). \tag{1.62}$$

Let us note the following error equations from (1.24)-(1.25), (1.39)-(1.40), and (1.36):

$$(A^{-1}\varepsilon_2, v_h) - (e_2, \operatorname{div}v_h) = -(g_1'(p) - g_1'(p_h(u)), v_h), \tag{1.63}$$
$$(\operatorname{div}\varepsilon_2, w_h) + (\Phi'(y)e_2, w_h) = (g_2'(y) - g_2'(y_h(u)), w_h) - (y_h(u)\tilde{\Phi}''(y)e_1, w_h), \tag{1.64}$$

for any $v_h \in V_h$ and $w_h \in W_h$.

Applying the assumption (**A3**) and using the argument similar to the proof of Lemma 3, we can also derive the following results:

Lemma 4. *Suppose that* (**A1**)-(**A6**) *are valid. There is a positive constant C independent of h such that*

$$\| z - z_h(u) \|_0 \leq Ch, \tag{1.65}$$

$$\| q - q_h(u) \|_0 \leq Ch, \tag{1.66}$$

$$\| \text{div}(q - q_h(u)) \|_0 \leq Ch. \tag{1.67}$$

With the intermediate errors, we can decompose the errors as following

$$p - p_h = p - p_h(u) + p_h(u) - p_h := \varepsilon_1 + \epsilon_1, \tag{1.68}$$

$$y - y_h = y - y_h(u) + y_h(u) - y_h := e_1 + r_1, \tag{1.69}$$

$$q - q_h = q - q_h(u) + q_h(u) - q_h := \varepsilon_2 + \epsilon_2, \tag{1.70}$$

$$z - z_h = z - z_h(u) + z_h(u) - z_h := e_2 + r_2. \tag{1.71}$$

By using the standard results of mixed finite element methods [4], we have the following results.

Lemma 5. *Suppose that* (**A1**)-(**A6**) *are valid. There is a positive constant C independent of h such that*

$$\| \varepsilon_1 \|_{\text{div}} + \| r_1 \|_0 \leq C \| u - u_h \|_U, \tag{1.72}$$

$$\| \varepsilon_2 \|_{\text{div}} + \| r_2 \|_0 \leq C \| u - u_h \|_U. \tag{1.73}$$

Proof. It follows from (1.22)-(1.25), (1.41)-(1.44), and (1.36), we have the error equations:

$$(A^{-1}\varepsilon_1, v_h) - (r_1, \text{div}v_h) = 0,$$

$$(\text{div}\varepsilon_1, w_h) + (\tilde{\Phi}'(y_h(u))r_1, w_h) = (u - u_h, w_h),$$

$$(A^{-1}\varepsilon_2, v_h) - (r_2, \text{div}v_h) = -(g_1'(p_h(u)) - g_1'(p_h), v_h),$$

$$(\text{div}\varepsilon_2, w_h) + (\Phi'(y_h(u))r_2, w_h) = (g_2'(y_h(u)) - g_2'(y_h), w_h) + (\tilde{\Phi}''(y_h)z_h r_1, w_h),$$

for any $v_h \in V_h$ and $w_h \in W_h$. Set

$$\varpi_1(v_h) = 0,$$

$$\psi_1(w_h) = (u - u_h, w_h),$$

$$\varpi_2(v_h) = -(g_1'(p_h(u)) - g_1'(p_h), v_h),$$

$$\psi_2(w_h) = (g_2'(y_h(u)) - g_2'(y_h), w_h) + (\tilde{\Phi}''(y_h)z_h r_1, w_h).$$

Since the terms $\varpi_1(v_h)$, $\psi_1(w_h)$, $\varpi_2(v_h)$, and $\psi_2(w_h)$ can be regarded as linear functionals of v_h and w_h defined on V_h and W_h, respectively, then we know from the stability results of [4] and the assumption (**A1**) that

$$\| \varepsilon_1 \|_{\text{div}} + \| r_1 \|_0 \leq C \left\{ \sup_{v_h \in V_h} \frac{|\varpi_1(v_h)|}{\| v_h \|_{\text{div}}} + \sup_{w_h \in W_h} \frac{|\psi_1(w_h)|}{\| w_h \|_0} \right\}, \tag{1.74}$$

$$\| \varepsilon_2 \|_{\text{div}} + \| r_2 \|_0 \leq C \left\{ \sup_{v_h \in V_h} \frac{|\varpi_2(v_h)|}{\| v_h \|_{\text{div}}} + \sup_{w_h \in W_h} \frac{|\psi_2(w_h)|}{\| w_h \|_0} \right\}. \tag{1.75}$$

Applying the assumption **(A3)** and the continuity of the linear operator B, we clearly see that

$$|\varpi_1(v_h)| = 0, \tag{1.76}$$

$$|\psi_1(w_h)| \leq C \parallel u - u_h \parallel_U \cdot \parallel w_h \parallel_0, \tag{1.77}$$

$$|\varpi_2(v_h)| \leq C \parallel \varepsilon_1 \parallel_{\text{div}} \cdot \parallel v_h \parallel_{\text{div}}, \tag{1.78}$$

$$|\psi_2(w_h)| \leq C \parallel r_1 \parallel_0 \cdot \parallel w_h \parallel_0. \tag{1.79}$$

It follows from (1.74) and (1.76)-(1.77) that

$$\parallel \varepsilon_1 \parallel_{\text{div}} + \parallel r_1 \parallel_0 \leq C \parallel u - u_h \parallel_U. \tag{1.80}$$

Then, by the estimates (1.75), (1.78)-(1.79), and (1.80) we obtain that

$$\parallel \varepsilon_2 \parallel_{\text{div}} + \parallel r_2 \parallel_0 \leq \parallel \varepsilon_1 \parallel_{\text{div}} + \parallel r_1 \parallel_0 \leq C \parallel u - u_h \parallel_U, \tag{1.81}$$

which completes the proof. □

Let $(p(u), y(u))$ and $(p_h(u), y_h(u))$ be the solutions of (1.12)-(1.13) and (1.41)-(1.42), respectively. Let $J(\cdot) : U \to \mathbb{R}$ be a G-differential convex functional near the solution u which satisfies the following form:

$$J(u) = g_1(p(u)) + g_2(y(u)) + j(u).$$

Then we have a sequence of convex functional $J_h : U \to \mathbb{R}$:

$$J_h(u) = g_1(p_h(u)) + g_2(y_h(u)) + j(u),$$
$$J_h(u_h) = g_1(p_h(u_h)) + g_2(y_h(u_h)) + j(u_h).$$

It can be shown that

$$(J'(u), v) = (j'(u) + z, v),$$
$$(J'_h(u), v) = (j'(u) + z_h(u), v),$$
$$(J'_h(u_h), v) = (j'(u_h) + z_h, v).$$

In the following we estimate $\parallel u - u_h \parallel_U$. A additional assumption is needed. We assume that the cost function J is strictly convex near the solution u, i.e.,

(A7) For the solution u there exists a neighborhood of u in L^2 such that J is convex in the sense that there is a constant $c > 0$ satisfying:

$$(J'(u) - J'(v), u - v) \geq c\|u - v\|_U^2, \tag{1.82}$$

for all v in this neighborhood of u. The convexity of $J(\cdot)$ is closely related to the second order sufficient optimality conditions of optimal control problems, which are assumed in many studies on numerical methods of the problem. For instance, in many references, the authors assume the following second order sufficiently optimality condition (see [3], [1]): there is $c > 0$ such that $J''(u)v^2 \geq c\|v\|_0^2$.

From the assumption (1.82), by the proof contained in [1], there exists a constant $c > 0$ satisfying

$$(J_h'(v) - J_h'(u), v - u) \geq c\|v - u\|_U^2, \quad \forall v \in U_h. \tag{1.83}$$

Now, we are able to derive our main result.

Theorem 1. *Suppose* (**A1**)-(**A7**) *are fulfilled. Let* $(p, y, q, z, u) \in (V \times W)^2 \times U$ *and* $(p_h, y_h, q_h, z_h, u_h) \in (V_h \times W_h)^2 \times U_h$ *be the solutions of (1.14)-(1.18) and (1.22)-(1.26), respectively. We assume that* $j'(u) + z \in H^1(\Omega_U)$. *Then, we have*

$$\| u - u_h \|_U \leq C(h + h_U), \tag{1.84}$$
$$\| p - p_h \|_{\mathrm{div}} + \| y - y_h \|_0 \leq C(h + h_U), \tag{1.85}$$
$$\| q - q_h \|_{\mathrm{div}} + \| z - z_h \|_0 \leq C(h + h_U). \tag{1.86}$$

Proof. We choose $\tilde{u} = u_h$ in (1.18) and $\tilde{u}_h = Q_h u$ in (1.26) to deduce that

$$(j'(u) + z, u_h - u) \geq 0, \tag{1.87}$$

and

$$(j'(u_h) + z_h, Q_h u - u_h) \geq 0. \tag{1.88}$$

Note that $Q_h u - u_h = Q_h u - u + u - u_h$ in (1.88) and add the two inequalities (1.87)-(1.88), we verify that

$$(j'(u_h) + z_h - j'(u) - z, u - u_h) + (j'(u_h) + z_h, Q_h u - u) \geq 0. \tag{1.89}$$

By using (1.83) and (1.89), we obtain

$$
\begin{aligned}
c \| u - u_h \|_U^2 &\leq (J_h'(u), u - u_h) - (J_h'(u_h), u - u_h) \\
&= (z_h(u) + j'(u), u - u_h) - (z_h + j'(u_h), u - u_h) \\
&\leq (z_h(u) - z, u - u_h) + (j'(u_h) - j'(u), Q_h u - u) \\
&\quad + ((z_h - z_h(u)), Q_h u - u) + ((z_h(u) - z), Q_h u - u) \\
&\quad + (j'(u) + z, Q_h u - u).
\end{aligned} \tag{1.90}
$$

Now, we estimate all terms at the right side of (1.90). By using Lemma 4, we deduce that

$$
\begin{aligned}
(z_h(u) - z, u - u_h) &\leq C \| z_h(u) - z \|_0 \cdot \| u - u_h \|_U \\
&\leq Ch \| u - u_h \|_U.
\end{aligned} \tag{1.91}
$$

By the local Lipschitz continuity of j', δ-Cauchy inequality, and the approximation property (1.35) of the projection Q_h, it is clear that

$$(j'(u_h) - j'(u), Q_h u - u) \leq Ch_U^2 + \delta \| u - u_h \|_U^2, \tag{1.92}$$

for any small $\delta > 0$.
From (1.35) and (1.49), we have

$$(z_h - z_h(u), Q_h u - u) \leq Ch_U^2 + \delta \parallel u - u_h \parallel_U^2, \tag{1.93}$$

and

$$(z_h(u) - z, Q_h u - u) \leq C(h^2 + h_U^2). \tag{1.94}$$

Furthermore, using the assumption in Theorem 1 and approximation property (1.35), we obtain

$$(j'(u) + z, Q_h u - u) \leq C \parallel j'(u) + z \parallel_{1,\Omega_U} \cdot \parallel Q_h u - u \parallel_{-1,\Omega_U} \leq Ch_U^2. \tag{1.95}$$

Finally, applying (1.91)-(1.95) to (1.90), we derive the results (1.84). By using (1.68)-(1.71) and Lemmas 3-5, we conclude the result (1.85)-(1.86).

□

1.5 Numerical tests

In this section, we are going to validate the priori error estimates for the error in the control, state, and co-state numerically. The optimal control problems were dealt numerically with codes developed based on AFEPACK. The package is freely available and the details can be found at [14]. The discretization was simplified: the control function u was discretized by piecewise constant functions, whereas the state (y, p) and the co-state (z, q) were discretized by the lowest order Raviart-Thomas mixed finite element functions.

In our numerical examples, we consider the following elliptic optimal control problem:

$$\min_{u \in K \subset U} \left\{ \frac{1}{2} \parallel p - p_0 \parallel^2 + \frac{1}{2} \parallel y - y_0 \parallel^2 + \frac{1}{2} \parallel u - u_0 \parallel^2 \right\} \tag{1.96}$$

$$\mathrm{div}\, p + y^5 = u + f, \qquad p = -A\nabla y, \qquad x \in \Omega, \quad y|_{\partial\Omega} = 0, \tag{1.97}$$

$$\mathrm{div}\, q + 5y^4 z = y - y_0, \qquad q = -A(\nabla z + p - p_0), \qquad x \in \Omega, \quad z|_{\partial\Omega} = 0. \tag{1.98}$$

We choose the domain $\Omega = \Omega_U = [0,1] \times [0,1]$ and $A = I$. We present below an example to illustrate the theoretical results of the elliptic optimal control problem. Let $\mathbf{T}_h$ be regular triangulations or rectangulations of Ω and Ω_U.

The convergence order is computed by the following formula:

$$\mathrm{order} \simeq \frac{log(E_i/E_{i+1})}{log(h_i/h_{i+1})},$$

where i responds to the spatial partition, and E_i denote the norm for the state, costate and control approximation.

Example   In this example, we assume that the data is as follows:

$$u_0 = 1 - \sin\frac{\pi x_1}{2} - \sin\frac{\pi x_2}{2},$$

$$y_0 = y - 2x_2(1-x_2) - 2x_1(1-x_1) - 5y^4 z,$$

$$y = z = x_1 x_2(1-x_1)(1-x_2),$$

$$u = \max(u_0 - z, 0),$$

$$f = 2x_2(1-x_2) + 2x_1(1-x_1) + y^5 - u,$$

$$q = -\left((1-2x_1)x_2(1-x_2), (1-2x_2)x_1(1-x_1)\right),$$

$$p = p_0 = -\left((1-2x_1)x_2(1-x_2), (1-2x_2)x_1(1-x_1)\right).$$

Firstly, the profile of the numerical solution on the 64×64 triangle mesh grids is presented in Fig. 1.1. From the error data on the uniform refined meshes, as listed in Table 1.1, it can be seen that the priori error estimates results remains in our data. Furthermore we also show the convergence orders by slopes in Fig. 1.2, where dofs represents the total number of nodal points in the finite element partition.

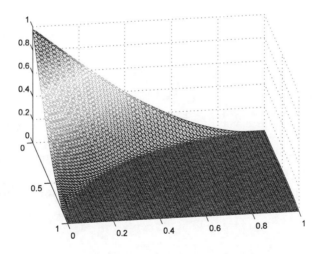

Fig. 1.1. The numerical solution on the 64×64 triangle mesh grids

Secondly, the errors obtained on a sequence of uniformly refined rectangle meshes are presented in Table 1.2. The theoretical results can be observed clearly from the data. The profile of the numerical solution on the 64×64 rectangle mesh

Table 1.1. The numerical errors on a sequence of uniformly triangle mesh grid

Resolution	$\|u-u_h\|$	$\|p-p_h\|$	$\|y-y_h\|$	$\|q-q_h\|$	$\|z-z_h\|$
16×16	1.874e-02	2.407e-02	2.192e-03	2.401e-02	2.196e-03
32×32	9.429e-03	1.204e-02	1.098e-03	1.202e-02	1.098e-03
64×64	4.728e-03	6.018e-03	5.490e-04	6.013e-03	5.490e-04
128×128	2.367e-03	3.008e-03	2.745e-04	3.007e-03	2.745e-04

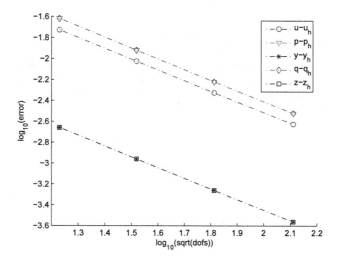

Fig. 1.2. Convergence orders of $u-u_h$, $p-p_h$, $y-y_h$, $q-q_h$, and $z-z_h$ on triangle mesh grids.

grids is plotted in Fig. 1.3. In order to show the convergence order clearly, we display the convergence orders by slopes in Fig. 1.4.

1.6 Conclusion and future works

The present chapter discussed the mixed finite element methods of the semilinear elliptic optimal control problems (1.1)-(1.3). By applying the priori error estimate results (see [13]) of the standard mixed finite element methods, we have established some error estimate results for both the state, the co-state and the control approximation with convergence order $h + h_U$. The priori error estimates for the general

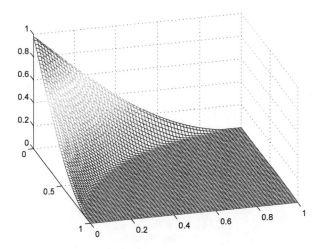

Fig. 1.3. The numerical solution on the 64 × 64 rectangle mesh grids

Table 1.2. The numerical errors on a sequence of uniformly rectangle mesh grid

Resolution	$\|u - u_h\|$	$\|p - p_h\|$	$\|y - y_h\|$	$\|q - q_h\|$	$\|z - z_h\|$
16×16	1.891e-02	2.947e-02	2.683e-03	2.940e-02	2.683e-03
32×32	9.521e-03	1.474e-02	1.344e-03	1.472e-02	1.344e-03
64×64	4.777e-03	7.369e-03	6.723e-04	7.365e-03	6.723e-04
128×128	2.393e-03	3.684e-03	3.362e-04	3.683e-03	3.362e-04

semilinear elliptic optimal control problems by mixed finite element methods seem to be new.

In our future work, we shall use the mixed finite element method to deal with the optimal control problems governed by nonlinear parabolic equations and convex boundary control problems. Furthermore, we shall consider a priori error estimates and superconvergence of optimal control problems governed by nonlinear parabolic equations or convex boundary control problems.

References

1. Arada N., Casas E., Tröltzsch F.: Error estimates for the numerical approximation of a semilinear elliptic control problem. Comp. Optim. appl. 23, 201–229 (2002)
2. Babuska I., Strouboulis T.: The finite element method and its reliability. Oxford University press, Oxford, (2001)

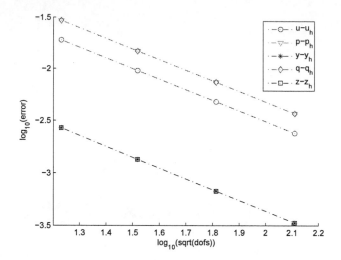

Fig. 1.4. Convergence orders of $u - u_h$, $p - p_h$, $y - y_h$, $q - q_h$, and $z - z_h$ on rectangle mesh grids.

3. Bonnans J. F.: Second-order analysis for control constrained optimal control problems of semilinear elliptic systems. Appl. Math. Optim. 38, 303–325 (1998)
4. Brezzi F., Fortin M.: Mixed and hybrid finite element methods. Springer, Berlin, (1991)
5. Chen Y.: Superconvergence of mixed finite element methods for optimal control problems. Math. Comp.. 77, 1269–1291 (2008)
6. Chen Y.: Superconvergence of quadratic optimal control problems by triangular mixed finite elements. Internat. J. Numer. Methods in Engineering. 75, 881–898 (2008)
7. Chen Y., Dai L., Lu Z.: Superconvergence of rectangular mixed finite element methods for constrained optimal control problem. Adv. Appl. Math. Mech. accepted
8. Chen Y., Liu W. B.: A posteriori error estimates for mixed finite element solutions of convex optimal control problems. J. Comp. Appl. Math. 211, 76–89 (2008)
9. Chen Y., Liu W. B.: Error estimates and superconvergence of mixed finite elements for quadratic optimal control. Internat. J. Numer. Anal. Modeling 3, 311–321 (2006)
10. Chen Y., Liu W. B.: Posteriori error estimates for mixed finite elements of a quadratic optimal control problem. Recent Progress comput. Appl. PDEs 2, 123–134 (2002)
11. Falk F. S.: Approximation of a class of optimal control problems with order of convergence estimates. J. Math. Anal. Appl. 44, 28–47 (1973)
12. Geveci T.: In the approximation of the solution of an optimal control problem governed by an elliptic equation. R.A.I.R.O. Numer. Anal. 13, 313–328 (1979)
13. Kwon Y., Milner F. A.: L^∞-error estimates for mixed methods for semilinear second-order elliptic equations. SIAM J. Numer. Anal. 25, 46–53 (1988)
14. Li R., Liu W. B.: http://circus.math.pku.edu.cn/AFEPack.
15. Li R., Liu W. B., Ma H. P., Tang T.: Adaptive finite element approximation for distributed convex optimal control problems. SIAM J. Control Optim. 41, 1321–1349 (2002)
16. Lions J. L.: Optimal control of systems governed by partial differential equations. Springer, Berlin, (1971)

17. Liu W. B., N. N. Yan: A posteriori error estimates for control problems governed by nonlinear elliptic equations. Appl. Numer. Math. 47, 173–187 (2003)
18. Liu W. B., Yan N. N.: A posteriori error estimates for control problems governed by Stokes equations. SIAM J. Numer. Anal. 40, 1850–1869 (2002)
19. Lu Z., Chen Y.: L^∞-error estimates of triangular mixed finite element methods for optimal control problem govern by semilinear elliptic equation. Numer. Anal. Appl. 12, 74–86 (2009)
20. Lu Z., Chen Y.: A posteriori error estimates of triangular mixed finite element methods for semilinear optimal control problems. Adv. Appl. Math. Mech. 1, 242–256 (2009)
21. Lu Z., Chen Y., Zhang H.: A priori error estimates of mixed finite element methods for nonlinear quadratic optimal control problems. Lobachevskii J. Math. 29, 164–174 (2008)
22. Miliner F. A.: Mixed finite element methods for quasilinear second-order elliptic problems. Math. Comp. 44, 303–320 (1985)
23. Mossino J.: An application of duality to distributed optimal control problems with constraints on the control and the state. J. Math. Anal. Appl. 50, 223–242 (1975)
24. Raviart P. A., Thomas J. M.: A mixed finite element method for 2nd order elliptic problems, Math. Aspects of the Finite Element Method. Lecture Notes in Math. 606, 292–315 (1977)
25. Xing X., Chen Y.: L^∞-error estimates for general optimal control problem by mixed finite element methods. Internat. J. Numer. Anal. Modeling 5, 441-456 (2008)

Sliding Mode Control Approach combined with Block Control for a Unified Power Flow Controller (UPFC) in a Small Power System*

Fidel Robles-Aguirre[1], Alexander Loukianov[1], Leonid Fridman[2], J.M. Cañedo[1]

[1]Cinvestav, Unidad Jalisco, México
[2]UNAM, Mexico City, México
fidelarturorobles@hotmail.com, louk@gdl.cinvestav.mx

Summary. In this work, 1st and 2nd order Sliding Mode (SM) controllers for the synchronous generator speed, UPFC series power flows, UPFC voltage magnitude and UPFC internal DC link voltage are designed. In order to design these controllers, combinations of Sliding Mode and Block Control (BC) techniques are constructed. These controllers are implemented in a small power system. The block control approach is used to design linear and nonlinear sliding manifolds with desired motions. 1st and 2nd order sliding mode control laws are applied to reject perturbations and avoid the so-called *chattering* effect. The effectiveness of the proposed algorithm is illustrated by means of digital simulations.

Keywords: Sliding Modes, Block Control, UPFC, FACTS.

2.1 Introduction

One major problem has been faced by modern power systems: the problem of loss of synchronism or angle instability. Therefore, two fundamental issues in the control design for power systems are: robust stabilization of generators speed and voltage magnitudes, and the need that the power system achieves a good level of transient performance. The operation limits and performance of a power network can be improved throughout the control of Flexible AC Transmission Systems (FACTS) technology. This family of devices has the ability to control different variables that govern the operation of transmission systems, such as power flows, phase angles and voltage magnitudes. In addition, these devices allow the introduction of a damping effect in the power system's oscillations. The FACTS technologies brings new possibilities of energy control in power networks at reasonable costs and allow the utilization of existing power facilities with high efficiency and maximum capacity [1]. The control schemes for power systems with FACTS are commonly based on linearized plant models and classical linear control algorithms. In [2], attention has been focused on

---

*This work was supported by Conacyt, Mexico, under grant 165253.

creating an energy function that describes the UPFC in order to include it on a larger function that preserves the structure of an electric network, although this approach is useful only for design purposes. In [3], a fuzzy control with a self adjustable factor is analyzed, but the model used for the synchronous generator is both reduced and linear. In [4, 5, 6] a double closed loop PI control, for the currents and voltages, in the Unified Power Flow Controller (UPFC) is discussed. In [7] a neural-network based adaptive controls, whose inputs are proportional to the error signals of each single neuron, is proposed to improve transient stability. In [8], applications of evolutionary optimization techniques for optimal locations and parameters for multiple UPFC devices are discussed. The application of these controllers has been done with reduced or linearized plant models.

2.2 Power System Description

In this work it's considered a small power network, which includes a synchronous generator with stator dynamics, two long transmission lines (with their capacitive shunt admittances), an UPFC - FACTS device, one short transmission line connected to the series branch of the UPFC, and a constant impedance load (Fig. 2.1). All the power system components are modeled in d-q reference frame. All system variables are expressed in pu units, excepts generator speed (rad/sec), machine angle (radians) and time (secs).

Based on the plant model, we shall resort sliding mode control technique [9, 10, 14] combined with block control [11, 12] to be able to control the generator speed, series power flows through the UPFC, voltage magnitude at the FACTS connection node, and the internal UPFC DC link voltage. The first control objective is generator's speed. The block control technique is applied to design a nonlinear sliding manifold, in such a way that the closed-loop generator's mechanical dynamics are represented by a linear system with desired eigenvalues. For a given bound on the control signal, sliding mode control laws are implemented to ensure stability. The second control objective is the magnitude voltage regulation at node 2 (where the UPFC shunt branch and local load are connected) close to a desired value. The voltage magnitude control in power systems is a non-linear task. The third control objective is to regulate the active and reactive power flows through the series branch of the UPFC towards the local load. This objective can be achieved through the independent current control in $d - q$ reference frame of the UPFC series branch. The power flow regulation is also a non-linear task. The fourth control objective is to regulate the internal DC link voltage of the UPFC near to a desired reference value. The second and fourth control objectives can be achieved through the independent currents control of the UPFC shunt branch (Fig. 2.1).

Based on block control technique, linear or non-linear manifolds are designed for each control objective. 1st and 2nd order sliding mode control laws are implemented to ensure closed-loop stability and tracking for the desired references. It is important that each proposed controllers be computationally low demanding and, furthermore, also takes into account bounded control inputs. The proposed control approaches enable the compensation of the inherent nonlinearities of the power system and the rejection of external disturbances. The simulation results reveal that the proposed combined control strategies provide good dynamic performance.

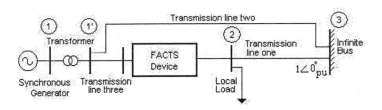

Fig. 2.1. Small power system for transient phenomena studies with UPFC (FACTS) device.

2.3 Dynamic Models of Power System Components

2.3.1 Synchronous Generator Dynamic Model

The synchronous generator model used in this work is the 8th order synchronous machine dynamic model, with the mechanical, electrical, rotor and stator dynamics. The model is presented in the $d-q$ reference frame obtained from Park's transformation [12]:

$$\dot{x}_1 = x_2 - \omega_b$$
$$\dot{x}_2 = a_1 \left[T_m - a_2 z_1 z_2 - a_3 z_2 x_3 - a_4 z_2 x_4 + a_5 z_1 x_5 + a_6 x_6 z_1 \right]$$
$$\dot{x}_3 = -b_1 \left[b_2 x_3 + b_3 z_1 + b_4 x_4 \right] + b_5 E_{fd}$$
$$\dot{x}_4 = -c_1 \left[x_4 + c_2 z_1 - c_3 x_3 \right]$$
$$\dot{x}_5 = -d_1 \left[d_2 z_2 + d_3 x_5 - d_4 x_6 \right]$$
$$\dot{x}_6 = -e_1 \left[x_6 + e_2 z_2 - e_3 x_5 \right]$$
$$\dot{z}_1 = h_1 x_4 + h_2 x_2 x_6 + h_3 x_2 x_5 + h_4 x_3 + h_5 z_1 + h_6 x_2 z_2 + h_7 v_{1d} + h_8 E_{fd}$$
$$\dot{z}_2 = k_1 x_2 x_4 + k_2 x_6 + k_3 x_5 + k_4 x_2 x_3 + k_5 x_2 z_1 + k_6 z_2 + k_7 v_{1q} \qquad (2.1)$$

where $x = \begin{bmatrix} x_1 & x_2 & x_3 & x_4 & x_5 & x_6 \end{bmatrix}^T = \begin{bmatrix} \delta & \omega_r & \psi_{fd} & \psi_{kd} & \psi_g & \psi_{kq} \end{bmatrix}^T$ is the synchronous generator state vector. x_1 represents the load angle of synchronous generator, expressed in radians. x_2 represents the generator rotor speed expressed in radians per second. x_3 represents the magnetic flux linkages of the rotor field winding in d axis, expressed in pu units. x_4 represents the magnetic flux linkages of the damping winding in rotor's d axis, in pu units. x_5, x_6 represents the magnetic flux linkages of the damping windings in rotor's q axis ψ_{kd}, ψ_{kq}, in pu units. $z = \begin{bmatrix} z_1 & z_2 \end{bmatrix}^T = \begin{bmatrix} i_{dg} & i_{qg} \end{bmatrix}^T$ are the stator currents in d and q axis, respectively, in pu units. ω_b is the synchronous speed in $rads/sec$, E_{fd} is the excitation input voltage in pu unit, T_m is the mechanical input torque in pu units, which in this work is assumed to be constant, thus $\dot{T}_m = 0$. The parameters $\{a_j, b_j, c_j, d_j, e_j, h_j\}$, and k_j, are constants. v_{1d} and v_{1q} are the generator's terminal voltage in the $d - q$ reference frame respectively, expressed in pu units.

2.3.2 UPFC's Dynamic Model

The dynamic model of the FACTS device (UPFC) after Park's transformation [1], in $d - q$ reference frame is described by

$$
\begin{aligned}
\dot{\eta}_1 &= -q_1\eta_1 + m_1\eta_2 + q_2\left[v_1 - v_3 - v_{serd}\right] \\
\dot{\eta}_2 &= -m_1\eta_1 - q_1\eta_2 + q_2\left[v_2 - v_4 - v_{serq}\right] \\
\dot{\eta}_3 &= -q_3\eta_3 + m_1\eta_4 + q_4\left[v_3 - v_{shd}\right] \\
\dot{\eta}_4 &= -m_1\eta_3 - q_3\eta_4 + q_4\left[v_4 - v_{shq}\right] \\
\dot{\eta}_5 &= \frac{q_5}{\eta_5}\left[v_{serd}\eta_1 + v_{serq}\eta_2 - v_{shd}\eta_3 - v_{shq}\eta_4\right]
\end{aligned}
\tag{2.2}
$$

where $\eta = \begin{bmatrix} \eta_1 & \eta_2 & \eta_3 & \eta_4 & \eta_5 \end{bmatrix}^T = \begin{bmatrix} i_{serd} & i_{serq} & i_{shd} & i_{shq} & v_{cd} \end{bmatrix}^T$ is the state vector of the UPFC. η_1 and η_2 are the currents flowing through the series branch of the UPFC from node 1 to node 2, in $d - q$ axis respectively. Expressed in pu units. η_3 and η_4 are the currents flowing through the shunt branch of the UPFC from node 0 to node 2, in $d - q$ axis respectively. Expressed in pu units also. η_5 is the internal DC link voltage of the UPFC, in pu units. q_j and m_j are positive constants (see Appendix). v_{serd} and v_{serq} are UPFC's series Variable Source Converter (VSC) voltages, in $d - q$ axis respectively, which represents the series control inputs, in pu units. v_{shd} and v_{shq} are UPFC's shunt Variable Source Converter (VSC) voltages, in $d - q$ axis respectively, which represents the shunt control inputs, in pu units. $v_1 = v_{1d}$, $v_2 = v_{1q}$, are the voltages at node 1 in $d - q$ reference frame respectively, in pu units. v_3 and v_4 are the voltages at node 2 in $d - q$ reference frame respectively, in pu units.

2.3.3 Large Transmission line's dynamic model

The three phase balanced dynamic models of the series elements of the two aerial transmission lines [13] in $d - q$ reference frame are represented by

$$
\begin{aligned}
\dot{\beta}_1 &= -\gamma_1\beta_1 + m_1\beta_2 + \gamma_2\left[v_3 - v_{infd}\right] \\
\dot{\beta}_2 &= -m_1\beta_1 - \gamma_1\beta_2 + \gamma_2\left[v_4 - v_{infq}\right] \\
\dot{\beta}_3 &= -\gamma_3\beta_3 + m_1\beta_4 + \gamma_4\left[v_1 - v_{infd}\right] \\
\dot{\beta}_4 &= -m_1\beta_3 - \gamma_3\beta_4 + \gamma_4\left[v_2 - v_{infq}\right]
\end{aligned}
\tag{2.3}
$$

where β_1 and β_2, are the $d - q$ current components flowing through transmission line 1, from node 2 to node 3, respectively. β_3 and β_4 are the $d - q$ current components flowing through transmission line 2 from node 1 to node 3, respectively.

Expressed in pu units. $v_{\mathrm{inf}d}$ and $v_{\mathrm{inf}q}$ are the components of node 3 voltage in $d-q$ reference frame, also in pu units. From the equivalent shunt capacitive admittances of the transmission lines at nodes 1 and 2, the dynamic behavior of the voltages at these nodes can be represented as

$$\dot{v}_1 = m_1 v_2 + m_2 [z_1 - \eta_1 - \beta_3]$$
$$\dot{v}_2 = -m_1 v_1 + m_2 [z_2 - \eta_2 - \beta_4], \tag{2.4}$$

$$\dot{v}_3 = m_1 v_4 + m_3 [\eta_1 - \eta_3 - \alpha_1 - \beta_1]$$
$$\dot{v}_4 = -m_1 v_3 + m_3 [\eta_2 - \eta_4 - \alpha_2 - \beta_2] \tag{2.5}$$

where v_1, and v_2, are the voltage components at node 1 in $d-q$ reference frame, respectively. v_3, and v_4, are the voltage components at node 2 in $d-q$ reference frame, respectively. Expressed in pu units. m_j, m_2, m_3, γ_1, γ_2, γ_3, and γ_4 are positive constants obtained from resistive, capacitive and inductive parameters of the transmission lines (see Appendix).

2.3.4 Local Load Dynamic Model

The dynamic model of a three phase balanced RL load in $d-q$ reference frame [13] can be represented by

$$\dot{\alpha}_1 = -p_1 \alpha_1 + m_1 \alpha_2 + p_2 v_3$$
$$\dot{\alpha}_2 = -m_1 \alpha_1 - p_1 \alpha_2 + p_2 v_4 \tag{2.6}$$

where α_1 and α_2 are current components in $d-q$ axis flowing from node 2 to node 0, Expressed in pu units. p_j are positive constants obtained from the resistive and inductive parameters (see Appendix).

2.4 Controller's Design

The main focus of this work is to study the interaction between the different Single Input Single Output (SISO) sliding mode and block control based controllers for the synchronous generator, and UPFC (FACTS) device. The five SISO controllers designed in this work operate together and at the same time on the small power system.

2.4.1 Nonlinear Block Integral Sliding Mode Control for the Synchronous Generator's Rotor Speed

The purpose of this controller is to provide large generator's speed stability during a transient phenomena scenario. The synchronous generator mechanical dynamics (2.1) are presented in the nonlinear controllable block form

$$\dot{x}_1 = x_2 - \omega_b$$
$$\dot{x}_2 = a_1 \left[T_m - a_2 z_1 z_2 - a_3 z_2 x_3 - a_4 z_2 x_4 + a_5 z_1 x_5 + a_6 x_6 z_1 \right]$$
$$\dot{x}_3 = -b_1 \left[b_2 x_3 + b_3 z_1 + b_4 x_4 \right] + b_5 E_{fd} \tag{2.7}$$

the rotor speed error is defined as

$$\sigma_r = x_2 - \omega_{ref} \tag{2.8}$$

In order to reject speed perturbations and estimate T_m, an integrer dynamic is introduced as

$$\dot{x}_0 = \sigma_r \tag{2.9}$$

Taking the derivative of (2.8) along the trajectories of (2.7)

$$\dot{\sigma}_r = a_1 \left[T_m - a_2 z_1 z_2 - a_3 z_2 x_3 - a_4 z_2 x_4 + a_5 z_1 x_5 + a_6 x_6 z_1 \right] \tag{2.10}$$

introducing the desired dynamics for σ_r (eq:eight:ctrl) on the form

$$\dot{\sigma}_r = a_1 T_m - k_g \sigma_r - k_1 x_0, \quad k_g > 0, \quad k_1 > 0 \tag{2.11}$$

the desired value for the virtual control x_3 is calculated from (2.10) and (2.11) as

$$x_{3des} = \frac{1}{a_3 z_2} \left(\frac{k_g}{a_1} \sigma_r + \frac{k_1}{a_1} x_0 + a_5 x_5 z_1 + a_6 x_6 z_1 \right) - \frac{a_2}{a_3} z_1 - \frac{a_4}{a_3} x_4 \tag{2.12}$$

Defining the control manifold as,

$$\sigma_g = x_3 - x_{3des} \tag{2.13}$$

Grouping (2.9) - (2.12), the mechanical dynamics (2.7) become

$$\dot{x}_0 = \sigma_r$$
$$\dot{x}_1 = x_2 - \omega_{ref}$$
$$\dot{\sigma}_r = a_1 T_m - k_g \sigma_r - k_1 x_0 + g_r \sigma_r \tag{2.14}$$

$$\dot{\sigma}_g = f_1 + g_1 E_{fd} \tag{2.15}$$

where $g_r = -a_1 a_3 z_2 g_1$ and $g_1 (\cdot) = \left(\frac{1}{a_3 z_2} \right) (+a_5 x_5 h_8 + a_6 x_6 h_8 + a_2 h_8 z_2 + a_3 z_2 b_5)$ is a positive function of time. Two sliding mode control laws can be used with relative degree one [10] related to x_2, the 1st order Sliding Mode Control law described by

$$E_{fd} = -V_{rmax} sign(\sigma_g) \tag{2.16}$$

or the 2nd Order Sliding Mode Control law known as *"Super-Twisting"* [10, 14] described by

$$E_{fd} = -M_{g1}\sqrt{|\sigma_g|}sign(\sigma_g) + u_{gen}$$
$$\dot{u}_{gen} = -M_{g2}sign(\sigma_g) \qquad (2.17)$$

Both control laws (2.16), (2.17) ensure the convergence of the closed-loop state vector to the sliding manifold $\sigma_g = 0$ in finite time. The sliding mode motion on this manifold is described by the third order system

$$\dot{x}_0 = \sigma_r$$
$$\dot{x}_1 = x_2 - \omega_{ref}$$
$$\dot{\sigma}_r = a_1 T_m - k_g \sigma_r - k_1 x_0 + g_r \sigma_r \qquad (2.18)$$

If $k_g > 0$ and $k_1 > 0$, then the system (2.18) is stable. Moreover, $\lim\limits_{t \to \infty} \sigma_r(t) = x_2(t) - \omega_b = 0$, the angle $x_1(t)$ tends to a steady state value x_{1ss} and $x_0(\infty) = \frac{a_1 T_m}{k_1}$.

2.4.2 Block Integral Sliding Mode Control for UPFC (FACTS) device

Control design for Voltage Magnitude at node 2

The purpose of this controller is to provide effective control of the voltage magnitude at node 2, where the shunt branch of the FACTS device is connected. This objective can be achieved using the current η_4 in the q axis in the shunt branch as the quasi control. The voltage magnitude error in power system's node 2 in $d - q$ reference frame can be represented by

$$\sigma_1 = |V_2| - v_{2ref} = v_3^2 + v_4^2 - v_{2ref} \qquad (2.19)$$

where V_{2ref} is a constant reference signal. Taking the derivative of (2.19) along the trajectories of (2.4), and introducing the desired dynamics for (2.19) as $\dot{\sigma}_1 = -k_{shq1}\sigma_1$, $k_{shq1} > 0$, it is obtained

$$\dot{\sigma}_1 = f_2 + f_3 \eta_4 = -k_{shq1}\sigma_1 \qquad (2.20)$$

where f_2 and f_3 are smooth functions. The desired value η_{4des} is calculated from (2.20) as

$$\eta_{4des} = f_4 + f_5 \sigma_1 \qquad (2.21)$$

with $f_4 = -\frac{f_2}{f_3}$ and $f_5 = -\frac{k_{shq1}}{f_3}$. Introducing the new variable

$$\sigma_2 = \eta_4 - \eta_{4des} \qquad (2.22)$$

Taking the derivative of (2.22) along the trajectories of (2.2) it is obtained

$$\dot{\sigma}_2 = -m_1\eta_3 - q_3\eta_4 + q_4v_4 - q_4v_{shq} - \dot{\eta}_{4des} = -k_{shq2}\sigma_2 \qquad (2.23)$$

With $k_{shq2} > 0$. The quasi control v_{shqdes} can be obtained from (2.23) as

$$v_{shqdes} = -\frac{1}{q_4}\left(m_1\eta_3 + q_3\eta_4 - q_4v_4 + \dot{\eta}_{4des} - k_{shq2}\sigma_2\right) \qquad (2.24)$$

In order to ensure that the voltage v_{shq} be a continuous function, it's taken into account the additional integral dynamic (linear actuator) described by

$$\dot{v}_{shq} = -v_{shq} + u_{shq} \qquad (2.25)$$

defining the new variable as the control manifold

$$\sigma_3 = v_{shq} - v_{shqdes} \qquad (2.26)$$

Taking the derivative of (2.26) along the trajectories of (2.24) and (2.25)

$$\dot{\sigma}_3 = f_6 + b_6 u_{shq} \qquad (2.27)$$

Grouping (2.20),(2.23),(2.27) it's obtained

$$\begin{aligned} \dot{\sigma}_1 &= -k_{shq1}\sigma_1 + g_2\sigma_2 \\ \dot{\sigma}_2 &= -k_{shq2}\sigma_2 - q_4\sigma_3 \\ \dot{\sigma}_3 &= f_6 + b_6 u_{shq} \end{aligned} \qquad (2.28)$$

with the function g_2 described by

$$g_2 = f_3 = -2m_3v_4, \quad b_6 = 1 \qquad (2.29)$$

To enforce the sliding mode motion on manifold $\sigma_3 = 0$ (2.26) it can either be chosen the 1st order sliding mode control law described by

$$u_{shq} = -M_{shq}\eta_5 sign(\sigma_3), \quad M_{shq}\eta_5 > \left|\frac{f_6}{b_6}\right|, \quad M_{shq}\eta_5 \leq V_{shq\ max} \qquad (2.30)$$

or the 2nd order sliding mode control law known as "*Super-Twisting*" [10, 14] described by

$$\begin{aligned} u_{shq} &= -M_0\eta_5\sqrt{|\sigma_3|}sign(\sigma_3) + u_1 \\ \dot{u}_1 &= -M_1\eta_5 sign(\sigma_3) \end{aligned} \qquad (2.31)$$

Where $v_{shq\ max}$ represents the maximum value of the series VSC in q axis. The sufficient convergence conditions for "*Super-Twisting*" control law are

$$\left| \frac{f_6}{b_6} \right| \leq \Phi_{shqst}, \quad \Phi_{shqst} > 0,$$

$$0 < \Gamma_{m_{shqst}} \leq |b_6| \leq \Gamma_{M_{shqst}}$$

$$M_1 \eta_5 > \frac{\Phi_{shqst}}{\Gamma_{m_{shqst}}}, \quad M_0^2 \eta_5 \geq \frac{4\Phi_{shqst}\Gamma_{M_{shqst}}(M_1\eta_5 + \Phi_{shqst})}{\Gamma_{m_{shqst}}^2 \Gamma_{m_{shqst}}(M_1\eta_5 - \Phi_{shqst})} \tag{2.32}$$

When a sliding mode occurs in (2.26) the sliding mode motion is reduced to the second order system described as

$$\dot{\sigma}_1 = -k_{shq1}\sigma_1 + g_2\sigma_2$$
$$\dot{\sigma}_2 = -k_{shq2}\sigma_2 \tag{2.33}$$

and the linear system (2.33) converges asymptotically to

$$\lim_{t\to\infty}\sigma_1(t) = 0, \quad \lim_{t\to\infty}\sigma_2(t) = 0, \quad \lim_{t\to\infty}V_2(t) = v_{2ref} \tag{2.34}$$

Control design for Active and Reactive Power, flowing through the series branch of the UPFC

The third control objective is to regulate the real P_{12} and reactive Q_{12} power flows from node 1 to node 2, which are expressed by

$$P_{12} = v_1\eta_1 + v_2\eta_2 \tag{2.35}$$

$$Q_{12} = v_2\eta_1 - v_1\eta_2 \tag{2.36}$$

The control errors of active and reactive power are expressed as

$$\sigma_P = P_{12} - P_{12ref} \tag{2.37}$$

$$\sigma_Q = Q_{12} - Q_{12ref} \tag{2.38}$$

Where P_{12ref} and Q_{12ref} are constant reference signals. Taking (2.35) – (2.38) and assuming that the errors σ_P and σ_Q in the steady state are equal to zero, the desired values for the series currents η_1 and η_2 can be obtained, respectively, as

$$\eta_{1des} = \frac{1}{v_1}(P_{12ref} - v_2\eta_2) = \frac{1}{v_1^2 + v_2^2}(v_1P_{ref} + v_2Q_{ref}) \tag{2.39}$$

$$\eta_{2des} = \frac{1}{v_1}(-Q_{12ref} + v_2\eta_1) = \frac{1}{v_2^2 + v_1^2}(v_2P_{ref} - v_1Q_{ref}) \tag{2.40}$$

Defining the new variables for series currents as

$$\sigma_4 = \eta_1 - \eta_{1des} \tag{2.41}$$

$$\sigma_5 = \eta_2 - \eta_{2des} \tag{2.42}$$

The desired dynamics for series currents η_1 and η_2 (2.2) are introduced with block control technique as

$$\dot{\eta}_1 = -q_1\eta_1 + m_1\eta_2 + q_2v_1 - q_2v_3 - q_2v_{serd} = -k_{dser}\sigma_4 \tag{2.43}$$

$$\dot{\eta}_2 = -m_1\eta_1 - q_1\eta_2 + q_2v_2 - q_2v_4 - q_2v_{serq} = -k_{qser}\sigma_5 \tag{2.44}$$

with $k_{dser} > 0$ and $k_{qser} > 0$. From (2.43) and (2.44) its is possible to define the quasi-controls v_{serd} and v_{serq}, as

$$v_{serddes} = \frac{1}{q_2}\left(-q_1\eta_1 + m_1\eta_2 + q_2v_1 - q_2v_3 + k_{dser}\sigma_4\right) \tag{2.45}$$

$$v_{serqdes} = \frac{1}{q_2}\left(-q_1\eta_2 - m_1\eta_1 + q_2v_2 - q_2v_4 + k_{qser}\sigma_5\right) \tag{2.46}$$

Substituting (2.39) and (2.40) in equations (2.45), (2.46) it's obtained

$$v_{serddes} = \frac{1}{q_2}\left(-q_1\eta_{1des} + m_1\eta_{2des} + q_2v_1 - q_2v_3 + k_{dser}\sigma_4\right) \tag{2.47}$$

$$v_{serqdes} = \frac{1}{q_2}\left(-q_1\eta_{2des} - m_1\eta_{1des} + q_2v_2 - q_2v_4 + k_{qser}\sigma_5\right) \tag{2.48}$$

Then, adding the first order dynamics (as linear actuators) described by

$$\dot{v}_{serd} = -v_{serd} + u_{serd} \tag{2.49}$$

$$\dot{v}_{serq} = -v_{serq} + u_{serq} \tag{2.50}$$

The series d and q control manifolds are defined as

$$\sigma_6 = v_{serd} - v_{serddes} \tag{2.51}$$

$$\sigma_7 = v_{serq} - v_{serqdes} \tag{2.52}$$

Taking the derivative of (2.51) and (2.52) along the trajectories of (2.47) - (2.50)

$$\dot{\sigma}_6 = f_7 + b_7 u_{serd} \tag{2.53}$$

$$\dot{\sigma}_7 = f_8 + b_8 u_{serq} \tag{2.54}$$

Grouping (2.43), (2.44), (2.53), and (2.54) it's obtained

$$\begin{aligned} \dot{\eta}_1 &= -k_{dser}\sigma_4 + g_3\sigma_6 \\ \dot{\sigma}_6 &= f_7 + b_7 u_{serd} \end{aligned} \tag{2.55}$$

$$\begin{aligned} \dot{\eta}_2 &= -k_{qser}\sigma_5 + g_4\sigma_7 \\ \dot{\sigma}_7 &= f_8 + b_8 u_{serq} \end{aligned} \tag{2.56}$$

With $g_3 = g_4 = -q_2$ and $b_7 = b_8 = 1$. To enforce the sliding mode motion on manifolds $\sigma_6 = 0$ (2.51) and $\sigma_7 = 0$ (2.52) it can either be chosen the 1st order sliding mode control laws described by

$$u_{serd} = -M_{serd}\eta_5 sign(\sigma_6), \quad M_{serd}\eta_5 > \left|\frac{f_7}{b_7}\right| \tag{2.57}$$

$$u_{serq} = -M_{serq}\eta_5 sign(\sigma_7), \quad M_{serq}\eta_5 > \left|\frac{f_8}{b_8}\right| \tag{2.58}$$

or the 2nd order sliding mode control laws known as *"Super-Twisting"* [10, 14] described by

$$u_{serd} = -M_2\eta_5\sqrt{|\sigma_6|}sign(\sigma_6) + u_2$$
$$\dot{u}_2 = -M_3\eta_5 sign(\sigma_6) \tag{2.59}$$

$$u_{serq} = -M_4\eta_5\sqrt{|\sigma_7|}sign(\sigma_7) + u_3$$
$$\dot{u}_3 = -M_5\eta_5 sign(\sigma_7) \tag{2.60}$$

Both control laws working independently, with the input signal bounds as

$$\left|\frac{f_7}{b_7}\right| \leq \Phi_{serdst}, \quad \Phi_{serdst} > 0,$$
$$0 < \Gamma_{m_{serdst}} \leq |b_7| \leq \Gamma_{M_{serdst}},$$
$$M_3\eta_5 > \frac{\Phi_{serdst}}{\Gamma_{m_{serdst}}} \quad M_2^2\eta_5 \geq \frac{4\Phi_{serdst}\Gamma_{M_{serdst}}(M_3\eta_5 + \Phi_{serdst})}{\Gamma_{m_{serdst}}^2\Gamma_{m_{serdst}}(M_3\eta_5 - \Phi_{serdst})} \tag{2.61}$$

$$\left|\frac{f_8}{b_8}\right| \leq \Phi_{serqst}, \quad \Phi_{serqst} > 0,$$
$$0 < \Gamma_{m_{serqst}} \leq |b_8| \leq \Gamma_{M_{serqst}},$$
$$M_5\eta_5 > \frac{\Phi_{serqst}}{\Gamma_{m_{serqst}}} \quad M_4^2\eta_5 \geq \frac{4\Phi_{serqst}\Gamma_{M_{serqst}}(M_5\eta_5 + \Phi_{serqst})}{\Gamma_{m_{serqst}}^2\Gamma_{m_{serqst}}(M_5\eta_5 - \Phi_{serqst})} \tag{2.62}$$

When sliding mode occurs the state vector converges to the manifolds $\sigma_6 = \sigma_7 = 0$ in different finite times, therefore the dynamics (2.55) and (2.56) are reduced to

$$\dot{\eta}_1 = -k_{dser}\sigma_4 \tag{2.63}$$

$$\dot{\eta}_2 = -k_{qser}\sigma_5 \tag{2.64}$$

Providing that the control errors σ_P (2.37) and σ_Q (2.38), converge asymptotically to

$$\lim_{t\to\infty}\sigma_P(t) = 0, \quad \lim_{t\to\infty}\sigma_Q(t) = 0 \tag{2.65}$$

Control design for internal DC Link Voltage regulation of UPFC (FACTS) device

The last control objective is to regulate the internal UPFC's DC link voltage η_5. Defining the error variables σ_8 and σ_9 as

$$\sigma_8 = \eta_5 - \eta_{5ref} \tag{2.66}$$

$$\sigma_9 = -v_{shdref} + \eta_{5ref} \tag{2.67}$$

Where η_{5ref} and v_{shdref} represent the constant reference values for η_5 and shunt VSC in d axis, respectively. The desired dynamics for η_5 (2.2) is given by

$$\dot{\eta}_5 = \frac{q_5}{\eta_5}\left[-v_{serd}\eta_1 - v_{serq}\eta_2 + v_{shq}\eta_4 + \eta_3(\eta_5 - \sigma_9 - \sigma_{cd})\right] \tag{2.68}$$

Therefore, the value of v_{shq} as quasi control is defined as

$$v_{shddes} = \eta_5 - \sigma_9 - \sigma_{cd} \tag{2.69}$$

Adding the linear dynamics v_{shd}, σ_{cd} described by

$$\dot{\sigma}_{cd} = (-\sigma_{cd} + u_{cd})k_{cd1}, \quad k_{cd1} > 0 \tag{2.70}$$

$$\dot{v}_{shd} = -v_{shd} + u_{shd} \tag{2.71}$$

Where the dynamic (2.71) represents a linear actuator. Defining the input u_{cd} as

$$u_{cd} = k_{cd2}\sigma_8, \quad k_{cd2} > 0 \tag{2.72}$$

Defining the control manifold as

$$\sigma_{10} = v_{shd} - v_{shddes} \tag{2.73}$$

Taking the derivative of (2.73) along the trajectories of (2.70) and (2.71) it's obtained

$$\dot{\sigma}_{10} = f_9 + b_9 u_{shd} \tag{2.74}$$

Where $b_9 = 1$ and f_9 is described by

$$f_9 = -v_{shd} - \frac{q_5}{\eta_5}\left(-v_{serd}\eta_1 - v_{serq}\eta_2 + v_{shd}\eta_3 + v_{shq}\eta_4\right) + k_{cd1}\left(\sigma_{cd} - u_{cd}\right) \tag{2.75}$$

Grouping η_3, η_5 from (2.2), (2.70) and (2.74), it's obtained

$$\dot{\eta}_5 = \frac{q_5}{\eta_5}\left[-v_{serd}\eta_1 - v_{serq}\eta_2 + v_{shq}\eta_4 + \eta_3(\eta_5 - \sigma_9 - \sigma_{cd} - \sigma_{10})\right]$$

$$\dot{\eta}_3 = -q_3\eta_3 + m_1\eta_4 + q_4 v_3 - q_4(\eta_5 - \sigma_9 - \sigma_{cd}) - q_4\sigma_{10}$$

$$\dot{\sigma}_{cd} = -k_{cd1}\left(\sigma_{cd} - k_{cd2}\sigma_8\right)$$

$$\dot{\sigma}_{10} = f_9 - b_9 u_{shd} \tag{2.76}$$

To enforce the sliding mode motion on manifolds $\sigma_{10} = 0$ (2.73), it can either be chosen the 1st order sliding mode control law described by

$$u_{shd} = -M_{shd}\eta_5 sign(\sigma_{10}), \quad M_{shd}\eta_5 > \left| \frac{f_9}{b_9} \right| \tag{2.77}$$

or the 2nd order sliding mode control law known as "*Super-Twisting*" [10, 14] described by

$$u_{shd} = -M_6\eta_5\sqrt{|\sigma_{10}|}sign(\sigma_{10}) + u_4$$
$$\dot{u}_4 = -M_7\eta_5 sign(\sigma_{10}) \tag{2.78}$$

with the input signal bounds as

$$\left| \frac{f_9}{b_9} \right| \leq \Phi_{shdst}, \quad \Phi_{shdst} > 0,$$

$$0 < \Gamma_{m_{shdst}} \leq |b_9| \leq \Gamma_{M_{shdst}},$$

$$M_7\eta_5 > \frac{\Phi_{shdst}}{\Gamma_{m_{shdst}}} \quad M_6^2\eta_5 \geq \frac{4\Phi_{shdst}\Gamma_{M_{shdst}}(M_7\eta_5 + \Phi_{shdst})}{\Gamma_{m_{shdst}}^2\Gamma_{m_{shdst}}(M_7\eta_5 - \Phi_{shdst})} \tag{2.79}$$

When sliding mode occurs the state vector converges to the manifolds σ_{10} in finite time, therefore the dynamics (2.76) are reduced to

$$\dot{\eta}_5 = \frac{q_5}{\eta_5}\left[-v_{serd}\eta_1 - v_{serq}\eta_2 + v_{shq}\eta_4 + \eta_3(\eta_5 - \sigma_9 - \sigma_{cd})\right]$$
$$\dot{\eta}_3 = -q_3\eta_3 + m_1\eta_4 + q_4v_3 - q_4(\eta_5 - \sigma_9 - \sigma_{cd})$$
$$\dot{\sigma}_{cd} = -k_{cd1}(\sigma_{cd} - k_{cd2}\sigma_8) \tag{2.80}$$

Equation (2.67) ensures stabilization of η_5 and regulation of η_3 close to the initial condition, while equation (2.72) achieves asymptotic regulation of η_5 to the initial value of η_{5ref}.

2.5 Power System Parameters

In order to simplify the power system, the transmission line three can be reduced by adding its resistance and reactance to the values of the transformer in the series branch of the FACTS device. Also the generator's transformer values of R and XL can be added to the stator parameters of the synchronous generator. The schematic diagram of the power system is shown in Fig. 2.2. The parameter values for the synchronous generator, transmission lines and FACTS device are shown in Tablesřeftabl:tabla:uno and 2.2.

The constant values of dynamic models (2.2) - (2.6) are calculated with parameters shown in Tables 2.1 and 2.2 (see Appendix). The control parameters and gains are shown in Table 2.3.

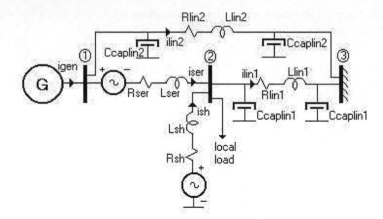

Fig. 2.2. Power system's reduced schematic diagram for transient phenomena studies with UPFC (FACTS) device.

Table 2.1. Synchronous generator's parameters (Fig. 2.2).

Parameter	Value	Parameter	Value
Num. of poles	2	Operation	24 KV
H	3.5250 sec	Nominal Power	550 MVA
Ld	1.8102 pu	Lq	1.76 pu
Ld'	0.3 pu	Lq'	0.65 pu
Ld''	0.23 pu	Lq''	0.25 pu
La	0.15 pu	ra	0.0006 pu
$\tau d'$	8 sec	$\tau q'$	1 sec
$\tau d''$	0.3 sec	$\tau q''$	0.7 sec

Table 2.2. Power system's parameters under 550 MVA power base (Fig. 2.2).

Element	Value	Element	Value
R_{lin1}	0.039 pu	L_{lin1}	0.17 pu
R_{lin2}	0.039 pu	L_{lin2}	0.17 pu
R_{ser}	0.0195 pu	L_{ser}	0.1475 pu
R_{sh}	0 pu	L_{sh}	0.0625 pu
$C_{caplin1}$	0.079 pu	$C_{caplin2}$	0.079 pu
R_{load}	1.8114 pu	L_{load}	0.7245 pu
C_{UPFC}	2 pu	ωb	376.9911 rad/sec

Table 2.3. Parameters and gains for the designed controls.

Parameter	Value	Parameter	Value
k_g	5	k_{shq2}	10,000
k_1	a_1	k_{cd1}	10
V_{rmax}	6	k_{cd2}	30
M_{g1}	$0.7\,V_{rmax}$ $M_{serd} = M_{serq} = M_{shd} = M_{shq}$		4
M_{g2}	$0.3\,V_{rmax}$ $M_0 = M_2 = M_4 = M_6$		$0.7\,M_{serd}$
$k_{dser} = k_{qser}$	3	$M_1 = M_3 = M_5 = M_7$	$0.3\,M_{serq}$
k_{shq1}	100	η_{5ref}	1 pu

2.6 Power system operating point and disturbances

In this section the operating point for the power system is described as well as the disturbances to be simulated.

2.6.1 Power system's operating point

The equilibrium is set such that the local load demand is $P =0.5$ and $Q =0.2$ pu. Power flows in the series branch of the UPFC are $P =0.5$ and $Q =0.2$ pu (from node 1 to node 2). The transmission line two power flows are $P =0.4$ and $Q =0$ pu (from node 1 to node 3). The transmission line one power flows are $P =0$ and $Q =0$ pu (from node 2 to node 3). The generator power supply is $P =0.9$ and $Q =0.2$ pu. The generator bus voltage magnitude is equal to 1.040 pu (node 1). The voltage magnitude at node 2 is equal to 1.025 pu. The node 3 (infinite bus) voltage magnitude is equal to 1.0 pu. The input torque of synchronous machine remains constant and equal to 0.9 pu.

2.6.2 Disturbances description

The first disturbance consists in a change of reference values for active and reactive power, flowing through the series branch of UPFC (from node 1 towards node 2). The power flows references are $P_{12ref} =0.5$ and $Q_{12ref} =0.2$ pu (from node 1 to node 2). At $t =1$ seconds, the power references are set as $P_{12ref} =1$ and $Q_{12ref} =0.5$ pu. At the time $t=4$ seconds, the power references are reestablished to $P_{12ref} =0.5$ and $Q_{12ref} =0.2$ pu. The second disturbance consists in a solid three phase short circuit at the infinite bus (node 3) with duration of 0.05 seconds (3 AC cycles). The disturbance initial time is set at $t =9$ seconds. At the time $t =9.05$ seconds, the voltage at node 3 is released.

2.7 Simulation results

In this section, the simulation results are shown. The disturbances described in Section 2.6.2 are introduced into the power system under the operating conditions described in Section 2.6.1. Figs. 2.3 – 2.7 show the behavior of the most important variables of the power system under the two disturbances described in Section 2.6.2.

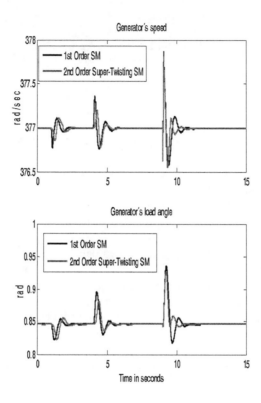

Fig. 2.3. Generator's speed x_2, and generator's load angle x_1, for the power system at operating point under disturbances one and two.

2.8 Conclusions

In this work, non-linear controllers were designed for a UPFC (FACTS) in a small power system. To design these controllers, combinations of Block Control (BC) with 1st order and 2nd order Sliding Mode control (SM) techniques were implemented. The linear dynamics included in the controllers design (integrators and 1st order dynamics as linear actuators) guarantees that the errors converge asymptotically to zero. Providing that the proposed control algorithms are continuous functions of the system state, thus the so-called *chattering effect* is considerably reduced. The designed controllers have been tested through simulations under many important disturbances in power systems (three phase faults, sudden changes of network parameters, down

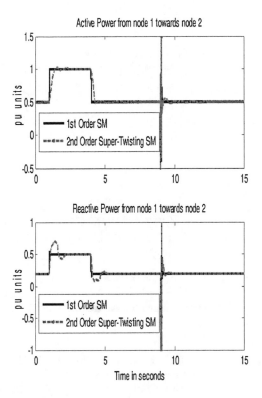

Fig. 2.4. Active and reactive power flows from node 1 towards node 2, (controlled by the series branch of the UPFC) under the disturbances one and two.

and up steps of torque input, load changes, etc) and multiple operating points. Due to number of page restrictions only two disturbances were shown here. The simulation results show that the proposed controllers were able to achieve good level of transient performance, and control goals are reached despite the presence of large disturbances.

References

1. Hingorani, N., Gyugyi, L. (2000), *Understanding FACTS: Concepts and Technology of Flexible AC Transmission Systems*. IEEE Press.
2. Valentin, A., Uros, G., Dusan, P., Mihalic, R. (2005), The Energy Function of a General Multimachine System with Unified Power Flow Controller, *IEEE Transactions on Power Systems*, vol. 20, no. 3.
3. Na, He., Liu, R., Xu, D. (2005), The Study of UPFC Fuzzy Control with self-adjustable Factor. In: *IEEE/PES Transmission and Distribution Conference & Exhibition*, Asia Pacific Dalian, China.

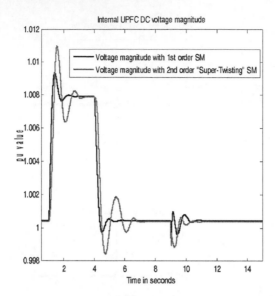

Fig. 2.5. UPFC's series VSC voltage magnitude and VSC shunt voltage magnitude, under disturbances one and two.

4. Zhang, Y., Liu, L., Zhu, P., Liu, X., Kang, Y. (2006), *Double Closed Loop Control and Analysis for Shunt Inverter of UPFC*. IEEE Press.
5. Chaudhuri, N., Kothari, M. (2006), *Optimum Design of UPFC Controllers Using GEA: Decoupled Real & Reactive Power Flow and Damping Controllers*. IEEE Press.
6. Liming, L., Pengcheng, Z., Yong, K., Jian, C. (2006), *Unified Power Flow Controller: Comparison of Two Advanced Control Schemes and Performance Analysis for Power Flow Control*, IEEE Press.
7. Mishra, S. (2006), Neural-Network-Based Adaptive UPFC for Improving Transient Stability Performance of Power System, *IEEE Transactions on Neural Networks*, Vol. 17, No. 2.
8. Shaheen, H., Rashed, G., Cheng, S. (2007), Application of Evolutionary Optimization Techniques for Optimal Location and Parameters Setting of Multiple UPFC Devices. In *3rd International Conference on Natural Computation (ICNC)*.
9. Utkin, V. (1992), *Sliding Modes in Control and Optimization*, Springer-Verlag, Berlin.
10. Fridman, L., Levant, A. (2002), Higher Order Sliding Modes in Sliding Mode Control Engineering. In Barbot, J. P., Perruguetti, W. (eds.) Marcel Dekker, pp. 53—101. New York.
11. Loukianov, A. (1998), Nonlinear Block Control with Sliding Mode, In *Automation and Remote Control*, Vol. 59, No. 7, pp. 916–933.
12. Loukianov, A., Cañedo, J., Utkin, V., Cabrera-Vazquez, J. (2004), Discontinuous Controller for Power Systems: Sliding Mode Block Control Approach, *IEEE Transaction on Power Systems*.
13. Krauss, Paul, Wasynczuk, O., Sudhoff, S. (1986), *Analysis of Electric Machinery*, IEEE press.
14. Fridman, L., Colet, E., Edwards, C. (2006), Advances in Variable Structure and Sliding Mode Control. Springer Verlag. LNCIS 334, ISBN-10 3-540-32800-9, Germany.

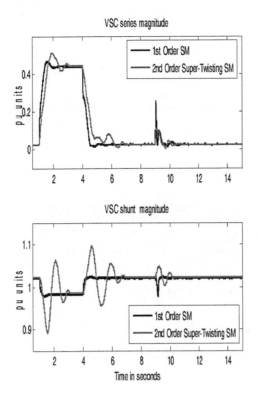

Fig. 2.6. UPFC's internal, DC link voltage.

Appendix

The constant parameters of dynamic models (2.2)-(2.6) are calculated as

$$m_1 = \omega_b, \qquad m_2 = \frac{\omega_b}{C_{caplin2}} \tag{2.81}$$

$$\gamma_1 = \frac{R_{lin1}\omega_b}{L_{lin1}}, \gamma_2 = \frac{\omega_b}{L_{lin1}}, \ \gamma_3 = \frac{R_{lin2}\omega_b}{L_{lin2}}, \gamma_4 = \frac{\omega_b}{L_{lin2}} \tag{2.82}$$

$$p_1 = \frac{R_{load}\omega_b}{L_{load}}, \qquad p_2 = \frac{\omega_b}{L_{load}} \tag{2.83}$$

$$q_1 = \frac{R_{ser}\omega_b}{L_{ser}}, q_2 = \frac{\omega_b}{L_{ser}}, \ q_3 = \frac{R_{ftesh}\omega_b}{L_{sh}}, q_4 = \frac{\omega_b}{L_{sh}}, \ q_5 = \frac{3}{2C_{upfc}} \tag{2.84}$$

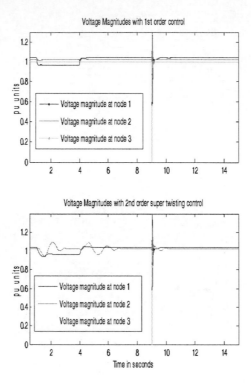

Fig. 2.7. Voltage magnitudes at the nodes of the power system under the 1st and 2nd order sliding mode control.

3

Closed-Loop Identification for a Velocity Controlled DC Servomechanism: Theory and Experiments

Rubén Garrido, Roger Miranda

Departamento de Control Automático, CINVESTAV
{garrido,rmiranda}@ctrl.cinvestav.mx

Summary. This chapter presents a method for DC servomotors working in closed loop. A Proportional Integral (PI) controller stabilizes the servomotor. Another PI controller closes the loop around a linear model of the servomotor. The error between the outputs of the servomotor and the model drives a gradient-like parameter estimation algorithm; afterwards, these parameter estimates update the servomotor model. Experiments using a laboratory prototype validate the proposed method.

3.1 Introduction

Servomotors are key elements in a great variety of industrial applications requiring speed and position control. The knowledge of the servomotor parameters is necessary to improve their transient, steady state and dynamic characteristics. For this reason, parameter identification techniques are a first step prior to setting up a model-based control law. A linear model of a servomotor considering its angular velocity as the controlled variable is open loop stable; therefore, many of the identification techniques in the literature are adequate for obtaining parameter estimates. However, in some cases, it would be interesting to identify a system in closed loop; for instance, if the open loop gain is high, then, small input values would produce large output values that would not be desirable for security reasons. This situation occurs in servomotors driven by current-controlled amplifiers; in this case the open loop gain is high. On the other hand, having a low open loop gain may produce a sluggish response and the identification procedure may slow down. A way to cope with the aforementioned problems is to modify the system gain using feedback. At this stage it is interesting to point out that velocity regulation of servomotors is accomplished in many industrial applications using Proportional Integral controllers [1].

There are several works dealing with parameter estimation of servomotors considering velocity as the output variable. Works in this vein are [2], [3], [4], [5], [6], [7], [8], [9], [10]. In [2] the authors propose a graphical method using velocity step responses. Frequency based methods are studied in [3], [4], [5], [6]. In [3] the authors employ an off-line Least Squares algorithm in the frequency domain, compare it against a standard time domain Least Squares algorithm, and show that the former has better performance. Velocity and armature voltage feed the estimation algorithms. An interesting feature is the multifrequency excitation signal used to perform the parameter estimation, however, the authors underline that the frequency approach is not suitable for on-line DC motor parameter identification. Reference [4] is an interesting survey about estimation of nonlinear models of DC motors. In this case, using armature current measurements and assuming that the motor works in open loop, the authors show that it is possible to estimate the servomotor parameters through the so-called Hartley functions. These functions allow converting a nonlinear differential equation into an algebraic relationship in the same way that the Laplace transform is applied to a linear differential equation with constant parameters. The off-line frequency-weighted Least Squares algorithm estimates the unknown parameters. The approach described in [6] also employs the Least Squares method where the output variable is the motor velocity; all the experiments are performed in open loop. In [7] an off-line Maximum Likelihood algorithm permits servomotor identification considering the presence of nonlinear friction. A feature of the method is the use of multifrequencial binary excitation signals that overcome the identification errors introduced by the presence of Coulomb friction. The authors employ open loop experiments and the measured variables are both the servomotor angular velocity and the rectilinear velocity of a table driven by the servomotor through a ball screw. Reference [8] describes another interesting approach. In this case, the servomotor operates in open loop and a recursive on-line Least Squares algorithm with forgetting factor estimates the servo parameters using a Hammerstein model. Reference [9] proposes a similar approach. It is worth mentioning the approach in [10] where the authors apply closed loop identification techniques to a rotating three-mass electromechanical system. A recursive Least Squares algorithm with forgetting factor performs parameter identification employing a discrete-time model. The servomotor speed and the armature voltage feed the identification algorithm. A Proportional Integral (PI) velocity controller controls the servomotor for the closed loop experiments. Convergence of the parameter estimates is rather slow in open and closed loop; however, it is experimentally shown that convergence rates in closed loop are faster than in open loop. Algebraic tools are another approach for parameter identification [11]. In that reference, the authors apply their methodology by means of numerical simulations to several plants. The plant is firstly identified then controlled using an algorithm computed employing the parameter estimates previously estimated. From the review it is clear that most of the approaches rely on open loop experiments; moreover, the Least Squares algorithm is in most cases the main tool for parameter identification.

This work presents some preliminary results concerning an identification methodology for a velocity-controlled servomotor. A depart from most previous approaches is the fact that parameter identification is performed while the motor works in closed loop under PI control. Moreover, the analysis of the identification algorithm takes into account the PI controller. Section 3.2.1 presents the proposed identification methodology. Section 3.3 shows some experimental results that sustain the proposed approach. Finally, the chapter ends with some concluding remarks.

3.2 Identification Algorithm

3.2.1 Preliminaries

The following paragraphs describe the proposed identification approach. Consider a servomotor driven by a current amplifier. Two identical PI controllers close the loop around the servomotor and a model as depicted in Fig. 3.1. The difference between the velocity of the servomotor and the estimated velocity generated by the model feeds the identifier; subsequently, the parameter estimates update the model. By using the PI controller, the servomotor has a stable behavior despite the amplifier gain. However, even when the closed-loop system associated to the servomotor is stable, it is not the case for the model; therefore, stability of the model needs further study.

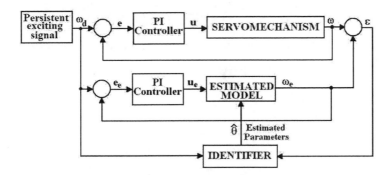

Fig. 3.1. Block diagram of the proposed identification mehtod.

The servomotor working in velocity mode driven by a current amplifier obeys the following model

$$J\dot{\omega}(t) + f\omega(t) = ku(t) \tag{3.1}$$

where $\omega(t)$ represents the servomotor shaft angular velocity, J and f are the servomotor inertia and viscous friction, $u(t)$ is the amplifier input voltage and k stands for the amplifier gain. Rewriting (3.1) gives

$$\dot{\omega}(t) = -a\omega(t) + bu(t) \tag{3.2}$$

with $a = f/J$, $b = k/j$ being positive constants. Consider the PI control law

$$u(t) = k_p e(t) + k_i \int_0^t e(\tau)\,d\tau \tag{3.3}$$

and the velocity error

$$e(t) = \omega_d - \omega(t) \tag{3.4}$$

with ω_d a reference. Consider now the following estimated model

$$\dot{\omega}_e(t) = -\hat{a}\omega_e(t) + \hat{b}u_e(t) \tag{3.5}$$

where $\hat{a}$ and $\hat{b}$ being estimates of a and b respectively, in closed-loop with the PI controller

$$u_e(t) = k_p e_e(t) + k_i \int_0^t e_e(\tau)\, d\tau \tag{3.6}$$

The next expression defines the velocity error associated to the model

$$e_e(t) = \omega_d - \omega_e(t) \tag{3.7}$$

Note that the control laws (3.3) and (3.6) uses the same gains. Substituting (3.3) into (3.2) and (3.6) into (3.5) yields

$$\dot{\omega}(t) = -a\omega(t) + bk_p e(t) + bk_i \int_0^t e(\tau)\, d\tau \tag{3.8}$$

$$\dot{\omega}_e(t) = -\hat{a}\omega_e(t) + \hat{b}k_p e_e(t) + \hat{b}k_i \int_0^t e_e(\tau)\, d\tau \tag{3.9}$$

Define the error between the outputs of the servomotor and its model as

$$\varepsilon(t) = \omega(t) - \omega_e(t) \tag{3.10}$$

Hence, employing (3.8) and (3.9), the time derivative of (3.10) is given by

$$\begin{aligned}
\dot{\varepsilon} &= \dot{\omega} - \dot{\omega}_e \\
&= -a\omega + bk_p e + bk_i \int_0^t e\, d\tau \\
&\quad + \hat{a}\omega_e - \hat{b}k_p e_e - \hat{b}k_i \int_0^t e_e\, d\tau
\end{aligned}$$

Adding and substracting $a\omega_e$, $bk_p e_e$ and $bk_i \int_0^t e_e\, d\tau$ from the right-hand side of this last expression leads to

$$\begin{aligned}
\dot{\varepsilon} &= -a\varepsilon - bk_p\varepsilon - bk_i \int \varepsilon\, d\tau \\
&\quad + (\hat{a} - a)\omega_e - (\hat{b} - b)\left[k_p e_e + k_i \int e_e\, d\tau \right]
\end{aligned}$$

Define

$$z(t) \; := \int_0^t \varepsilon(\tau)\, d\tau, \quad \alpha = a + bk_p > 0 \tag{3.11}$$

$$\tilde{\theta} = \hat{\theta} - \theta = \begin{bmatrix} \hat{a} - a \\ \hat{b} - b \end{bmatrix}$$

$$\phi(t) = \begin{bmatrix} \omega_e(t) \\ -k_p e_e(t) - k_i \int_0^t e_e(\tau)\, d\tau \end{bmatrix} = \begin{bmatrix} \omega_e(t) \\ -u_e(t) \end{bmatrix}$$

then, $\dot{\varepsilon}$ becomes

$$\dot{\varepsilon} = -\alpha\varepsilon - bk_i z + \tilde{\theta}^T \phi(t) \tag{3.12}$$

3.2.2 Stability Analysis

This section deals with the stability analysis of (3.12) which allows concluding stability of the model (3.9). To this end consider the following Lyapunov function candidate

$$V(\varepsilon,z) = \frac{1}{2}\begin{bmatrix} \varepsilon & z \end{bmatrix}\begin{bmatrix} 1 & \mu \\ \mu & bk_i \end{bmatrix}\begin{bmatrix} \varepsilon \\ z \end{bmatrix} + \frac{1}{2}\tilde{\theta}^T\Gamma^{-1}\tilde{\theta} \tag{3.13}$$

where $\Gamma = \Gamma^T \in \mathbb{R}^{2\times 2}$ is positive definite and $\mu \in \mathbb{R}^+$. It is clear that (3.13) is positive definite if the next inequality holds

$$\mu < \sqrt{bk_i} \tag{3.14}$$

Taking the time derivative of (3.13) along the trajectories of (3.12) yields

$$\begin{aligned}
\dot{V}(\varepsilon,z) &= \varepsilon\dot{\varepsilon} + \mu z\dot{\varepsilon} + \mu\dot{z}\varepsilon + bk_i z\dot{z} + \tilde{\theta}^T\Gamma^{-1}\dot{\tilde{\theta}} \\
&= \varepsilon\left[-\alpha\varepsilon - bk_i z + \tilde{\theta}^T\phi\right] + \mu z\left[-\alpha\varepsilon - bk_i z + \tilde{\theta}^T\phi\right] \\
&\quad + \mu\varepsilon^2 + bk_i z\varepsilon + \tilde{\theta}^T\Gamma^{-1}\dot{\tilde{\theta}} \\
&= -(\alpha - \mu)\varepsilon^2 - \mu bk_i z^2 - \mu\alpha z\varepsilon + \tilde{\theta}^T\left[\phi\varepsilon + \phi\mu z + \Gamma^{-1}\dot{\tilde{\theta}}\right]
\end{aligned}$$

Since $\dot{\hat{\theta}} = \dot{\tilde{\theta}}$, choosing the update law

$$\dot{\hat{\theta}} = -\Gamma\phi(\varepsilon + \mu z) \tag{3.15}$$

and defining $\gamma = (\alpha - \mu) > 0$ allows writing $\dot{V}$ as

$$\dot{V}(\varepsilon,z) = -\gamma\varepsilon^2 - \mu bk_i z^2 - \mu\alpha z\varepsilon \tag{3.16}$$

$$\leq -\left[\gamma - \frac{\mu\alpha^2}{4bk_i}\right]\varepsilon^2$$

If μ fulfills the following inequality

$$\mu < \frac{4\alpha bk_i}{4bk_i + \alpha^2} \tag{3.17}$$

then, (3.16) is negative semidefinite. Consequently, ε and $\tilde{\theta}$ are bounded. Define $\beta = \gamma - \mu\alpha^2/(4bk_i)$, thus, from (3.16) it follows that

$$\int_0^t \varepsilon^2(\tau)\,d\tau \leq \frac{V(0) - V(t)}{\beta} < \infty \tag{3.18}$$

so $\varepsilon \in L_2 \cap L_\infty$ and $\dot{\varepsilon}(t) \in L_\infty$. Therefore, from Barbalat's lemma [12] it follows that $\varepsilon(t) \to 0$ as $t \to \infty$. Hence, $\omega_e(t) \in L_\infty$ and stability of (3.9) follows. Note however, that the above analysis does not allow concluding convergence of $\tilde{\theta}$ to zero.

3.2.3 Parameter Convergence

Using (3.12) and (3.15) and defining $y(t) := \mu z + \varepsilon$, it is possible to obtain the following state-space representation

$$\dot{x} = Ax + Bu \tag{3.19}$$
$$y = Cx$$

$$\dot{\tilde{\theta}} = -\Gamma \phi y \tag{3.20}$$

where

$$A = \begin{bmatrix} 0 & 1 \\ -bk_i & -\alpha \end{bmatrix}, B = \begin{bmatrix} 0 \\ 1 \end{bmatrix} \tag{3.21}$$

$$C = \begin{bmatrix} \mu & 1 \end{bmatrix}, u = \tilde{\theta}^T \phi, x(t) = \begin{bmatrix} z \\ \varepsilon \end{bmatrix}$$

The corresponding transfer function for (3.19) is given by

$$G(s) = C(sI - A)^{-1}B = \frac{s + \mu}{s^2 + \alpha s + bk_i} \tag{3.22}$$

It is not difficult to obtain an expression for $G(s)$ evaluated in $j\omega$ in terms of its real and imaginary parts

$$G(j\omega) = \mathrm{Re}\,[G(j\omega)] + j\mathrm{Im}\,[G(j\omega)]$$
$$\mathrm{Re}\,[G(j\omega)] = \frac{bk_i\mu + \omega^2(\alpha - \mu)}{(bk_i - \omega^2)^2 + (\alpha\omega)^2}$$
$$\mathrm{Im}\,[G(j\omega)] = \frac{(bk_i - \omega^2)\omega - \mu\alpha\omega}{(bk_i - \omega^2)^2 + (\alpha\omega)^2} \tag{3.23}$$

Note that

$$\lim_{\omega \to \infty} \omega^2 \mathrm{Re}\,[G(j\omega)] = \lim_{\omega \to \infty} \frac{bk_i\mu\omega^2 + \omega^4(\alpha - \mu)}{(bk_i - \omega^2)^2 + \alpha^2\omega^2}$$

$$= \lim_{\omega \to \infty} \left\{ \left(1 + \frac{\alpha^2 - 2bk_i}{\omega^2} + \frac{b^2 k_i^2}{\omega^4}\right)^{-1} \left((\alpha - \mu) + \frac{bk_i\mu\omega^2}{\omega^4}\right) \right\}$$

$$= \alpha - \mu$$

Then, according to standard results about Strictly Positive Real (SPR) functions, transfer function (3.22) is SPR if

$$\mu < \alpha \tag{3.24}$$

The following definition about a Persistently Exciting (PE) signal [12] will be useful for stating parameter convergence.

Definition 1. *A vector* $\phi : \mathbb{R}_+ \rightarrow \mathbb{R}^{2n}$ *is PE if there exist* α_1, α_2, γ *such that*

$$\alpha_1 I \leq \int_{t_0}^{t_0+\delta} \phi(\tau)\phi^T(\tau)\, d\tau \leq \alpha_2 I, \forall t_0 \geq 0 \qquad (3.25)$$

Considering the updating equation (3.20) and using standard results on parameter convergence of adaptive identifiers using SPR equations [12], if the regressor vector $\phi(t)$ is PE, then $\tilde{\theta}$ converges exponentially to zero. It is worth to note that the signals in $\phi(t)$ belong to a time-varying system, namely, the estimated system (3.9). However, from the stability analysis performed in Subsection 3.2.2, it follows that $\phi(t)$, corresponding to the estimated model, converges to a vector $\phi_r(t)$ related to the servomotor. Then, if $\phi_r(t)$ fulfills the PE condition so does $\phi(t)$. This observation allows concluding that it is only neccessary to establish the relationship between Sufficiently Richness (SR) of the reference signal ω_d and the PE conditions on ϕ_r, that corresponds to the input and output of a linear time-invariant system respectively. Now, consider the following definition about SR signals [12].

Definition 2. *A stationary signal* $r(t) : \mathbb{R}_+ \rightarrow \mathbb{R}^n$ *is called SR of order n if the support of the spectral density of* $r(t)$, *namely,* $S_r(d\omega)$, *contains at least n points.*

The following proposition will allow obtaining the main result of this subsection.

Proposition 1. *Let* $\phi \in \mathbb{R}^n$ *the output of a stable linear time-invariant system with transfer function* $L(s)$ *and input* $r(t)$ *SR of order n. Assume that* $L(j\omega_1), ..., L(j\omega_n)$ *are linearly independent on* $\mathbb{C}^n$ *for all* $\omega_1, ..., \omega_n \in \mathbb{R}$. *Then* ϕ *is PE.* $\square$

Regressor vector $\phi_r(t)$ is given by

$$\phi_r(t) = \begin{bmatrix} \phi_{r,1} \\ \phi_{r,2} \end{bmatrix} = \begin{bmatrix} \omega(t) \\ -k_p e(t) - k_i \int_0^t e(\tau)\, d\tau \end{bmatrix} \qquad (3.26)$$

To prove that $\phi_r(t)$ is PE, consider the following transfer function related to the signals belonging to $\phi_r(t)$ defined in (3.26)

$$\begin{aligned} L(s) &= \begin{bmatrix} L_1(s) \\ L_2(s) \end{bmatrix} = \begin{bmatrix} \Phi_{r,1}(s)/Q_d(s) \\ \Phi_{r,2}(s)/Q_d(s) \end{bmatrix} \\ &= \begin{bmatrix} (bk_p s + bk_i)/(s^2 + (a+bk_p)s + bk_i) \\ -((k_p s + k_i)(s+a))/(s^2 + (a+bk_p)s + bk_i) \end{bmatrix} \end{aligned} \qquad (3.27)$$

where $\Phi_{r,1}(s)$, $\Phi_{r,2}(s)$ and $Q_d(s)$ are the Laplace transforms of $\phi_{r,1}$, $\phi_{r,2}$ and q_d, respectively. We have that $(a+bk_p) > 0$ and $bk_i > 0$, then (3.27) is stable. To apply Proposition 1, assume that $q_d(t)$ is SR and consider the following complex matrix

$$M = \begin{bmatrix} L(j\omega_1) & L(j\omega_2) \end{bmatrix} = \begin{bmatrix} \kappa_{11} & \kappa_{12} \\ \kappa_{21} & \kappa_{22} \end{bmatrix}$$

with

$$\kappa_{11} = \frac{bk_p(j\omega_1) + bk_i}{(j\omega_1)^2 + (a + bk_p)(j\omega_1) + bk_i}$$

$$\kappa_{12} = \frac{bk_p(j\omega_2) + bk_i}{(j\omega_2)^2 + (a + bk_p)(j\omega_2) + bk_i}$$

$$\kappa_{21} = -\frac{[k_p(j\omega_1) + k_i][(j\omega_1) + a]}{(j\omega_1)^2 + (a + bk_p)(j\omega_1) + bk_i}$$

$$\kappa_{22} = -\frac{[k_p(j\omega_2) + k_i][(j\omega_2) + a]}{(j\omega_2)^2 + (a + bk_p)(j\omega_2) + bk_i}$$

whose determinat is

$$\det(M) = \frac{(\omega_2 - \omega_1)[bk_ik_p(\omega_2 + \omega_1)] + j(\omega_2 - \omega_1)(bk_p^2\omega_1\omega_2 - bk_i^2)}{\left[(j\omega_1)^2 + (a + bk_p)(j\omega_1) + bk_i\right]\left[(j\omega_2)^2 + (a + bk_p)(j\omega_2) + bk_i\right]}$$

Then, $\det(M) \neq 0$ if $\omega_1 \neq \omega_2$, which implies that $L(j\omega_1)$, $L(j\omega_2)$ are linearly independent. Consequently, ϕ_r is PE. Finally, since vector ϕ converges to ϕ_r, thus ϕ converges to a vector satisfying a PE condition, therefore, $\tilde{\theta}$ converges exponentially to zero.

3.3 Experimental Results

The servomechanism employed for the experiments consists of a DC brushed motor controlled through a Copley Controls power amplifier, model 413, configured in current mode. A Servotek tachometer model SA-7388-1 furnishes velocity measurements. A MultiQ-3 card from Quanser Consulting performs data adquisition. The card has $12\,bits$ analog-to-digital and digital-to-analog converters with a voltage range of $\pm 5\,volts$. The Matlab-Simulink software operating with the WINCON software from Quanser Consulting serves as a programming platform. Sampling period was set to $1\,ms$. For the proposed identification method, parameters were set to $k_p = 5$, $k_i = 2$, $\Gamma = diag\left[5000\ 100000\right]$, $\mu = 10$. The servo was excited using the block Band-Limited White Noise from Matlab-Simulink. The output of this block was subsequently filtered by a low pass filter with a cutoff frequency of $5Hz$. Fig. 3.2 shows the excitation signal. Fig. 3.3 depicts the signals ω and ω_e and parameter estimates $\hat{a}$ and $\hat{b}$ are shown in Fig. 3.4 and Fig. 3.5, respectively. Finally, Fig. 3.6 shows the identification error ε.

It can be seen that whitin a few seconds parameter estimates settles around constant values. It is also worth mentioning that the proposed parameter identification method worked out in spite of measurement noise and unmodelled dynamics. In the case of measurement noise, the tachometer employed during the experiments has a voltage ripple of $300mV$. The unmodelled dynamics correspond to the electrical dynamics associated to the servomotor and the current amplifier.

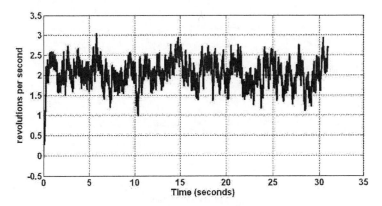

Fig. 3.2. Excitation signal ω_d.

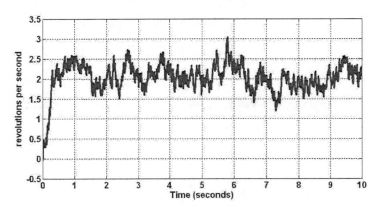

Fig. 3.3. Signals ω and ω_e.

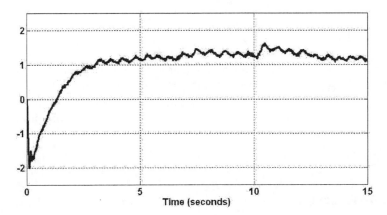

Fig. 3.4. Parameter estimated $\hat{a}$.

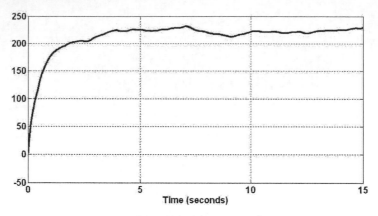

Fig. 3.5. Parameter estimated $\hat{b}$.

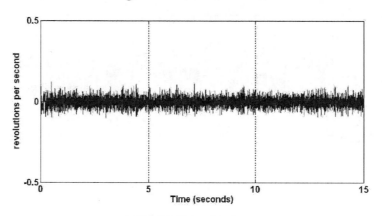

Fig. 3.6. Error signal ε.

3.4 Conclusion

This chapter presents results concerning a new methodology for closed-loop identification of servomotors. Compared with previous developments, the proposed approach allows estimating the servomotor parameters when it works in closed loop
under velocity PI control. Experiments using a laboratory prototype shows the performance of the method.

Acknowledgements

The authors would like to thank Gerardo Castro and Jesús Meza for their help in
setting up the laboratory prototype.

References

1. G. Ellis, *Control System Design Guide*, San Diego, USA: Academic Press, 2000.
2. W. Lord, J. H. Hwang, "DC Servomotors: Modeling and parameter determination", *IEEE Trans. Industrial Applications*, vol. 1A-13, no. 3, pp. 234-243, 1977.
3. W. Lord, J. H. Hwang, "DC Servomotors: Modeling and parameter determination", *IEEE Trans. Industrial Applications*, vol. 1A-13, no. 3, pp. 234-243, 1977.
4. S. Daniel-Berhe, H. Unbehauen, "Experimental physical parameter estimation of a thyristor driven DC motor using the HMF method", *Control Engineering Practice*, vol. 6, pp. 615-626, 1998.
5. S. Daniel-Berhe, H. Unbehauen, "Physical parameters estimation of the nonlinear continuous-time dynamics of a DC motor using Hartley modulating functions", *Journal of the Franklin Institute*, vol. 336, pp. 481-501, 1999.
6. S. A. Soliman, A. M. Al-Kandari, M. E. El-Hawari, "Parameter identification method of a separately excited DC motor for speed control", *Electric Machines and Power Systems*, vol. 26, no. 8, pp. 831-838, 1998.
7. Y. Y. Chen, P. Y. Huang, J. Y. Yen, "Frequency domain identification algorithms for servo systems with friction", *IEEE Trans. on Control Systems Technology*, vol. 10, no. 5, pp. 654-665, 2002.
8. T. Kara, I. Eker, "Nonlinear closed loop identification of DC motor with load for low speed two directional operation", *Electrical Engineering*, vol. 86, no. 2, pp. 87-96, 2004.
9. A. Krneta, S. Antić, Danilo Stojanović, "Recursive Least Squares Method in Parameter Identification of DC Motor Models", *Facta Universitatis, Series: Electronics and Energetics*, Vol. 18, N° 3, 2005.
10. I. Eker, "Open loop and closed loop experimental on-line identification of a three mass electromechanical system", *Mecatronics*, vol. 14, pp. 549-565, 2004.
11. M. Fliess, H. Sira-Ramirez, "Closed-loop parametric identification for continuous-time linear systems via new algebraic techniques. In *Continuous-Time Model Identification from Sampled Data*, H. Garnier & L. Wang Eds. Springer Verlag, 2008.
12. S. Sastry, M. Bodson, *Adaptive Control, Stability, Convergence and Robustness*, Prentice-Hall, 1989.

4

Practical stability of neutral and retarded type time delay systems: LMI's approach

R. Villafuerte and S. Mondié

Department of Automatic Control. CINVESTAV,
Av. Instituto Politécnico Nacional 2508. 07300 México D.F., México
{rvillafuerte,smondie}@ctrl.cinvestav.mx

Summary. In this chapter we discuss the practical stability of a class of neutral and retarded type time delay systems. First we extend the definition of practical stability, a concept closely related to ultimate boundedness, for time delay systems subject to noise or disturbances. We establish the properties a Lyapunov-Krasovskii functional must satisfy to be employed in the proof of practical stability. Then, we obtain sufficient conditions based on Lyapunov-Krasovskii functionals stated in terms of the feasibility of a set of linear matrix inequalities. A "practical" exponential estimate of the solution is also obtained. The approach is validated with some illustrative examples.

4.1 Introduction

In practice, systems usually exhibit nonlinear characteristics and are subject to various forms of disturbances. As a consequence the uncertainty in the model needs to be taken into account when designing a model-based feedback controller for the process. Furthermore, there is often an interval of time between the application of a stimulus and the system's response, hence problems due to the presence of time delays must also be addressed [6].

The stability of systems, with or without delays, is one of the main research topics for the control community. The theoretical definitions of asymptotic stability and of stability in the sense of Lyapunov used in most contributions is too restrictive when considering problems in the real world. Sometimes the state of a system may be mathematically unstable and yet the system may oscillate sufficiently near this state so that its performance is acceptable. It is clear that another notion of stability that is more suitable than Lyapunov stability is needed in such situations. The definition of Practical Stability introduced in La Salle et al. [11] and Lakshmikantham et al. [8] provides indeed an appropriate performance specification from the engineering point of view.

In the above mentioned publications the definitions and the analysis is restricted to delay free systems. Practical stability conditions for time delay systems were derived in a number of contributions: in Ghunyu Yang et al. [6] Lyapunov-Krasovskii functionals are employed to study a class of time delay systems of neutral type, in Qingling Zhang Chunyu et al. [3] the comparison principle is used for a class of descriptor systems. An analysis based on the fundamental matrix of the system is employed for linear systems with a single delay [4] and for a class of retarded linear nonautonomous systems [5].

In this chapter, following the ideas presented in the work of Poznyak [10] on the inequality the Lyapunov function associated to a system should satisfy for achieving practical stability in the case of delay free systems, we propose a new approach for the analysis of the case of neutral and retarded type time delay systems. The definition of practical stability of systems with delays is introduced in section 4.2. The case of neutral time delay systems is addressed in section 4.3: The results for delay free systems introduced in [10] are extended and technical results linking the stability of the difference operator of the neutral system with that of the state are developed in section 4.3.1. Next, sufficient practical stability conditions in the Lyapunov-Krasovskii stability framework are obtained in section 4.3.2: the characterization is given in terms of the feasibility of linear matrix inequality derived from the Lyapunov-Krasovskii functional proposed in Kharitonov et al. [7]. It should be mentioned that because of the type of functional we use, the feasibility of the LMI's implies a "practical" exponential convergence of the solutions of the neutral type time delay system. Section 4.4 on retarded time delay system is organized similarly and it is completed with stabilization results. The approach is validated with some examples and a case study in sections 4.5 and 4.6, respectively. The contribution ends with some concluding remarks.

4.2 Definition

In order to introduce the formal definition of practical stability, we consider general time delay systems of the form

$$\dot{x}(t) = f(x_t, \dot{x}_t) + n(t)$$
$$x(\theta) = \phi(\theta), \quad \theta \in [-h, 0],$$

where f is a function satisfying a Lipchitz condition with respect to its arguments, $\phi \in C[-h, 0]$ is the initial condition, and $n(t)$ is an external signal such that $\|n(t)\| \leq \gamma$, $t \geq 0$. For $t \geq 0$ we denote by $x(t, \phi)$ the solution of the system with initial condition ϕ, by $x_t(\phi) = \{x(t + \theta, \phi) : \theta \in [-h, 0]\}$ the segment of trajectory of the system and by $\mathfrak{C} := C([-h, 0], \mathbb{R}^n)$ the Banach space with norm $\|\phi\|_h := \max_{\theta \in [-h, 0]} \|\phi(\theta)\|$.where and $n(t)$ is an external signal such that $\|n(t)\| \leq \gamma$, $t \geq 0$.

Definition 1. *The system (4.1) is said to be μ-practically stable if for some $\mu > 0$ there exists $T(\mu, \phi)$ such that $\|x(t)\| \leq \mu$ for $t \geq T(\mu, \phi)$.*

4.3 Neutral type time delay systems

We consider neutral type time delay systems of the form

$$\dot{x}(t) + D\dot{x}(t-h) = A_0 x(t) + A_1 x(t-h) + n(t)$$
$$x(\theta) = \phi(\theta), \quad \theta \in [-h, 0], \tag{4.1}$$

where $A_0, A_1 \in \mathbb{R}^{n \times n}$, $D \in \mathbb{R}^{n \times n}$ is Schur stable, $h \geq 0$ is the time delay and $n(t)$ is an external signal such that $\|n(t)\| \leq \gamma$, $t \geq 0$.

4.3.1 Preliminary results

Next, we introduce a preliminary result in the Lyapunov-Krasovskii functional framework for the practical stability of neutral time delay system of the form (4.1).

Lemma 1. *Let a time delay system (4.1) be given. If there exists a functional $v(x_t)$ such that*

$$\alpha_1 \|x(t) + Dx(t-h)\|^2 \leq v(x_t) \leq \alpha_2 \|x_t\|_h^2, \ \forall t \geq 0, \tag{4.2}$$

and

$$\frac{d}{dt} v(x_t) \leq -2\sigma v(x_t) + \kappa \sqrt{v(x_t)}, \quad \forall t \geq 0, \tag{4.3}$$

for positive constants α_1, α_2, σ and κ. Then for any initial condition $\phi \in \mathfrak{C}$, the solution of system (4.1) satisfies:

$$\|x(t) + Dx(t-h)\| \leq \sqrt{\frac{\alpha_2}{\alpha_1}} e^{-\sigma t} \|\phi\|_h + \frac{\kappa}{2\sigma\sqrt{\alpha_1}} (1 - e^{-\sigma t}), \ \forall t \geq 0. \tag{4.4}$$

Proof. Premultiplication of (4.3) by $e^{2\sigma\theta}$ yields

$$\frac{\frac{d}{d\theta}\left(e^{2\sigma\theta} v(x_\theta)\right)}{\sqrt{e^{2\sigma\theta} v(x_\theta)}} \leq \kappa e^{\sigma\theta}.$$

Integration from 0 to t gives

$$\sqrt{e^{2\sigma t} v(x_t)} - \sqrt{v(\phi)} \leq \frac{\kappa}{2\sigma}(e^{\sigma t} - 1)$$

hence

$$\sqrt{v(x_t)} \leq \frac{\kappa}{2\sigma}(1 - e^{-\sigma t}) + e^{-\sigma t}\sqrt{v(\phi)}.$$

and it follows that

$$\|x(t) + Dx(t-h)\| \leq \sqrt{\frac{\alpha_2}{\alpha_1}} e^{-\sigma t} \|\phi\|_h + \frac{\kappa}{2\sigma\sqrt{\alpha_1}} (1 - e^{-\sigma t}), \ \forall t \geq 0.$$

The above result allows to conclude on the response of the linear combination $x(t) + Dx(t-h)$ and not, as wished, on $x(t)$. The following Lemma, which extends results presented in [7], is instrumental in establishing the practical stability of the variable $x(t)$.

Lemma 2. *Consider the system*

$$x(t) + Dx(t-h) = f(t), \quad t \geq 0, \tag{4.5}$$

where the matrix $D \in \mathbb{R}^{n \times n}$ is Schur stable, and

$$\|f(t)\| \leq \delta e^{-\sigma t} + \beta, \quad t \geq 0. \tag{4.6}$$

Then, for any initial condition $\phi \in \mathfrak{C}$, the solution of (4.1) satisfies the inequality

$$\|x(t,\phi)\| \leq \left[\zeta \|\phi\|_h + \delta + \frac{\zeta}{\varepsilon h r e}(\delta + \beta) \right] e^{-(1-\varepsilon)rt}$$
$$+ \beta, \quad t \geq 0, \tag{4.7}$$

with $\zeta = \sqrt{\lambda_{\max}(s)/\lambda_{\min}(s)}$, $r = \min\{\lambda/h, \sigma\}$, $\lambda = -\ln(\varsigma)$, $\varsigma \in (0,1)$, and $\varepsilon \in (0,1)$.

Proof. Let $t = kh + \xi$, $\xi \in [-h,0)$ and $k = 1, \ldots, n$. Iterating k times (4.5), we have that

$$x(t,\phi) = (-D)^k \phi(\xi) + f(t) + \sum_{i=1}^{k-1} (-D)^i f(t-ih). \tag{4.8}$$

Now, the Schur stability of matrix D implies that there exists a positive definite matrix S and a positive scalar $\varsigma \in (0,1)$ such that

$$D^T S D - \varsigma^2 S < 0,$$

hence, the matrix D satisfies the bound

$$\|D^i\| \leq \zeta e^{-\lambda i}, \tag{4.9}$$

with $\zeta = \sqrt{\lambda_{max}(S)/\lambda_{min}(S)}$ and $\lambda = -\ln(\varsigma)$.

Therefore, substituting (4.9) into (4.8) we obtain that

$$\|x(t,\phi)\| \leq \zeta e^{-\lambda k} \|\phi\|_h + \delta e^{-\sigma t} + \beta + \sum_{i=1}^{k-1} \zeta e^{-\lambda i} (\delta e^{-\sigma(t-ih)} + \beta)$$

$$= \zeta e^{-\lambda k} \|\phi\|_h + \delta e^{-\sigma t} + \delta \zeta \sum_{i=1}^{k-1} e^{-\lambda i} e^{-\sigma(t-ih)}$$

$$+ \beta \zeta \sum_{i=1}^{k-1} e^{-\lambda i} + \beta, \quad \forall t \geq 0.$$

Choosing $r = \min\{\lambda/h, \sigma\}$, we see that $e^{-\lambda k} \leq e^{-rt}$, $e^{-\sigma t} \leq e^{-rt}$, $e^{\lambda i} e^{-\sigma(t-ih)} \leq e^{-rt}$, $i = 1, \ldots, k-1$ and $(k-1) \leq t/h$, then the following inequality holds:

$$\|x(t,\phi)\| \leq \zeta \|\phi\|_h e^{-rt} + \delta e^{-rt} + \zeta \frac{t}{h}(\beta + \delta) e^{-rt} + \beta, \quad \forall t \geq 0.$$

Now, using simple calculations one can verify that for any $\varepsilon \in (0,1)$,

$$\max_{t \geq 0}\{te^{-\varepsilon rt}\} = \frac{1}{\varepsilon re},$$

therefore,

$$\|x(t,\phi)\| \leq \zeta\|\phi\|_h e^{-(1-\varepsilon)rt} + \delta e^{-(1-\varepsilon)rt} + \frac{\zeta}{h}(\beta+\delta)te^{-rt} + \beta$$

$$= \zeta\|\phi\|_h e^{-(1-\varepsilon)rt} + \delta e^{-(1-\varepsilon)rt} + \frac{\zeta}{h}(\beta+\delta)te^{-rt}\left(\frac{e^{\varepsilon rt}}{e^{\varepsilon rt}}\right) + \beta$$

$$= \left[\zeta\|\phi\|_h + \delta + \frac{\zeta}{h}(\beta+\delta)te^{-\varepsilon rt}\right]e^{-(1-\varepsilon)rt} + \beta$$

$$\leq \left[\zeta\|\phi\|_h + \delta + \frac{\zeta}{\varepsilon hre}(\beta+\delta)\right]e^{-(1-\varepsilon)rt} + \beta.$$

Remark 1. For $\|\phi\| \neq 0$, $\varepsilon \in (0,1)$, and $\theta = (1-\varepsilon)r$, the bound in (4.5) may be rewritten as

$$\|x(t,\phi)\| \leq \left[\zeta + \frac{\delta}{\|\phi\|_h} + \frac{\zeta}{\varepsilon hre\|\phi\|_h}(\beta+\delta)\right]\|\phi\|_h e^{-\theta t} + \beta, \quad \forall t \geq 0.$$

Combining Lemma 1 and Lemma 2 leads to sufficient conditions for the practical stability in terms of the existence of an appropriate functional $v(x_t)$.

Theorem 1. *Given system (4.1) let a positive definite matrix S and $\varsigma \in (0,1)$ exist such that the following matrix inequality holds:*

$$\varsigma^2 S - D^T SD > 0.$$

If there exists a functional $v(x_t)$ such that

$$\alpha_1 \|x(t) + Dx(t-h)\|^2 \leq v(x_t) \leq \alpha_2 \|x_t\|_h^2, \quad \forall t \geq 0, \tag{4.10}$$

and

$$\frac{d}{dt}v(x_t) \leq -2\sigma v(x_t) + \kappa\sqrt{v(x_t)}, \quad \forall t \geq 0, \tag{4.11}$$

where α_1, α_2, σ and κ are positive constants. Then, for any initial condition $\phi \in \mathfrak{C}$ the following "practical" exponential estimate of the response holds:

$$\|x(t,\phi)\| \leq \rho\|\phi\|_h e^{-\theta t} + \beta(1 - e^{-\theta t}), \ \forall t \geq 0,$$

and the system (4.1) is μ-practically stable for all $t \geq T(\mu,\phi)$, where

$$T(\mu,\phi) = \begin{cases} 0, & \text{if } \|\phi\|_h \leq \frac{\mu}{\rho}; \\ \frac{1}{\theta}\ln\left(\frac{\rho\|\phi\|_h - \beta}{\mu - \beta}\right), & \text{elsewhere.} \end{cases}$$

Here, $\mu > \beta$, $\rho = \left[\zeta + \sqrt{\frac{\alpha_2}{\alpha_1}} + \frac{\zeta}{\varepsilon hre}\sqrt{\frac{\alpha_2}{\alpha_1}}\right]$, $\theta = (1-\varepsilon)r$ $\beta = \frac{\kappa}{2\sigma\sqrt{\alpha_1}}$, $\zeta = \sqrt{\lambda_{max}(S)/\lambda_{min}(S)}$, $r = \min\{\lambda/h, \sigma\}$, $\lambda = -\ln(\varsigma)$, and $\varepsilon \in (0,1)$.

Proof. Inequality (4.4) of Lemma 1 implies that the solution of system (4.1) satisfies

$$\|x(t) + Dx(t-h)\| \le \delta e^{-\sigma t} + \beta,$$

where $\delta = \left[\sqrt{\frac{\alpha_2}{\alpha_1}} \|\phi\|_h - \beta \right]$, and $\beta = \frac{\kappa}{2\sigma\sqrt{\alpha_1}}$.

Therefore, using Lemma 2 and substituting δ into (4.7) we obtain that

$$\|x(t,\phi)\| \le \left[\zeta\|\phi\|_h + \sqrt{\frac{\alpha_2}{\alpha_1}}\|\phi\|_h - \beta + \frac{\zeta}{\varepsilon hre}(\beta + \sqrt{\frac{\alpha_2}{\alpha_1}}\|\phi\|_h - \beta) \right] e^{-(1-\varepsilon)rt} + \beta$$
$$= \rho\|\phi\|_h e^{-\theta t} + \beta(1 - e^{-\theta t}).$$

where $\rho = \left[\zeta + \sqrt{\frac{\alpha_2}{\alpha_1}} + \frac{\zeta}{\varepsilon hre}\sqrt{\frac{\alpha_2}{\alpha_1}} \right]$ and $\theta = (1-\varepsilon)r$, equivalently,

$$\|x(t,\phi)\| \le (\rho\|\phi\|_h - \beta) e^{-\theta t} + \beta. \tag{4.12}$$

It follows from (4.12) that for any initial conditions $\phi \in \mathfrak{C}$ such that $\|\phi\|_h \le \frac{\beta}{\rho}$ and $\beta < \mu$ the solution of system (4.1) satisfies the inequality

$$\|x(t,\phi)\| \le \mu, \ \forall\, t \ge 0.$$

Now, for initial conditions $\phi \in \mathfrak{C}$ such that $\frac{\beta}{\rho} < \|\phi\|_h < \frac{\mu}{\rho}$ we have that

$$\|x(t,\phi)\| \le \mu, \ \forall\, t \ge 0.$$

Finally, for initial conditions $\phi \in \mathfrak{C}$ such that $\frac{\mu}{\rho} < \|\phi\|_h$ it follows that

$$\|x(t,\phi)\| \le \mu \ \forall\, t \ge T(\mu,\phi),$$

where $\mu > \beta$, and the time $T(\mu,\phi)$ follows from the condition

$$0 < e^{-\theta t}(\rho\|\phi\|_h - \beta) \le \mu - \beta,$$

hence

$$T(\mu,\phi) \ge \frac{1}{\theta} \ln\left(\frac{\rho\|\phi\|_h - \beta}{\mu - \beta} \right).$$

As $\|\phi\|_h > \frac{\mu}{\rho}$ and $\mu > \beta$, then $\rho\|\phi\|_h - \beta > \mu - \beta > 0$, therefore $T(\mu,\phi)$ exists.

4.3.2 Main results: Practical stability of neutral type delay systems

In this section we present our main result: sufficient conditions for the practical stability of neutral type time delay system of the form (4.1).

Theorem 2. *Consider a linear time delay system of the form* (4.1). *If there exist positive definite matrices $P, Q \in R^{n \times n}$ and a positive constant σ such that the following matrix inequality holds:*

$$\mathscr{M}(P,Q) + 2\sigma\mathscr{N}(P) < 0, \tag{4.13}$$

where

$$\mathscr{M}(P,Q) = \begin{bmatrix} PA_0 + A_0^T P + Q & PA_1 + A_0^T PD \\ A_1^T P + D^T PA_0 & -e^{-2\sigma h}Q + D^T PA_1 + A_1^T PD \end{bmatrix}, \tag{4.14}$$

$$\mathscr{N}(P) = \begin{bmatrix} P & PD \\ D^T P & D^T PD \end{bmatrix}, \tag{4.15}$$

then the following "practical" exponential estimate of the response is

$$\|x(t,\phi)\| \leq \rho \|\phi\|_h e^{-\theta t} + \beta(1 - e^{-\theta t}), \quad \forall\, t \geq 0, \tag{4.16}$$

and system (4.1) *is μ-practically stable with*

$$\mu > \beta, \tag{4.17}$$

and

$$T(\mu,\phi) = \begin{cases} 0, & \text{if } \|\phi\|_h \leq \frac{\mu}{\rho}; \\ \frac{1}{\theta}\ln\left(\frac{\rho\|\phi\|_h - \beta}{\mu - \beta}\right), & \text{elsewhere.} \end{cases} \tag{4.18}$$

Here $\beta = \frac{\gamma \lambda_{\max}(P)}{\sigma \lambda_{\min}(P)}$, $r = \min\{\lambda/h, \sigma\}$, $\lambda = -\ln(\varsigma)$, $\varepsilon \in (0,1)$, $\varsigma \in (0,1)$, $\theta = (1-\varepsilon)r$,

$$\rho = \sqrt{\frac{\lambda_{max}(S)}{\lambda_{min}(S)}} + \sqrt{\frac{\lambda_{\max}(P)(1 + \|D\|)^2 + h\lambda_{\max}(Q)}{\lambda_{\min}(P)}}$$
$$+ \sqrt{\frac{\lambda_{max}(S)[\lambda_{\max}(P)(1 + \|D\|)^2 + h\lambda_{\max}(Q)]}{(\varepsilon h r e)^2 \lambda_{min}(S)\lambda_{\min}(P)}}, \tag{4.19}$$

and S is a positive definite matrix such that

$$\varsigma^2 S - D^T SD > 0. \tag{4.20}$$

Proof. We consider the Lyapunov-Krasovskii functional proposed in [7]:

$$v(x_t) = [x(t) + Dx(t-h)]^T P[x(t) + Dx(t-h)]$$
$$+ \int_{-h}^{0} x^T(t+\tau)e^{2\sigma\tau}Qx(t+\tau)d\tau, \tag{4.21}$$

where P and Q are the positive definite matrices of (4.13).

First, it is straightforward to see that the functional (4.21) satisfies

$$\alpha_1 \|x(t) + Dx(t-h)\|^2 \le v(x_t) \le \alpha_2 \|x_t\|_h^2, \quad \forall\, t \ge 0, \tag{4.22}$$

where α_1 and α_2 are respectively given by

$$\alpha_1 = \lambda_{\min}(P),$$
$$\alpha_2 = \lambda_{\max}(P)(1 + \|D\|)^2 + h\lambda_{\max}(Q).$$

Next, the time derivative of $v(x_t)$ along the trajectories of system (4.1) is

$$\begin{aligned}
\frac{d}{dt} v(x_t) =& 2[x(t) + Dx(t-h)]^T P[A_0 x(t) + A_1 x(t-h) + n(t)] \\
&+ x^T(t) Q x(t) - x^T(t-h) e^{-2\sigma h} Q x(t-h) \\
&- 2\sigma \int_{-h}^0 x^T(t+\tau) e^{2\sigma\tau} Q x(t+\tau) d\tau, \quad \forall\, t \ge 0.
\end{aligned}$$

Now, observe that

$$2[x(t) + Dx(t-h)]^T Pn(t) \le 2\|x(t) + Dx(t-h)\| \|P\| \|n(t)\|.$$

The first inequality in (4.22) implies that $\|x(t) + Dx(t-h)\| \le \frac{1}{\sqrt{\alpha_1}}\sqrt{v(x_t)}$. As $\|n(t)\| \le \gamma$ it follows that

$$2[x(t) + Dx(t-h)]^T Pn(t) \le 2\frac{\gamma\|P\|}{\sqrt{\alpha_1}}\sqrt{v(x_t)},$$

therefore

$$\begin{aligned}
\frac{d}{dt} v(x_t) \le& \begin{bmatrix} x(t) \\ x(t-h) \end{bmatrix}^T \mathscr{M}(P,Q) \begin{bmatrix} x(t) \\ x(t-h) \end{bmatrix} - 2\sigma \int_{-h}^0 x^T(t+\tau) e^{2\sigma\tau} Q x(t+\tau) d\tau \\
&+ 2\frac{\gamma\|P\|}{\sqrt{\alpha_1}}\sqrt{v(x_t)}, \quad \forall\, t \ge 0,
\end{aligned}$$

where $\mathscr{M}(P,Q)$ is given in (4.14).

Observe that the functional (4.21) can be rewritten as

$$\begin{aligned}
v(x_t) =& \begin{bmatrix} x(t) \\ x(t-h) \end{bmatrix}^T \begin{bmatrix} P & PD \\ D^T P & D^T PD \end{bmatrix} \begin{bmatrix} x(t) \\ x(t-h) \end{bmatrix} \\
&+ \int_{-h}^0 x^T(t+\tau) e^{2\sigma\tau} Q x(t+\tau) d\tau.
\end{aligned}$$

Clearly, we have that

$$\begin{aligned}
&\frac{d}{dt} v(x_t) + 2\sigma v(x_t) - 2\frac{\gamma\|P\|}{\sqrt{\alpha_1}}\sqrt{v(x_t)} \\
&\le \begin{bmatrix} x(t) \\ x(t-h) \end{bmatrix}^T \{\mathscr{M}(P,Q) + 2\sigma\mathscr{N}(P)\} \begin{bmatrix} x(t) \\ x(t-h) \end{bmatrix},
\end{aligned}$$

with $\mathscr{M}(P,Q)$ and $\mathscr{N}(P)$ given by (4.14) and (4.15), respectively.

We can conclude that if the condition (4.13) holds, then

$$\frac{d}{dt}v(x_t) \leq -2\sigma v(x_t) + 2\frac{\gamma\|P\|}{\sqrt{\alpha_1}}\sqrt{v(x_t)}, \quad \forall\, t \geq 0.$$

Now, the Schur stability of matrix D implies that there exist a positive definite matrix S and a positive scalar $\varsigma \in (0,1)$ such that

$$D^T SD - \varsigma^2 S < 0.$$

Clearly, the result follows straightforwardly from Theorem 1 by substituting $k = 2\frac{\gamma\lambda_{\max}(P)}{\sqrt{\lambda_{\min}(P)}}$, $\beta = \frac{\gamma\lambda_{\max}(P)}{\sigma\lambda_{\min}(P)}$, $r = \min\{\lambda/h, \sigma\}$, $\lambda = -\ln(\varsigma)$, $\varepsilon \in (0,1)$, and

$$\rho = \sqrt{\frac{\lambda_{max}(S)}{\lambda_{min}(S)}} + \sqrt{\frac{\lambda_{\max}(P)(1+\|D\|)^2 + h\lambda_{\max}(Q)}{\lambda_{\min}(P)}}$$
$$+ \sqrt{\frac{\lambda_{max}(S)(\lambda_{\max}(P)(1+\|D\|)^2 + h\lambda_{\max}(Q))}{(\varepsilon h r e)^2 \lambda_{min}(s)\lambda_{\min}(P)}}.$$

4.4 Retarded type time delay system

In this section, we consider retarded type time delay system of the form

$$\dot{x}(t) = A_0 x(t) + A_1 x(t-h) + n(t),$$
$$x(\theta) = \varphi(\theta), \ \theta \in [-h, 0], \tag{4.23}$$

where $A_0, A_1 \in R^{n\times n}$, $h \geq 0$ is the time-delay, $n(t) \in \mathbb{R}^n$, $\|n(t)\| \leq \gamma \in \mathbb{R}$, is an external signal such that the existence and the uniqueness of the solution of system (4.23) holds, and φ is the initial condition.

4.4.1 Preliminary results

Next, we present general conditions a Lyapunov-Krasovskii functional must satisfy to insure practical stability.

Lemma 3. *Let a time-delay system of the form (4.23) be given. If there exists a functional $v(x_t)$ such that*

$$\alpha_1 \|x(t)\|^2 \leq v(x_t) \leq \alpha_2 \|x_t\|_h^2 \tag{4.24}$$

and that

$$\frac{d}{dt}v(x_t) \leq -2\sigma v(x_t) + \kappa\sqrt{v(x_t)} \tag{4.25}$$

with the positive constants α_1, α_2, σ and κ, then for a given initial condition φ, the solution of system (4.23) satisfies:

$$\|x(t)\| \le \frac{\sqrt{\alpha_2}}{\sqrt{\alpha_1}} e^{-\sigma t} \|\varphi\|_h + \frac{\kappa}{2\sigma\sqrt{\alpha_1}} (1 - e^{-\sigma t}). \qquad (4.26)$$

Furthermore, the system (4.23) is μ-practically stable with $\mu > \dfrac{\kappa}{2\sigma\sqrt{\alpha_1}}$, and

$$T = \begin{cases} 0, & if \|\varphi\|_h \le \dfrac{k}{2\sigma\sqrt{\alpha_2}}; \\[3mm] \dfrac{1}{\sigma} \ln \left(\dfrac{2\sigma\sqrt{\alpha_2}\|\varphi\|_h - \kappa}{2\sigma\sqrt{\alpha_1}\mu - \kappa} \right), & elsewhere. \end{cases} \qquad (4.27)$$

Proof. Premultiplication of (4.25) by $e^{2\sigma\theta}$ yields

$$\frac{\frac{d}{d\theta}(e^{2\sigma\theta} v(x_\theta))}{\sqrt{e^{2\sigma\theta} v(x_\theta)}} \le \kappa e^{\sigma\theta}.$$

Integration from 0 to t gives

$$\sqrt{e^{2\sigma t} v(x_t)} - \sqrt{v(\varphi)} \le \frac{\kappa}{2\sigma}(e^{\sigma t} - 1)$$

hence

$$\sqrt{v(x_t)} \le \frac{\kappa}{2\sigma}(1 - e^{-\sigma t}) + e^{-\sigma t}\sqrt{v(\varphi)}.$$

Now, it follows from (4.24) that

$$\|x(t)\| \le \frac{\kappa}{2\sigma\sqrt{\alpha_1}} + e^{-\sigma t} \left(\frac{\sqrt{\alpha_2}}{\sqrt{\alpha_1}} \|\varphi\|_h - \frac{\kappa}{2\sigma\sqrt{\alpha_1}} \right).$$

Observe that for an initial conditions φ such that $\|\varphi\|_h \le \dfrac{\kappa}{2\sigma\sqrt{\alpha_2}}$ we have that

$$\|x(t)\| \le \frac{\kappa}{2\sigma\sqrt{\alpha_1}}, \ \forall t \ge 0.$$

For an initial conditions φ such that $\|\varphi\|_h > \frac{\kappa}{2\sigma\sqrt{\alpha_2}}$ it follows that

$$\|x(t)\| \le \mu, \ \forall t \ge T(\mu, \varphi)$$

where $\mu > \dfrac{\kappa}{2\sigma\sqrt{\alpha_1}}$ and the time $T(\mu, \varphi)$ is obtained from the condition

$$0 < e^{-\sigma t} \left(\frac{\sqrt{\alpha_2}}{\sqrt{\alpha_1}} \|\varphi\|_h - \frac{\kappa}{2\sigma\sqrt{\alpha_1}} \right) \le \mu - \frac{\kappa}{2\sigma\sqrt{\alpha_1}}$$

hence,

$$T \ge \frac{1}{\sigma} \ln \left(\frac{2\sigma\sqrt{\alpha_2}\|\varphi\|_h - \kappa}{2\sigma\sqrt{\alpha_1}\mu - \kappa} \right).$$

We observe that if $\|\varphi\|_h > \dfrac{\kappa}{2\sigma\sqrt{\alpha_2}}$ and $\mu > \dfrac{\kappa}{2\sigma\sqrt{\alpha_1}}$, then $2\sigma\sqrt{\alpha_2}\|\varphi\|_h - \kappa > 0$, $2\sigma\sqrt{\alpha_1}\mu - \kappa > 0$, and $2\sigma\sqrt{\alpha_2}\|\varphi\|_h - \kappa > 2\sigma\sqrt{\alpha_1}\mu - \kappa$, hence T exists.

4.4.2 Main results: practical stability of retarded type delay systems

In this section, practical stability sufficient conditions for retarded time delay systems are presented. They are followed by practical stabilizability conditions.

Theorem 3. *If there exist positive definite matrices $P, Q \in R^{n \times n}$ and a positive constant σ such that the inequality*

$$\mathcal{M}(P,Q) + 2\sigma \mathcal{N}(P) < 0 \tag{4.28}$$

holds, where

$$\mathcal{M}(P,Q) = \begin{bmatrix} PA_0 + A_0^T P + Q & PA_1 \\ A_1^T P & -e^{-2\sigma h} Q \end{bmatrix}, \tag{4.29}$$

$$\mathcal{N}(P) = \begin{bmatrix} P & 0 \\ 0 & 0 \end{bmatrix}. \tag{4.30}$$

Then the solution of (4.23) satisfies the following "practical exponential estimate"

$$\|x(t)\| \leq \frac{\sqrt{\lambda_{max}(P) + h\lambda_{max}(Q)}}{\sqrt{\lambda_{min}(P)}} e^{-\sigma t} \|\varphi\|_h + \frac{\gamma \lambda_{max}(P)}{\sigma \lambda_{min}(P)} (1 - e^{-\sigma t}), \tag{4.31}$$

thus, we conclude that the system (4.23) is μ-practically stable with

$$\mu > \frac{\gamma \lambda_{max}(P)}{\sigma \lambda_{min}(P)}, \tag{4.32}$$

and

$$T = \begin{cases} 0, & if \quad \|\varphi\|_h \leq \dfrac{\gamma \lambda_{max}(P)}{\sigma \sqrt{\lambda_{min}(P)[\lambda_{max}(P) + h\lambda_{max}(Q)]}} \\[2em] \dfrac{1}{\sigma} \ln \left(\dfrac{\sigma \sqrt{\lambda_{min}(P)[\lambda_{max}(P) + h\lambda_{max}(Q)]} \|\varphi\|_h - \gamma \lambda_{max}(P)}{\sigma \mu \lambda_{min}(P) - \gamma \lambda_{max}(P)} \right), & elsewhere. \end{cases} \tag{4.33}$$

Proof. Consider the Lyapunov-Krasovskii functional

$$v(x_t) = x^T(t) Px(t) + \int_{-h}^{0} x^T(t + \theta) e^{2\sigma \theta} Qx(t + \theta) d\theta \tag{4.34}$$

where P and Q are the positive definite matrices of Theorem 3.
First, it is straightforward to verify that the functional (4.34) satisfies

$$\alpha_1 \|x(t)\|^2 \leq v(x_t) \leq \alpha_2 \|x_t\|_h^2, \tag{4.35}$$

where $\alpha_1 = \lambda_{min}(P)$, $\alpha_2 = \lambda_{max}(P) + h\lambda_{max}(Q)$.

Next, the time derivative of $v(x_t)$ along the trajectories of system (4.23) is

$$\frac{d}{dt}v(x_t) = 2x^T(t)P\left[A_0x(t) + A_1x(t-h) + n(t)\right]$$
$$+ x^T(t)Qx(t) - x^T(t-h)e^{-2\sigma h}Qx(t-h)$$
$$- 2\sigma \int_{-h}^{0} x^T(t+\theta)e^{2\sigma\theta}Qx(t+\theta)d\theta.$$

The majorization of the term $2x^T(t)Pn(t)$ gives

$$2x^T(t)Pn(t) \le 2\|x(t)\|\,\|P\|\,\|n(t)\|.$$

Now, it follows from the first inequality in (4.35) that $\|x(t)\| \le \dfrac{1}{\sqrt{\alpha_1}}\sqrt{v(x_t)}$. There-
fore, $\|n(t)\| \le \gamma$ implies

$$2x^T(t)Pn(t) \le 2\frac{\gamma\|P\|}{\sqrt{\alpha_1}}\sqrt{v(x_t)}.$$

Thus, we have that

$$\frac{d}{dt}v(x_t) \le \begin{bmatrix} x(t) \\ x(t-h) \end{bmatrix}^T \mathscr{M}(P,Q)\begin{bmatrix} x(t) \\ x(t-h) \end{bmatrix}$$
$$- 2\sigma \int_{-h}^{0} x^T(t+\theta)e^{2\sigma\theta}Qx(t+\theta)d\theta + 2\frac{\gamma\|P\|}{\sqrt{\alpha_1}}\sqrt{v(x_t)}.$$

Observe that the functional (4.34) can be rewritten as

$$v(x_t) = \begin{bmatrix} x(t) \\ x(t-h) \end{bmatrix}^T \begin{bmatrix} P & 0 \\ 0 & 0 \end{bmatrix}\begin{bmatrix} x(t) \\ x(t-h) \end{bmatrix} + \int_{-h}^{0} x^T(t+\theta)e^{2\sigma\theta}Qx(t+\theta)d\theta.$$

Clearly, we have that

$$\frac{d}{dt}v(x_t) + 2\sigma v(x_t) - 2\frac{\gamma\|P\|}{\sqrt{\alpha_1}}\sqrt{v(x_t)} \le \begin{bmatrix} x(t) \\ x(t-h) \end{bmatrix}^T \{\mathscr{M}(P,Q) + 2\sigma\mathscr{N}(P)\}\begin{bmatrix} x(t) \\ x(t-h) \end{bmatrix},$$

with $\mathscr{M}(P,Q)$ and $\mathscr{N}(P)$ given by (4.29) and (4.30), respectively and we can conclude that if the condition (4.28) holds, then

$$\frac{d}{dt}v(x_t) \le -2\sigma v(x_t) + 2\frac{\gamma\|P\|}{\sqrt{\alpha_1}}\sqrt{v(x_t)}.$$

Clearly, $v(x_t)$ satisfies the conditions of Lemma 3 and it follows that the system is
μ-practically stable with

$$\mu > \frac{2\frac{\gamma\|P\|}{\sqrt{\alpha_1}}}{2\sigma\sqrt{\alpha_1}} = \frac{\gamma\|P\|}{\sigma\alpha_1} = \frac{\gamma\lambda_{\max}(P)}{\sigma\lambda_{\min}(P)},$$

and

$$T = \frac{1}{\sigma} \ln \left(\frac{\sigma \sqrt{\lambda_{min}(P)[\lambda_{max}(P) + h\lambda_{max}(Q)]} \|\varphi\|_h - \gamma \lambda_{max}(P)}{\sigma \mu \lambda_{min}(P) - \gamma \lambda_{max}(P)} \right).$$

Now, it follows from (4.26) that the solution of system (4.23) satisfies:

$$\|x(t)\| \leq \frac{\sqrt{\lambda_{max}(P) + h\lambda_{max}(Q)}}{\sqrt{\lambda_{min}(P)}} e^{-\sigma t} \|\varphi\|_h + \frac{\gamma \lambda_{max}(P)}{\sigma \lambda_{min}(P)} (1 - e^{-\sigma t}), \ \forall \, t \geq 0.$$

It is straightforward to extend this result to the practical stabilization of non autonomous systems with delay.

Corollary 1. *Consider a system of the form*

$$\dot{x}(t) = A_0 x(t) + A_1 x(t - h) + Bu(t) + n(t) \tag{4.36}$$

where A_0, $A_1 \in R^{n \times n}$, $B \in R^{n \times m}$, $u(t) \in R^m$, $h \in \mathbb{R}^+$, $n(t) \in \mathbb{R}^n$ with $\|n(t)\| \leq \gamma \in \mathbb{R}^+$. If there exist positive definite matrices Q_0, $Q_1 \in R^{n \times n}$, a positive constant σ and a matrix Y_0, $Y_1 \in R^{m \times n}$ such that

$$\tilde{\mathcal{M}}(Q_0, Q_1) + 2\sigma \tilde{\mathcal{N}}(Q_0) < 0 \tag{4.37}$$

holds, where

$$\tilde{\mathcal{M}}(Q_0, Q_1) = \begin{bmatrix} A_0 Q_0 + Q_0 A_0^T + BY_0 + Y_0^T B^T + Q_1 & A_1 Q_0 + BY_1 \\ Q_0 A_1^T + Y_1^T B^T & -e^{-2\sigma h} Q_1 \end{bmatrix}, \tag{4.38}$$

$$\tilde{\mathcal{N}}(Q_0) = \begin{bmatrix} Q_0 & 0 \\ 0 & 0 \end{bmatrix}, \tag{4.39}$$

then the feedback control law

$$u(t) = Y_0 Q_0^{-1} x(t) + Y_1 Q_0^{-1} x(t - h) \tag{4.40}$$

μ-practically stabilizes the system (4.36) with

$$\mu > \frac{\gamma \lambda_{max}(Q_0^{-1})}{\sigma \lambda_{min}(Q_0^{-1})}, \tag{4.41}$$

and

$$T = \begin{cases} 0, & \text{if } \|\varphi\|_h \leq c_1 \\ \dfrac{1}{\sigma} \ln(c_2), & \text{elsewhere}, \end{cases} \tag{4.42}$$

where

$$c_1 = \frac{\gamma \lambda_{max}(Q_0^{-1})}{\sigma \sqrt{\lambda_{min}(Q_0^{-1})[\lambda_{max}(Q_0^{-1}) + h\lambda_{max}(Q_0^{-1} Q_1 Q_0^{-1})]}},$$

$$c_2 = \frac{\sigma \sqrt{\lambda_{min}(Q_0^{-1})[\lambda_{max}(Q_0^{-1}) + h\lambda_{max}(Q_0^{-1} Q_1 Q_0^{-1})]} \|\varphi\|_h - \gamma \lambda_{max}(Q_0^{-1})}{\sigma \mu \lambda_{min}(Q_0^{-1}) - \gamma \lambda_{max}(Q_0^{-1})}.$$

Furthermore,

$$\|x(t)\| \leq \frac{\sqrt{\lambda_{\max}(Q_0^{-1}) + h\lambda_{\max}(Q_0^{-1}Q_1Q_0^{-1})}}{\sqrt{\lambda_{\min}(Q_0^{-1})}} e^{-\sigma t} \|\varphi\|_h$$

$$+ \frac{\gamma\lambda_{max}(Q_0^{-1})}{\sigma\lambda_{min}(Q_0^{-1})}(1 - e^{-\sigma t}), \ \forall\, t \geq 0. \tag{4.43}$$

Proof. We consider a feedback control law of the form $u(t) = K_0 x(t) + K_1 x(t-h)$, $K_0, K_1 \in R^{m \times n}$. The closed-loop system (4.36) is:

$$\dot{x}(t) = (A_0 + BK_0)x(t) + (A_1 + BK_1)x(t-h) + n(t).$$

Using the result of Theorem 3 we have that if there exist positive definite matrices P, $Q \in R^{n \times n}$ and a positive constant σ such that

$$\mathcal{M}(P,Q) + 2\sigma\mathcal{N}(P) < 0 \tag{4.44}$$

holds, where

$$\mathcal{M}(P,Q) = \begin{bmatrix} P(A_0 + BK_0) + (A_0 + BK_0)^T P + Q & P(A_1 + BK_1) \\ (A_1 + BK_1)^T P & -e^{-2\sigma h}Q \end{bmatrix},$$

$$\mathcal{N}(P) = \begin{bmatrix} P & 0 \\ 0 & 0 \end{bmatrix},$$

then the closed loop system is μ-practically stable. Left and right multiplication of (4.44) by $\mathrm{diag}\{P^{-1}, P^{-1}\}$ and setting $Q_0 = P^{-1}$, $Q_1 = P^{-1}QP^{-1}$, $Y_0 = K_0 P^{-1}$ and $Y_1 = K_1 P^{-1}$ we have

$$\tilde{\mathcal{M}}(Q_0, Q_1) + 2\sigma\tilde{\mathcal{N}}(Q_0) < 0,$$

where $\tilde{\mathcal{M}}(Q_0, Q_1)$ and $\tilde{\mathcal{N}}(Q_0)$ are given by (4.38) and (4.39). Moreover,

$$u(t) = Y_0 Q_0^{-1} x(t) + Y_1 Q_0^{-1} x(t-h).$$

Finally, (4.41), (4.42) and (4.43) follow from (4.32), (4.33) and (4.31), respectively.

4.5 Illustrative examples

Unlike previous contributions on the topic, our results allows as to actually compute the constant μ and the time $T(\mu, \phi)$ that characterize the practical stability and to determine a "practical" exponential estimate of the system response. The above is illustrated in the following examples:

Example 1. Consider the system

$$\dot{x}(t) + D\dot{x}(t-h) = A_0 x(t) + A_1 x(t-h) + n(t) \tag{4.45}$$

where $h = 0.5$, $\|n(t)\| \leq \gamma = 0.1$,

$$A_0 = \begin{pmatrix} -4 & -1 \\ 0 & -3 \end{pmatrix}, A_1 = \begin{pmatrix} 2 & 0 \\ 1 & 1 \end{pmatrix} \text{ and } D = \begin{pmatrix} -0.2 & 0 \\ -0.1 & -0.2 \end{pmatrix}.$$

Choosing $\varsigma = 0.33$ the LMI (4.20) remain feasible for

$$S = \begin{pmatrix} 1.3025 & -0.019 \\ -0.019 & 1.3026 \end{pmatrix},$$

which implies that $\lambda_{min}(S) = 1.3022$ and $\lambda_{max}(S) = 1.3029$. The LMI (4.13) is feasible for $\sigma = 1.1$ and the positive definite matrices

$$P = \begin{pmatrix} 3.395 & -0.001 \\ -0.001 & 3.394 \end{pmatrix}, Q = \begin{pmatrix} 8.62 & -1.4 \\ -1.4 & 2.32 \end{pmatrix},$$

which implies that $\lambda_{min}(P) = 3.393$, $\lambda_{max}(P) = 3.396$, $\lambda_{min}(Q) = 2.019$ and $\lambda_{max}(Q) = 8.913$. Substituting into (4.19) for $\varepsilon = 0.01$ we obtain $\rho = 1.1648$.

Thus, it follows from (4.17) that the system (4.45) is μ-practically stable with $\mu > \beta = 0.0909$. For a choice of $\mu = 0.0959$ and $\|\phi\|_h = 0.15$ we obtain from (4.18) that $T(\mu, \phi) = 7.48$, as shows on Fig. 4.1.

Finally, the solution of system (4.45) satisfy the following "practical" exponential estimate of the form (4.16):

$$\|x(t, \phi)\| \leq 116.5 \|\phi\|_{0.5} e^{-1.089t} + 0.09(1 - e^{-1.089t}), \forall t \geq 0.$$

Remark 2. Observe on Fig. 4.2 that the norm of the solution is bounded by $\mu = 0.095$ which is actually smaller than the norm $\gamma = 0.1$ of the external signal $n(t)$.

Remark 3. Observe that to obtain a μ of smaller value in (4.32), it is possible to minimize the condition number of the positive define matrix P, i.e. minimize $\lambda_{max}(P)/\lambda_{min}(P)$. This problem can be reformulated as a GEVP (Generalized eigenvalue problem) [2]: minimize $\kappa > 0$ subject to $\alpha I_n < P < \kappa \alpha I_n$, $\alpha > 0$. Thus, this additional restriction should be included to achieve better results.

Example 2. Consider the model of a mechanical system with self-excited oscillations with a retarded action [9],

$$\ddot{\theta}(t) + 2\delta\omega\dot{\theta}(t) + \rho\theta(t - h) + \omega^2\theta(t) = M\sin(kt), \tag{4.46}$$

where δ, ρ and ω are well-known physical constants, θ is the angle of oscillation, h is the time lag which we assume to be constant, and $M\sin(kt)$ is the external moment.
By introducing the new variable $x(t) = [\theta(t)\ \dot{\theta}(t)]^T$, system (4.46) can be rewritten as:

$$\dot{x}(t) = \begin{pmatrix} 0 & 1 \\ -\omega^2 & -2\delta\omega \end{pmatrix} x(t) + \begin{pmatrix} 0 & 0 \\ \rho & 0 \end{pmatrix} x(t - h) + \begin{pmatrix} 0 \\ M\sin kt \end{pmatrix}. \tag{4.47}$$

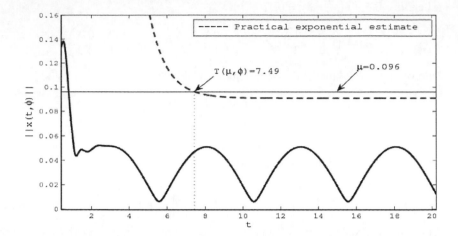

Fig. 4.1. Practical stability of system (4.45)

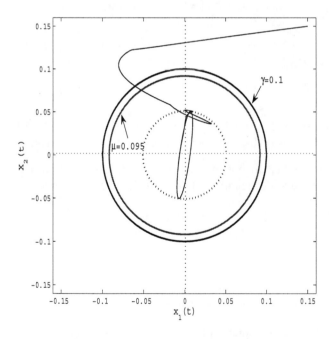

Fig. 4.2. Phase diagram of system (4.45)

For $\omega = 3.1321$, $\delta = 1.6762$, $\rho = 0.32$, $h = 0.5$, $M = 0.1$, and $k = 1$. Notice that $\|n(t)\| \leq 0.1$. The LMI conditions (4.28) is feasible for $\sigma = 0.94$ and

$$P = \begin{pmatrix} 0.4917 & 0.1833 \\ 0.1833 & 0.1983 \end{pmatrix}, \quad Q = \begin{pmatrix} 1.1160 & 1.2871 \\ 1.2871 & 1.4942 \end{pmatrix}.$$

It follows from (4.31) in Theorem 3 that the solution of system (4.47) satisfies:

$$\|x(t)\| \le 4.134 e^{-0.9t} \|\varphi\|_{0.5} + 0.585(1 - e^{-0.9t}).$$

Furthermore, (4.32) and (4.33) imply that

$$\|x(t)\| \le 0.585, \quad \forall\, t > 10.9.$$

Now, we consider parameter values from which the system (4.47) is unstable and we apply the results of Theorem 3 to obtain a control law that μ-practically stabilizes it.

Example 3. Consider now a nonautonomous system of the form

$$\dot{x}(t) = \begin{pmatrix} 0 & 1 \\ -\omega^2 & -2\delta\omega \end{pmatrix} x(t) + \begin{pmatrix} 0 & 0 \\ \rho & 0 \end{pmatrix} x(t-h) + \begin{pmatrix} 2 & 0 \\ 0 & 2 \end{pmatrix} u(t) + \begin{pmatrix} 0 \\ M\sin(kt) \end{pmatrix}.$$
(4.48)

First, we observe that for $\omega = 3.1321$, $\delta = 0.006$, $\rho = 0.9$, $h = 1.5$, $M = 0.1$, and $k = 1$, $\|n(t)\| \le 0.1$, the system (4.48) is unstable (see Fig. 4.3).

Now, applying the results of Theorem 3, the LMI conditions (4.28) is feasible for $\sigma = 1.1$ and

$$Q_0 = \begin{pmatrix} 0.5482 & 0 \\ 0 & 0.55 \end{pmatrix}, \quad Q_1 = \begin{pmatrix} 0.5126 & 0 \\ 0 & 0.5147 \end{pmatrix}$$
$$Y_0 = \begin{pmatrix} -5.3811 & 1.207 \\ 1.207 & -5.8427 \end{pmatrix}, \quad Y_1 = \begin{pmatrix} 0 & 0 \\ 0 & 0 \end{pmatrix}$$

It follows from (4.40) that the feedback control law

$$u(t) = \begin{pmatrix} -9.8159 & 2.1945 \\ 2.2017 & -10.6232 \end{pmatrix} x(t) \tag{4.49}$$

μ-practically stabilizes the system (4.48) with $\mu = 0.0909$, and $T = 5.7896$. Furthermore, (4.31) implies that the solution of the system satisfies:

$$\|x(t)\| \le 1.4 e^{-1.1t} \|\varphi\|_{1.5} + 0.09(1 - e^{-1.1t}), \quad \forall\, t > 0.$$

The practical stabilization of the system (4.48) with the control law (4.49) is shown on Fig. 4.4. The phase diagram of the system is shown on Fig. 4.5.

Remark 4. We observe that the value $\mu = 0.0909$ achieved for the controlled system is smaller than the noise bound, $\|n(t)\| \le 0.1$.

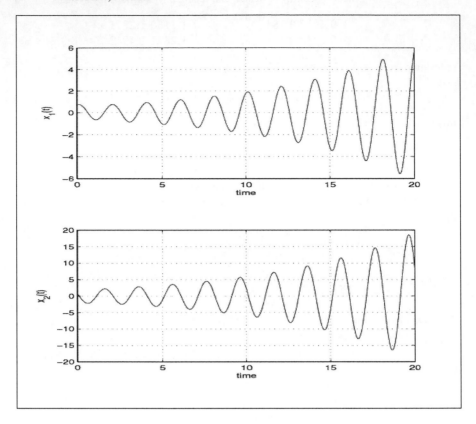

Fig. 4.3. Instability of system (4.48).

4.6 Case study

In [1] the system of the form $\ddot{\theta}(t) + \omega^2\theta(t) = u(t)$ is stabilized with an output feedback control law $u(t) = K\theta(t-h)$ using a frequency analysis approach. Here, we stabilize the system of the form (4.50) with feedback control $u(t) = k_1\theta(t) + k_2\theta(t-h)$ using the time domain approach of Theorem 3.

Example 4. Now, we consider the system with bounded additive uncertainty $n(t) = M\sin(kt)$:

$$\ddot{\theta}(t) + \omega^2\theta(t) + n(t) = u(t), \tag{4.50}$$

whose state space model is

$$\dot{x}(t) = \begin{pmatrix} 0 & 1 \\ -\omega^2 & 0 \end{pmatrix} x(t) + \begin{pmatrix} 0 \\ -n(t) \end{pmatrix} + \begin{pmatrix} 0 \\ 1 \end{pmatrix} u(t).$$

It is clear that this system is unstable for $u(t) = 0$.

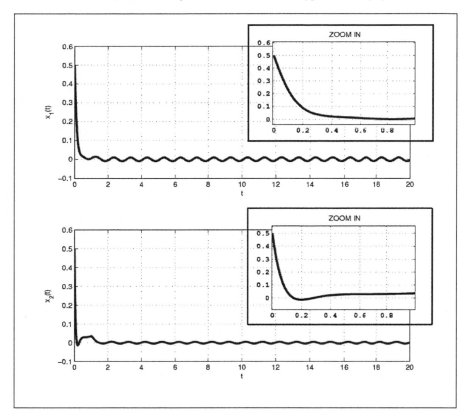

Fig. 4.4. μ-practical stabilization of system (4.48).

Applying the results of Theorem 3, we have that the LMI conditions (4.28) is feasible for $\sigma = 1.5$, $h = 0.5$, and

$$Q_0 = \begin{pmatrix} 0.23 & -0.46 \\ -0.46 & 1.61 \end{pmatrix}, \ Q_1 = \begin{pmatrix} 0.2 & 1.7 \times 10^{-5} \\ 1.7 \times 10^{-5} & 1.91 \end{pmatrix}$$
$$Y_0 = \begin{pmatrix} 0.5 & -5.51 \end{pmatrix}, \qquad Y_1 = \begin{pmatrix} -6.9 \times 10^{-4} & 0.21 \end{pmatrix}.$$

It follows from (4.40) that the feedback control law

$$u(t) = \begin{pmatrix} -9.73 & -6.18 \end{pmatrix} x(t) + \begin{pmatrix} 0.54 & 0.28 \end{pmatrix} x(t - 0.5) \tag{4.51}$$

μ-practically stabilizes the system (4.50) with $\mu = 1.16$, and $T = 7.2$ (see, Fig. 4.6). Furthermore, we derive from (4.31) that the solution of the system satisfies:

$$\|x(t)\| \leq 5.20 e^{-1.5t} \|\varphi\|_{0.5} + 1.16(1 - e^{-1.5t}), \quad \forall \, t > 0.$$

Notice that $x(t) = (\theta(t) \ \dot{\theta}(t))$, hence the control law (4.51) depends on $\dot{\theta}$. Moreover, the LMI's (4.37)-(4.39) do not admit as a solution an output control law of the

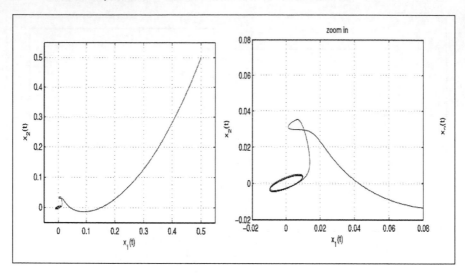

Fig. 4.5. Phase diagram of system (4.48).

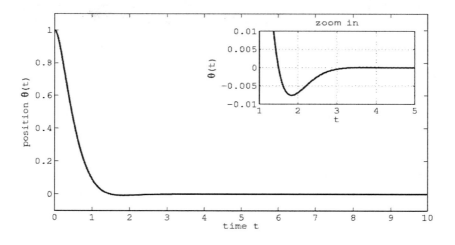

Fig. 4.6. μ-practical stabilization of system (4.50) with feedback control (4.51).

form $u(t) = k_1\theta(t) + k_2\theta(t-h)$, k_1, k_2, $h \geq 0$, due to the fact that the zero delay system is unstable. To overcome this problem, we substitute into (4.51) the approximation of the time derivative $\dot{\theta}(t) \approx \frac{\theta(t)+\theta(t-r)}{r}$ where r is sufficiently small. For $r = 0.1$, we obtain the control law

$$u(t) = -71.53\theta(t) + 61.8\theta(t-0.1) + 3.35\theta(t-0.5) - 2.82\theta(t-0.6) \quad (4.52)$$

Now, we observe that the feedback control (4.52) μ-practically stabilizes the system (4.50) as depicted on Fig. 4.7.

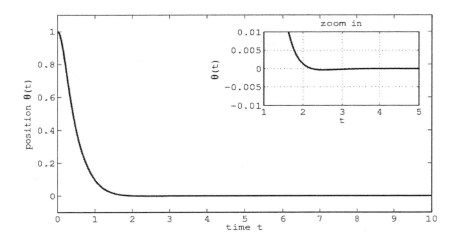

Fig. 4.7. μ-practical stabilization of system (4.50) with feedback control (4.52).

As shown on Fig. 4.8 the new output feedback control (4.52) converges more quickly to the equilibrium and its behavior is improved, specially in the context of a practical implementation because the efforts imposed on the actuators are substantially reduced.

4.7 Concluding remarks

A constructive approach for the determination of the practical stability of neutral and retarded type time delay systems is presented. The sufficient conditions we obtain for the practical stability analysis are derived in the framework of the Lyapunov-Krasovskii approach. The characterization is given in terms of easy to verify LMI conditions.

Notice that our conditions imply a "practically exponential convergence" of the solutions of the time delay system.

References

1. C. Abdallah, P. Dorato, J. Benitez-Read, and Byrne R. Delayed positive feedback can stabilize oscillatory system. *Proceedings of the American Control Conference*, 3106-3107, 1993.

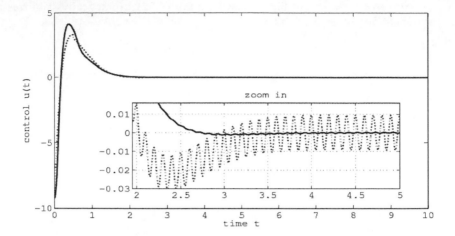

Fig. 4.8. The dotted line $(\cdots)$ is the control signal (4.51) and the solid line $(-)$ is the control signal (4.52).

2. S. Boyd, L. El Ghaoui, E. Feron, and Balakrishnan V. *Linear matrix inequalities in system and control theory.* SIAM Studies in Applied Mathematics.Philadelphia,, 1994.
3. Qingling Zhang Chunyu Yang and Linna Zhou. Practical stability of descriptor systems with time delay in terms of two measurements. *Journal of Control and Information,* 18:1-18, 2001.
4. D. Lj. Debeljkovic and S. A. Milinkovic. On practical stability of time delay systems. *Proceedings of the American Control Conference,* pg. 3235-3236, 1997.
5. D. Lj. Debeljkovic and S. A. Milinkovic. Further results on the stability of linear nonautonomous systems with delayed state defined over finite time interval. *Proceedings of the American Control Conference,* pg. 1450-1451, 2000.
6. Hu Guang-Di and Hu Guang-Da. Stabilization of an uncertain large-scale time-dependent bilinear neutral differential system by memory feedback control. *IMA Journal of Mathematical Control and Information,* 18:1-18, 2001.
7. V. Kharitonov, S. Mondié, and J. Collado. Exponential estimates for neutral time delay systems: an LMI approach. *IEEE Trans. on Autom. Contr.,* 50:666-670, 2005.
8. V. Lakshmikantham, S. Leela, and Martynyuk A.A. *Practical Stability Of Nonlinear Systems.* World Scientific Publishing Co. Pte. Ltd., 1990.
9. N. Minorsky. Self-excited mechanical oscillations. *Journal of Applied Physics,* 19:332-338, 1947.
10. A. Poznyak. *Deterministic noise effects in sliding mode observation, In: variable structure systems: from principle to implementation.* IEE-Pres (London), Chapter 3, pg. 45-80, 2004.
11. J. La Salle and S. Lefschetz. *Stability by Lyapunov's Direct Method: with applications.* Academic Press Inc (London) Ltd., 1961.

5

Relationship Between Dynamic Programming and the Maximum Principle for Impulsive Hybrid LQ Optimal Control Problems*

R. Galvan-Guerra and V. Azhmyakov

Departamento de Control Automatico, CINVESTAV
A.P. 14-740, Av. Instituto Politecnico Nacional No. 2508, C.P. 07360
Mexico D.F., Mexico
rgalvan@ctrl.cinvestav.mx

Summary. This chapter deals with optimization techniques for linear impulsive hybrid systems (LIHSs), i.e., systems with continuous dynamics and jumps in the continuous state, where the transitions between different discrete locations occur autonomously when the continuous state intersects given switching surfaces, and whose jumps can be controlled to improve the behavior of the system. We consider the LQ (linear quadratic) impulsive hybrid optimal control problem (OCP) and we apply the corresponding Pontryagin-type Maximum Principle (MP). Our aim is to investigate the natural relationship between the above MP and the Bellman Dynamic Programming (DP) approach to the impulsive hybrid OCP under consideration. We use a simple transformation of the original impulsive control system, and formulate necessary optimality conditions for the above-mentioned LQ problem, as a consequence we derive the associated Riccati-type equation, whose give us an implementable numerical algorithm, and we discuss some related numerical schemes.

5.1 Introduction

This chapter addresses optimization problems for a class of linear impulsive hybrid systems, i.e. hybrid systems with continuous dynamics and jumps in transitions times. Several impulsive dynamical systems have been studied due to their wide applications in impact mechanics, biological population management, quality control, and non–smooth optimization problems, (see e.g. [9, 15, 16]); involving many applications such as sample data control, switching power converters and intelligent vehicle highway systems.

It is well-known that the ability to operate a hybrid control system in an optimal way remains a challenging theoretical task(see e.g., [22, 17, 20, 14, 3, 11, 12, 1, 2, 4, 6, 8]). Indeed, for the general setting of hybrid systems, one has to deal not only with the infinite dimensional optimization problems related to the continuous dynamics,

---

*This work was sponsored by CONACyT

but also with a potential combinatorial explosion related to the discrete part (see e.g., [11]). In this context and with focus on particular classes, many schemes have been proposed to tackle the problem. For a deeper discussion on the main theoretical and computational results see e.g. [22, 17, 20, 14, 12, 1].

In the last years, there has been a revival of the first-order optimization techniques and related numerical schemes based on the corresponding hybrid version of the Pontryagin type MP (see e.g. [1, 6]). This fact is due to their numerical robustness, reliability and the existence of well established convergence results. On the other hand, the above-mentioned approaches are not sufficiently advanced to the LQ-type OCPs governed by LIHSs, and do not give us an easy implementable tool to compute the optimal control.

For a classical OCP governed by a closed-loop control system, one of the main tools toward the construction of optimal trajectories is the celebrated Bellman DP method. It is also well-known, that for a conventional OCP the DP approach is equivalent to the techniques based on the usual Pontryagin Maximum Principle(see e.g., [10, 13]).

The class of hybrid systems considered in this contribution involves systems driven by continuous control inputs where switching is accompanied by a jump in the state. A similar class has been considered in [21] where the authors focus their attention on state delayed systems with controlled switches and in [6] where the authors consider a general class of impulsive hybrid system and develop the corresponding hybrid MP.

The aim of this contribution is to study a possible relationship between the DP and MP in the case of a LQ impulsive hybrid OCPs. We use a simple transformation that relates the optimal control problem for the impulsive class of systems to a non–impulsive optimal control problems, for which necessary conditions of optimality have been previously derived by one of the authors [7]. These results make it possible to use conceptual algorithms and their corresponding convergence results, see e.g., [4]. Note that using transformations is a standard approach in optimal control theory and has been used extensively in the past to formulate different results.

Moreover, following this approach we are able to deduce the corresponding Riccati-formalism (similarly to the conventional LQ-theory [10, 13]), proving that this Riccati equation is a discontinuous equation in contrast to the classic case, but whose discontinuities are governed by an algebraic Riccati Equation, giving us a useful tool to the design of a hybrid linear optimal control. Using all these results we discuss some numerical schemes and present some theoretical examples to show the effectiveness of our approach.

The remainder of our chapter is organized as follows. Section 5.2 contains some preliminary facts, basic concepts and the problem formulation. We show some concepts for the case of impulsive and non–impulsive systems. Moreover, is in this part where we present the transformation that we will use in the remain of the chapter.

We also present the two forms of the different definitions to emphasize the difference between the impulsive and the non–impulsive case.

Section 5.3 is devoted to an equivalent representation of the LIHSs and to the Riccati-formalism for the LQ problems in the hybrid setting. In this section we prove the discontinuity of the Riccati equation. Section 5.4 contains the develop of the Algebraic Riccati equation that govern the jumps in the Riccati differential equation, also in this section we propose an implementable algorithm and some theoretical examples with the aim of shown the effectiveness of the proposed algorithm, and we also discuss shortly some general numerical aspects in the context of impulsive hybrid OCPs. Section 5.5 summarizes the chapter.

5.2 Optimization of Linear Impulsive Hybrid Systems

Let us start with the definition of the kind of LIHS used in this chapter. A LIHS is a linear hybrid system where the transitions between different discrete locations occur autonomously when the continuous state intersects given switching surfaces, and that can have discontinuities ("jumps") in the switching times (see Fig. 5.1); this "jumps" can be controlled to improve the behavior of the system. Now lets introducing a variant of the formal general concept of a LIHS (see e.g., [6],[1] for details).

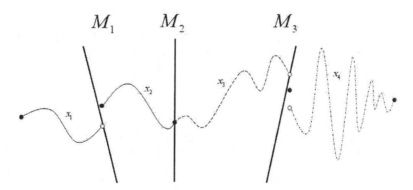

Fig. 5.1. Linear Impulsive Hybrid System (LIHS)

Definition 1. *An LIHS is an 8-tuple*

$$\chi = \{Q, X, U, A, B, \mathcal{U}, \Theta, S\}$$

where

- *Q is a finite set of locations;*

- $X = \{X_q\}, q \in Q$, where $X_q \subseteq \mathbb{R}^n$, is a family of the state spaces;
- $U \subseteq \mathbb{R}^m$ is a set of admissible control input values;
- $A = \{A_q(\cdot)\}, B = \{B_q(\cdot)\}, q \in Q$ are families of continuously matrix-functions

$$A_q : \mathbb{R} \to \mathbb{R}^{n \times n}, \ B_q : \mathbb{R} \to \mathbb{R}^{n \times m};$$

- $\mathscr{U} := \{u(\cdot) \in \mathbb{L}_m^\infty(0, t_f) : u(t) \in U \text{ a. e. on } [0, t_f]\}$ is a set of all admissible control functions;
- $\Theta = \{\Theta_q\}, q \in Q$ is a collection of maximal constant amplitudes (state jumps);
- S is a subset of Ξ, where

$$\Xi := \{(q, x, q', x') : q, q' \in Q, x \in X_q, x' \in X_{q'}\}.$$

In the following, we assume that the control set U is compact and convex. Moreover, we suppose that smooth functions $m_{q,q'} : \mathbb{R}^n \to \mathbb{R}$, $q, q' \in Q$,

$$m_{q,q'}(x) = b_{q,q'}^T x + c_{q,q'}$$

are given such that the hyperplanes are defined as

$$M_{q,q'} := \{x \in \mathbb{R}^n : m_{q,q'}(x) = 0\}.$$

The condition that the manifolds be linear can eliminated and all the results presented in this chapter remain the same, for simplicity we will assume that the manifolds are hyperplanes.

Here $b_{q,q'} \in \mathbb{R}^n$ and $c_{q,q'} \in \mathbb{R}$ for every $q, q' \in Q$. The given hyperplanes $M_{q,q'}$ represents the affine switching sets at which a switch from location q to location q' can take place. In our chapter we consider LIHSs with $r \in \mathbb{N}$ switchings. The switching times are given by the following sequence: $\{t_i\}, i = 1, \ldots, r$, where

$$0 = t_0 < t_1 < \cdots < t_{r-1} < t_r = t_f.$$

Note that the above sequence of switching times $\{t_i\}$ is not defined a priori, and the hybrid control system remains in location $q_i \in Q$ for all $t \in [t_{i-1}, t_i), i = 1, \ldots, r$.

Let us note that the pair $(q, x(t)), q \in Q, x \in \mathbb{R}^n$, represents the hybrid state at time t. Now let us recall some standard spaces, namely, the space $\mathbb{C}_0^\infty(0, t_f)$ of all C^∞ functions that vanish outside a compact subset of $(0, t_f)$ and the space $D'(o, t_f)$ of generalized functions (Schwartz distributions). It is well-known that $D'(o, t_f)$ can be considered as a space of linear, sequentially continuous functionals with respect to the convergence on the space $\mathbb{C}_0^\infty(0, t_f)$. We are ready to introduce the notion of a linear hybrid trajectory to the LIHS under consideration (see e.g., [6]).

Definition 2. *A hybrid trajectory of linear IHS is a triple $X = (x(\cdot), \{q_i\}, \tau)$, where $x(\cdot) \in D'(0, t_f)$ is a discontinuous trajectory, $\{q_i\}, i = q, \ldots, r$ is a finite sequence of locations and τ is the corresponding sequence of switching times such that for each $i = 1, \ldots, r$ there exist $u(\cdot) \in \mathscr{U}$ such that:*

– $x(0) = x_0 \notin \bigcup_{q,q' \in Q} M_{q,q'}$ and $x_i(\cdot) = x(\cdot)|_{(t_{i-1},t_i)}$ *is an absolutely continuous func-*
tion on (t_{i-1}, t_i);
– $x_i(t_i) \in M_{q_i, q_{i+q}}$ *for i=1,...,r-1;*
– $\dot{x}_i(t) = A_i(t)x_i(t) + B_i(t)u_i(t) + \theta_{q_i}\delta(t - t_i)$ *for almost all* $t \in [t_{i-1}, t_i]$, *where*
δ *is the Dirac function and* $\|\theta_{q_i}\| \leq \Theta_{q_i}$ *and* $u_i(\cdot)$ *is a restriction of the chosen*
control function $u(\cdot)$ *on the time interval* $[t_{i-1}, t_i]$.

The derivative $\dot{x}_i(\cdot)$ in Definition 2 is considered as a weak derivative of the generalized function $x_i(\cdot)$ defined on the full interval $[t_{i-1}, t_i]$. It is also evident that a function $x(\cdot)$ from definition 2 consists of absolutely continuous parts defined on the open interval (t_{i-1}, t_i) and involves jumps of magnitude θ_{q_i} at the switching times t_i. The global evolution equation of the given LIHS can also be represented as follows.

$$\dot{x}(t) = \sum_{i=1}^{r} \beta_{[t_{i-1},t_i)}(t)\left[A_i(t)x_i(t) + B_i(t)u_i(t)\right] + \sum_{i=1}^{r} \theta_{q_i}\delta(t - t_i) \text{ a.e. on } [0, t_f],$$

$$x(0) = x_0$$

$$(5.1)$$

where $\beta_{[t_{i-1},t_i)}(\cdot)$ is the characteristic function of the interval $[t_{i-1}, t_i)$

$$\beta_{[t_{i-1},t_i)}(t) = \begin{cases} 1 \text{ if } t \in [t_{i-1}, t_i) \\ 0 \quad \text{otherwise} \end{cases}$$

for $i = 1, \ldots, r$.

Note that the initial value problem (5.1) is also considered in the sense of weak derivatives on the space $D'(0, t_f)$ and for each $u(\cdot) \in \mathcal{U}$ and all $\|\theta_{q_i}\| \leq \Theta_{q_i}$, $i = 1, \ldots, r$, the initial value problem (5.1) has a unique solution in $D'(0, t_f)$.

Let $S_f : \mathbb{R} \to \mathbb{R}^{n \times n}$, $S_q : \mathbb{R} \to \mathbb{R}^{n \times n}$ and $R_q : \mathbb{R} \to \mathbb{R}^{m \times m}$, where $q \in Q$. Assume that S_f is symmetric and positive semidefinite, and that for every time instant $t \in [0, t_f]$ and every $q \in Q$, $S_q(t)$ is also a symmetric and positive semidefinite matrix. Moreover, let $R_q(t)$ be a symmetric and positive definite for every $t \in [0, t_f]$ and every $q \in Q$. We also assume that the given matrix-functions $S_q(\cdot)$, $R_q(\cdot)$ are continuous. Given an LIHS we consider the following LQ-type problem:

minimize

$$J(u, \theta, (\cdot), x(\cdot)) := \frac{1}{2}\left(x^T(t_f)S_fx(t_f)\right) + \sum_{i=1}^{r}\frac{1}{2}\int_{t_{i-1}}^{t_i}\left(x^T(t)S_{q_i}(t)x(t) + u^T(t)R_{q_i}(t)u(t)\right) dt$$

over all admissible trajectories X of LIHS

$$(5.2)$$

Evidently (5.1) is the problem of minimizing the quadratic Bolza cost functional J over all trajectories of the given linear hybrid system. Note that we study the impulsive hybrid OCP (5.1) in the absence of possible target and state constraints.

Throughout this chapter we assume that the LQ problem (5.2) has an optimal solution

$$\left(u^{opt}\left(\cdot\right),\theta^{opt},X^{opt}\left(\cdot\right)\right) \in \mathscr{C} := \mathscr{U} \times \mathbb{R}^{n \times r} \times Q^r \times \left[0,t_f\right]^r,$$

where $\theta^{opt} := \left(\theta_{q_1}^{opt} \ldots \theta_{q_r}^{opt}\right)$ is a matrix representing the optimal jumps. The optimal control problem 5.2 is an optimization problem formulated on the space $\mathscr{C}$ which involves the space of generalized functions $D'\left(0,t_f\right)$. Our aim is to simplify the initial problem 5.2. Considering the following auxiliary initial value problem (see [6])

$$\dot{y}\left(t\right) = \sum_{i=1}^{r} \beta_{[t_{i-1},t_i)}\left(t\right)\left[A_i\left(t\right)\left(y_i\left(t\right) + \theta_{q_i}\eta\left(t - t_i\right)\right) + B_i\left(t\right)u_i\left(t\right)\right] \text{ a.e. on } \left[0,t_f\right],$$

$$y\left(0\right) = x_0,$$

(5.3)

where $i = 1,\ldots,r$ and $\eta\left(\cdot\right)$ is a Heaviside step-function, we could formulate an auxiliary hybrid OCP governed by a hybrid system with autonomous location transitions (without jumps in the continuous state, see [6]). Evidently, the initial value problem (5.3) has a unique absolutely continuous solution for each $u\left(\cdot\right) \in \mathscr{U}$. We could consider $y\left(\cdot\right)$ as an element of the Sobolev space $x\left(\cdot\right) \in \mathbb{W}_n^{1,\infty}\left(0,t_f\right)$, i.e, the space of absolutely continuous functions with essentially bounded derivatives. We have the following equivalence result.

Theorem 1. *Under the assumptions of this Section, the (unique) solution $x\left(\cdot\right) \in D'\left(0,t_f\right)$ of the initial value problem (5.1) can be represented in the following form:*

$$x\left(t\right) = y\left(t\right) + \sum_{i=1}^{r} \theta_{q_i}\eta\left(t - t_i\right),$$

where $y\left(\cdot\right) \in \mathbb{W}_n^{1,\infty}\left(0,t_f\right)$ is a (unique) solution to the auxiliary initial value problem (5.3)

We refer to [6] for the proof of this result. Using Theorem 1 we could study an auxiliary hybrid system with autonomous location transitions (see [4]). Recall the corresponding concept.

Definition 3. *A hybrid control system with autonomous location transitions is the following 7-tuple*

$$\zeta = \{Q,X,U,A,B,\mathscr{U},S^a\}$$

where $Q,X,U,\mathscr{U},A,B$ are from Definition 1 and

$$S^a \subset \varXi^a := \{(q,y,q',y') \, , \, q,q' \in Q, y \in X_q, y' \in X_{q'}\}.$$

Moreover a hybrid trajectory of ζ is a triple

$$Y = \left(y(\cdot), \{q_i\}^a, \tau^a \right),$$

where $y(\cdot) : [0, T] \to \mathbb{R}^n$, and for each $i = 1, \ldots, r$ there exist $u(\cdot) \in \mathcal{U}$ such that:

- $y(0) = y_0$ and $y_i(\cdot) = y(\cdot)|_{(t_{i-1}, t_i)}$ is an absolutely continuous function on (t_{i-1}, t_i) continuously prolongable to $[t_{i-1}, t_i]$, $i = 1, \ldots, r$;
- $\dot{y}_i(t) = A_i(t) y_i(t) + B_i(t) u_i(t)$ for almost all $t \in [t_{i-1}, t_i]$, $i = 1, \ldots, r$

The switching manifolds $M_{q_i, q_{i+1}}$ are now characterized by the following equations:

$$m_{q, q'} \left(y(t) + \sum_{i=1}^{r} \theta_{q_i} \eta(t - t_i) \right) = 0, \quad i = 1, \ldots, r.$$

Now we could formulate the auxiliary OCP using the above transformation and the initial value problem 5.3

$$\text{minimize} \ \frac{1}{2} \left[\left(\left(y(t_f) + \theta_{q_r} \right)^T S_f \left(y(t_f) + \theta_{q_i} \right) \right) \right.$$

$$\left. + \sum_{i=1}^{r} \int_{t_{i-1}}^{t_i} \left(\left(y + \theta_{q_i} \eta(t - t_i) \right)^T S_{q_i} \left(y \theta_{q_i} \eta(t - t_i) \right) \right) + \sum_{i=1}^{r} \int_{t_{i-1}}^{t_i} \left(u^T R_{q_i} u \right) dt \right] \tag{5.4}$$

over all admissible trajectories Y of ζ.

It is necessary to stress that an optimal solution $(v^{opt}(\cdot), Y^{opt}(\cdot))$, where $v := (u, \theta)$, of the auxiliary OCP (5.4) defines the corresponding optimal solution $(u^{opt}(\cdot), \theta^{opt}, X^{opt}(\cdot))$ for the problem (5.2). We formulate this relationship more precisely.

Theorem 2. *Suppose that problems (5.2) and (5.4) have both optimal solutions. Under the assumptions of this Section, every optimal solution $(v^{opt}(\cdot), Y^{opt})$ of problem (5.4) defines the corresponding optimal solution $(v^{opt}(\cdot), X^{opt}(\cdot))$ for problem (5.2), where*

$$x^{opt}(t) = y^{opt}(t) + \sum_{i=1}^{r} \theta_{q_i}^{opt} \eta(t - t_i^{opt}).$$

The proof of this result follows from the following fact: the transformation from Theorem 1 is a bijective transformation (see [6] for theoretical details). Let us now formulate the hybrid MP presented in [5] for the case of LQ hybrid OCP (5.4).

Theorem 3. *Let the matrices A, B be continuous and the optimal control problem (5.4) be regular. Then there exist a function $\psi_i(\cdot)$ from $\mathbb{W}_m^{1, \infty}(0, t_f)$ and a non-zero vector $a = (a_1, \ldots, a_{r-1})^T \in \mathbb{R}^{r-1}$ such that*

$$\frac{d}{dt} \psi_i(t) = -A_{q_i}^T(t) \psi_i(t) + S_{q_i} \left(y(t) + \theta_{q_i} \eta(t - t_i) \right) \quad a.e. \ on \ \left[t_{i-1}^{opt}, t_i^{opt} \right]$$

$$\psi_r(t_f) = -S_f \left(y^{opt}(t_f) + \theta_{q_r} \right) \tag{5.5}$$

and $\psi_i\left(t_i^{opt}\right) = \psi_{i+1}\left(t_i^{opt}\right) + a_i b_{q_i,q_{i+1}}$, where $i = 1,\dots,r-1$. Moreover, for every admissible control $u(\cdot) \in \mathscr{U}$ the partial Hamiltonian

$$H_{q_i}(t,Y,v,\psi) = \langle \psi_i, A_{q_i}(t)(y_i + \theta_{q_i}\eta(t-t_i)) + B_{q_i}(t)u_i \rangle - \frac{1}{2}(y_i + \theta_{q_i}\eta(t-t_i))^T S_{q_i}(y_i + \theta_{q_i}\eta(t-t_i)) - \frac{1}{2}u_i^T R_{q_i}u_i$$

satisfies the following maximization conditions

$$\max_{v \in \mathscr{U} \times \Theta} H_{q_i}\left(t, Y^{opt}, v, \psi(t)\right) = H_{q_i}\left(t, y^{opt}(t) + \sum_{i=1}^{r}\theta_{q_i}^{opt}\eta(t-t_i), u^{opt}(t), \psi(t)\right).$$

$$(5.6)$$

Here are $t \in \left[t_{i-1}^{opt}, t_i^{opt}\right)$, $i = 1,\dots,r$. Moreover, we use the notation $\psi(t) := \sum_{i=1}^{r}\beta_{\left[t_{i-1}^{opt},t_i^{opt}\right)}\psi_i(t)$ for all $t \in [0,t_f]$.

To avoid that a trajectory of the given LHS (5.3) has a sliding mode behavior (zeno behavior) with respect to a switching hyperplane, we will assume that the system fulfils the additional condition:

$$\alpha \dot{y}_i(t_i) + (1-\alpha)\dot{y}_{i+1}(t_i) \notin M_{q_i,q_{i+1}}; \quad \forall \alpha \in \mathbb{R}, \ i = 1,\dots,r$$

The last relation is equivalent to the following

$$b[(\alpha A_i(t_i) + (1-\alpha)A_{i+1}(t_i))(y_i(t_i) + \theta_{q_i}) + \alpha B_i u_i(t_i) + (1-\alpha)B_{i+1}u_{i+1}] + c \neq 0$$

where the value of y_i is determined by the Cauchy formula

$$y_i(t_i) = \Phi(t_i,t_0)x_0 + \sum_{j=2}^{i}\Phi(t_i,t_j)\theta_{j-1} + \sum_{j=1}^{i}\int_{t_{j-1}}^{t_j}\Phi(t_i,t_j)\Phi_j(t_j,\tau)B_j(\tau)u_j(\tau)\mathrm{d}\tau,$$

where $\Phi(\cdot)$ is a joint "hybrid fundamental matrix" (in the sense of system (5.3)) defined by

$$\Phi(t,t_j) = \Phi_i(t,t_i)\prod_{j=1}^{i}\Phi_j(t_j,t_{j-1}) := \Phi_i(t,t_i)\Phi_{i-1}(t_{i-1},t_{i-2})\Phi_{i-2}(t_{i-2},t_{i-3})\dots\Phi_1(t_1,t_0)$$

and $\Phi_i(\cdot)$ is the transfer function of the subsystem from the location q_i.

Using the correspondence between the solutions $x^{opt}(\cdot)$ and $y^{opt}(\cdot)$ of the initial value problems 5.1 and 5.3, see [6] we are now able to formulate the necessary optimality conditions for the original problem 5.2.

Theorem 4. *Under the conditions of Theorem 1 there exist a function $p_i(\cdot)$ from $\mathbb{W}_m^{1,\infty}(0,t_f)$ and a non-zero vector $d = (d_1,\ldots,d_{r-1})^T \in \mathbb{R}^{r-1}$ such that*

$$\frac{d}{dt}p_i(t) = -A_{q_i}^T(t)\,p_i(t) + S_{q_i}x^{opt}(t) \ \text{ a.e. on } \left[t_{i-1}^{opt},t_i^{opt}\right]$$
$$p_r(t_f) = -S_f x^{opt}(t_f)$$

(5.7)

and $p_i\left(t_i^{opt}\right) = p_{i+1}\left(t_i^{opt}\right) + a_i b_{q_i,q_{i+1}}$, where $i = 1,\ldots,r-1$. Moreover, for every admissible control $u(\cdot) \in \mathscr{U}$ the Hamiltonian

$$H_{q_i}(t,X,u,p) = \left\langle p_i, A_{q_i}(t)x + B_{q_i}(t)u \right\rangle - \frac{1}{2}x^T S_{q_i}x - \frac{1}{2}u^T R_{q_i}u$$

satisfies the following maximization conditions

$$\max_{v \in \mathscr{U} \times \Theta} H_{q_i}\left(t,x^{opt}(t),v^{opt}(t),p(t)\right),\ t \in \left[t_{i-1}^{opt},t_i^{opt}\right),$$

(5.8)

where $i = 1,\ldots,r$ and $p(t) := \sum_{i=1}^{r} \beta_{\left[t_{i-1}^{opt},t_i^{opt}\right)}p_i(t)$ for all time instants $t \in \left[0,t_f\right]$.

Note that the function H is a real Hamiltonian in the sense of the auxiliary OCP (5.4). On the other side this function does not represent the Hamiltonian of the original OCP (5.2). As shown in [19] the "full" Hamiltonian (in the sense of the auxiliary problem (5.4)), namely, the function

$$\tilde{H}^{opt}(t) := \sum_{i=1}^{r} \beta_{\left[t_{i-1}^{opt},t_i^{opt}\right)}(t)H_{q_i}(t,y^{opt}(t),v^{opt}(t),\psi(t))$$

computed for optimal pair $(v^{opt}(\cdot),X^{opt}(\cdot))$ and for the corresponding adjoint variable $\psi(\cdot)$ is a continuous function. Using the equivalent representation of the initial impulsive hybrid system from (2), we can write this fact in the following form:

$$\left\langle \psi_i(t), A_{q_i}(t)x_i^{opt}(t) + B_{q_i}(t)u_i^{opt}(t) \right\rangle - \left\langle \psi_{i+1}(t), A_{q_{i+1}}(t)x_{i+1}^{opt}(t) + B_{q_{i+1}}(t)u_{i+1}^{opt}(t) \right\rangle$$
$$= \frac{1}{2}x_i^T(t)S_{q_i}x_i(t) + \frac{1}{2}u_i^T(t)R_{q_i}u_i(t) - \frac{1}{2}x_{i+1}^T(t)S_{q_{i+1}}x_{i+1}(t) - \frac{1}{2}u_{i+1}^T(t)R_{q_{i+1}}u_{i+1}(t)$$

(5.9)

for all $i = 1,\ldots r-1$ and $t \in [0,t_f]$. Finally, note that the correct Hamiltonian in the sense of the original OCP (5.2) does not posses this continuity properties.

5.3 The Riccati-Formalism for Impulsive Hybrid Linear Quadratic OCPs

In this section we extend the well know Riccati techniques to the LQ hybrid OCP of the type (5.2). First Let us consider the linear boundary value problem (5.3), (5.5) for $U \equiv \mathbb{R}^m$. The maximization condition (5.6) from the above MP (Theorem 3) with respect to the first variable u of the full control vector v implies

$$u_i^{opt}(t) = R_{q_i}^{-1}(t) B_{q_i}^T(t) \psi_i(t), t \in \left[t_{i-1}^{opt}, t_i^{opt}\right] \qquad (5.10)$$

Using this representation of an optimal control and the basic facts from the theory of linear differential equations, we now compute (similar to [10], [13]) an optimal control $u^{opt}(\cdot)$ for (5.4) in the form of a partially linear feedback control law

$$u^{opt}(t) = -\sum_{i=1}^{r} \beta_{[t_{i-1}, t_i)}(t) C_i(t) \left(y_i(t) + \theta_{q_i} \eta(t - t_i)\right) \qquad (5.11)$$

where $C_i(t) = R_{q_i}^{-1}(t) B_{q_i}^T(t) P_i(t)$ is a partial gain matrix. For a LQ hybrid OCP governed by a linear hybrid system with autonomous location transitions we can find the matrix P_i as a solution to the differential equation of the Riccati type. This equation (written for every current location q_i) gives a rise to the dynamical behavior of this matrix. As shown in [7] the global Riccati matrix given by the following relation

$$P(t) = \sum_{i=1}^{r} \beta_{[t_{i-1}, t_i)}(t) P_i(t)$$

is a discontinuous matrix-function. Analogously to the above-mentioned case of a hybrid LQ problem in the absence of state jumps, for every location $q_i \in Q$ and for almost all $t \in \left(t_{i-1}^{opt}, t_i^{opt}\right)$ we obtain the following Riccati-type equation

$$\dot{P}_i(t) + P_i(t) A_{q_i}(t) + A_{q_i}^T(t) P_i(t) - P_i(t) B_{q_i}(t) R_{q_i}^{-1}(t) B_{q_i}^T(t) P_i(t) + S_{q_i} = 0. \qquad (5.12)$$

Note that the main equation (5.12) is satisfied only on the open intervals $\left(t_{i-1}^{opt}, t_i^{opt}\right)$. For the time instants $t_i \in \tau$ we have the following jump condition (see [7]):

$$\left[P_{i+1}(t_i^{opt}) - P_i(t_i^{opt})\right]\left(y_i(t_i^{opt}) + \theta_i\right) = a_i b_{q_i, q_{i+1}}. \qquad (5.13)$$

A symmetric discontinuous global Riccati matrix, see [7],

$$P(t) = \sum_{i=1}^{r} \beta_{[t_{i-1}, t_i)}(t) P_i(t)$$

which satisfies all equations (5.12) and the boundary (terminal) condition $P(t_f) = S_f$ determines the optimal feedback dynamics of (5.3). Note that the corresponding global Riccati matrix-function is in general a discontinuous function even in the case of a LQ-type OCP governed by hybrid systems with autonomous location transitions (see [7]). In the case of a LIHS the discontinuity of the global matrix $P(\cdot)$ is also caused by the state jump of the amplitude θ_i in the location $q_i \in Q$. Clearly, the above partially linear feedback control function u^{opt} is a strongly discontinuous (piecewise continuous) function of x. Using the relation between variables x and y, we now are able to rewrite the optimal feedback control in the sense of the original impulsive hybrid OCP (5.2)

$$u^{opt}(t) = -C(t)x(t) = -\sum_{i=1}^{r} \beta_{[t_{i-1},t_i)}(t)C_i(t)x_i(t)$$

where C_i is the same partial gain matrix as in the case of problem (5.4) and $P_i(\cdot)$ is the solution to the above Riccati matrix equation (5.12). Finally, note that similarly to (5.4) the optimal feedback control strategy for the original impulsive hybrid OCP (5.2) is an piecewise continuous function of x. The discontinuity points of this function are determined by switching times $t_i \in \tau$ as well as by the state jumps $\theta_i \in \Theta$.

The maximization condition (5.6) from Theorem 3 with respect to the second variable θ of the full control vector v are equivalent to the following sequence of finite dimensional maximization problems

$$\max_{\theta} H_{q_i}\left(t_i^{opt}, Y^{opt}, (u^{opt}(t_i^{opt}), \theta), \psi(t_i^{opt})\right)$$
$$\text{subject to } \|\theta_{q_i}\| \le \Theta_{q_i}, \ i = 1, ..., r-1$$
(5.14)

Since the functions H_{q_i} are a concave (quadratic with respect to θ) functions, the optimal solution for every problem (5.14) can belongs to the interior $\text{int}\{F\}$ of the admissible set

$$F := \{\xi_{q_i} \in \mathbb{R}^n : \|\xi_{q_i}\| \le \Theta_{q_i}\}$$

or satisfies the condition $\|\theta_{q_i}^{opt}\| = \Theta_{q_i}$.

In the scalar case the optimal $\theta_{q_i}^{opt}$ can be found by the following heuristic rules:

- if $i \le r - 2$ and

$$\mu_i := x(t) \bigcap M_{q_i, q_{i+1}},$$
$$\mu_{i+1} := x(t) \bigcap M_{q_{i+1}, q_{i+2}}$$

  then we have $\theta_{q_i}^{opt} = \mu_{i+1} - \mu_i$;
- if $i = r - 1$, then $\theta_{q_i}^{opt} = -\mu_i$

In general the quadratic optimization problem (5.14) can be solved by well-known effective algorithms from the theory of the quadratic programming (see e.g., [18] for details). Note that the maximization problem from Theorem 3 is determined as a minimization problem on the Cartesian product $\mathscr{U} \times \Theta$. Since the first variable of the control vector v takes the value in full space $\mathbb{R}^m$ and the second value is a finite dimensional component, we can separate the above maximization problem for the Hamiltonian H and consider two independent maximization procedures.

5.4 Numerical Approach to Impulsive Hybrid Optimization

Relations (5.13), (5.12) and the affine restrictions

$$m_{q_i, q_{i+1}}(x) = b_{q_i, q_{i+1}}^T x + c_{q_i, q_{i+1}} = 0,$$

provide a basis for an effective numerical treatment for the LQ problem (5.2). As evident the Riccati-type equations (5.12) determine the Riccati matrix $P_i(t)$ for every t from open time intervals $(t_{i-1}^{opt}, t_i^{opt})$. The continuity property of the function $\check{H}^{opt}(\cdot)$ introduced above makes it possible to derive the necessary conditions for computing the value $P_i(t_i^{opt})$ for $i = 1, ..., r - 1$. Using the continuity formula (5.9), we obtain the following relation

$$P_i(t_i^{opt}) D_{q_i}(t_i^{opt}) P_i(t_i^{opt}) - 2P_i(t_i^{opt}) A_{q_i}(t_i^{opt}) - F_{q_i, q_{i+1}}(t_i^{opt}) = 0, \qquad (5.15)$$

where

$$F_{q_i, q_{i+1}}(t) = 2\big(S_{q_{i+1}}(t) - S_{q_i}(t)\big) + P_{i+1}(t) D_{q_{i+1}}(t) P_{i+1}(t) + 2P_{i+1}(t) A_{q_{i+1}}(t)$$

and

$$D_{q_i}(t) = B_{q_i}(t) R_{q_i}^{-1}(t) B_{q_i}^T(t).$$

We now are able to summarize a general conceptual computational algorithm for the numerical treatment of the optimal partially linear feedback control in the auxiliary hybrid LQ problem (5.4). Note that in the algorithm presented below an approximating control $v^{appr}(\cdot)$ to $v^{opt}(\cdot)$, the corresponding trajectory $y^{appr}(\cdot)$ and the sequence τ^{appr} to τ^{opt} are assumed to be given. The elements of τ^{appr} approximate the optimal switching times $t_i^{opt} \in \tau^{opt}$ for every $i = 1, ..., r - 1$. The control $v^{appr}(\cdot)$ the trajectory $y^{appr}(\cdot)$, the sequence τ^{appr} and the associated sequence of the corresponding locations can be obtained in various ways, for instance, with help of the gradient-based algorithms proposed in [4], or using the optimality zone algorithms from [19].

Example 1. Let us consider a LIHS with one switching. The dynamics of the system is given by two linear equations associated with the corresponding location

$$\begin{aligned}
\dot{x}_1 &= x_1(t) + u_1(t) + \theta \delta(t - t_1) \ \forall t \in [0, t_1] \\
\dot{x}_2 &= -x_2(t) + 2u_2(t) \qquad\qquad \forall t \in [t_1, t_f]
\end{aligned}$$

where t_1 is a switching time and $x(0) = -0.04$. Lets considered various switching manifolds which are an affine-linear manifolds to show the different values that can take θ

$$M_{1,2}^{(1)}(x) = x + 0.3, \ M_{1,2}^{(2)}(x) = x + 0.1, \ M_{1,2}^{(3)}(x) = x + 0.05$$

and the value of the state jump is determined by the condition $||\theta|| \leq 0.1$. Our aim is to minimize the quadratical cost function

$$J(u(\cdot), x(\cdot)) = \frac{1}{2} \sum_{i=1}^{2} \int_{t_{i-1}}^{t_i} (x_i^2(t) + u_i^2(t)) dt$$

1) Consider an initial pair $(v^0(\cdot), y^0(\cdot))$, where $v^0 \equiv v^{appr}$ and $y^0 \equiv y^{appr}$, an initial sequence $\tau^0 \equiv \tau^{appr}$, the corresponding sequence of locations and the terminal condition $P(t_f) = S_f$ for a given LIHS. Set $k = 1$ and $l = 1$.

2) With help of the inverted-time integrating procedure compute the value $P_r(t_{r-1}^{opt})$ of the partial Riccati matrix P_r as a solution of (5.12). Using (5.15) we calculate the Riccati matrix $P_{r-1}(t_{r-1}^{opt})$.

3) By the inverted-time integrating solution define $P_{r-k}(t_{r-k-1}^{opt})$, increase k by one. If $k = r - 1$, then go to Step 4. Otherwise go to Step 2.

4) Complete all partial Riccati matrices $P_i(\cdot)$ and define the corresponding partial gain matrices

$$C_i(t) = R_{q_i}^{-1}(t) B_{q_i}^T(t) P_i(t).$$

Compute the quasi-optimal (in the sense of the above approximations) piecewise feedback control function from (5.11).

5) Complete partial Hamiltonian functions

$$H_{q_i}(t_i^{l-1}, Y^{l-1}, (u^l(t_i^{l-1}), \theta), \psi(t_i^{l-1})), \quad i = 1, ..., r;$$

where $\psi(\cdot)$ is a solution of the corresponding adjoint system (5.5). Solve the auxiliary maximization problem (5.14) and compute θ^l as a solution to (5.14) with

$$H_{q_i}(t_i^{l-1}, Y^{l-1}, (u^l(t_i^{l-1}), \theta), \psi(t_i^{l-1})).$$

6) Using the obtained quasi-optimal feedback control low and the solution to (5.14), compute the corresponding trajectory $y^l(\cdot)$ and define the hybrid trajectory pair $Y^l(\cdot)$ for problem (5.4). Determine the new approximating sequence τ^l from the conditions

$$t_i^l := \min\{t \in [0, t_f] : y^l(t) \bigcap M_{q_i, q_{i+1}} \neq \emptyset\}.$$

where $i = 1, ..., r - 1$. Finally, increase l by one and go to Step 2.

Since the presented example deals with the one-dimensional model, the Algorithm 1 can be sufficiently simplified. Note that in this one-dimensional situation a solution of (5.14) can be found from the condition

$$\max_{\theta \in [-0.1, 0.1]} H_{q_i}\left(t_1, Y^{opt}, (u^{opt}(t_1), \theta), \psi(t_1)\right).$$

Hence $\theta^{opt} \in [-0.1, 0.1]$ and an optimal value can be found by an easy sorting procedure. Applying the simplified variant of the proposed Algorithm 1 we obtain the next optimal cost values for each switching rule

$$J^{opt^{(1)}} = 0.0791, \ J^{opt^{(2)}} = 0.0054, \ J^{opt^{(3)}} = 4.8455 \times 10^{-4}$$

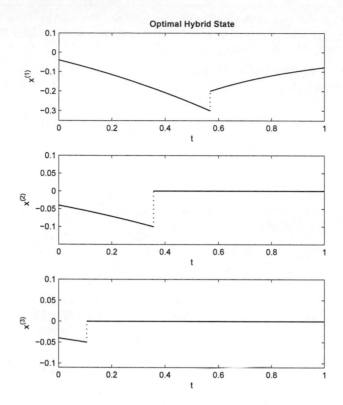

Fig. 5.2. Optimal trajectory for 3 different switching rules

with an optimal jump of $\theta^{opt(1)} = 0.1$, $\theta^{opt(2)} = 0.1$, $\theta^{opt(3)} = 0.05$, respectively. Comparison between each of the trajectories is shown in the Fig. 5.2 and the optimal control for the first switching rule is presented in Fig. 5.3.

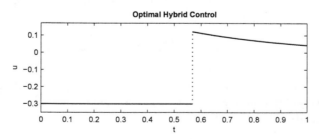

Fig. 5.3. Optimal Control for $M_{1,2}^{(1)}$

Example 2. Let us consider a LIHS with one switching. The dynamics of the system is given by two linear equations associated with the corresponding location

$$\dot{x}_1 = \begin{bmatrix} 1.5 & 0 \\ 0 & 1 \end{bmatrix} x_1(t) + \begin{bmatrix} 1 \\ 1 \end{bmatrix} u_1(t) + \theta \delta(t - t_1)$$

$$\forall t \in [0, t_1]$$

$$\dot{x}_2 = \begin{bmatrix} -0.5 & 0 \\ 0 & -3 \end{bmatrix} x_2(t) + \begin{bmatrix} 1 \\ 1 \end{bmatrix} u_2(t)$$

$$\forall t \in [t_1, t_f]$$

where t_1 is a (unknown) switching time and $x(0) = \begin{bmatrix} 1 \\ -1 \end{bmatrix}$. The switching manifold is an affine-linear manifold

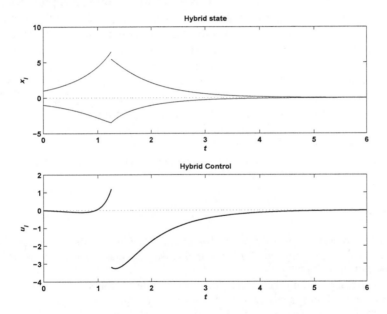

Fig. 5.4. Example 2: Optimal trajectory for the *LIHS*

$$M_{1,2}(x) = \begin{bmatrix} 1 & 1 \end{bmatrix} x - 3$$

and the value of the state jump is determined by the condition $||\theta|| \leq 1$. Our aim is to minimize the quadratical cost function

$$J(u(\cdot),x(\cdot)) = \frac{1}{2}\sum_{i=1}^{2}\int_{t_{i-1}}^{t_i}(x_i^2(t)+u_i^2(t))dt$$

Applying the proposed Algorithm 1 we obtain an optimal cost of $J^{opt} = 0.3153$ and a optimal jump of $\theta^{opt} = \begin{bmatrix} -1 \\ 0 \end{bmatrix}$. The behavior of the system is shown in Fig 5.4.

5.5 Conclusion

In this contribution, we establish the natural relationship between the hybrid MP and the Bellman DP approach for a family of impulsive hybrid LQ problems. Using a simple transformation of the original impulsive control system, we formulate necessary optimality conditions for the above-mentioned LQ problem and also derive the associated Riccati-type equation. This Riccati-type formalism provides a basis for an implementable numerical algorithm which can be applied to the original impulsive hybrid LQ problem under consideration. In different to the conventional LQ-based computational schemes, the proposed numerical method for the hybrid LQ-optimization uses the hybrid trajectory information. In some specific cases one can apply a simple version of Algorithm 1 and omit the forward integration procedure. Note that a simplified version of the proposed algorithm was applied to an illustrative example. The consistency analysis of the proposed numerical method and the corresponding (theoretical or implementable) convergence results belong to further works. Finally, note that the theoretical and computational approaches presented in this chapter can be applied to some other classes of hybrid LQ-type OCPs.

References

1. Attia, S.A. and Azhmyakov, V. and Raisch, J., "State jump optimization for a class of hybrid autonomous systems", *IEEE International Conference on Control Applications*, pp. 1408-1413, 2007.
2. S.A. Attia, V. Azhmyakov and J. Raisch, On an optimization problem for a class of impulsive hybrid systems, *Discrete Event Dynamic Systems*, accepted
3. H. Axelsson, H. Boccardo, M. Egerstedt, M. Valigi and Y. Wardi, Optimal mode-switching for hybrid systems with varying initial states, *Nonlinear Analysis: Hybrid Systems*, vol. 2, 2008, pp. 765 – 772.
4. V. Azhmyakov and J. Raisch, "A gradient-based approach to a class of hybrid optimal control problems", In *Proceedings of the 2nd IFAC Conference on Analysis and Design of Hybrid Systems*, Alghero, 2006, pp. 89-94.
5. Azhmyakov, V., "Optimal control of hybrid and switched systems", In *Proceedings of the IX chetaev Conference "Analytical Mechanics, Stability and Control of Motion"* Irkutsk, 2007, pp.308-317.
6. Azhmyakov, V. Attia, S.A. and Raisch, J., "On the Maximum Principle for impulsive hybrid systems", *Lecture Notes in Computer Science*, vol. 4981, pp. 30-42, Springer, Berlin, 2008.

7. Azhmyakov, V., Galvan-Guerra, R. and Polyakov, A., "On the Dynamic Programming method for LQ hybrid optimal control problems" *Automation and Remote Control* (submitted).

8. V. Azhmyakov, V. Boltyanski, A. Poznyak, Optimal control of impulsive hybrid systems, *Nonlinear Analysis: Hybrid Systems*, vol. 2, 2008, pp. 1089 – 1097.

9. D.D. Bainov, P.S. Simeonov, *Systems with Impulse Effects: Stability Theory and Applications*, Halsted, New York, 1989.

10. Bryson, A.E., Ho, Y.C., *Applied Optimal Control: Optimization, Estimation and Control*, Hemisphere Publising Corp., New York, 1975.

11. Cassandras, C., Pepyne, D.L. and Wardi, Y., "Optimal control of class of hybrid systems", *IEEE Transactions on Automatic Control*, vol. 46, 2001, pp.398-415.

12. Egerstedt, M., Wardi, Y., and Axelsson, H., "Transition-time optimization for switched-mode dynamical systems", *IEEE Transactions on Automatic Control*, vol. 51, 2006, pp. 110-115.

13. Fattorini, H.O., *Infinite-Dimensional Optimization and Control Theory*, Cambridge University Press, Cambridge, UK, 1999.

14. Garavello, M. and Piccoli B., "Hybrid necessary principle", *SIAM Journal on Control and Optimization*, **43**, 2005, pp.1867-1887.

15. V. Lakshmikantham, D.D. Bainov, P.S. Simeonov, *Theory of Impulsive Differential Equations*, World Scientific, Singapore, 1989.

16. B. Liu, X. Liu, X.X. Liao, "Robust stability analysis of uncertain impulsive systems", J. Math. Anal. Appl., vol. 290, pp. 519533, 2004.

17. Piccoli, B., "Necessary conditions for hybrid optimization", In *Proceedings of the 38th IEEE Conference on Decision and Control*, Phoenix, 1999, pp.410-415

18. E. Polak, *Optimization*, Springer-Verlag, New York, 1997.

19. M.S. Shaikh and P. E. Caines, "On the hybrid optimal control problem: theory and algorithms", *IEEE Transactions on Automatic Control*, vol. 52, 2007, pp. 1587 – 1603.

20. Sussmann, H. J., "A maximum principle for hybrid optimization", In *Proceedings of 38th IEEE Conference on Decision and Control*, Phoenix, 1999, pp. 425-430.

21. Verriest, E., Delmotte, F. and Egerstedt, "Optimal impulsive control of point delay systems with refractory period", In *Proceedings of the 5th IFAC workshop on Time Delay Systems*, Luven, Belgium, 2004.

22. Xu, X., and Antsaklis, P.J., "Optimal control of hybrid autonomous systems with state jumps", In *Proceedings of the American Control Conference*, Denver, 2003, pp. 5191-5196.

Task Space Robot Control Using Joint Proportional Action

Rubén Garrido, E. Alberto Canul, Alberto Soria

Departamento de Control Automático, CINVESTAV
{garrido,ecanul,soria}@ctrl.cinvestav.mx

Summary. This work presents a task-space controller for robot manipulators. A key feature of the proposed approach is the simultaneous use of two proportional actions at joint and task levels. This topology exploits the fact that in general, measurements at the task level are noisier than the measurements at the joint level. Moreover, task measurements are obtained at a lower sampling rate than joint level measurements. Therefore, it is possible to use high proportional gains at the joint level and low proportional gains at the task level. The Lyapunov method permits assessing closed-loop stability. A visual servoing application allows evaluating the performance of the proposed approach.

6.1 Introduction

Robot manipulator controllers generally fall into two categories, namely joint or task space-based designs. However, since a task is specified in the robot task space, it seems more natural to develop control algorithms for this case.

One of the first works was the seminal paper of Takegaki and Arimoto [1]. The authors assumed that the robot is not redundant and its Jacobian matrix is nonsingular at the desired position. This result was subsequently reported in [2]. References [3],[4] proposed a controller under the assumptions that the robot Jacobian matrix is full rank and the robot is non-redundant. The approach in [5] uses an energy shaping plus damping injection approach and achieves local asymptotic stability despite of the singularities on the task function Jacobian. Further contributions can be found in [6], [7] where the authors consider uncertainties on the robot Jacobian matrix and gravity torques. Asymptotic stability is assured provided that there exists a bound on the difference between the real and the estimated Jacobian matrix. The approach proposed in [7] uses a control law endowed with an integral action to compensate for the robot uncertain gravitational torques. Later contributions [8],[9] propose task space feedback controllers using the transpose robot Jacobian matrix instead of its generalized inverse.

As a particular case of the task space approach, visual servoing is considered for solving the position measurement problem in task space; therefore, the position is obtained directly from task space without requiring the robot kinematic inverse mapping. This approach is used in [8], [10]. In [10] the author proves asymptotic stability and robustness of the proposed controller under uncertainties in camera orientation and lens radial distortion. It is worth remarking that the aforementioned approaches rely on joint velocity and position task feedback and do not take into account joint position measurements. Considering robot joint position as feedback in task space controllers has several advantages. Firstly, all the industrial robots are controlled using joint proportional integral derivative (PID) controllers. In order to put to work most of the task space controllers mentioned above, the joint PID position controllers should be reconfigured as joint velocity controllers and the loop is closed using task space position measurements, for instance, measurements from a vision system. In this case, failure of the task space sensor would produce unpredictable movement since the robot is controlled only by the joint velocity loops. Secondly, having a PID or proportional derivative (PD) joint controller allows compensating for disturbances and nonlinear phenomena as stiction at the joint level at a higher bandwidth. Note that task space sensors have poor bandwidth compared with joint sensors. Typically, a vision system using a high-speed digital camera is able to capture 500 images per second and a data acquisition card may sample an optical encoder at 10 Khz. Moreover, it is not always possible to reconfigure the joint controller in industrial robots.

This chapter proposes a task-space robot controller using simultaneously proportional actions at the joint and the task levels. The Lyapunov method allows concluding stability of the proposed approach. Experimental results using a robot controlled through visual feedback asses the performance of the proposed approach.

6.2 Preliminaries

6.2.1 Review of Task Space controllers

Consider the controller proposed in [1]

$$\tau = J^T(\mathbf{q}) K_p \tilde{\mathbf{y}} - K_v \dot{\mathbf{q}} + g(\mathbf{q}) \tag{6.1}$$

where τ is the $n \times 1$ vector of applied joint torques, $J(\cdot) \in \mathbb{R}^{m \times n}$ is the robot Jacobian matrix, $K_p \in \mathbb{R}^{m \times m}$, $K_v \in \mathbb{R}^{n \times n}$ are positive definite feedback proportional and derivative gains respectively, $\tilde{\mathbf{y}} = \mathbf{y}_d - \mathbf{y} \in \mathbb{R}^m$ is the task space error where $\mathbf{y}_d$ and $\mathbf{y}$ are the desired and measured positions respectively, expressed in task space coordinates, $\dot{\mathbf{q}} \in \mathbb{R}^n$ is the measured joint velocity, and $\mathbf{g}(\mathbf{q}) \in \mathbb{R}^n$ is the gravity compensation. The first element of the right hand side of equation (1) and the gravity compensation modify the potential energy in the robot and the term $K_v \dot{\mathbf{q}}$ injects damping. Notice that the control law (6.1) can be interpreted as an inner velocity loop connected in cascade with an external position loop (*see Fig. 6.1*).

Task Space controllers in [5], [8] and [10] share the structure of controller (6.1). Note that if the task space position sensor fails, the proportional action $J^T(\mathbf{q}) K_p \tilde{\mathbf{y}}$ is not longer active and the robot motion may get unbounded.

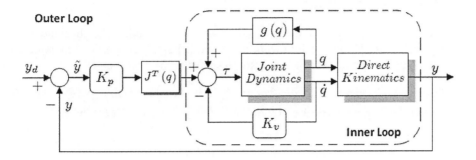

Fig. 6.1. Block diagram of controller (6.1)

6.2.2 Robot Dynamics

In absence of friction and some other disturbances, the dynamics of a serial n-link rigid robot manipulator can be written as [11],[12]:

$$M(\mathbf{q})\,\ddot{\mathbf{q}} + C(\mathbf{q},\dot{\mathbf{q}})\,\dot{\mathbf{q}} + g(\mathbf{q}) = \tau \tag{6.2}$$

Vectors q, $\dot{q}$ and $\ddot{q}$ correspond to the joint position, velocity and acceleration respectively, $M(\mathbf{q}) \in \mathbb{R}^{n \times n}$ is the robot inertia matrix, $C(\mathbf{q},\dot{\mathbf{q}})\,\dot{\mathbf{q}} \in \mathbb{R}^{n}$ corresponds to the *Coriolis* and centripetal forces vector, $\mathbf{g}(\mathbf{q}) \in \mathbb{R}^{n}$ is the gravity torque vector and τ the input torque vector. Throughout this chapter, the use of the notation $\lambda_{\min}\{\cdot\}$ and $\lambda_{\max}\{\cdot\}$ will indicate the smallest and largest eigenvalues of a symmetric positive bounded given matrix respectively, $\|\cdot\|$ is the norm of a given matrix or vector and $\langle \cdot,\cdot\rangle$ is the scalar product.

Model (6.2) considering revolute joint robots has the following properties [13].

1. $M(\mathbf{q}) = M^{T}(\mathbf{q})$
2. $M(\mathbf{q}) > 0$
3. $\lambda_{\min}\{M(\mathbf{q})\}\|\mathbf{x}\|^{2} \leq \mathbf{x}^{T}M(\mathbf{q})\mathbf{x} \leq \lambda_{\max}\{M(\mathbf{q})\}\|\mathbf{x}\|^{2}$ where $\lambda_{\min}\{M(\mathbf{q})\}$, $\lambda_{\max}\{M(\mathbf{q})\}$ denotes the minimum and maximum eigenvalues of $M(\mathbf{q})$ respectively.
4. The matrix $N(\mathbf{q},\dot{\mathbf{q}}) = \dot{M}(\mathbf{q}) - 2C(\mathbf{q},\dot{\mathbf{q}})$ is skew-symmetric and $\dot{M}(\mathbf{q}) = C(\mathbf{q},\dot{\mathbf{q}}) + C^{T}(\mathbf{q},\dot{\mathbf{q}})$.
5. There exists a constat k_{C} such that the following inequality holds:

$$\|C(\mathbf{q},\dot{\mathbf{q}})\,\dot{\mathbf{q}}\| \leq k_{C}\|\dot{\mathbf{q}}\|^{2}$$

6. The robot potential energy $U(\mathbf{q})$ and gravity vector $g(\mathbf{q})$ are bounded, i.e. $\|U(\mathbf{q})\| \leq k_{u}$ and $\|g(\mathbf{q})\| \leq k_{g}$. Furthermore, there exists a constant k' such that $\|g(\mathbf{x}) - g(\mathbf{y})\| \leq k'\|\mathbf{x} - \mathbf{y}\|$ where $k' \geq n\left[\max_{ij}\left|\frac{\partial g_{i}(\mathbf{q})}{\partial q_{j}}\right|\right]$.

6.2.3 Saturation Functions.

In recent years the use of a special kind of saturation functions has become more common [2],[14]-[17]. The following definition [2] states properties of saturation functions useful for the stability analysis presented in the next section.

Definition 1. *Consider de scalar function $S_i(\theta_i)$ and its derivative $\sigma_i(\theta_i)$ where θ_i is a real number. Functions $S_i(\theta_i)$ and $\sigma_i(\theta_i)$ fulfill the following properties:*

1. *$S_i(\theta) > 0$ for $\theta \neq 0$ and $S_i(0) = 0$*
2. *$S_i(\theta)$ is at least two times differentiable and the derivative $\sigma_i(\theta_i) = \frac{d}{d\theta_i}[S_i(\theta_i)]$ is strictly increasing in θ for $|\theta| < \gamma_i$ for some γ_i, and saturates for $|\theta| \geq \gamma_i$, that is:*

$$\sigma_i(\theta) = \pm\sigma_i, |\gamma_i| < \theta$$

   with σ_i a positive constant.
3. *There exists a constant $\bar{c}_i > 0$, such that:*

$$S_i(\theta) \geq \bar{c}_i \sigma_i^2(\theta)$$

for $\theta \neq 0$.

6.3 Task-Space Controller with an Inner PD Loop

Consider the following control law

$$\tau = -K_d \dot{\mathbf{q}} + K_p(\mathbf{u} - \mathbf{q}) \tag{6.3}$$

$$\mathbf{u} = \tilde{J}^T(\mathbf{q}) L_p \sigma(\mathbf{e}) + \mu L_i \int_0^t \tilde{J}^T(\mathbf{q}) L_p \sigma(\mathbf{e}) \, d\xi \tag{6.4}$$

Equation (6.3) corresponds to the inner PD control law and (6.4) to the outer PI controller. $K_p = k_p I \in \mathbb{R}^{n \times n}$, $k_p > 0$, $K_d > 0 \in \mathbb{R}^{n \times n}$ are the proportional and derivative gain matrices respectively with I the identity matrix, $L_p > 0 \in \mathbb{R}^{m \times m}$ and $L_i > 0 \in \mathbb{R}^{n \times n}$ are the proportional and integral gain matrices for the outer loop, $\mathbf{e}$ is the task space error defined as $\mathbf{e} = \mathbf{X}_d - \mathbf{X} \in \mathbb{R}^m$; where $\mathbf{X}$ and $\mathbf{X}_d$ are the measured and desired positions and $\mu > 0$ is an arbitrary constant. Fig. 6.2 shows the block diagram for this controller.

Let $S_i(e_i)$ and $\sigma_i(\theta_i)$ a scalar function and its time derivative respectively, according to *Definition 1* $\sigma(\mathbf{e}) = \left(\sigma_1(e_1) \ldots \sigma_m(e_m) \right)^T$ is the saturation error vector. Assume that an estimate $\tilde{J}^T(\mathbf{q})$ of the robot Jacobian matrix is available such that

$$\left\| J(\mathbf{q}) - \hat{J}(\mathbf{q}) \right\| \leq r \tag{6.5}$$

with r a positive constant.

Let $\mathbf{w}(t)$ be defined as

Outer Loop

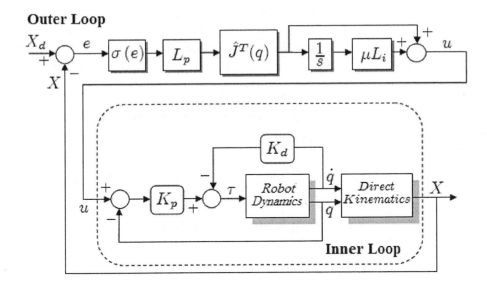

Fig. 6.2. Block diagram of control law (6.3),(6.4)

$$\mathbf{w}(t) = K_p^{-1}\left[g(\mathbf{q}_d) + K_p\mathbf{q}(0)\right] + \int_0^t \mathbf{y}(\xi)\,d\xi \qquad (6.6)$$

where:

$$\mathbf{y} = \dot{\mathbf{q}} - \mu L_i \hat{J}^T(\mathbf{q})L_p\sigma(\mathbf{e}) \qquad (6.7)$$

From (6.6) it follows that $\dot{\mathbf{w}} = \mathbf{y}$. Then, using (6.6), control law (6.3,6.4) can be rewritten as

$$\tau = K_p\hat{J}^T(\mathbf{q})L_p\sigma(\mathbf{e}) - K_d\dot{\mathbf{q}} + g(\mathbf{q}_d) - K_p\mathbf{w} \qquad (6.8)$$

Substituting (6.8) into (6.2) yields the closed loop system equation:

$$M(\mathbf{q})\ddot{\mathbf{q}} + C(\mathbf{q},\dot{\mathbf{q}})\dot{\mathbf{q}} + g(\mathbf{q}) = K_p\hat{J}^T(\mathbf{q})L_p\sigma(\mathbf{e}) - K_d\dot{\mathbf{q}} + g(\mathbf{q}_d) - K_p\mathbf{w} \qquad (6.9)$$

Now, consider the following Lyapunov function candidate

$$\begin{aligned}
V(\mathbf{w},\sigma(\mathbf{e}),\dot{\mathbf{q}}) &= \frac{1}{2}\dot{\mathbf{q}}^T M(\mathbf{q})\dot{\mathbf{q}} - \mu\dot{\mathbf{q}}^T M(\mathbf{q})L_i\hat{J}^T(\mathbf{q})L_p\sigma(\mathbf{e}) \\
&\quad + U(\mathbf{q}) - U(\mathbf{q}_d) + \blacksquare\mathbf{q}^T g(\mathbf{q}_d) \\
&\quad + k_p\langle \mathbf{S}(\mathbf{e}),\mathcal{L}_p\rangle + \frac{1}{2}\mathbf{w}^T K_p\mathbf{w}
\end{aligned} \qquad (6.10)$$

where $\langle \mathscr{L}_p, \mathbf{S}(\mathbf{e}) \rangle = \mathscr{L}_p^T \mathbf{S}(\mathbf{e}) = \sum_{j=1}^m l_{pj} S_j(e_j)$, $\mathbf{S}^T(\mathbf{e}) = \begin{bmatrix} S_1(\mathbf{e}_1) & \dots & S_m(\mathbf{e}_m) \end{bmatrix}^T$, $\mathscr{L}_p = \text{diag}\{L_p\}$ and $\blacksquare \mathbf{q} = \mathbf{q}_d - \mathbf{q}$, with $\mathbf{q}_d$ and $\mathbf{q}$ the desired and measured joint position respectively. First, to verify that (6.10) is positive definite consider the following expression

$$\frac{1}{2}\mathbf{y}^T M(\mathbf{q})\mathbf{y} = \frac{1}{2}\dot{\mathbf{q}}^T M(\mathbf{q})\dot{\mathbf{q}} - \mu\dot{\mathbf{q}}^T M(\mathbf{q})L_i\hat{J}^T(\mathbf{q})L_p\sigma(\mathbf{e})$$
$$+ \frac{1}{2}\mu^2\sigma^T(\mathbf{e})L_p\hat{J}(\mathbf{q})L_iM(\mathbf{q})L_i\hat{J}^T(\mathbf{q})L_p\sigma(\mathbf{e}) \qquad (6.11)$$

where y is given in (6.7). Therefore, the Lyapunov function candidate (6.10) can be rewritten as

$$V(\mathbf{w},\sigma(\mathbf{e}),\dot{\mathbf{q}}) = \frac{1}{2}\mathbf{y}^T M(\mathbf{q})\mathbf{y} + U(\mathbf{q}) - U(\mathbf{q}_d)$$
$$- \frac{1}{2}\mu^2\sigma^T(\mathbf{e})\left[L_p\hat{J}(\mathbf{q})L_iM(\mathbf{q})L_i\hat{J}^T(\mathbf{q})L_p\right]\sigma(\mathbf{e})$$
$$+ k_p\langle \mathbf{S}(\mathbf{e}),\mathscr{L}_p \rangle + \frac{1}{2}\mathbf{w}^T K_p\mathbf{w} + \Delta\mathbf{q}^T g(\mathbf{q}_d) \qquad (6.12)$$

Since the inertia matrix is positive definite the following inequality holds:

$$-\lambda_2\|\sigma(\mathbf{e})\|^2 \leq -\sigma^T(\mathbf{e})\left[L_p\hat{J}(\mathbf{q})L_iM(\mathbf{q})L_i\hat{J}^T(\mathbf{q})L_p\right]\sigma(\mathbf{e})$$

where $\lambda_2 = \lambda_{\max}\{L_p\}^2 \lambda_{\max}\{L_i\}^2 \lambda_{\max}\{M(\mathbf{q})\}\lambda_{\max}\{\hat{J}(\mathbf{q})\hat{J}^T(\mathbf{q})\}$. As a consequence of the robot model properties, for k_p large enough the following inequality is fulfilled:

$$\frac{1}{2}k_p\|\sigma(\mathbf{e})\|^2 \leq U(\mathbf{q}) - U(\mathbf{q}_d) + \Delta\mathbf{q}^T g(\mathbf{q}_d) + \frac{1}{2}k_p\langle \mathbf{S}(\mathbf{e}),\mathscr{L}_p \rangle$$

Then, (6.10) is lower bounded as follows

$$V(\mathbf{w},\sigma(\mathbf{e}),\dot{\mathbf{q}}) \geq \frac{1}{2}\lambda_{\min}\{M(\mathbf{q})\}\|\mathbf{y}\|^2$$
$$+ \frac{1}{2}[k_p - \mu^2\lambda_2]\|\sigma(\mathbf{e})\|^2 + \frac{1}{2}k_p\|\mathbf{w}\|^2$$

Hence, $V(\mathbf{w},\sigma(\mathbf{e}),\dot{\mathbf{q}})$ is positive definite if

$$\sqrt{\frac{k_p}{\lambda_2}} > \mu \qquad (6.13)$$

Taking the time derivative of (6.10) and using (6.9) yields

$$\dot{V}\left(\mathbf{w},\sigma\left(\mathbf{e}\right),\dot{\mathbf{q}}\right) = -k_p\dot{\mathbf{q}}^T\left[J^T\left(\mathbf{q}\right)-\hat{J}^T\left(\mathbf{q}\right)\right]L_p\sigma\left(\mathbf{e}\right)$$
$$-\mu k_p\sigma\left(\mathbf{e}\right)^T L_p\hat{J}\left(\mathbf{q}\right)L_i\hat{J}^T\left(\mathbf{q}\right)L_p\sigma\left(\mathbf{e}\right)$$
$$-\mu\sigma\left(\mathbf{e}\right)^T L_p\hat{J}\left(\mathbf{q}\right)L_i\left[g\left(\mathbf{q}_d\right)-g\left(\mathbf{q}\right)\right]$$
$$-\mu Z_2\left(\mathbf{q},\dot{\mathbf{q}},\sigma\left(\mathbf{e}\right)\right)-\dot{\mathbf{q}}^T K_d\dot{\mathbf{q}} \tag{6.14}$$

where

$$Z_2\left(\mathbf{q},\dot{\mathbf{q}},\sigma\left(\mathbf{e}\right)\right) = \dot{\sigma}\left(\mathbf{e}\right)^T L_p\hat{J}\left(\mathbf{q}\right)L_i M\left(\mathbf{q}\right)\dot{\mathbf{q}}$$
$$+\sigma\left(\mathbf{e}\right)^T L_p\dot{\hat{J}}\left(\mathbf{q}\right)L_i M\left(\mathbf{q}\right)\dot{\mathbf{q}}$$
$$+\sigma\left(\mathbf{e}\right)^T L_p\hat{J}\left(\mathbf{q}\right)L_i\left\{C^T\left(\mathbf{q},\dot{\mathbf{q}}\right)-K_d\right\}\dot{\mathbf{q}} \tag{6.15}$$

Since the gravity torques are bounded, there exists a positive constant α_3 [7], [8] such that for k_p large enough the following inequality holds

$$-\frac{1}{2}\mu k_p\sigma\left(\mathbf{e}\right)^T L_p\hat{J}\left(\mathbf{q}\right)L_i\hat{J}^T\left(\mathbf{q}\right)L_p\sigma\left(\mathbf{e}\right)$$
$$-\mu\sigma\left(\mathbf{e}\right)^T L_p\hat{J}\left(\mathbf{q}\right)L_i\left[g\left(\mathbf{q}_d\right)-g\left(\mathbf{q}\right)\right]\leq -\frac{1}{2}k_p\alpha_3\left\|\sigma\left(\mathbf{e}\right)\right\|^2$$

Finally, since the robot Jacobian matrix is bounded, it follows that

$$\mu\left|Z_2\left(\mathbf{q},\dot{\mathbf{q}},\sigma\left(\mathbf{e}\right)\right)\right|\leq \mu c_0\left\|\dot{\mathbf{q}}\right\|^2,\forall c_0>0 \tag{6.16}$$

As a consequence, $\dot{V}$ is upper bounded as follows

$$\dot{V}\left(\mathbf{w},\sigma\left(\mathbf{e}\right),\dot{\mathbf{q}}\right)\leq k_p r\lambda_{\max}\left\{L_p\right\}\left\|\dot{\mathbf{q}}\right\|\left\|\sigma\left(\mathbf{e}\right)\right\|$$
$$-\left(\lambda_{\min}\left\{K_d\right\}-\mu c_0\right)\left\|\dot{\mathbf{q}}\right\|^2-\frac{1}{2}k_p\alpha_3\left\|\sigma\left(\mathbf{e}\right)\right\|^2 \tag{6.17}$$

where $\alpha_3 = \lambda_{\min}\left\{L_p\right\}^2\lambda_{\min}\left\{L_i\right\}\lambda_{\min}\left\{\hat{J}\left(\mathbf{q}\right)\hat{J}^T\left(\mathbf{q}\right)\right\}$ and r is defined in (6.5). Inequality (6.17) can be further rewritten as:

$$\dot{V}\left(\mathbf{w},\sigma\left(\mathbf{e}\right),\dot{\mathbf{q}}\right)\leq -\frac{1}{2}\left(\lambda_{\min}\left\{K_d\right\}-\mu c_0\right)\left\|\dot{\mathbf{q}}\right\|^2-\frac{1}{2}k_p\alpha_3\left\|\sigma\left(\mathbf{e}\right)\right\|^2$$
$$-\frac{1}{2}\left[\sqrt{\left(\lambda_{\min}\left\{K_d\right\}-\mu c_0\right)}\left\|\dot{\mathbf{q}}\right\|^2\right.$$
$$\left.-\frac{\left(rk_p\lambda_{\min}\left\{L_p\right\}\right)}{\sqrt{\lambda_{\min}\left\{K_d\right\}-\mu c_0}}\left\|\sigma\left(\mathbf{e}\right)\right\|\right]^2$$
$$-\frac{1}{2}\left[k_p\alpha_3-\frac{\left(rk_p\lambda_{\min}\left\{L_p\right\}\right)^2}{\lambda_{\min}\left\{K_d\right\}-\mu c_0}\right]\left\|\sigma\left(\mathbf{e}\right)\right\|^2 \tag{6.18}$$

Then, $\dot{V}\left(\mathbf{w},\sigma\left(\mathbf{e}\right),\dot{\mathbf{q}}\right)\leq 0$ if:

$$\frac{\lambda_{\min}\{K_d\}}{c_0} > \mu \qquad (6.19)$$

$$\alpha_3\left(\lambda_{\min}\{K_d\} - \mu c_0\right) > k_p\left(r\lambda_{\min}\{L_p\}\right)^2 \qquad (6.20)$$

The above inequalities hold for μ and r small enough. Asymptotic stability follows directly by applying La Salle's invariance theorem [11],[12],[18].

Proposition 1. *Closed loop of dynamic system (6.2) and control law (6.3), (6.4) has a unique equilibrium point at $\dot{q} = 0, \sigma(e) = 0, w = 0$, and it is asymptotically stable if $\hat{J}(q)$ fulfills (6.5) and K_p, K_d, L_p, L_i and μ are chosen such that inequalities (6.13), (19) and (6.20) are fulfilled.*

Note that small values of μ and r allow increasing the value of k_p without violating inequalities (6.13), (19) and (6.20)

6.4 Experimental Results

6.4.1 Experimental test bed

To show the performance of the proposed controller, it was tested using a planar 2-link revolute joint robot under visual feedback. The robot is driven by DC brushed motors through timing belts. The motors are subsequently driven by power amplifiers from Copley Controls model 413, working in current mode. Joint position feedback is performed by incremental encoders with 10,000 pulses per revolution. Data acquisition was performed through a MultiQ 3 card from Quanser Consulting. Image acquisition for task space position feedback was performed using a CCD-Dalsa camera connected to a National Instruments PCI-1422 card through an RS-422 protocol. The visual sample time for the outer loop was 7 *ms* and the inner sample time for the joint position loop was 1 *ms*. The experimental test bed is shown in Fig. 6.3.

6.4.2 Performance of the proposed controller

The manipulator Jacobian matrix for the visual servoing application is given by

$$\hat{J}(\mathbf{q}) = R(\theta)\hat{J}_a(\mathbf{q}) \qquad (6.21)$$

where $R(\theta)$ is the rotation matrix between the cartesian and the visual space and is taken as the identity and $\hat{J}_a(\mathbf{q})$ is an estimate of the robot analytic Jacobian matrix. For the robot used in the experiments $\hat{J}(\mathbf{q})$ is expressed as

$$\hat{J}(\mathbf{q}) = \begin{bmatrix} \hat{l}_1\cos q_1 & \hat{l}_2\cos q_2 \\ \hat{l}_1\sin q_1 & \hat{l}_2\sin q_2 \end{bmatrix} \qquad (6.22)$$

q_1 and q_2 are the robot angular positions, l_1 and l_2 denote the lengths of the first and second robot links. Their approximate values are $\hat{l}_1 = \hat{l}_2 = 0.21$ m. The gains of the proposed controller (6.3), (6.4) were chosen as follows[1]

---

[1]x and y in pixels

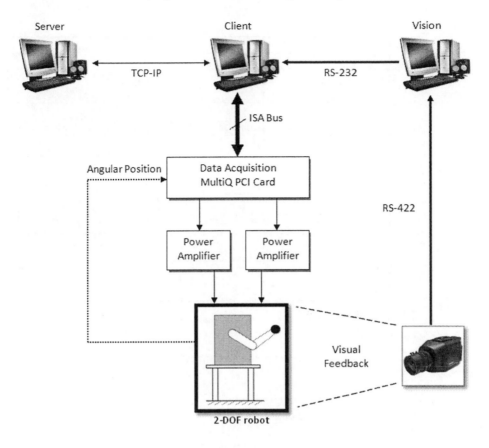

Fig. 6.3. Experimental Test Bed

$$K_p = \begin{bmatrix} 10 & 0 \\ 0 & 10 \end{bmatrix}, K_d = \begin{bmatrix} 0.24 & 0 \\ 0 & 0.24 \end{bmatrix}$$

$$L_p = \begin{bmatrix} 0.15 & 0 \\ 0 & 0.1 \end{bmatrix}, L_i = \begin{bmatrix} 1.6 & 0 \\ 0 & 1.5 \end{bmatrix}$$

$$\mu = 1 \text{ and sat}(\mathbf{e}) = \pm 8 \tag{6.23}$$

Fig. 6.4 shows the robot displacement on the x axis and Fig. 6.5 depicts the robot displacements on the y axis.

6.4.3 Comparative Study.

The performance of the proposed controller (6.3), (6.4) is compared with itself but without the inner joint proportional action. Therefore, (6.3) becomes:

$$\tau = -K_d \dot{\mathbf{q}} + \mathbf{u} \tag{6.24}$$

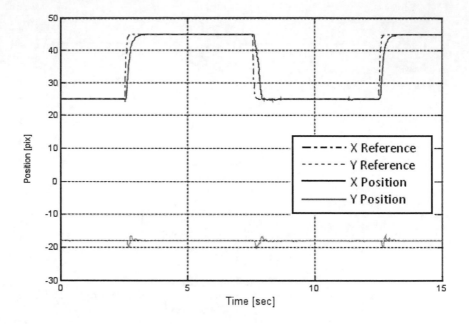

Fig. 6.4. Desired position versus actual position in x

Fig. 6.5 depicts the block diagram with this modification.
The gains of the modified controller (6.24) were set up as follows:

$$K_d = \begin{bmatrix} 0.15 & 0 \\ 0 & 0.18 \end{bmatrix}$$

$$L_p = \begin{bmatrix} 1 & 0 \\ 0 & 1 \end{bmatrix}, L_i = \begin{bmatrix} 0.3 & 0 \\ 0 & .05 \end{bmatrix}$$

$$\mu = 1 \text{ and } \text{sat}(\mathbf{e}) = \pm 8 \tag{6.25}$$

Fig. 6.7 shows the measurement noise with the set of parameters (6.23) for the proposed controller and Fig. 6.8 for the set of parameters (6.25) for the modified controller.

One can see that due to the use of higher proportional gains at the task level in the modified controller (6.24) the noise coming from the vision system is amplified; on the other hand, noise from the task level sensor under controller (6.3) is not greatly amplified. Note also that the L_p gain in controller (6.24) is seven times higher than the corresponding gain in controller (6.3), (6.4).

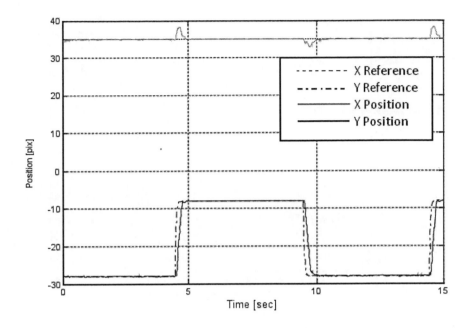

Fig. 6.5. Desired position versus actual position in y

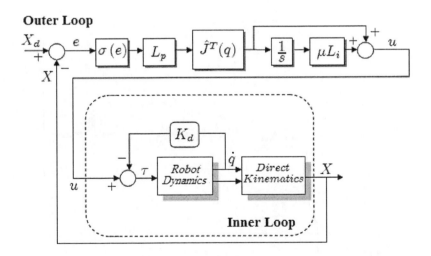

Fig. 6.6. Proposed controller without inner joint position measurements

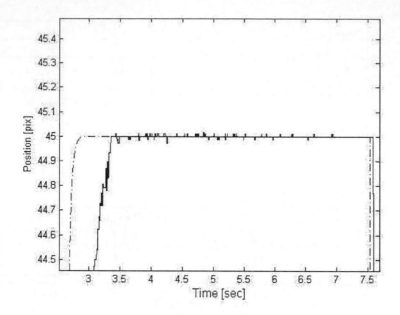

Fig. 6.7. Measurement noise with the proposed controller (6.3)

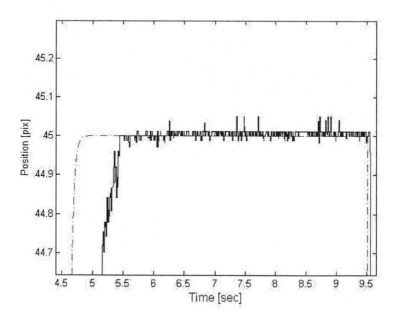

Fig. 6.8. Measurement noise with the modified controller (6.24)

6.5 Conclusion

This work presents an approach for robot manipulator control in task space. The proposed control law makes use of proportional actions at the joint and the task levels. Experiments show that the performance of the proposed approach and a controller without joint proportional action are similar. However, this last controller uses a high gain proportional gain at the task level thus amplifying the measurement noise. On the other hand, the proposed controller employs a smaller proportional gain at the task level and the measurement noise is not greatly amplified.

Acknowledgment. The Authors would like to thank Gerardo Castro and Jesús Meza for their support during the experiments.

References

1. M. Takegaki, S. Arimoto. *A new feedback method for dynamic control of manipulators.* Trans. ASME; J. Dyn. Syst., Meas., Ctrl., vol.102, pp.119-125, June 1981.
2. S. Arimoto. *Control Theory on Non-Linear Mechanical Systems.* Oxford, U.K.: Claredon, 1996.
3. L. Sciavicco and B. Siciliano, *Modeling and Control of Robot Manipulators.* New York: McGraw-Hill, 1996.
4. C. Canudas de Wit, B. Siciliano and G. Bastin, *Theory of Robot Control.* New York: Springer-Verlag, 1996.
5. R. Kelly. *Regulation of manipulators in generic task space: An energy shaping plus damping injection approach.* IEEE Trans. on Robotics and Automation. vol.15, no.2, 1999.
6. C. C. Cheah, S. Kawamura and S. Arimoto. *Feedback control for robotic manipulators with uncertain kinematics and dynamics.* Proc. IEEE Int. Conf. Robotics and Automation, Leuven, Belgium, 1998, pp. 3607-3612.
7. C. C. Cheah, S. Kawamura, S. Arimoto and K. Lee. *PID control for robotic manipulator with uncertain Jacobian matrix.* Proc. IEEE Int. Conf. Robotics and Automation, Detroit, MI, May 1999, pp. 494-499.
8. C. C. Cheah, M. Hirano, S. Kawamura and S. Arimoto. *Approximate Jacobian control for Robots with uncertain Kinematics and Dynamics.* IEEE Trans. on Robotics and Automation, vol. 19, no. 4, August 2003.
9. C.Q. Huang, X.G. Wang and Z.G. Wang. *A class of transposed Jacobian-based NPID regulators for robot manipulators with an uncertain Kinematics.* Journal of Robotic Systems, 2002, pp.527-539.
10. R. Kelly. *Robust Asymptotically stable visual servoing of planar robots.* IEEE Trans. on Robotics and Automation. vol.12, no.5, October 1996.
11. M. Spong and M. Vidyasagar. *Robot Dynamics and control.* New York: Wiley 1989
12. J.J. Craig. *Introduction to Robotics: Mechanics and Control.* New York: Addison-Wesley 2000.
13. R. Kelly and V. Santibáñez. *Control of Robot manipulators in Joint Space.* Springer-Verlag London 2005
14. R. Kelly. *Global positioning of robot manipulators via PD control plus a class of nonlinear integral actions.* IEEE Transactions on Automatic Control, Vol 43, pp. 934-938, July 1998.
15. R. Kelly and V. Santibáñez. *Global convergence of the adaptive PD controller with computed feedforward for robot manipulators.* in Proc. IEEE Int. Conf. Robotics and Automation, Detroit, MI, May 1999, pp. 1831-1836
16. R. Kelly and V. Santibañez. *A class of nonlinear PID global regulators for robot manipulators.* in Proc. IEEE Int. Conf. Robotics and Automat., Leuven Bélgica, May 1998, pp. 3601-3606.
17. J. Álvarez, R. Kelly and I. Cervantes. *Semiglobal stability of saturated linear PID control for robot manipulators.* Elsevier science Ltd. Automática no. 39, pp. 989-995, June 2003.
18. H. Khalil, *Nonlinear Systems.* 3rd. edition, Prentice Hall. USA c.2002

7

Further Properties of Carleman Linearization

Irving Sánchez, Joaquín Collado

Automatic Control Dept.
CINVESTAV
Av. IPN 2508, Col. San Pedro Zacatenco
07360 México, D. F., MEXICO
{jcollado,isanchez}@ctrl.cinvestav.mx

Summary. This chapter deals with the Standard Truncated Carleman Linearization as well proposes a Modified version. The Carleman Linearization states that every analytic n-dimensional nonlinear system is equivalent to an infinite dimensional bilinear system. As a result, the new system is made up of a state linear, a control linear and a bilinear terms in the state space format. In this work is shown that the controllable part coincides with Taylor Linearization when the control part is constant. Then we introduce some modification which reduces the error in the truncation process. We applied this Modified Truncated Carleman Linearization to two van der Pol oscillators with slight different frequencies. Each one is approximate by a 14th order linear system; thus we coupled this two linear oscillators and look for a synchronization. We conclude that even the approximation is very good, is not possible synchronize high order linear oscillators as the nonlinear oscillators do.

Keywords - Carleman Linearization, bilinear systems, controllability, synchronization, van der Pol oscillators.

7.1 Introduction.

The original Carleman Linearization [1] applies to systems of the form

$$\dot{x} = f(x) \tag{7.1}$$

where $x \in \mathbb{R}^n$, and $f(\cdot)$ is an analytic vector-valued function and states that ODE (7.1) is equivalent to

$$\dot{z}(t) = Az(t) \tag{7.2}$$

where $\dim(z(t)) = \infty$.

The new system has a infinite dimension and it is necessary truncate the Taylor series at some point in order to have an approximate but finite dimensional system.

When system (7.2) is truncated to dimension n, Carleman Linearization coincides with Taylor Linearization.

If the original nonlinear system is controlled, affinely in control, i. e.

$$\dot{x}(t) = f(x) + g(x)u(t) \qquad (7.3)$$
$$y(t) = Cx(t)$$

then, the Truncated Carleman Linearization applied to (7.3) give a bilinear system as

$$\dot{z}(t) = Az(t) + u(t)Fz(t) + Bu(t) \qquad (7.4)$$

Systems (7.3) and (7.4) are equivalent only when $\dim(z) = \infty$. Any finite dimensional truncation of (7.4) introduces an error between the systems (7.3) and (7.4). Bellamn [1] proposed to add some extra terms (see equation (7.26)) to reduce the truncation error. In this chapter we use least square method [2] to choose the parameters introduced in equation (7.26) in order to reduce the truncation error. To approximate non-linear oscillators, we introduce a second modification, which consists in relocate the eigenvalues of the truncated system, in order that all the eigenvalues be in the $j\omega$ axis and all of them be a multiple of a ω_0, the natural frequency of the oscillator.

An example of two coupled oscillators, which approximate van der Pol oscillators is presented. It is concluded that is not possible to synchronize linear oscillators with linear coumplings.

7.2 Preliminaries

In this section we introduce some definitions and preliminaries required in the sequel.

Let $x = \begin{bmatrix} x_1 & x_2 & \dots & x_n \end{bmatrix}^T$ be an n-dimensional vector, where $x_i \in \mathbb{R}$, then $x^{(2)}$ is defined as follows

$$x^{(2)} = x \otimes x$$

$$= \left[\underbrace{x_1^2, x_1 x_2, x_1 x_3, \dots, x_1 x_n}_{n \text{ terms}}, \underbrace{x_2^2, x_2 x_3, \dots, x_2 x_n}_{(n-1) \text{ terms}}, \right.$$

$$\left. \underbrace{x_3^2, x_3 x_4, \dots, x_3 x_n}_{(n-2) \text{ terms}}, \dots, \underbrace{x_n^2}_{1 \text{ term}} \right]^T \in R^{n(n+1)/2}$$

where $\otimes^1$ denotes the Reduced Kronecker Product [3] [4] [8] [17] and [23]. In this way

$$x^{(p)} = x \otimes x \otimes ...x \otimes x \tag{7.5}$$

where there are $p - 1$ Reduced Kronecker products.

Remark 1. Notice that in the Reduced Kronecker Product all the redundant products have been removed.

The matrix case is the following, if $y = Ax$ then $A^{(p)}$ is such that the following equation

$$y^{(p)} = A^{(p)}x^{(p)} \tag{7.6}$$

is satisfied.

Now consider a nonlinear dynamic system described by (7.3), where we assumed that $f(x)$ and $g(x)$ are analytic in $\mathbb{R}^n$ and $f(0) = 0$, thus we can write the Taylor series of $f(x)$ and $g(x)$ [18] as:

$$f(x) = A_1 x(t) + A_2 x^{(2)}(t) + \cdots \tag{7.7}$$
$$g(x) = B_0 + B_1 x(t) + B_2 x^{(2)}(t) + \cdots$$

where $A_k, B_k \in \mathbb{R}^{n \times m_k}$ are constant matrices, $k = 1, 2, ...$, and $x^{(k)} \in \mathbb{R}^{m_k}$ means the reduced Kronecker Product of x with itself as it was defined above with $m_k \triangleq \begin{pmatrix} n+k-1 \\ k \end{pmatrix}$.

Remark 2. Note that the index in A_k matrices corresponds to the degree in x of the Taylor series and m_k is the number of different elements in $x^{(k)}$.

Then dynamic system described in (7.3) may be rewritten as

$$\dot{x}(t) = \sum_{k=1}^{\infty} A_k x^{(k)}(t) + \sum_{k=0}^{\infty} B_k x^{(k)}(t)u(t) \tag{7.8}$$

is an equivalent representation of (7.3), see [5] and [18]. Again, if we truncate (7.8) to $k = 1$, we obtain the Taylor Linearization of the controlled system (7.3).

Remark 3. Notice that equation (7.8) is not in the state-space format, because the RHS contains products of the elements of $x(t)$.

---

[1] The symbol $\otimes$ is normally used by the standard Kronecker Product, which is defined as
$\begin{bmatrix} x_1 x \\ x_2 x \\ \vdots \\ x_n x \end{bmatrix} \in \neg \mathbb{R}^{n^2}$, but here is used for this reduced Kronecker Product.

7.3 Carleman Linearization

In this section we describe the procedure of Carleman Linearization applied to a nonlinear-analytic state equation that can be represented in a Taylor series form around the origin.

From (7.8) we can write

$$\dot{x}(t) = \sum_{k=1}^{N} A_k x^{(k)}(t) + \sum_{k=0}^{N-1} B_k x^{(k)}(t) u(t) + ... \tag{7.9}$$

where we have specifically retained the terms with order equal or less than N in the expansion of $f(x)$ and $N-1$ for $g(x)$.

Now we obtain a differential equation for $x^{(2)}(t)$, neglecting the terms with order higher than N^2.

$$\frac{d}{dt}\left[x^{(2)}\right] = \frac{d}{dt}[x \otimes x] \tag{7.10}$$

$$= \dot{x} \otimes x + x \otimes \dot{x}$$

$$= \left[\sum_{k=1}^{N} A_k x^{(k)} + \sum_{k=0}^{N-1} B_k x^{(k)} u(t)\right] \otimes x$$

$$+ x \otimes \left[\sum_{k=1}^{N} A_k x^{(k)} + \sum_{k=0}^{N-1} B_k x^{(k)} u(t)\right] + ...$$

$$= \sum_{k=1}^{N-1} [A_k \otimes I_n + I_n \otimes A_k] x^{(k+1)}$$

$$+ \sum_{k=0}^{N-2} [B_k \otimes I_n + I_n \otimes B_k] x^{(k+1)} u(t) + ...$$

$$x^{(2)}(0) = x_0^{(2)}$$

Thus $x^{(2)}(t)$ also satisfies a differential equation with the same general form than $x(t)$ at (7.9). Then, in this way, we can write a differential equation for $x^{(j)}(t)$ with order N as

$$\frac{d}{dt}\left[x^{(j)}\right] = \sum_{k=1}^{N-j+1} A_{j,k} x^{(k+j-1)} + \sum_{k=0}^{N-j} B_{j,k} x^{(k+j-1)} + ... \tag{7.11}$$

$$x^{(j)}(0) = x_0^{(j)}, \qquad j = 1, ..., N$$

---

[2]For Reduced Kronecker Product $\otimes$ as for the normal Kronecker Product it is valid the Mixed Product property, namely: $(AB) \otimes (CD) = (A \otimes C)(B \otimes D)$.

where $A_{1,k} = A_k$, and for $j > 1$,

$$A_{j,k} = A_k \otimes I_n \otimes ...I_n + I_n \otimes A_k \otimes I_n \otimes ... \otimes I_n \qquad (7.12)$$
$$+ ... + I_n \otimes ... \otimes I_n \otimes A_k$$

there are $j - 1$ reduced Kronecker products at each term and j-terms. Likewise for $B_{j,k}$.

Now, we define

$$\tilde{x}(t) \triangleq \left[\left[x^{(1)}(t) \right]^T \left[x^{(2)}(t) \right]^T \cdots \left[x^{(N)}(t) \right]^T \right]$$

Here the dimension of $\tilde{x}$ is

$$\dim(\tilde{x}) = M = n + m_2 + ... + m_N$$

It is possible to write a collection of differential equations for (7.11) as a single state equation in the following way

$$\frac{d}{dt}\tilde{x} = \begin{bmatrix} A_{11} & A_{12} & ... & A_{1N} \\ 0 & A_{21} & ... & A_{2,N-1} \\ 0 & 0 & ... & A_{3,N-2} \\ \vdots & \vdots & \ddots & \vdots \\ 0 & 0 & ... & A_{N1} \end{bmatrix} \tilde{x} \qquad (7.13)$$

$$+ u \begin{bmatrix} B_{1,1} & B_{1,2} & \cdots & B_{1,N-1} & 0 \\ B_{2,0} & B_{2,1} & \cdots & B2 & 0 \\ 0 & B_{3,0} & \cdots & B_{3,N-3} & 0 \\ \vdots & \vdots & \ddots & \vdots & \vdots \\ 0 & 0 & \cdots & B_{N,0} & 0 \end{bmatrix} \tilde{x} + \begin{bmatrix} B_{1,0} \\ 0 \\ \vdots \\ 0 \end{bmatrix} u$$

$$\frac{d}{dt}\tilde{x} = A\tilde{x} + uF\tilde{x} + Bu$$

$$\tilde{x}(0) = \left[x_0 \; x_0^{(2)} \; ... \; x_0^{(N)} \right]^T$$

This is called a Truncated Carleman Linearization of a non-linear analytic state equation.

7.4 Controllability of Linearized Systems by using Carleman Linearization.

In this section we deal with the controllability of the resultant bilinear system obtained through Truncated Carleman Linearization procedure.

Firstly   we will explain a necessary condition for controllability of bilinear systems then we will analyze systems obtained from linear, non-linear with constant control field, bilinear and non-linear systems which we have applied the Carleman technique.

7.4.1 Controllability of Bilinear Systems.

This section deals with controllability for bilinear systems and its application in (7.13). In [14] some necessary conditions are shown to establish controllability of bilinear systems. In our case consider the system

$$\frac{dz(t)}{dt} = Az(t) + Fv(t)z(t) + Bv(t) \tag{7.14}$$

$A, F \in \mathbb{R}^{n \times n}, \quad B \in \mathbb{R}^n, z \in \mathbb{R}^n$ and $v \in \mathbb{R}$.

For linear time-invariant systems a convenient rank test for the pair (A, B) is available to determinate controllability [6]. On the other hand Bilinear systems has the possibility of controllability through the pairs (A, B), (A, F) and (B, F) by appropriate system design. [14].

For system (7.14) a pair (z_e, v_e) is called an equilibrium pair, if

$$Az_e + Fv_e z_e + Bv_e = 0. \tag{7.15}$$

A necessary condition in order to this pair be said controllable [14] [7] is that

$$(A + Fv_e, Fz_e + B) \tag{7.16}$$

is a controllable pair.

Remark 4. Notice that for $(z_e, v_e) = (0, 0)$, the condition (7.16) reduces to our familiar (A, B) pair.

7.4.2 Linear case.

This section deals with Carleman Linearization applied to a lineal system

$$\dot{x} = A_1 x + B_0 u \tag{7.17}$$

where the matrices $A \in \mathbb{R}^{n \times n}$ and $B \in \mathbb{R}^n$ are real and constant, and the controllability of the resulting system.

Using (7.11), (7.12) and (7.13) on (7.17) we obtain

$$\frac{d}{dt}\tilde{x} = \begin{bmatrix} A_{11} & 0 & \cdots & 0 \\ 0 & A_{21} & \cdots & 0 \\ 0 & 0 & \cdots & 0 \\ \vdots & \vdots & \ddots & \vdots \\ 0 & 0 & \cdots & A_{N1} \end{bmatrix}\tilde{x} + u\begin{bmatrix} 0 & 0 & \cdots & 0 & 0 \\ B_{2,0} & 0 & \cdots & 0 & 0 \\ 0 & B_{3,0} & \cdots & 0 & 0 \\ \vdots & \vdots & \ddots & \vdots & \vdots \\ 0 & 0 & \cdots & B_{N,0} & 0 \end{bmatrix}\tilde{x} + \begin{bmatrix} B_{1,0} \\ 0 \\ \vdots \\ 0 \end{bmatrix}u \qquad (7.18)$$

Now, if we use the condition from the previous section it is easy to verify that the last bilinear system is not completely controllable around the origin, i. e., $(\tilde{x}_e, u_e) = (0,0)$, in fact the number of controllable state variables in x_e is at most n, the original dimension of x if the pair (A_1, B_0) is a controllable pair.

In this way we can say that we can apply the Truncated Carleman Linearization to a linear system as (7.17) we obtain again a linear system regarding the controllable variables and the control field.

7.4.3 Non-Linear case with constant control field.

It is known if $g(x)$ in (7.3) is "almost" constant or constant with respect to $x(t)$, i. e.

$$m_i \leq g_i(x) \leq m_i + \varepsilon \qquad (7.19)$$

$$\left|\frac{dg_i(x)}{dx}\right|_{x=x_e} \leq \delta \ll 1$$

for $i = 1, 2, ..., n$ and some $\varepsilon \ll 1$ and $\delta \ll 1$, the Taylor series elements B_k are almost zero for $k \geq 1$, then

$$g(x) \approx B_0 \qquad (7.20)$$

Furthermore (7.8) has the form

$$\dot{x} = \sum_{K=1}^{\infty} A_k x^{(k)}(t) + B_0 u(t) \qquad (7.21)$$

and

$$\frac{d\tilde{x}(t)}{dt} = A\tilde{x}(t) + u(t)F\tilde{x}(t) + Bu(t) \qquad (7.22)$$

where A and B are as in (7.13) and F reduces to the following due to $B_{j,k} = 0$ for $k > 0$, i. e.,

$$F = \begin{bmatrix} 0 & 0 & \cdots & 0 & 0 \\ B_{2,0} & 0 & \cdots & 0 & 0 \\ 0 & B_{3,0} & \cdots & 0 & 0 \\ \vdots & \vdots & \vdots & \vdots & \vdots \\ 0 & 0 & \cdots & B_{N,0} & 0 \end{bmatrix}$$

In (7.22) the controllability matrix $\mathscr{C}$ from (7.16) at equilibrium point $\tilde{x}_e = 0$ and $u_e = 0$.

$$\mathscr{C} = \begin{bmatrix} B, AB, \ldots, A^{M-1}B \end{bmatrix} \tag{7.23}$$

but due to

$$B = \begin{bmatrix} B_0 & 0 & \cdots & 0 \end{bmatrix}^T$$

we have

$$A^k B = \begin{bmatrix} A_1^k B_0 & 0 & \cdots & 0 \end{bmatrix}^T$$

.

Thus

$$rank(\mathscr{C}) \leq rank(A_1) \leq n \leq M. \tag{7.24}$$

This last, as in the previous section, is a consequence of the fact that $\tilde{x}$ is an array of the original n-dimensional vector and powers of the same, thus only the first n elements of $\tilde{x}$ can move freely, i. e. $rank(\mathscr{C}) = n$, if and only if pair (A_1, B_0) is a controllable pair [10] and [6], whereas remainder elements depend on powers of these "n" first components.

Therefore (7.22) is not completely controllable by using scalar control but at most only the first n original vector state elements.

7.4.4 Bilinear case.

In this section we show the case when we have originally a bilinear system as

$$x = A_1 x + B_1 x u + B_0 u$$

with $A_1, B_1 \in \mathbb{R}^{n \times n}$ and $B_0 \in \mathbb{R}^n$ and u an scalar. In this case it is obvious the difference regarding a linear system is the order where we truncate the Taylor series in (7.3) with respect to control field, at the linear case this order is zero and at bilinear case is one. Thus after apply Carleman Linearization we have the system

$$\frac{d\tilde{x}(t)}{dt} = A\tilde{x}(t) + u(t)F\tilde{x}(t) + Bu(t)$$

where

$$F = \begin{bmatrix} B_{1,1} & 0 & \cdots & 0 & 0 \\ B_{2,0} & B_{2,1} & \cdots & 0 & 0 \\ 0 & B_{3,0} & \cdots & 0 & 0 \\ \vdots & \vdots & \vdots & \vdots & \vdots \\ 0 & 0 & \cdots & B_{N,0} & 0 \end{bmatrix}$$

In this case there exists a slight difference with respect to the two previous cases, where the controllability was restricted to the original controllable part of linear and nonlinear with constant control field systems, but it is easy to verify the new bilinear system is as before only controllable on the first n variable state elements as well through the pairs (A_1, B_0), (A_1, B_1) and (B_1, B_0).

7.4.5 Non-linear case.

Finally the non-linear case

$$\dot{x} = f(x) + g(x)u$$

where f and g are as in (7.7) from section 2. The resulting truncated linearized system is exactly (7.13) and as well as the linear, constant field and bilinear systems, its controllable part are the first n state elements corresponding to the original number of elements in the state vector space $\mathbb{R}^n$.

This last result is easily to prove by applying the sufficient condition explained above when we take the equilibrium at the origin, where all linearized systems have as controllable part the n first original state elements if the original systems are controllable.

7.5 Modified Carleman Linearization.

The Truncated Carleman Linearization neglects high order terms from the Taylor Series in (7.7) and (7.8) therefore the most part of the times some important dynamics are omitted after linearization process.

In order to avoid this Bellman [1] proposed to make an approximation of the deleted terms by a weighted linear terms that depends on the variables in $\tilde{x}$, this improves in general the linear approximation of (7.3), thus we can write the state equation of $\tilde{x}$ when $g(x) = 0$, as

$$\frac{d\tilde{x}}{dt} = A\tilde{x} + S(\tilde{x}) \tag{7.25}$$

where $S(\tilde{x}) = \begin{bmatrix} s_1(\tilde{x}) \\ s_2(\tilde{x}) \\ \vdots \\ s_N(\tilde{x}) \end{bmatrix}$ contains the terms of order higher than N, and each $s_i(\tilde{x})$

is a polynomial of degree $\geq N + 1$. Partitions on $S(\tilde{x})$ are compatible with partitions in A.

If we make approximations of $S_i(\tilde{x})$ by $\sum_{k=1}^{N} \rho_{i,k} x^{(k)}$, where $\rho_{i,k} \in \mathbb{R}^{m_i \times m_k}$, see [1]; then the finite order linearization of degree M of (7.3) becomes:

$$\frac{d\tilde{x}}{dt} = (A + \mathscr{R})\tilde{x} \tag{7.26}$$
$$= A\tilde{x}$$

where

$$\mathscr{R} = \begin{bmatrix} \rho_{1,1} & \rho_{1,2} & \cdots & \rho_{1,N} \\ \rho_{2,1} & \rho_{2,2} & \cdots & \rho_{2,N} \\ \vdots & \vdots & \ddots & \vdots \\ \rho_{N,1} & \rho_{N,2} & \cdots & \rho_{N,N} \end{bmatrix}, \quad \rho_{i,k} \in \mathbb{R}^{m_i \times m_k} \tag{7.27}$$

Not necessarily all the row blocks $\rho_{i,*}$ for $i = 1, \ldots, N$ need to be different from zero.

There are two forms to determine this set of parameters:

I) Choosing a region $D \subset \mathbb{R}^n$ of the original state space where our system is know to be operating, sample enough number of points in D and minimize

$J = \left\| s_i(\tilde{x}) - \sum_{k=1}^{N} \rho_{i,k} x^{(k)} \right\|$ using least squares [2], [19], [20] and [21] to calculate the

free parameters introduced.

II) If we want to approximate a particular solution, as the limit cycle in a van der Pol oscillator, sample this trajectory equidistantly and again apply least squares to minimize the same criteria J.

In this way, the equation (7.26) represents a better finite order linearization of degree M of (7.3). Any of the two modifications of the Truncated Carleman Linearization improves its performance.

In this chapter we are going to apply the first **I)** procedure described.

7.6 Linearized van der Pol Oscillator

The Modified Carleman Linearization can be used for any non-linear systems whose right hand side of defining equations be analytic; we are going to apply the above procedure to the well known van der Pol Oscillator, see [15] and [22], the equation that describes this system is

$$\ddot{z} + \varepsilon(z^2 - 1)\dot{z} + \omega^2 z = 0 \tag{7.28}$$

where the term with ε represents a non-linear damping term (when $\varepsilon = 0$ the van der Pol oscillator reduces to an harmonic linear oscillator, see [22]), and ω is the frequency in radians.

We define $x_1 \triangleq z$, $x_2 \triangleq \dot{z}$ and $x \triangleq [x_1 \ x_2]^T$, then rewriting (7.28) in Taylor Series form and using the Modified Kronecker Product we have

$$\dot{x} = A_1 x + A_2 x^{(2)} + A_3 x^{(3)} + A_4 x^{(4)} \tag{7.29}$$

$$A_1 = \begin{bmatrix} 0 & 1 \\ -\omega^2 & \varepsilon \end{bmatrix}, \quad A_3 \begin{bmatrix} 0 & 0 & 0 & 0 \\ 0 & -\varepsilon & 0 & 0 \end{bmatrix}, \quad A_2 = A_4 = 0$$

Then using (7.12) and (7.13) the linearization of van der Pol Oscillator for $N = 4$ is

$$\frac{d\tilde{x}}{dt} = \begin{bmatrix} A_{1,1} & 0 & A_{1,3} & 0 \\ 0 & A_{2,1} & 0 & A_{2,3} \\ 0 & 0 & A_{3,1} & 0 \\ 0 & 0 & 0 & A_{4,1} \end{bmatrix} \tilde{x} \tag{7.30}$$

$$= A\tilde{x}$$

the vector state $\tilde{x} \in \mathbb{R}^{14}$, and

$$A_{1,1} = \begin{bmatrix} 0 & 1 \\ -\omega^2 & \varepsilon \end{bmatrix}, \quad A_{1,3} = \begin{bmatrix} 0 & 0 & 0 & 0 \\ 0 & -\varepsilon & 0 & 0 \end{bmatrix}, \quad A_{2,1} = \begin{bmatrix} 0 & 2 & 0 \\ -\omega^2 & \varepsilon & 1 \\ 0 & -2\omega^2 & 2\varepsilon \end{bmatrix}$$

$$A_{2,3} = \begin{bmatrix} 0 & 0 & 0 & 0 & 0 \\ 0 & -\varepsilon & 0 & 0 & 0 \\ 0 & 0 & -2\varepsilon & 0 & 0 \end{bmatrix}, \quad A_{3,1} = \begin{bmatrix} 0 & 3 & 0 & 0 \\ -\omega^2 & \varepsilon & 2 & 0 \\ 0 & -2\omega^2 & 2\varepsilon & 1 \\ 0 & 0 & -3\omega^2 & 3\varepsilon \end{bmatrix}$$

$$A_{4,1} = \begin{bmatrix} 0 & 4 & 0 & 0 & 0 \\ -\omega^2 & \varepsilon & 3 & 0 & 0 \\ 0 & -2\omega^2 & 2\varepsilon & 2 & 0 \\ 0 & 0 & -3\omega^2 & 3\varepsilon & 1 \\ 0 & 0 & 0 & -4\omega^2 & 4\varepsilon \end{bmatrix}$$

However, as we saw in the previous section, (7.30) is a truncated linearization, where the higher order terms have been deleted, in fact there exists a residual $S(\tilde{x})$ of the form

$$S(x) = \begin{bmatrix} 0 & 0 & 0 & 0 & 0 & 0 & s_1(\tilde{x}) & s_2(\tilde{x}) & s_3(\tilde{x}) \\ 0 & s_4(\tilde{x}) & s_5(\tilde{x}) & s_6(\tilde{x}) & s_7(\tilde{x}) \end{bmatrix}^T$$

where

$$s_1(\tilde{x}) = -\varepsilon x_1^4 x_2, \quad s_2(\tilde{x}) = -2\varepsilon x_1^3 x_2^2, \quad s_3(\tilde{x}) = -3\varepsilon x_1^2 x_2^3$$
$$s_4(\tilde{x}) = -\varepsilon x_1^5 x_2, \quad s_5(\tilde{x}) = -2\varepsilon x_1^4 x_2^2, \quad s_6(\tilde{x}) = -3\varepsilon x_1^3 x_2^3$$
$$s_7(\tilde{x}) = -4\varepsilon x_1^2 x_2^4$$

Then the equation that describes the real dynamics of $\tilde{x}$ is

$$\frac{d\tilde{x}}{dt} = A\tilde{x} + S(\tilde{x})$$

Now, we can calculate $\rho_{i,k}$, for which we use least square method for each $s_i(\tilde{x})$ above, the following example shows a particular case of this linearization for the van der Pol Oscillator.

Example 1. From the van der Pol Oscillator equation in (7.28) when $\omega = \varepsilon = 1$, we calculate the value of $\rho_{i,k}$ using the least squares method as described above. As example $s_1(\tilde{x})$ is approximate by

$$s_1(\tilde{x}) \approx \rho\tilde{x}$$

$$\rho_1 = \begin{bmatrix} -7.9848 & 1.3921 & -0.0137 & -0.0011 & -0.004 & 1.9669 & -4.4836 & 2.3702 \\ -0.2828 & 0.004 & -0.0021 & 0.0056 & -0.0019 & 0.0009 \end{bmatrix}$$

In this form we obtain (7.27) for the van der Pol oscillator, then we have a finite order linearization of the van der Pol Oscillator, which is an unstable 14th order system, because some eigenvalues of the matrix $A = A + \mathcal{R}$ are on the right hand side of the complex plane.

For the linear oscillator to be marginally stable and have a periodic solution, the eigenvalues of the state matrix must lie on the imaginary axis and they should be a multiple of some fundamental frequency in conjugated complex pairs, i. e., let the spectra of A be $\sigma(A) = \{\lambda_1, \lambda_2, \ldots, \lambda_{14}\}$ then

$$\text{Re}(\lambda_i) = 0 \quad \text{and} \quad \lambda_i = \pm jm\omega_0 \qquad (7.31)$$
$$m = 1, \ldots, 7 \quad i = 1, \ldots, 14$$

To make this we use a control signal such that

$$\frac{d\tilde{x}}{dt} = A\tilde{x} + bv$$

where

$$b \in \mathbb{R}^{14}, \quad b = \begin{bmatrix} 0 & 0 & \dots & 1 \end{bmatrix}^T$$
$$v = -K\tilde{x}, \quad K \in \mathbb{R}^{1 \times 14}$$

The b used above is such that the pair (A, b) must be controllable [6]. Then K must be such that the matrix $\mathbb{A} = A - bK$ has its eigenvalues on the vertical axis of the complex plane and they satisfy the multiplicity described in (7.31), see [6] and [10]. The effect of this state feedback is exemplified in Fig. 7.1

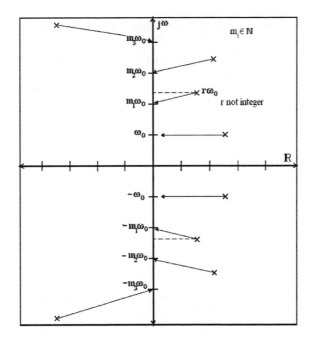

Fig. 7.1. Eigenvalue assignment in order to satisfy the constraint in (7.31).

Remark 5. The simple projection of the eigenvalues to the imaginary axis does not guarantee that the eigenvalues be multiple of the lowest one, then we may interpret as a *nonlinear projection* as shown in Fig. 7.1.

Hence K must be such that one of the eigenvalues for the matrix $\mathbb{A} = A - bK$ must be the corresponding to the natural frequency of (7.28) on the vertical axis and the rest, multiples of this one. This fact makes that the linearized van der Pol systems oscillates. Then

$$\sigma\left(\mathbb{A}\right) = \{\lambda_i \mid \lambda_i = \pm m\lambda_0, \quad m \in \mathbb{N}, \quad m = 1,...,7\}$$
$$\lambda_0 = \omega_0 j$$

A way to obtain ω_0 when ε is large is in [9], otherwise we can the less precise formula in (7.28).

In this way the system

$$\frac{d\tilde{x}}{dt} = \mathbb{A}\tilde{x} \tag{7.32}$$

is an stable, but not asymptotically stable, finite order linearization of (7.28).

The Fig. 7.2 shows the waveforms of the original and the linearized van der Pol oscillators. Notice how well approximates the linear system to the van der Pol oscillator. In Fig. 7.3 we show the error between the original non-linear van der Pol oscillator and the linearized one.

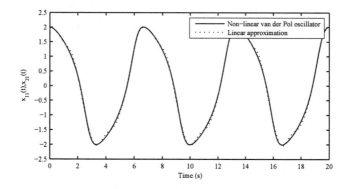

Fig. 7.2. First state of the original van der Pol oscillator compared with the first component of the 14th order linear approximation.

7.6.1 Symmetric coupled linearized van der Pol Oscillators

In (7.32) we have a stable finite 14th order linearization of the van der Pol Oscillator for $\omega = \varepsilon = 1$. Now apply our procedure for two original van der Pol oscillators with slightly different frequencies as

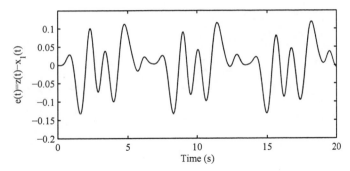

Fig. 7.3. Error between non-linear van der Pol oscillator and its linear approximation.

$$\ddot{z}_1 + (z_1^2 - 1)\dot{z}_1 + z_1 = 0 \qquad z_1(0) = 2, \ \dot{z}_1(0) = 0$$

$$\ddot{z}_2 + (z_2^2 - 1)\dot{z}_2 + (1.01)^2 z_2 = 0 \quad z_2(0) = 2, \ \dot{z}_2(0) = 0$$

i. e., $\omega_1 = \varepsilon_1 = \varepsilon_2 = 1$ and $\omega_2 = 1.01$, we can use $\mathbb{A}_1$ from example above and obtain $\mathbb{A}_2$ using the method described in the previous section.

For coupling the two oscillators, we use the diagram in Fig. 7.4, see [13], [16] and [12]. Then we add a control signal in (7.32) for both oscillators such that.

$$\frac{d\tilde{x}_i}{dt} = \mathbb{A}_i \tilde{x}_i + \mathbb{B} u_i \tag{7.33}$$

$$y_i = \mathbb{C} \tilde{x}_i$$

$$y_{i0} = \tilde{x}_{i,1}$$

$$\mathbb{B} \in \mathbb{R}^{14 \times 1}, \quad \mathbb{C} \in \mathbb{R}^{1 \times 14}, \quad u_i \in \mathbb{R}$$

$$u_i = -k(y_i - y_j), \quad i \neq j, \quad k \in \mathbb{R}, \quad k > 0$$

$$\mathbb{B} = \begin{bmatrix} 01\vdots010\vdots0100\vdots01000 \end{bmatrix}^T$$

$$\mathbb{C} = \begin{bmatrix} 01\vdots010\vdots0100\vdots01000 \end{bmatrix} \tag{7.34}$$

Thus

$$\frac{d\tilde{x}_i}{dt} = (\mathbb{A}_i - k\mathbb{B}\mathbb{C})\tilde{x}_i - k\mathbb{B}\tilde{x}_j$$

These values for $\mathbb{C}$ and $\mathbb{B}$ are been chosen in order to couple both oscillators through the element $\tilde{x}_{i,2}(t)$ and its powers per each linearized oscillator corresponding to the velocity element $\dot{z}$ in the original van der Pol oscillator, see [13], [16] and [12].

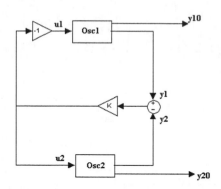

Fig. 7.4. Symmetric coupled linearized van der Pol oscillators.

In this case both pairs $(\mathbb{A}_i, \mathbb{B})$ $i = 1, 2$; are controllable and both pairs $(\mathbb{C}, \mathbb{A}_i)$ are observable.

The behavior of $\tilde{x}_1$ and $\tilde{x}_2$ depends directly from the value of the scalar k, in fact both systems become unstable for any value of k as it is shown in the Fig. 7.5.

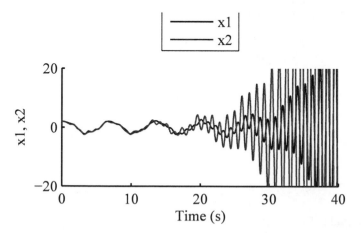

Fig. 7.5. The coupled oscillators cannot synchronize.

The fact that the amplitude increases is due to during the state feedback through $u_i = -k(y_i - y_j)$ modifies the eigenvalues of the matrix $(\mathbb{A}_i - k\mathbb{B}\mathbb{C})$ to the left and to the right side on the complex plane in general. The Fig. 7.6 shows the root locus of the eigenvalues of the matrix $(\mathbb{A}_1 - k\mathbb{B}\mathbb{C})$.

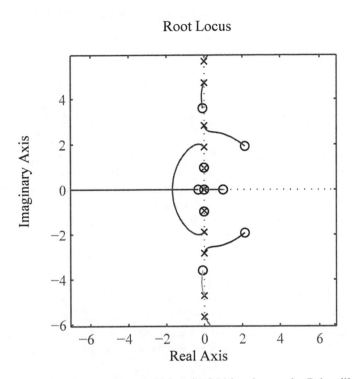

Fig. 7.6. Root Locus of coupled Linearized 14th order van der Pol oscillators.

This shows that the eigenvalues turn on the right side at the same moment when k is different of zero, thus we can say it is no possible to synchronize the linear oscillators with diffusively symmetric coupling. Notice this behavior is not for the coupling proposed, but for any linear coupling, because to synchronize linear oscillator is required that eigenvalues coincide on the imaginary axis, that keeps the multiplicity property and that these conditions holds for a range of values of k greater that a threshold [16]. The Fig. 7.7 shows how two original nonlinear van der Pol oscillators can be synchronized with diffusively symmetric coupling, the natural frequencies of the oscillators are the same that the example with linear oscillators above.

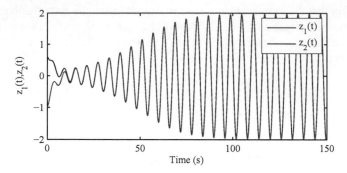

Fig. 7.7. Two nonlinear van der Pol oscillators synchronized.

7.7 Conclusions.

In the section 4 we analyzed the effect of Truncated Carleman Linearization explained in section 3 on the controllability of a set of different systems. We found that the controllability of the linearized systems is the same that the original ones, i. e., the Truncated Carleman procedure does not affect this property without matter what kind of original system be. After being linearized the new system is only controllable trough the original number of state elements. In the case of a lineal system the new system is virtually the same.

In section 5 we proposed a Modified version of the Truncated Carleman Linearization method, then in section 6 we applied this novel procedure to the van der Pol oscillator and used to try to synchronize to linearized van der Pol oscillators with coupling proportional to velocity. We conclude that it is not possible to synchronize them as they do in the original systems using the same coupling.

References

1. Bellman R. and Richardson J.M., "On some questions arising in the approximative solution of nonlinear differential equations", Quart. Appl. Math, Vol. 20, pp. 333-339, 1963.
2. Björck A., *Numerical Methods for Least Squares Problems*, SIAM, Philadelphia, 1996.
3. Brewer J., "Kronecker products and matrix calculus in system theory". IEEE Trans. on Circuits and Systems, Volume CAS-25, Issue 9, pp. 772 - 781, Sep 1978.
4. Brockett R. "Lie Algebras and Lie Groups in Control Theory". In *Geometric Methods In Systems Theory*. D. Mayne and R. Brockett (Eds.), Reidel Dordrecht, 1973.
5. Carleman T., "Application de la Théorie des Équations Intégrales Linéares aux Systémes D' Équations Différentialles Non Linéaires", Acta. Math., Vol. 59, pp. 63-87, 1932.
6. Chen C. T., *Linear System Theory and Design*, Third Edition, Oxford University Press, 1999.
7. Cheng D. and Liu J., "On Controllability of Switched Bilinear Systems", American Control Conference Vol. 7, pp. 5109-5114, 2005.

8. Collado J. and Sánchez I., "Carleman Linearization and its Use in Oscillators", 5th International Conference on Electrical Engineering, Computing Science and Automatic Control (ICEEE), November 2008, pp. 13-19.
9. Dorodnicyn A. A., "Asymptotic Solution of van der Pol's Equation", Engl. transl. in Amer. Math. Soc. Transl., Series 1, 88, pp. 1-24, 1953.
10. Kailath T., *Linear Systems*, Prentice-Hall, Inc., Englewood Cliffs, N. J. 07632, 1980.
11. Khalil H. K., *Nonlinear Systems*, 3rd. Ed. Upper Saddle River:Prentice-Hall, USA, 2002.
12. Kuramoto Y., *Chemical Oscillations, Waves, and Turbulence*, New York: Springer, 1984.
13. Mimila P. O., *Sincronización de Osciladores*, Tesis de maestría, Departamento de Control Automático CINVESTAV-IPN, Sep. 2006.
14. Mohler R. R. *Nonlinear Systems Vol. II Aplications to Bilinear Control,* Prentice-Hall Inc., 1991.
15. Nayfeh A. H., *Nonlinear Oscillations*, John Wiley & Sons Inc., 1979.
16. Pikovsky A., Rosenblum M., and J. Kurths. *Synchronization: A universal concept in nonlinear sciences*, Cambridge University Press, Cambridge, 2001.
17. Qiu L. and Davison E. J., "The Stability Robustness Determination of State Space Models with Real Unstructured Uncertainty". Math. of Contr., Signals and Systems, Vol. 4, pp. 247-267, 1991.
18. Rugh W. J., *Non Linear System Theory The Volterra/Wiener Approach*. The Johns Hopkins University Press, 1981.
19. Stewart G. W., *Introduction to Matrix Computations*, Academic Press, 1973.
20. Stewart G. W., *Matrix Algorithms: Basic Decompositions*, SIAM, Philadelphia, 1998.
21. Trefethen L. N. and Bau D. III, *Numerical Linear Algebra*, SIAM, Philadelphia, 1997.
22. van der Pol B. L., "The nonlinear theory of electric oscillators", Proc of the IRE, Vol. 22, Num. 9, pp. 1051-1056, September 1934.
23. van Loan C. F., "The ubiquitous Kronecker product". Journal of Computational and Applied Mathematics, 123 (2000) 85–100.

8

Finite Element Modelling and Unbalance Compensation for an Asymmetrical Rotor-Bearing System with Two Disks

M. Arias-Montiel and G. Silva-Navarro

Centro de Investigacion y de Estudios Avanzados del I.P.N.
Departamento de Ingenieria Electrica - Seccion de Mecatronica
A.P. 14-740, C.P. 07360 Mexico, D.F., MEXICO
mam7915@yahoo.com.mx, gsilva@cinvestav.mx

Summary. In this work a LQR control scheme for unbalance compensation using an active magnetic bearing in a rotor-bearing system is presented. The rotor-bearing system has two disks in an asymmetrical configuration along the shaft and its mathematical model is obtained by applying finite element techniques. A reduced order model of seven degrees of freedom is experimentally validated for the design of an active LQR control law, using the output feedback of only one disk displacement, a Luenberger-type asymptotic observer and an estimated state feedback controller, synthesized by means of a convenient LQ criterion. Thus, the active balancing control is tuned to attenuate those vibrations caused by the rotor imbalance in the two disks, using only an actuator (active magnetic bearing) and one proximitor at each rotor plane. Some simulation and experimental results are included to show the transient and steady-state behavior of the overall closed-loop system.

8.1 Introduction

The control of vibrations in rotating machinery is very important to improve the machines performance and to extend their useful life. The main source of vibrations in rotating machinery is attributed to the unbalance (Vance [10]). Many control schemes to reduce the unbalance response in the classical Jeffcott rotor have been proposed in the literature (Zhou and Shi [12]).

Recently, the analysis and control of rotating systems have been extended to systems with more than one disk and finite element techniques are now widely used to model these systems. The finite element approach was applied to rotors for the first time by Ruhl and Booker [8], with subsequent contributions by Nelson and Mcvaugh [7]. For more details, in Genta [3] is presented a survey about finite element types used in rotordynamics.

On the other hand, the Active Magnetic Bearings (AMB) present some advantages over most of the conventional journal bearings. For instance, the complete absence of mechanical contact and no wear due to mechanical friction make to the lubricant unnecessary. Therefore, the number of AMB applications is growing very fast due to these advantages, to reduce the unbalance response using modern feedback control techniques. Some researchers have considered reduced order finite element models to design algorithms for active vibration control in rotor-bearing systems with magnetic bearings, such as Tian [9], where is used a discrete-time sliding-mode control with a variable structure system and a disturbance observer to estimate the state vector and disturbance in order to attenuate the vibrations in an AMB system. Couzon and Der Hagopian [2] combine neural networks and fuzzy logic to control the flexible modes of a rotor suspended on AMB and working under the first critical speed. Lewis et. al. [6] present a sliding-mode control with output feedback to control a flexible Jeffcott rotor and numerical simulations to get some validation. Moreover, in Yu et al. [11] is proposed a robust modal control for the vibration suppression in a flexible Jeffcott rotor, and the validation is done via numerical simulations. All these works propose the use of two magnetic bearings to control the unbalance response.

In the present chapter, a finite element model of a rotor-bearing system, with two disks on the shaft in an asymmetrical configuration, based on the so-called Euler beam is developed and experimentally validated. The reduced order model is used to synthesize an active vibration control law. First, we design an asymptotic observer to estimate those not measurable states, via the knowledge of only one disk displacement, and then, a LQR state feedback controller is properly tuned to attenuate the unbalance response in the overall rotor-bearing system. This control scheme employs only a displacement measurement in one disk and one AMB to reduce the vibrations caused by the unbalance in two balancing planes (disks). In addition , some numerical and experimental results are presented to illustrate the transient and steady-state behavior of the closed-loop system.

8.2 Modelling and analysis of the rotor-bearing system

8.2.1 Finite element model of the asymmetrical rotor

First, we consider a rotor-bearing system, which consists of a flexible steel shaft, supported by two conventional journal bearings and two disks placed asymmetrically along the shaft, as shown in Fig. 8.1.

For modelling purposes the shaft is divided in three finite elements and four nodes. The finite elements have different lengths L_1, L_2 and L_3. A finite element of the Euler beam is considered with the following assumptions: i) the system is isotropic, such that we can analyze only a rotor plane of motion, ii) the shaft exhibits a linear elastic behavior, iii) the disk masses are lumped in the corresponding nodes, and iv) the gyroscopic effects and shear deformations can be neglected.

The Euler beam element is illustrated in Fig. 8.2. The element has two degrees of freedom in each node, that is, displacement and angular deflection (see, e.g., Genta [3]).

The four generalized coordinates in each element are given by

$$u_1 = R_{X1}, u_2 = \beta_{Y1}, u_3 = R_{X2}, u_4 = \beta_{Y2} \tag{8.1}$$

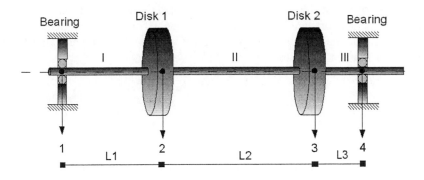

Fig. 8.1. Schematic diagram of an asymmetrical rotor-bearing system with two unbalanced disks.

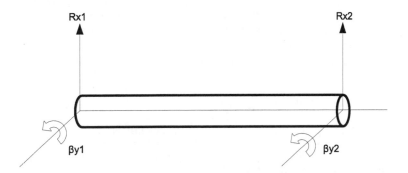

Fig. 8.2. A finite element of the Euler-type beam.

The mass and stiffness matrices for this type of element are detailed in Genta [3]. In order to assembly the mass and stiffness matrices, we employ the consistent approach presented in Genta [4]. Thus, by considering the vector of generalized displacements

$$\mathbf{u} = \begin{bmatrix} R_{x1} & \beta_{y1} & R_{x2} & \beta_{y2} & R_{x3} & \beta_{y3} & R_{x4} & \beta_{y4} \end{bmatrix} \tag{8.2}$$

we can obtain the global matrices for the rotor-bearing system, which are provided in Appendix A.

8.2.2 Natural frequencies

The natural frequencies can be easily obtained from the undamped system dynamics in free vibration, that is, the eight degree-of-freedom system

$$\mathbf{M}\ddot{\mathbf{u}} + \mathbf{K}\mathbf{u} = \mathbf{0}, \quad \mathbf{u} \in \mathbb{R}^8 \tag{8.3}$$

To find the natural frequencies, first it is necessary to apply the boundary conditions to system. In this case, the displacements in nodes 1 and 4 are fixed to zero, because the supports are considered rigid enough. Therefore, the global system matrices are reduced as is shown in Appendix B and the vector of generalized displacements (8.2) is simplified to

$$\mathbf{u} = \begin{bmatrix} \beta_{y1} & R_{x2} & \beta_{y2} & R_{x3} & \beta_{y3} & \beta_{y4} \end{bmatrix} \tag{8.4}$$

and the six degree-of-freedom system dynamics, for only one rotor plane, is described by

$$\mathbf{M}_{red}\ddot{\mathbf{u}} + \mathbf{K}_{red}\mathbf{u} = \mathbf{0}, \quad \mathbf{u} \in \mathbb{R}^6 \tag{8.5}$$

In this case, the characteristic equation for system (8.5) is given by

$$\det\left\{-\omega^2\mathbf{M}_{red} + \mathbf{K}_{red}\right\} = \mathbf{0} \tag{8.6}$$

By solving (8.6), we are able to find the natural frequencies of the rotor-bearing system over rigid supports.

The experimental plataform to validate the above mathematical model is a rotor kit by *Bently Nevada*, which has a steel shaft with diameter 10 mm, total length between bearings of 450 mm and two disks with unbalance of diameter 75 mm, width 25 mm and mass 0.806 kg. The system parameters used to carry out the numerical computations and to obtain the simulation results are given in Table 8.1.

Table 8.1. Physical system parameters of the experimental setup

Parameter	Numerical value
Shaft density, ρ	$7850\frac{kg}{m^3}$
Shaft diameter, d	0.010m
Mass per length unit, m	$0.6165\frac{kg}{m}$
Disk mass, m_{d1}, m_{d2}	0.806kg
Length, L_1	0.150m
Length, L_2	0.250m
Length, L_3	0.050m
Young's modulus, E	205GPa
Moment of inertia, I	4.9×10^{-10}m^4

The natural frequencies are computed as follows

$$\omega_1 = 8931.3\text{Hz} = 535881.06\text{rpm}$$
$$\omega_2 = 3306.7\text{Hz} = 198401.01\text{rpm}$$
$$\omega_3 = 1847.3\text{Hz} = 110842.3\text{rpm}$$
$$\omega_4 = 778.1\text{Hz} = 46685.7\text{rpm}$$
$$\omega_5 = 123.7\text{Hz} = 7427.5\text{rpm}$$
$$\omega_6 = 40.43\text{Hz} = 2425.8\text{rpm}$$

Note that, there are two natural frequencies into the operation range of the experimental platform (i.e., ω_5 and ω_6), that is, rotor speeds up to 10000 rpm. Therefore, we are able to analyze two critical speeds.

8.2.3 Model validation

In order to validate the proposed model, we now proceed to compare the simulation and experimental results with respect to the displacements in both disks (nodes 2 and 3). First of all, the modal damping coefficients were obtained experimentally by applying the so-called *Peak Picking Method*, and then, for numerical simulations, we adjust these values to satisfy the assumption of proportional damping into the model.

In Figs. 8.3 and 8.4 the time domain results are presented. Two constant angular speeds are tested, that is, below the first critical speed and, later on, between the first and the second critical speed.

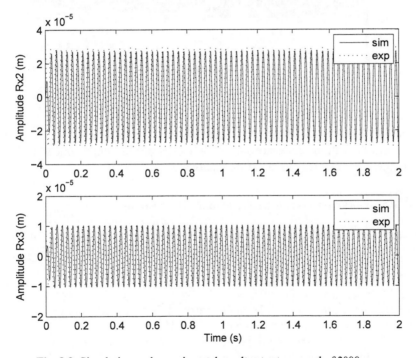

Fig. 8.3. Simulation and experimental results at rotor speed of 2000 rpm.

Here, we can observe that the numerical simulations approximately match to the experimental system responses.

In Figs. 8.5 and 8.6 the frequency response function (FFT) of each disk is shown. In these figures we can note that, the resonant frequencies in simulation and experimental are sufficiently close. It is important to point out that, in this case, we only

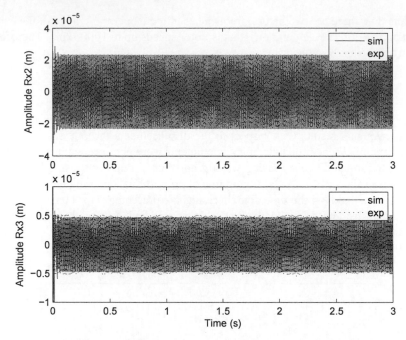

Fig. 8.4. Simulation and experimental results at rotor speed of 7000 rpm.

compared the numerical values of the resonant frequencies. The shape of the experimental and simulation curves are not the same because the sample time was different during the experiments than in the simulations.

8.2.4 Rotor-bearing system with one magnetic bearing

The proposed reduced order model for the rotor-bearing system depicted in Fig. 8.7, without unbalance forces, is described by the equation of motion

$$\mathbf{M}_{cl}\ddot{\mathbf{u}} + \mathbf{D}_{cl}\dot{\mathbf{u}} + \mathbf{K}_{cl}\mathbf{u} = \mathbf{b}f, \ \mathbf{u} \in \mathbb{R}^7, f \in \mathbb{R} \qquad (8.7)$$

where $\mathbf{D}_{cl}$ is the proportional damping matrix, $\mathbf{b}$ is a constant vector and f is the control input (lateral force) provided by the AMB in one direction.

In this case it is important to remark that, when a conventional journal bearing is replaced by a magnetic bearing at the right end of the shaft, results in a different system dynamics due to changes in the boundary conditions. Therefore, we consider an additional degree of freedom to the system, such that, the matrices $\mathbf{M}_{cl}$, $\mathbf{D}_{cl}$ and $\mathbf{K}_{cl}$ are all of dimensions 7×7.

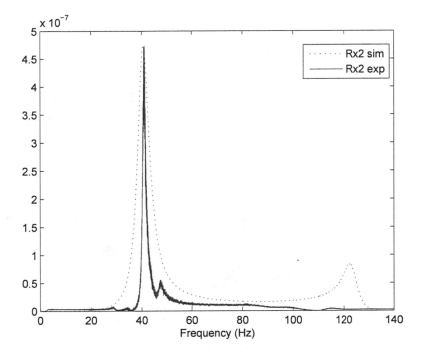

Fig. 8.5. Frequency response function (FFT) in Disk 1.

8.2.5 Controllability and observability properties

To check the controllability and observability properties of the rotor-bearing system with an AMB, the system dynamics is transformed to a state space form, by defining the state vector $\mathbf{z} = \begin{bmatrix} u & \dot{u} \end{bmatrix}^T$

Thus, system (8.7) becomes

$$\dot{\mathbf{z}} = \mathbf{A}\mathbf{z} + \mathbf{B}f, \quad \mathbf{z} \in \mathbb{R}^{14}, f \in \mathbb{R} \tag{8.8}$$

$$y = \mathbf{C}\mathbf{z}, \quad y \in \mathbb{R} \tag{8.9}$$

where

$$\mathbf{A} = \begin{bmatrix} \mathbf{0} & \mathbf{I} \\ -\mathbf{M}_{cl}^{-1}\mathbf{K}_{cl} & -\mathbf{M}_{cl}^{-1}\mathbf{D}_{cl} \end{bmatrix}, \quad \mathbf{B} = \begin{bmatrix} \mathbf{0} \\ \mathbf{M}_{cl}^{-1}\mathbf{b} \end{bmatrix}$$

$$\mathbf{C} = \begin{bmatrix} 0 & 1 & 0 & 0 & 0 & 0 & 0 & 0 & 0 & 0 & 0 & 0 & 0 & 0 \end{bmatrix}$$

From the matrices $\mathbf{A}$, $\mathbf{B}$ and $\mathbf{C}$ in system (8.8)-(8.9), we can easily compute the controllability and observability matrices (see Kailath [5]):

$$\mathbf{C}_{con} = \begin{bmatrix} \mathbf{B} & \mathbf{AB} & \mathbf{A}^2\mathbf{B} & \cdots & \mathbf{A}^{13}\mathbf{B} \end{bmatrix} \tag{8.10}$$

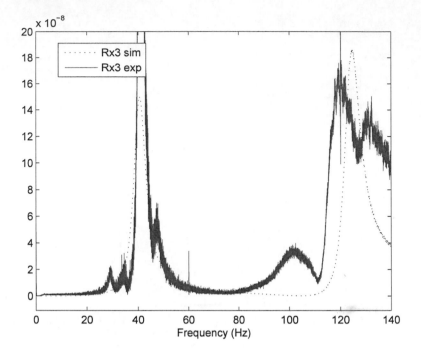

Fig. 8.6. Frequency response function (FFT) in Disk 2.

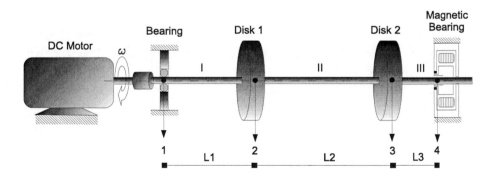

Fig. 8.7. Schematic diagram of the rotor-bearing system with an AMB in the right end.

$$\mathbf{O}_{obs} = \begin{bmatrix} \mathbf{C} \ \mathbf{AC} \ \mathbf{A}^2\mathbf{C} \cdots \mathbf{A}^{13}\mathbf{C} \end{bmatrix}^T \qquad (8.11)$$

which are both of full rank, thus proving that system (8.8)-(8.9) is completely controllable from the lateral control force f and observable from the output $y = z_2$ (displacement in Disk 1). The 14×14 controllability and observability matrices (8.10) and (8.11) were numerically processed in *Matlab* and, due to space limitations, these are omitted here.

8.3 Active unbalance compensation using a LQR control scheme

First, we propose a static output feedback to stabilize the system, because the open-loop system is inherently unstable. To do this, we synthesize the control gains using a LQ criterion using the only output $y = z_2$. Then, we design a full order asymptotic (Luenberger) observer to estimate the state vector in order to apply a LQR controller based on a full state feedback.

8.3.1 State estimation using an asymptotic observer

The dynamic behavior of the perturbed system is given by

$$\dot{z} = Az + Bf + E\varpi, \quad z \in \mathbb{R}^{14}, f \in \mathbb{R}, \varpi \in \mathbb{R} \tag{8.12}$$

$$y = Cz, \quad y \in \mathbb{R} \tag{8.13}$$

where E is the coefficient disturbance matrix and ϖ is the disturbance associated to the existing rotor imbalance.

To estimate the state vector, we apply a Luenberger-type asymptotic observer (Kailath [5]). We consider that only the displacement in the Disk 1 (i.e., $z_2 = R_{x2}$) is available for measurement, as well as the control force f; hence, we need to estimate the remaining displacements and velocities.

The asymptotic observer is described by the dynamics

$$\dot{\hat{z}} = A\hat{z} + Bf(t) + E\varpi(t) + L(y - \hat{y}), \quad \hat{z} \in \mathbb{R}^{14}, f \in \mathbb{R}, \varpi \in \mathbb{R} \tag{8.14}$$

where $y = z_2$, $\hat{z}$ is the estimated state vector and L is some feedback gain matrix. By defining the estimation error as

$$\tilde{z} = z - \hat{z} \tag{8.15}$$

and properly selecting the gain matrix L, we can guarantee that the estimation error asymptotically converges to zero, that is,

$$\lim_{t \longrightarrow \infty} \tilde{z}(t) = 0 \tag{8.16}$$

which is equivalent to that the estimated state $\hat{z}$ converges to the actual state z. In order to satisfy this condition, the matrix L is computed via a pole placement technique, such that the observer dynamics (8.14) be sufficiently faster than the rotor-bearing system dynamics (8.8)-(8.9).

In fact, the rotor speed in the experimental platform can reach up to 10000 rpm (166.7 Hz) and, therefore, the observer has to be tuned to robustly estimate the fastest physical vibration mode.

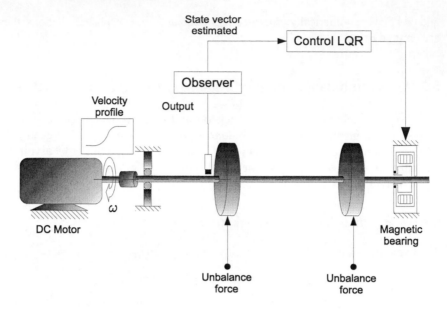

Fig. 8.8. Schematic diagram of the overall rotor-bearing system with controller-observer.

8.3.2 LQR control-observer scheme

In Fig. 8.8 we present a schematic diagram of the overall rotor-bearing system, including the observer-controller structure.

Now, using the estimated state vector, we propose a LQR control based on the full state feedback for the perturbed system (8.12), as follows

$$f = -\mathbf{G}\hat{\mathbf{z}} \tag{8.17}$$

where the state regulation problem is solved by computing a 1×14 gain matrix $\mathbf{G}$ that minimizes a conventional LQR performance index with a positive semi-definite 14×14 matrix $\mathbf{Q}$ and constant $r > 0$ (see, e.g., Kailath [5] and Albertos et al. [1]).

It is important to note that, the implemented unbalance compensation scheme in Fig. 8.8 also includes a local PI speed controller for typical run-up or coast-down operations to/from some desired rotor speed. The selected speed profile is based on Bézier polynomials such that the acceleration/deceleration scheduling be smooth enough to aim at reductions on the vibration amplitudes, specially when the rotor-bearing system goes through its critical speeds.

8.4 Simulation results

To carry out the numerical simulations of the closed-loop system (8.12)-(8.13) with the control law (8.17), we propose the angular speed profile (variable speed) as shown in Fig. 8.9.

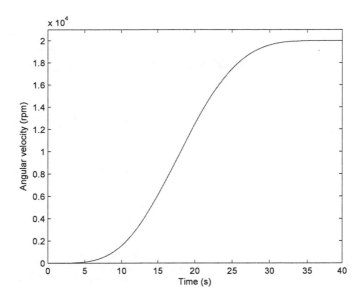

Fig. 8.9. Rotor speed profile based on Bézier polynomials.

In Fig. 8.10 the simulation results for the displacements in the Disks 1 and 2, as well as the the control effort (lateral force) are shown.

The overall LQR control scheme was also proved with constant speeds, which correspond to the first two critical speeds. These frequencies were obtained by solving the eigenproblem, considering a rigid left support and free the right support (AMB). These values are 4300 and 18250 rpm and the results are presented in Figs. 8.11 and 8.12.

The closed-loop system performance using the LQR control scheme is also compared with the (open-loop) stabilizing controller (Figs. 8.11 and 8.12). Here, we can observe that the overall LQR unbalance controller improves the system behavior, with important vibration attenuation, with constant and variable rotor speeds, even at the first and second critical speeds.

8.5 Conclusions

This work deals with the active vibration control of a flexible rotor system with two disks in an asymmetrical configuration. The main contribution of this chapter is the reduction of the vibration amplitude in the two disks using only an actuator and considering only one measurable state (the displacement in the second disk). Because the system dynamics is of 14th order and there are states that are not measurable, we designed an asymptotic observer to estimate these states from one disk displacement. We have proposed a LQR control scheme with state feedback to attenuate the vibrations caused by the mass unbalance and numerical simulations have

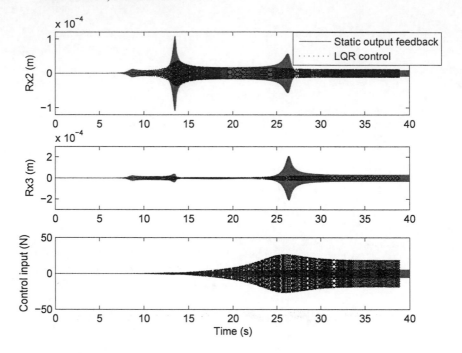

Fig. 8.10. Closed-loop system response with the speed profile as input.

been presented in order to validate the dynamic behavior of the proposed control scheme. The numerical results demonstrate reductions in the overall unbalance response about 70% and 78% in the first disk and about 77% and 85% for the second disk. Furthermore, the applied control efforts acquire reasonable values even when the system goes through its critical speeds. Experimental work is being performed to validate the proposed LQR control scheme and theoretical effort in order to improve the robust dynamic behavior against parametric uncertainties.

References

1. Albertos, P. and Sala, A.: Multivariable control systems: an engineering approach. Springer-Verlag, London (2004).
2. Couzon, P.Y. and Der Hagopian, J.: Neuro-fuzzy active control of rotor suspended on active magnetic bearings, Journal of Vibration and Control, Vol. 13, No. 4, 365–384 (2007).
3. Genta, G.: Dynamics of rotating systems. Springer, New York, NY (2005).
4. Genta, G.: Consistent matrices in rotordynamics, Meccanica, Vol. 20, 235–248 (1985).
5. Kailath, T.: Linear systems. Prentice-Hall. Englewood, NJ (1980).
6. Lewis. A. S., Sinha, A. and Wang, K. W.: Sliding mode output feedback control of a flexible rotor supported by magnetic bearings, International Journal of Rotating Machinery, Vol.7, No. 2, 117–129 (2001).

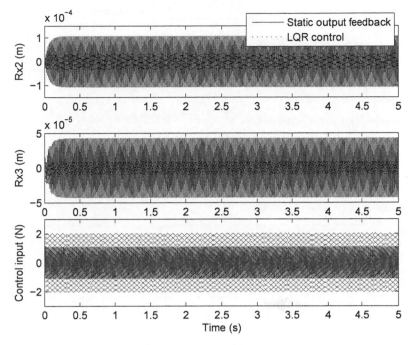

Fig. 8.11. Closed-loop system response for the displacements in Disks 1 and 2, and control force at the first critical speed (4300 rpm).

7. Nelson, H. and McVaugh, J.: The dynamics of rotor-bearing system using finite elements, Journal of Engineering for Industry, 593–600 (1976).
8. Ruhl, R. and Booker, J.: A finite element model for distributed parameter turborotor system, Journal of Engineering for Industry, 126–132 (1972).
9. Tian, H.: Robust control of a spindle-magnetic bearing system using sliding-mode control and variable structure systems disturbance observer, Journal of Vibration and Control, Vol. 5, 277–298 (1999).
10. Vance, J.M.: Rotordynamics of Turbomachinery. John Wiley & Sons, NY (1988).
11. Yu, H-C., Lin, Y-H. and Chu, C-L.: Robust modal vibration suppression of a flexible rotor, Mechanical Systems and Signal Processing, Vol. 21, 334–347 (2007).
12. Zhou, S. and Shi, J.: Active balancing and vibration control of rotating machinery: A survey, The Shock and Vibration Digest, Vol. 33, No. 4, 361–371 (2001).

Appendix A: Global matrices

The global mass matrix is described by

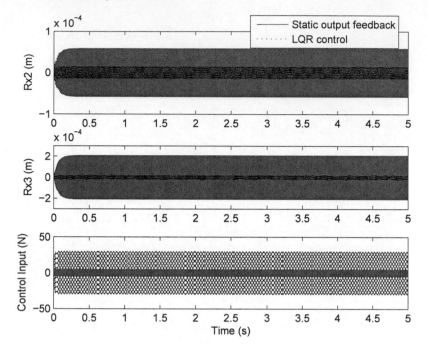

Fig. 8.12. Closed-loop system response for the displacements in Disks 1 and 2, and control force at the second critical speed (18250 rpm).

$$\mathbf{M} = \begin{bmatrix} m_{11} & m_{12} & m_{13} & m_{14} & 0 & 0 & 0 & 0 \\ & m_{22} & m_{23} & m_{24} & 0 & 0 & 0 & 0 \\ & & m_{33} & m_{34} & m_{35} & m_{36} & 0 & 0 \\ & & & m_{44} & m_{45} & m_{46} & 0 & 0 \\ & & & & m_{55} & m_{56} & m_{57} & m_{58} \\ & \text{sym} & & & & m_{66} & m_{67} & m_{68} \\ & & & & & & m_{77} & m_{78} \\ & & & & & & & m_{88} \end{bmatrix}$$

with

$$m_{11} = \frac{156mL_1}{420}, \qquad m_{12} = \frac{22mL_1^2}{420}, \qquad m_{13} = \frac{54mL_1}{420},$$

$$m_{14} = -\frac{13mL_1^2}{420}, \qquad m_{22} = \frac{4mL_1^3}{420}, \qquad m_{23} = \frac{13mL_1^2}{420},$$

$$m_{24} = -\frac{3mL_1^3}{420}, \qquad m_{33} = \frac{156m(L_1+L_2)}{420} + m_{d1}, \quad m_{34} = \frac{22m(L_2^2-L_1^2)}{420},$$

$$m_{35} = \frac{54mL_2}{420}, \qquad m_{36} = -\frac{13mL_2^2}{420}, \qquad m_{44} = \frac{4m(L_1^3+L_2^3)}{420},$$

$$m_{45} = \frac{13mL_2^2}{420}, \qquad m_{46} = -\frac{3mL_2^3}{420}, \qquad m_{55} = \frac{156m(L_2+L_3)}{420} + m_{d2},$$

$$m_{56} = \frac{22m(L_3^2-L_2^2)}{420}, \quad m_{57} = \frac{54mL_3}{420}, \qquad m_{58} = -\frac{13mL_3^2}{420},$$

$$m_{66} = \frac{4m(L_2^3+L_3^3)}{420}, \qquad m_{67} = \frac{13mL_3^2}{420}, \qquad m_{68} = -\frac{3mL_3^3}{420},$$

$$m_{77} = \frac{156mL_3}{420}, \qquad m_{78} = -\frac{22mL_3^2}{420}, \qquad m_{88} = \frac{4mL_3^3}{420}$$

where m is the mass per unit length and L is the length of the element.

The global stiffness matrix is then obtained as follows

$$\mathbf{K} = \begin{bmatrix} k_{11} & k_{12} & k_{13} & k_{14} & 0 & 0 & 0 & 0 \\ & k_{22} & k_{23} & k_{24} & 0 & 0 & 0 & 0 \\ & & k_{33} & k_{34} & k_{35} & k_{36} & 0 & 0 \\ & & & k_{44} & k_{45} & k_{46} & 0 & 0 \\ & & & & k_{55} & k_{56} & k_{57} & k_{58} \\ & \text{sym} & & & & k_{66} & k_{67} & k_{68} \\ & & & & & & k_{77} & k_{78} \\ & & & & & & & k_{88} \end{bmatrix}$$

with

$$k_{11} = \frac{12EI}{L_1^3}, \quad k_{12} = \frac{6EI}{L_1^2}, \quad k_{13} = -\frac{12EI}{L_1^3}, \quad k_{14} = \frac{6EI}{L_1^2},$$

$$k_{22} = \frac{4EI}{L_1}, \quad k_{23} = -\frac{6EI}{L_1^2}, \quad k_{24} = \frac{2EI}{L_1}, \quad k_{33} = \frac{12EI}{L_1^3+L_2^3},$$

$$k_{34} = \frac{6EI}{L_2^2-L_1^2}, \quad k_{35} = -\frac{12EI}{L_2^3}, \quad k_{36} = \frac{6EI}{L_2^2}, \quad k_{44} = \frac{4EI}{L_1+L_2},$$

$$k_{45} = -\frac{6EI}{L_2^2}, \quad k_{46} = \frac{2EI}{L_2}, \quad k_{55} = \frac{12EI}{L_2^3+L_3^3}, \quad k_{56} = \frac{6EI}{L_2^2-L_3^2},$$

$$k_{57} = -\frac{12EI}{L_3^3}, \quad k_{58} = \frac{6EI}{L_3^2}, \quad k_{66} = \frac{4EI}{L_2+L_3}, \quad k_{67} = -\frac{6EI}{L_3^2},$$

$$k_{68} = \frac{2EI}{L_3}, \quad k_{77} = \frac{12EI}{L_3^3}, \quad k_{78} = -\frac{6EI}{L_3^2}, \quad k_{88} = \frac{4EI}{L_3}$$

where E is the module of elasticity of the steel shaft and I is the moment of inertia.

Appendix B: Reduced matrices

The mass and stiffness reduced matrices are obtained by applying the corresponding boundary conditions (rigid supports) to the finite element model. These are expressed by

$$\mathbf{M_{red}} = \begin{bmatrix} m_{22} & m_{23} & m_{24} & 0 & 0 & 0 \\ & m_{33} & m_{34} & m_{35} & m_{36} & 0 \\ & & m_{44} & m_{45} & m_{46} & 0 \\ & & & m_{55} & m_{56} & m_{58} \\ & \text{sym} & & & m_{66} & m_{68} \\ & & & & & m_{88} \end{bmatrix}$$

$$\mathbf{K_{red}} = \begin{bmatrix} k_{22} & k_{23} & k_{24} & 0 & 0 & 0 \\ & k_{33} & k_{34} & k_{35} & k_{36} & 0 \\ & & k_{44} & k_{45} & k_{46} & 0 \\ & & & k_{55} & k_{56} & k_{58} \\ & \text{sym} & & & k_{66} & k_{68} \\ & & & & & k_{88} \end{bmatrix}$$

9

Semiactive Control for the Unbalance Compensation in a Rotor-Bearing System

A. Cabrera-Amado and G. Silva-Navarro

Centro de Investigacion y de Estudios Avanzados del I.P.N.
Departamento de Ingenieria Electrica - Seccion de Mecatronica
A.P. 14-740, C.P. 07360 Mexico, D.F., MEXICO
alvaroca_1@hotmail.com, gsilva@cinvestav.mx

Summary. The problem of unbalance compensation in a rotor-bearing system is addressed by applying a semiactive balancing control scheme based on two radial MR dampers mounted in one of the supports. For the analysis and control synthesis is employed a mathematical model of a Jeffcott-like rotor of 7 degrees-of-freedom for one rotor disk, two supports, the speed dynamics, and the dynamics associated to the MR dampers, whose viscoelastic properties depend on the current inputs (control actions). The semiactive control scheme consists of a rotating speed controller based on a fast PID controller, thus enabling a proper speed profile to pass over the first critical speeds (run-up or coast-down), and a balancing controller for the overall system response, associated to the disk with unbalance and both supports, one of which is semiactively controlled by two independent radial MR dampers, providing the damping forces to reduce the system response. The two radial forces for the unbalance compensation are computed by means of sliding-mode control techniques, which are designed to push the system response to a small orbit about an specific equilibrium. For simplicity during the control design and physical implementations we use the Choi-Lee-Park polynomial model for both MR dampers, which can adequately describe the highly nonlinear and hysteretic behavior. Some important aspects of the mechatronic integration are discussed in order to establish that the unbalance response is reduced up to 95% with respect to the open-loop behavior, employing less than 0.75% of the total power in the rotor-bearing system, and the system response can be stabilized and reduced above the first critical speed. Finally, some numerical simulations and experimental results on a physical platform are presented to validate the dynamic and robust performance of the overall control system.

9.1 Introduction

In rotating machinery the unbalance is a common problem that occurs when the principal axis of inertia of the rotor does not agree with its geometric axis. In general, this is the result of inevitable imperfections in manufacturing and assembly of rotors. These defects can cause high levels of vibrations, noise and wear, and they may lead to failures or lost of the machine. The balancing of rotors is therefore crucial during the normal operation of low and high speed rotating machines. See, e.g., Dimarogonas [3], Vance [13], Zhou and Shi [14].

Active and semiactive vibration control have deserved much attention in theoretical and experimental aspects in rotating machinery, providing many advantages for the attenuation of vibration amplitudes, during run-up and coast-down through critical speeds, and minimization of sudden transient behavior due to rotor unbalance or parametric uncertainty. This problem has been investigated using different devices such as magnetic bearings, active squeeze film dampers, lateral force actuators, pressurized or hybrid bearings, etc. (see, e.g., Carmignani et al. [1], El-Shafei et al. [4], Guozhi et al. [6], He et al. [7], Lum et al. [9], Zhou and Shi [14]).

In the last years the electrorheological (ER) and, more recently, the magnetorheological (MR) fluids have been employed in active or semiactive bearings, used for vibration attenuation on rotor-bearing systems, specially when the rotor goes through their critical speeds (Carmignani et al. [1], El-Shafei et al. [4], Forte et al. [5], Guozhi et al. [6], Silva–Navarro and Cabrera-Amado [10]). The ER and MR fluids are smart materials, based on suspensions of micron-sized dielectric/ferromagnetic particles, respectively, that can be excited by electric/magnetic fields to yield reversible variations in their rheological properties like viscosity and stiffness. Different hysteresis models for ER or MR dampers have been proposed in the literature (Bingham, Bouc-Wen, Spencer, Choi-Lee-Park, etc.), most of them theoretically and experimentally validated. In practice the MR fluids are more attractive than ER fluids, mainly because of the employment of low voltages and inherent higher yield strength (Spencer et al. [11]).

In this chapter we propose a simple semiactive balancing control scheme for a rotor-bearing system, mounted on journal bearings, one of them supported on two radial and controllable MR dampers. The mathematical model of the rotor-bearing system results from a 7 degrees of freedom Jeffcott-like model and the dynamics associated to the MR dampers, whose mechanical properties depend on the induced electromagnetic field by current inputs (control actions). The damping force provided by the MR dampers is then modelled using the Choi-Lee-Park polynomial model (Choi et al. [2]), which is quite consistent with the nonlinear and complex hysteresis damper models and also simplifies the controller design as well as the physical implementations. The semiactive control scheme is used to compensate the unbalance response of the rotor-bearing system, by means of a proper modification of the rotordynamics coefficients (damping and stiffness), which is based on sliding-mode control techniques. The actual rotor speed is also regulated with the application of an industrial PID controller, combined with trajectory planning, or acceleration scheduling, thereby reducing the amplitude of synchronous vibrations when passing through the first critical speed. The proposed control scheme is also validated on an experimental setup.

9.2 System description

In practice it is very difficult to find rotors supported by identical journal bearings. In addition, any machine has imperfections and, therefore, it is difficult to obtain journal bearings with the same dynamical properties in both directions in the supports.

The rotor-bearing system is described as a nonorthotropic system with different dynamics in the journal bearings housings (see Dimarogonas [3]). The rotor in Fig. 9.1 consists of a plane and rigid disk with mass m_r mounted at the midspan of a rigid shaft; this is supported in the left side by a journal bearing with stiffness (k_{ix}, k_{iy}) and the right support is a journal bearing mounted on a suspension based on two radial MR dampers and compression springs (k_{dx}, k_{dy}), providing damping forces (F_{MRx}, F_{MRy}). Because of the existing rotor imbalance, the mass center is not located at the geometric center of the disk O' but at the point G and the distance between these two points is denoted by the eccentricity e.

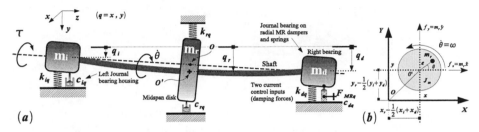

Fig. 9.1. Schematic diagrams: (a) Rotor-bearing system supported by two radial MR dampers and springs; (b) Disk with unbalance.

A 3-phase AC induction motor, coupled to the rotor-bearing system, provides the torque (τ) to regulate the angular speed of the rotor $(\omega = \dot{\theta})$, during run-up or coast-down, which is controlled by an industrial speed driver (local controller). In Fig. 9.1a a front view of the Jeffcott-like rotor-bearing system is shown, it is a system of 7 degrees of freedom, with two MR dampers on the right journal bearing housing. The MR dampers can attenuate the vibrations due to the unbalance in the disk or bearings. In Fig. 9.1b is also shown a view of the whirling disk, with coordinates (x, y) describing the horizontal and vertical motions, respectively.

The experimental platform of the rotor-bearing system with the MR suspension (right side) is illustrated in Fig. 9.2. Several non contact sensors (proximitors) are used to obtain the displacements in different locations.

9.2.1 The rotor-bearing model

The seven degrees of freedom rotor-bearing system is modeled with two degrees of freedom for each (left and right) journal bearing, two degrees for the disk and one degree of freedom for the angular speed. We assume a nonorthotropic reduced order model that does not consider the gyroscopic effects. The angular speed $\omega = \dot{\theta}$ is time-varying and controlled during any run-up and coast-down operation, in contrast to most of Jeffcott-like models. In addition. it is assumed that both journal bearings have different dynamics (mass, stiffness and damping). In fact, this is evident in the right journal bearing, because this is composed by a traditional journal bearing mounted

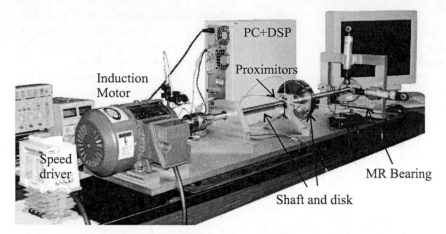

Fig. 9.2. The experimental setup.

on two radial MR dampers and compression springs. The left journal bearing housing is rigid enough and, hence, it is approximated with large stiffness values.

The mathematical model for the 7 degrees of freedom rotor-bearing system with the MR suspension is described as follows:

$$m_i\ddot{x}_i + c_{ix}\dot{x}_i + k_{ix}x_i - \frac{1}{2}k_{rx}\left[x_r - \frac{1}{2}(x_i + x_d)\right] = 0 \tag{9.1}$$

$$m_i\ddot{y}_i + c_{iy}\dot{y}_i + k_{iy}y_i - \frac{1}{2}k_{ry}\left[y_r - \frac{1}{2}(y_i + y_d)\right] = 0 \tag{9.2}$$

$$m_r\ddot{x}_r + c_{rx}\dot{x}_r + k_{rx}\left[x_r - \frac{1}{2}(x_i + x_d)\right] = f_x(t) \tag{9.3}$$

$$m_r\ddot{y}_r + c_{ry}\dot{y}_r + k_{ry}\left[y_r - \frac{1}{2}(y_i + y_d)\right] = f_y(t) \tag{9.4}$$

$$m_d\ddot{x}_d + c_{dx}\dot{x}_d + k_{dx}x_d + F_{MRx} - \frac{1}{2}k_{rx}\left[x_r - \frac{1}{2}(x_i + x_d)\right] = 0 \tag{9.5}$$

$$m_d\ddot{y}_d + c_{dy}\dot{y}_d + k_{dy}y_d + F_{MRy} - \frac{1}{2}k_{ry}\left[y_r - \frac{1}{2}(y_i + y_d)\right] = 0 \tag{9.6}$$

$$\left(J_m + m_r e^2\right)\ddot{\theta} + c_t\dot{\theta} = \tau + f_\theta(t) \tag{9.7}$$

where

$$f_x(t) = m_r e\dot{\theta}^2 \cos(\theta) + m_r e\ddot{\theta}\sin(\theta)$$
$$f_y(t) = m_r e\dot{\theta}^2 \sin(\theta) - m_r e\ddot{\theta}\cos(\theta)$$
$$f_\theta(t) = m_r e\sin(\theta)\left[\ddot{x}_r - \frac{1}{2}(\ddot{x}_i + \ddot{x}_d)\right] - m_r e\cos(\theta)\left[\ddot{y}_r - \frac{1}{2}(\ddot{y}_i + \ddot{y}_d)\right]$$

Here (x_r, y_r) denote the radial displacements of the disk, m_r is the unbalance mass, (c_{rx}, c_{ry}) are radial viscous dampings, (k_{rx}, k_{ry}) are the radial shaft stiffnesses ($k_r =$

$48EI/L^3$) and e is the disk eccentricity. The left journal bearing is characterized by radial displacements (x_i, y_i), mass m_i, stiffnesses (k_{ix}, k_{iy}) and viscous dampings (c_{ix}, c_{iy}). The right journal bearing has radial displacements (x_d, y_d) and its parameters are the mass m_d, two linear compression springs (k_{dx}, k_{dy}), connected in parallel with the MR dampers, and radial viscous dampings (c_{dx}, c_{dy}). The rotational motion considers torsional viscous damping c_t and its inertial mass J_m, corresponding to the disk mass and shaft mass. Finally, the MR damping forces are denoted by F_{MRx} and F_{MRy}, each corresponding to individual and independent radial control forces, which will be controlled through a proper manipulation of their own electrical currents (control actions). The physical parameters for the overall rotor-bearing system are given in Table 9.1.

Table 9.1. Rotor-bearing system parameters

Left journal-bearing	Rotor (disk)	Right journal-bearing
$m_i = 0.9\text{kg}$	$m_r = 5.4\text{kg}$	$m_d = 0.8\text{kg}$
$k_{ix} = 2.319 \times 10^8 \frac{\text{N}}{\text{m}}$	$k_{rx} = 2.009 \times 10^5 \frac{\text{N}}{\text{m}}$	$k_{dx} = 5.082 \times 10^3 \frac{\text{N}}{\text{m}}$
$k_{iy} = 2.368 \times 10^8 \frac{\text{N}}{\text{m}}$	$k_{ry} = 2.050 \times 10^5 \frac{\text{N}}{\text{m}}$	$k_{dy} = 5.270 \times 10^3 \frac{\text{N}}{\text{m}}$
$c_{ix} = 12 \frac{\text{Ns}}{\text{m}}$	$c_{rx} = 10.20 \frac{\text{Ns}}{\text{m}}$	$c_{dx} = 11 \frac{\text{Ns}}{\text{m}}$
$c_{iy} = 13 \frac{\text{Ns}}{\text{m}}$	$c_{ry} = 10.25 \frac{\text{Ns}}{\text{m}}$	$c_{dy} = 11.5 \frac{\text{Ns}}{\text{m}}$

The material shaft properties are the following: Young's modulus $E = 2.11 \times 10^{11} \text{N/m}^2$, shaft length $L = 0.7293\text{m}$, material density $\rho = 7800\text{kg/m}^3$ and shaft diameter $D = 0.02\text{m}$. Additional system parameters are the torsional viscous damping $c_t = 6 \times 10^{-3}\text{Nms/rad}$, inertia mass $J_m = 0.00113\text{kgm}^2$ and disk eccentricity $e = 100\mu\text{m}$. The mass m_r includes also the weight corresponding to the shaft $m_f = 1.7871\text{kg}$.

9.3 MR damper model

The MR fluids are *smart materials* that respond well to an applied magnetic field, leading to an important change in their rheological behavior (viscosity and stiffness). The viscosity and stiffness changes are continuous and reversible, which makes possible the application of MR dampers for vibration control (see Spencer et al. [11]). The passive nature of the MR dampers limits their practical use to *semiactive* vibration control, although this is sufficient to improve the rotor-bearing system response and extend the stability thresholds. Specifically, via the application of feedback control to manipulate the electrical currents of two radial dampers (directions x and y) one can control the two independent damping forces to attenuate the unbalance response in the rotor-bearing system.

For simplicity in the control synthesis, we consider the Choi-Lee-Park polynomial model for the MR dampers (see Choi et al. [2] and references therein). The Choi-Lee-Park polynomial model is able to predict the field-dependent damping force and hysteresis behavior of MR dampers. This simple model splits the hysteresis loop in two regions, for positive acceleration (lower loop) and negative acceleration (upper loop), and then the lower and upper loops are fitted by polynomials to experimental results. The schematic diagram of the polynomial model is shown in Fig. 9.3.

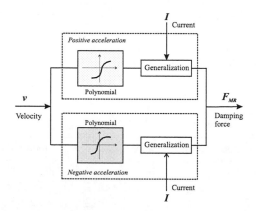

Fig. 9.3. Schematic diagram of the polynomial model for MR dampers.

In a previous work we have used the polynomial model for two identical MR dampers RD-1097-01, both manufactured by *Lord Corporation* (see Silva-Navarro and Cabrera-Amado [10]), whose damping forces are expressed by

$$F_{MRx}(\dot{x}_d, I_x) = \sum_{j=0}^{n=2} (\hat{b}_j + \hat{c}_j I_x) \dot{x}_d^j \tag{9.8}$$

$$F_{MRy}(\dot{y}_d, I_y) = \sum_{j=0}^{n=2} (\hat{b}_j + \hat{c}_j I_y) \dot{y}_d^j \tag{9.9}$$

where $(\dot{x}_d, \dot{y}_d)$ are the piston velocity for the horizontal and vertical MR dampers in the right bearing, respectively, (I_x, I_y) are the current control inputs for the dampers, which can be controlled independently. The general coefficients $(\hat{b}_j, \hat{c}_j)$ depend discontinuously on the piston accelerations $(\ddot{x}_d, \ddot{y}_d)$ for the horizontal and vertical MR dampers, respectively. Specifically, they can be expressed as

$$\hat{b}_j = \frac{(b_j^+ + b_j^-) + \left| b_j^+ - b_j^- \right| \text{sign}(\dot{v})}{2} \tag{9.10}$$

$$\hat{c}_j = \frac{(c_j^+ + c_j^-) + \left| c_j^+ - c_j^- \right| \text{sign}(\dot{v})}{2} \tag{9.11}$$

$$j \in \{0, 1, ..., n\}, \dot{v} \in \{\ddot{x}_d, \ddot{y}_d\}$$

where (b_j^+, c_j^+) and (b_j^-, c_j^-) denote the coefficients for positive and negative acceleration, respectively.

The polynomials with best fitting to the experimental results are of order $n = 2$ (see Silva-Navarro and Cabrera-Amado [10]), whose coefficients are given in Table 9.2.

Table 9.2. Polynomial coefficients b_i and c_i

Index	Positive acceleration $\dot{v} > 0$		Negative acceleration $\dot{v} < 0$	
i	b_i^+	c_i^+	b_i^-	c_i^-
0	0.403	2.928	0.5426	−3.105
1	−18.3	1156	−18.549	1161
2	19.01	−561.3	8.6212	−372.5

9.4 System properties and steady-state behavior

The overall 7 degree-of-freedom rotor-bearing system dynamics (9.1)-(9.7) considers two identical MR dampers F_{MRx} and F_{MRy} in the right journal bearing, which are modeled by (9.8) and (9.9), respectively. From the switching characteristics into the coefficients $(\hat{b}_j, \hat{c}_j)$, it is evident that the control system (9.1)-(9.7) for the radial directions (x, y) are highly nonlinear and nonsmooth, which complicates the synthesis of vibration controllers. Via these coefficients there exists certain acceleration feedback influencing the effective damping in the rotor-bearing system. This phenomenon, however, can be ignored, because the dynamics of the MR dampers is much slower than the rotor dynamics.

It is possible to prove that the control system (9.1)-(9.7) is locally controllable about equilibrium positions from the two current inputs (I_x, I_y). This dynamics also reveals that the average equilibrium displacements (without unbalance), for constant currents inputs, can be obtained by

$$\left(x_i = 0, x_r = -\frac{\hat{b}_0 + \hat{c}_0 I_x}{2k_{dx}}, x_d = -\frac{\hat{b}_0 + \hat{c}_0 I_x}{k_{dx}} \right)$$

$$\left(y_i = 0, y_r = -\frac{\hat{b}_0 + \hat{c}_0 I_y}{2k_{dy}}, y_d = -\frac{\hat{b}_0 + \hat{c}_0 I_y}{k_{dy}} \right) \tag{9.12}$$

This steady-state behavior has to be considered during the application of a feedback control law, because in any case the ideal equilibrium position should be the origin.

9.4.1 Speed regulation for acceleration scheduling

In this case we use a local PID control scheme to manipulate the rotor speed $\omega(t)$ in the overall system dynamics (9.7). Thus, the control objective is the asymptotic tracking of a sufficiently smooth speed profile to reduce the transient behavior of the unbalance response, via an acceleration scheduling policy.

For the speed regulation of the rotor-bearing system in (9.7), it is implemented a PID controller, whose control law is given by

$$\tau(t) = K_P \omega_e(t) + K_I \int_0^t \omega_e(s)ds + K_D \dot{\omega}_e(t) \tag{9.13}$$

where $\omega_e(t) = \omega^* - \omega(t)$ is the tracking error with respect to some desired speed profile ω^*. The desired speed trajectories $\omega^*(t)$, for run-up and coast-down operations, are based on the well-known Bézier polynomials.

The control parameters are selected as $K_P = 0.04$, $K_I = 1 \times 10^{-6}$ and $K_D = 10$, which guarantee the robust tracking in presence of the endogenous perturbation $f(t)$ resulting from the inertial couplings in (9.7).

9.5 Open-loop simulation results

Consider the rotor-bearing system (9.1)-(9.7) with MR dampers, whose parameters are given in Tables 9.1 and 9.2. The initial conditions are $x_i(0) = y_i(0) = 0$ m, $x_r(0) = -6 \times 10^{-5}$ m, $y_r(0) = -5.5 \times 10^{-5}$ m, $x_d = -8.5 \times 10^{-5}$ m, $y_d(0) = -9.2 \times 10^{-5}$ m, $\dot{x}_i(0) = \dot{x}_r(0) = \dot{x}_d(0) = \dot{y}_i(0) = \dot{y}_r(0) = \dot{y}_d(0) = 0$ m/s.

Fig. 9.4 shows the dynamic behavior of the open-loop left journal bearing, in case of three different constant current inputs $I \in \{0, 0.1, 0.55\}$ A and angular speed $\dot{\theta}(t)$, during a run-up from 0 to 3200 rpm. Similarly, Fig. 9.4 depicts the radial displacements of the disk for three current inputs. Note that, when the currents are increased, the vibration amplitudes are attenuated to small values. In addition, Fig. 9.4 shows the dynamic behavior for the right journal bearing, using similar currents and run-up. For the rotor speed regulation in Fig. 9.4 is used a smooth speed profile, based on Bézier polynomials.

It is evident how the dynamics of the semiactive suspension change with the current inputs, mainly that, its rheological properties contribute to increase the equivalent stiffness and damping in the right support and, as a consequence, the overall system dynamics is modified. In fact, higher current inputs lead to a quasi-rigid support (see Fig. 9.4). Above $I = 0.4$ A the absorption capacity of the MR suspension is slightly diminished in the disk (see Fig. 9.4). In addition, the vibration amplitudes in the right journal bearing are almost cancelled when $I = 0.55$ A, although this yields a bigger response in the disk (see Fig. 9.4).

On the other hand, the current inputs also modify the natural frequencies in the system. For instance, when $I = 0$ A the critical speed associated to the disk is 46.13 Hz and when $I = 0.55$ A this is reduced to 31.66 Hz. The steady-state attenuation of the system response becomes obvious when the current inputs are increased and, hence, this property enables the application of semiactive balancing control schemes to the overall rotor-bearing system (9.1)-(9.7).

The experimental rotor displacements (only direction x) for a similar run-up and constant current inputs $I_x \in \{0, 0.2, 0.55\}$ A are described in Fig. 9.5. Moreover, the experimental hysteresis (direction x) is depicted in Fig. 9.6.

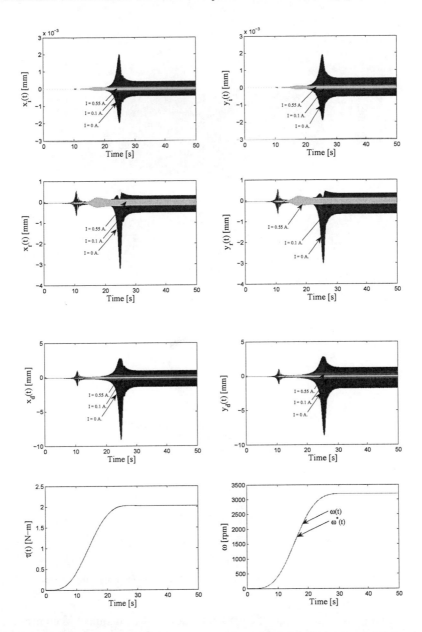

Fig. 9.4. Simulation results in the left journal bearing, right bearing and disk (open-loop response) with different current inputs and rotor run-up from 0 to 3200 rpm using the PID speed control.

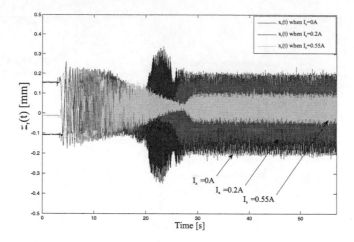

Fig. 9.5. Experimental results for the open-loop behavior for different current inputs $I_x \varepsilon \{0, 0.2, 0.55\}$ A and run-up from 0 to 1100 rpm.

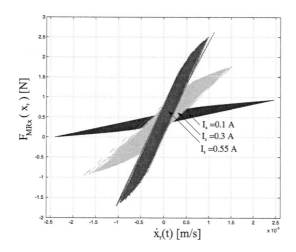

Fig. 9.6. Experimental hysteresis in the MR damper (only direction x_r).

The above results validate the application of the MR dampers to control the unbalance response of the rotor-bearing system.

9.6 Semiactive balancing control

For control purposes it is considered the inverse model for the MR damping forces in (9.8)-(9.9), such that the current control inputs can be expressed by (see Choi et al. [2])

$$I_x = \frac{u_x - \sum_{j=0}^{n=2} \hat{b}_j \dot{x}_d^j}{\sum_{j=0}^{n=2} \hat{c}_j \dot{x}_d^j} \tag{9.14}$$

$$I_y = \frac{u_y - \sum_{j=0}^{n=2} \hat{b}_j \dot{y}_d^j}{\sum_{j=0}^{n=2} \hat{c}_j \dot{y}_d^j} \tag{9.15}$$

where u denotes the desired damping forces (u_x, u_y) for the horizontal and vertical displacements, respectively. It is important to remark that, in general, there is no singularity in (9.14) because $\hat{c}_0 \neq 0$ and the actual acceleration and velocity ranges are far from any singularity.

Because the rotor-bearing system (9.1)-(9.6) is controllable from the two independent MR damping forces F_{MRx} and F_{MRy}, we can synthesize a simple balancing control scheme to reduce the unbalance response even when passing trough its critical speeds. The semiactive vibration control scheme is summarized in the following steps (see Fig. 9.7):

- It is designed a stabilizing vibration control law via the two independent MR damping forces, F_{MRx} and F_{MRy} for each radial direction (x, y) of the rotor-bearing system (9.1)-(9.6).
- Computed control forces are considered as the desired damping forces, $u_x = F_{MRDx}$ and $u_y = F_{MRDy}$, in (9.8)-(9.9). Then, the current control inputs (I_x, I_y) are calculated from the inverse dynamics (9.14), leading to an internal loop that linearizes the slow dynamics of both MR dampers.
- Due to physical limitations in both MR dampers, the computed current inputs (I_x, I_y) are saturated between 0A and 0.55 A as well as the switching frequencies are restricted to below 1 KHz. These saturated currents (I_{sx}, I_{sy}) are then applied to the MR dampers to generate the actual damping forces.
- This strategy is repeated until the unbalance response converges into a pre-specified region.

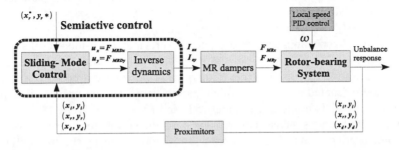

Fig. 9.7. Schematic diagram of the semiactive balancing control scheme using sliding-mode control.

Two sliding-mode controllers are applied to achieve the vibration attenuation of the unbalance response (see Utkin [12]). This is a robust control strategy against exogenous perturbations and parameter uncertainties. Thus, the two sliding surfaces are defined by

$$\sigma_x = \beta_{x3}\left[x_r^{(3)} - x_r^{(3)*}\right] + \beta_{x2}(\ddot{x}_r - \ddot{x}_r^*) + \beta_{x1}(\dot{x}_r - \dot{x}_r^*) + \beta_{x0}(x_r - x_r^*) \quad (9.16)$$

$$\sigma_y = \beta_{y3}\left[y_r^{(3)} - y_r^{(3)*}\right] + \beta_{y2}(\ddot{y}_r - \ddot{y}_r^*) + \beta_{y1}(\dot{y}_r - \dot{y}_r^*) + \beta_{y0}(y_r - y_r^*) \quad (9.17)$$

where x_r^* and y_r^* denote desired references for the rotor positions, which typically are selected as constant values $(x_r^* = y_r^* = 0)$, in order to center the orbit at the origin, although these can be used to compensate the average equilibria (9.12). The design parameters $\beta_{j0}, \beta_{j1}, \beta_{j2}, \beta_{j3}$, $j \in \{x, y\}$ are used to get some pre-specified dynamic behavior.

Finally, the sliding-mode controllers are synthesized as follows

$$u_x = \left(\frac{2m_r m_d \beta_{2x}}{\beta_{3x}k_{rx}} - \frac{2m_d c_{rx}}{k_{rx}}\right)x_r^{(3)} + \left(\frac{2m_r m_d \beta_{1x}}{\beta_{3x}k_{rx}} - 2m_d\right)\ddot{x}_r + \frac{2m_r m_d \beta_{0x}}{\beta_{3x}k_{rx}}\dot{x}_r$$

$$+ \frac{k_{rx}x_r}{2} - \frac{4k_{dx} + k_{rx}}{4}x_d + m_d\ddot{x}_i - c_{dx}\dot{x}_d - \frac{k_{rx}}{4}x_i + \frac{2m_r m_d}{\beta_{3x}k_{rx}}W_x \mathrm{sign}\sigma_x \quad (9.18)$$

$$u_y = \left(\frac{2m_r m_d \beta_{2y}}{\beta_{3y}k_{ry}} - \frac{2m_d c_{ry}}{k_{ry}}\right)y_r^{(3)} + \left(\frac{2m_r m_d \beta_{1y}}{\beta_{3y}k_{ry}} - 2m_d\right)\ddot{y}_r + \frac{2m_r m_d \beta_{0y}}{\beta_{3y}k_{ry}}\dot{y}_r$$

$$+ \frac{k_{ry}}{2}y_r - \frac{4k_{dy} + k_{ry}}{4}y_d + m_d\ddot{y}_i - c_{dy}\dot{y}_d - \frac{k_{ry}}{4}y_i + \frac{2m_r m_d}{\beta_{3y}k_{ry}}W_y \mathrm{sign}\sigma_y \quad (9.19)$$

where W_x and W_y are design parameters for the discontinuous control actions. For the speed control of the motor, the PID controller (9.13) is used to get the run-up (or coast-down) of the rotor.

9.6.1 Closed-loop simulation results

Now consider the system parameters in Tables 9.1 and 9.2. The design parameters for the sliding-mode controllers (9.18, 9.19) are: $\beta_{0x} = \beta_{0y} = 13.08 \times 10^6, \beta_{1x} = \beta_{1y} = 1.636 \times 10^5, \beta_{2x} = \beta_{2y} = 700, \beta_{3x} = \beta_{3y} = 1$ and $W_x = W_y = 40$.

The initial conditions are identical to the previous open-loop results. Fig. 9.8 shows the radial displacements of the rotor disk (x_r, y_r), with amplitudes smaller than 0.1 mm for a run-up from 0 to 3200 rpm. The control currents are saturated to 0.55 A, here denoted as (I_{sx}, I_{sy}), which lead to stabilized and attenuated responses employing small damping forces about 15 N and 10 N, respectively.

Fig. 9.9 illustrates the closed-loop and open loop orbits for the rotor disk, when the rotor-bearing system is started from 0 to 3200 rpm using the local PID speed control. Moreover, the steady-state responses, when the rotor speed remains at the (open-loop) critical speed of 3084 rpm, are described in Fig. 9.10. It is evident the effectiveness of the semiactive balancing control scheme, based on the MR dampers

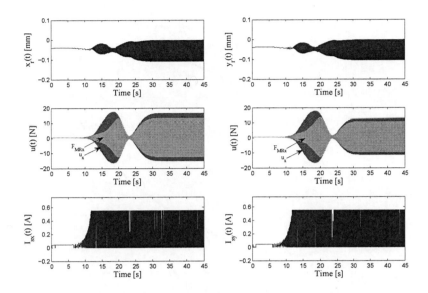

Fig. 9.8. Closed-loop radial displacements in the disk (x_r, y_r), computed damping forces (u_x, u_y), applied damping forces (F_{MRx}, F_{MRy}), and saturated control currents (I_{sx}, I_{sy}).

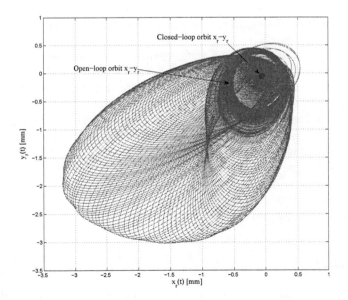

Fig. 9.9. Closed-loop and open-loop orbits in the disk $(y_r$ vs $x_r)$ during a run-up from 0 to 3200 rpm.

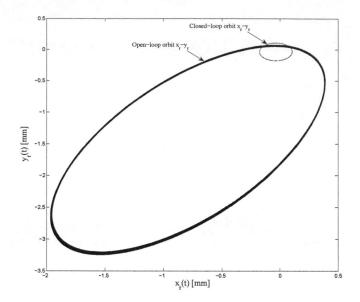

Fig. 9.10. Closed-loop and open-loop orbits in the disk (y_r vs x_r) in steady-state behavior at 3084 rpm.

and sliding-mode control techniques combined with the acceleration scheduling via a local PID speed control.

A comparison of the open-loop and closed-loop FRF unbalance response in the disk is presented in Fig. 9.11. It is clear that current inputs $I_x = I_y = 0$ A lead to an open-loop response with two resonant peaks, one corresponding to the right journal bearing (8.705 Hz or 54.69 rad/s) and the other associated to the disk (51.4 Hz or 322.95 rad/s). In contrast, when the sliding-mode controllers are applied, the overall system response is highly attenuated, with resonant peaks almost cancelled.

The steady-state radial displacements for the left journal bearing are very small, about 1.3×10^{-4} mm, which means that this rigid support does not have a significant influence in the rotordynamics.

Finally, one can conclude that the application of the above semiactive balancing control scheme is quite satisfactory to attenuate the unbalance response.

9.7 Conclusions

A rotor-bearing system with a MR suspension, modelled with a 7 degrees-of-freedom model and a semiactive balancing control scheme, is addressed. The MR dampers allow the effective attenuation of the unbalance response in different locations of the system. A simple and robust semiactive balancing control scheme, based on sliding-mode control techniques, for the rotor-bearing system with MR dampers is proposed.

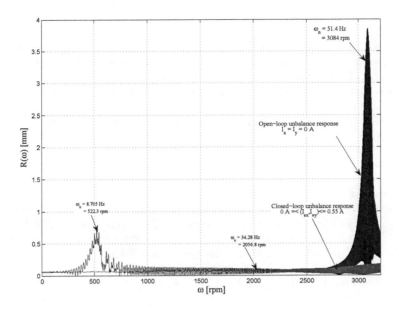

Fig. 9.11. Comparison of the open-loop and closed-loop total unbalance response, in the disk $R = \sqrt{x_r^2 + y_r^2}$.

The control scheme combines two radial MR dampers and springs to support the right journal bearing. The rheological properties (viscosity and stiffness) are then controlled through current control inputs, using sliding-mode controllers, to get the desired dynamic performance and FRF on the rotor-bearing system. The semiactive controllers reduce the steady-state vibration amplitudes in the rotor disk up to 87%, with respect to the open-loop behavior, as well as the system response when passing through the first critical speeds (disk), with reductions up to 95%.

References

1. Carmignani, C., Forte, P., Rustighi, E.: Design of a novel magneto-rheological squeeze-film damper, Smart Materials and Structures, Vol. 15, 164–170 (2006).
2. Choi, S.B., Lee, S.K., Park, Y.P.: A hysteresis model for the field-dependent damping force of a magnetorheological damper, Journal of Sound and Vibration, Vol. 245, No. 2, 375–383 (2001).
3. Dimarogonas, A.D.: Vibration for Engineers. Prentice Hall, NJ (1995).
4. El-Shafei, A.: Active control algorithms for the control of rotor vibrations using HSFDS, Proc. of the ASME TURBOEXPO, Munich, Germany, 1–16 (2000).
5. Forte, P., Paterno, M., Rustighi, E.: A magnetorheological fluid damper for rotor applications, International Journal of Rotating Machinery, Vol. 10, No. 3, 175–182 (2004).

6. Guozhi, Y., Fah, Y.F., Guang, C., Guang, M., Tong, F., Yang, Q.: Electro-rheological multi-layer squeeze film damper and its application to vibration control of rotor system, Journal of Vibration and Acoustics, Vol. 122, 7–11 (2000).

7. He, Y.Y., Oi, S., Chu, F.L., Li, H.X.: Vibration control of a rotor-bearing system using shape memory alloy: I. Theory, Smart Materials and Structures, Vol. 16, 114–121 (2007).

8. He, Y.Y., Oi, S., Chu, F.L., Li, H.X.: Vibration control of a rotor-bearing system using shape memory alloy: II. Experimental study, Smart Materials and Structures, Vol. 16, 122–127 (2007).

9. Lum, K.Y., Coppola, V.T., Bernstein, D.S.: Adaptive autocentering control for an active magnetive bearing supporting a rotor with unknown mass imbalance, IEEE Trans. on Control Systems Technology, Vol. 4, 587–597 (1996).

10. Silva-Navarro, G., Cabrera-Amado, A.: Semiactive sliding-mode control of the unbalance response in a rotor-bearing system supported on MR dampers, Proc. IEEE Conference on Decision and Control, New Orleans, USA, 4513–4518 (2007).

11. Spencer Jr, B.F., Dyke, S., Sain, M., Carlson, J.D.: Phenomenological model of a magne-toreological damper, ASCE Journal of Engineering Mechanics, Vol. 123, 230–238 (1997).

12. Utkin, V.I.: Sliding Modes in Control and Optimization. Springer-Verlag, Berlin (1992).

13. Vance, J.M.: Rotordynamics of Turbomachinery. John Wiley & Sons, NY (1988).

14. Zhou, S., Shi, J.: Active balancing and vibration control of rotating machinery: a survey, The Shock and Vibration Digest, Vol. 33, No. 4, 361–371 (2001).

Design of a Passive/Active Autoparametric Pendulum Absorber for Damped Duffing Systems

G. Silva-Navarro[1], L. Macias-Cundapi[1], and B. Vazquez-Gonzalez[2]

[1] Centro de Investigacion y de Estudios Avanzados del I.P.N.,
   Departamento de Ingenieria Electrica - Seccion de Mecatronica,
   Av. IPN 2508, Col. S.P. Zacatenco, CP 07360, Mexico, D.F., Mexico
   gsilva@cinvestav.mx, lilianamacu@hotmail.com
[2] Universidad Autonoma Metropolitana – Azcapotzalco, Departamento de Energia,
   Av. San Pablo No. 180, Col. Reynosa Tamaulipas, CP 02200, Mexico, D.F., Mexico
   bvg@correo.azc.uam.mx

Summary. A passive/active autoparametric pendulum-type absorber is designed to control the resonant vibrations in damped Duffing systems. The primary system is a damped Duffing system, with cubic stiffness function, affected by exogenous forces with excitation frequencies close to the principal parametric resonance. The passive design of the autoparametric pendulum absorber is obtained by using an approximation of the nonlinear frequency response, computed via the multiple scales method. Then, in order to improve the overall system performance against slow variations on the amplitude and excitation frequency in the external force, it is incorporated a servomechanism to manipulate the pendulum length and, therefore, the autoparametric pendulum-type absorber can be automatically tuned into a given frequency bandwidth, by means of the application of a nonlinear control law combining feedback and feedforward compensation terms. The design of the autoparametric absorber, approximate frequency analysis, synthesis of control laws, stability analysis and closed-loop system performance are discussed. Some simulations results are included to illustrate the dynamic performance of the overall system.

10.1 Introduction

Autoparametric vibration systems have an interesting dynamic that results from at least two non-linear subsystems coupled to interact in a way where one of them transfers the exogenous perturbation energy to the other (vibration absorber). Thus, the primary system can be externally excited by some harmonic force and, when it is connected to the secondary system (absorber), it can be verified the so-called parametric excitation, that is, a mechanism that transfers the exogenous energy to the vibration absorber. In case the primary system is excited exactly or near its linear resonant frequency, it is possible to get the principal parametric resonance for the absorber (autoparametric interaction) and, hence, the response of the primary system can be attenuated. See, e.g., Cartmell [3] and Tondl et al. [8].

The autoparametric interaction has been analyzed in the literature (Haxton and Barr [4], Cartmell [3], Korenev and Reznikov [5], Nayfeh and Mook [7]). There are many vibrating structures and machinery, where the pendulum-type absorbers can be used.

This work deals with the attenuation problem of harmonic vibrations affecting an inherently nonlinear mechanical system, using a hybrid passive/active vibration control scheme. The mechanical system consists of a perturbed Duffing primary system (machine) coupled to a secondary system, composed by a nonlinear autoparametric vibration absorber using a pendulum (Vazquez and Silva [9]; Macias-Cundapi [6]). The undesirable resonant vibration is a harmonic force with variations on the excitation frequency, which is applied to the primary Duffing system.

Our main goal is the synthesis of a feedback and feedforward control law to automatically tune the overall system to get the minimal attenuation gain. The approximate frequency response function of the passive system (two degree of freedom nonlinear system) is obtained via perturbation methods, thus guaranteeing a stable behavior and the resonance conditions for some specific excitation frequency. In case of frequency variations we add an extra degree of freedom to the system in order to hold the autoparametric interaction condition. This new coordinate is the position of the pendulum mass, implying the addition of a controlled third degree of freedom and enabling the energy dissipation over a frequency band. The application of a robust control law that asymptotically achieves an optimal position can yield indirectly the attenuation on the primary Duffing system. It is important to remark that similar results are reported in Cartmell et al. [1, 2] for linear primary systems, employing an on/off controller and analyzing only the dynamics associated to the linear primary system and the passive pendulum absorber (two degrees of freedom system).

10.2 Passive/Active Pendulum-Type Absorber

A damped Duffing system is attached to a pendulum vibration absorber (see Fig. 10.1), with motion restricted to a horizontal plane (no gravity effects). The primary system is a nonlinear vibrating system with mass m_1, a nonlinear spring with cubic stiffness $k_1 x + k_2 x^3$ and linear viscous damping c_1. This system is affected by an external force $F(t) = F_0 \cos(\Omega t)$, with amplitude F_0 and excitation frequency Ω tuned at the principal parametric resonance associated to the primary system.

In order to attenuate the harmonic vibrations $F(t)$, it is used a pendulum-type vibration absorber, composed by a mass m_3 and a rigid bar with mass m_2 and moment of inertia I_2, with respect to its center of mass. The primary and secondary subsystems are coupled by means of a linear torsional spring k_3 and linear viscous damping c_2. The lengths l_b and l denote the total pendulum length (bar) and the length measured along the bar from the position of the mass m_3 to the pivot point, respectively. A third degree of freedom is considered to include the position of the mass m_3, which is controlled via a dc motor with a ball-screw.

The equations of motion for the passive/active autoparametric pendulum vibration absorber are obtained as follows,

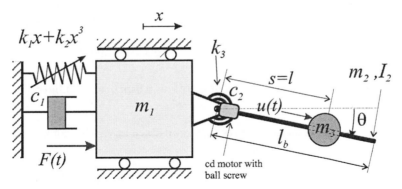

Fig. 10.1. Schematic diagram of the autoparametric pendulum vibration absorber.

$$M(q)\ddot{q} + C(q,\dot{q})\dot{q} + G(q) = \tau(t) \qquad (10.1)$$

where $q = [x, \theta, s]^T$ is the vector of generalized coordinates, that is, with the horizontal displacement of the primary system, the angular displacement of the pendulum and the pendulum length, respectively; $\tau = [F(t), 0, u(t)]^T$ is the vector of exogenous forces, with an external harmonic force $F(t) = F_0 \cos \Omega t$ affecting the primary system with an excitation frequency Ω close to the primary resonant frequency, and the control force $u(t)$ to be synthesized to get a proper pendulum length $s = l$. Moreover, $M(q)$ represents to the inertia matrix, $C(q,\dot{q})$ is the vector of centrifugal and Coriolis terms, also including dissipative forces, and $G(q)$ denotes the vector of gravitational forces. These are described by

$$M(q) = \begin{bmatrix} m_1 + m_2 + m_3 & -\frac{1}{2}m_2 l_b \sin\theta - m_3 s \sin\theta & m_3 \cos\theta \\ -\frac{1}{2}m_2 l_b \sin\theta - m_3 s \sin\theta & \frac{7}{12}m_2 l_b^2 + m_3 s^2 & 0 \\ m_3 \cos\theta & 0 & m_3 \end{bmatrix}$$

$$C(q,\dot{q})\dot{q} = \begin{bmatrix} c_1\dot{x} - 2m_3\dot{s}\dot{\theta}\sin\theta - \frac{1}{2}m_2 l_b \dot{\theta}^2 \cos\theta - m_3 s\dot{\theta}^2 \cos\theta \\ c_2\dot{\theta} + 2sm_3\dot{s}\dot{\theta} \\ -m_3 s\dot{\theta}^2 \end{bmatrix}$$

$$G(q) = \begin{bmatrix} k_1 x + k_2 x^3 \\ k_3 \theta \\ 0 \end{bmatrix}$$

The control system (10.1) is strongly nonlinear, underactuated and the output to be controlled x is not controllable from the input u, exactly at the equilibrium points of interest. The states that are controllable from u are only $(s, \dot{s})$ and, therefore, we propose to control the primary Duffing system response via an optimization approach and an indirect control method, first computing the optimal position $s = l^*$ leading to the minimal steady-state amplitude, and then applying a position control law to quickly achieve such a point.

Note that, when the position $s = l$ is forced to be constant, with $u(t) \equiv 0$, the passive/active pendulum-type vibration absorber (10.1) becomes the passive autoparametric pendulum vibration absorber (see Vazquez and Silva [9]).

10.2.1 The Passive Autoparametric Pendulum Absorber for Damped Duffing Systems

The equations of motion for the two degrees-of-freedom passive autoparametric vibration absorber are described by

$$(m_1 + m_2 + m_3)\ddot{x} + c_1\dot{x} + k_1 x + k_2 x^3 +$$

$$- \left(\frac{1}{2}m_2 l_b + m_3 l\right)\left(\ddot{\theta}\sin\theta + \dot{\theta}^2\cos\theta\right) = F_0\cos(\Omega t) \quad (10.2)$$

$$\left(\frac{1}{4}m_2 l_b^2 + I_2 + m_3 l^2\right)\ddot{\theta} + k_3\theta + c_2\dot{\theta} - \left(\frac{1}{2}m_2 l_b + m_3 l\right)\ddot{x}\sin\theta = 0 \quad (10.3)$$

By defining representative parameters and assuming small oscillations, one can transform the system (10.2)-(10.3) to get an approximate analytical solution for the nonlinear frequency response. This procedure results in the following two coupled and nonlinear differential equations,

$$\ddot{x} + 2\varepsilon\zeta_1\omega_1\dot{x} + \omega_1^2 x + \varepsilon\alpha x^3 - \varepsilon h\left(\ddot{\theta}\theta + \dot{\theta}^2\right) = \varepsilon f\cos(\Omega t) \quad (10.4)$$

$$\ddot{\theta} + 2\varepsilon\zeta_2\omega_2\dot{\theta} + \omega_2^2\theta - \varepsilon g\ddot{x}\theta = 0 \quad (10.5)$$

where the system parameters are defined by

$$M = m_1 + m_2 + m_3, \quad J = \tfrac{1}{4}m_2 l_b^2 + I_2 + m_3 l^2, \quad \omega_1^2 = \frac{k_1}{M}$$

$$\varepsilon\alpha = \frac{k_2}{M}, \qquad \varepsilon f = \frac{F}{M}, \qquad \varepsilon\zeta_2 = \xi_2 = \frac{c_2}{2\omega_2 J}$$

$$\varepsilon\zeta_1 = \xi_1 = \frac{c_1}{2\omega_1 M}, \quad \varepsilon h = \tfrac{1}{M}\left(\tfrac{1}{2}m_2 l_b + m_3 l\right), \quad \varepsilon g = \frac{1}{J}\left(\frac{1}{2}m_2 l_b + m_3 l\right)$$

$$\varepsilon = \frac{1}{2}m_2 l_b + m_3 > 0$$

Here the small perturbation parameter ε considers the internal couplings between the pendulum absorber and the Duffing primary system, viscous dampings, nonlinearities and external force into the system. These perturbed equations include the *cubic nonlinearity* in the restoring force for the primary system, which is multiplied by the small perturbation parameter ε, that is, $\varepsilon\alpha$ with a constant $\alpha > 0$ or $\alpha < 0$ corresponding to a hardening or softening spring, respectively.

10.2.2 Approximate Frequency Analysis

The method of multiple scales is used to compute an approximate solution for the perturbed system Eqs. (10.4)-(10.5) (Cartmell [2] and Nayfeh [4]). The perturbed solutions are expressed by $x = x_0(T_0, T_1) + \varepsilon x_1(T_0, T_1) + \dots$ and $\theta = \theta_0(T_0, T_1) + \varepsilon \theta_1(T_0, T_1) + \dots$, where $T_0 = t$ is the fast time scale, $T_1 = \varepsilon t$ is the slow time scale and the remaining time scales are related by the perturbation as $T_n = \varepsilon^n t$, with $n = 0, 1, 2, \dots$ Time derivatives along different time scales lead to differential operators $d/dt = D_0 + \varepsilon D_1 + \dots$ and $d^2/dt^2 = D_0^2 + 2\varepsilon D_0 D_1 + \dots$.

The external and internal resonance conditions, characterizing the autoparametric interaction between the two-degrees-of-freedom, are described by

$$\Omega = \omega_1 + \varepsilon \rho_1 \tag{10.6}$$

$$\omega_1 = \omega_2 + 2\varepsilon \rho_2 \tag{10.7}$$

where $\varepsilon \rho_1$ and $\varepsilon \rho_2$ define the external and internal detuning parameters, respectively. Substitution of the proposed first order solutions $x(T_0, T_1)$ and $\theta(T_0, T_1)$ into (10.4)-(10.5) and grouping the zero and first order terms in ε, yields the set of partial differential equations

$$\varepsilon^0 : \quad D_0^2 x_0 + \omega_1^2 x_0 = 0 \tag{10.8}$$

$$\varepsilon^1 : \quad D_0^2 x_1 + \omega_1^2 x_1 = -2\zeta_1 \omega_1 D_0 x_0 - 2D_0 D_1 x_0 - \alpha x_0^3 + h(D_0 \theta_0)^2$$
$$+ h\theta_0(D_0^2 \theta_0) + f\cos(\Omega T_0) \tag{10.9}$$

$$\varepsilon^0 : \quad D_0^2 \theta_0 + \omega_2^2 \theta_0 = 0 \tag{10.10}$$

$$\varepsilon^1 : \quad D_0^2 \theta_1 + \omega_2^2 \theta_1 = g(D_0^2 x_0)\theta_0 - 2D_0 D_1 \theta_0 - 2\zeta_2 \omega_2 D_0 \theta_0 \tag{10.11}$$

Now, the proposed solutions in their polar forms are given as,

$$x_0 = A(T_1) e^{i\omega_1 T_0} + \bar{A}(T_1) e^{-i\omega_1 T_0} \tag{10.12}$$

$$\theta_0 = B(T_1) e^{i\omega_2 T_0} + \bar{B}(T_1) e^{-i\omega_2 T_0} \tag{10.13}$$

where the amplitudes depend on the fast time scale T_1 and the oscillations on the time scale T_0. Here $\bar{A}(T_1)$ and $\bar{B}(T_1)$ denote complex conjugates of the amplitudes $A(T_1)$ and $B(T_1)$, respectively.

Substituting the proposed solutions in equations (10.9) and (10.11), cancelling secular terms and using the polar forms $A(T_1) = \frac{1}{2} a(T_1) e^{i\delta(T_1)}$ and $B(T_1) = \frac{1}{2} b(T_1) e^{i\gamma(T_1)}$, leads to

$$-i\zeta_1 \omega_1^2 a - i\omega_1 a' + \omega_1 a\delta' - \frac{3}{8}\alpha a^3 - \frac{1}{2}h\omega_2^2 b^2 e^{i\phi_2} + \frac{1}{2}f e^{i\phi_1} = 0 \quad (10.14)$$

$$-\frac{1}{4}g\omega_1^2 abe^{-i\phi_2} - i\omega_2 b' + \omega_2 b\gamma' - i\zeta_2 \omega_2^2 b = 0 \quad (10.15)$$

where

$$\phi_1 = \rho_1 T_1 - \delta \tag{10.16}$$

$$\phi_2 = 2\gamma - \delta - 2\rho_2 T_1 \tag{10.17}$$

Here a', b', δ' and γ' denote differentiation with respect to the slow time scale T_1.

10.2.3 Steady State Solutions

The steady state responses of the overall system are computed for $a' = 0$, $b' = 0$, $\delta' = \rho_1$ and $\gamma' = \rho_1/2 + \rho_2$. The steady state responses are obtained by taking real and imaginary parts in (10.14)-(10.17) for the steady-state conditions, hence, by solving these equations the approximate amplitude responses for the primary and secondary subsystems are given by,

$$a = \frac{4\omega_2^2}{(\varepsilon g)\,\omega_1^2} \sqrt{\left(\frac{\varepsilon\rho_1 + \omega_1}{2\omega_2} - 1\right)^2 + (\varepsilon\zeta_2)^2} \qquad (10.18)$$

$$b^4 + Qb^2 + R = 0 \qquad (10.19)$$

where

$$Q = \frac{12\omega_2(\varepsilon\alpha)\,(\Omega - 2\omega_2)^3}{(\varepsilon g)^3(\varepsilon h)\omega_1^6} + \frac{48\omega_2^3\,(\varepsilon\alpha)\,(\varepsilon\zeta_2)^2\,(\Omega - 2\omega_2)}{(\varepsilon h)\,(\varepsilon g)^3\,\omega_1^6}$$
$$- \frac{8\,(\Omega - 2\omega_2)\,(\Omega - \omega_1)}{(\varepsilon h)\,(\varepsilon g)\,\omega_1\,\omega_2} + \frac{16\,(\varepsilon\zeta_1)\,(\varepsilon\zeta_2)}{(\varepsilon h)\,(\varepsilon g)} \qquad (10.20)$$

$$R = \frac{2304\omega_2^8\,(\varepsilon\alpha)^2}{(\varepsilon h)^2\,(\varepsilon g)^6\,\omega_1^{12}} \left[\left(\frac{\Omega}{2\omega_2} - 1\right)^2 + (\varepsilon\zeta_2)^2\right]^3$$
$$- \frac{768\omega_2^4\,(\varepsilon\alpha)\,(\Omega - \omega_1)}{(\varepsilon h)^2\,(\varepsilon g)^4\,\omega_1^7} \left[\left(\frac{\Omega}{2\omega_2} - 1\right)^2 + (\varepsilon\zeta_2)^2\right]^2$$
$$+ \frac{64\left[(\Omega - \omega_1)^2 + \omega_1^2\,(\varepsilon\zeta_1)^2\right]}{(\varepsilon h)^2\,(\varepsilon g)^2\,\omega_1^2} \left[\left(\frac{\Omega}{2\omega_2} - 1\right)^2 + (\varepsilon\zeta_2)^2\right] - \frac{(\varepsilon f)^2}{(\varepsilon h)^2\,\omega_2^4} \qquad (10.21)$$

It is important to note how the primary Duffing system response (10.18) does not depend on the cubic stiffness and the external force. In fact, this expression coincides with that reported by Cartmell et al. [1, 2, 3], where there are only linear elements in the primary system. The secondary system response, however, is certainly influenced by the cubic nonlinearity through the parameters Q and R. For more details we refer to the previous work by Vazquez and Silva [9].

10.2.4 Simulation Results

Consider the system parameters given in Table 10.1 for the primary Duffing system and passive pendulum vibration absorber vibration absorption (10.2)-(10.3).

The approximate frequency response for the passive Duffing system and its dynamic behavior, without autoparametric interaction (i.e., $\theta(t) \equiv 0$), are described in Fig. 10.2. In both graphics the steady state amplitude $a = 0.05925$ m is similar.

The approximate frequency response for the primay and secondary systems, under autoparametric interaction, are illustrated in Fig. 10.3, when $l = 0.155$ m (constant) and considering the system parameters shown in Table 10.1. The transient

Table 10.1. System parameters

$m_1 = 2.962$ kg	$m_2 = 2.590$ kg	$m_3 = 1.070$ kg	$M = 6.622$ kg
$k_1 = 835$ N/m	$\omega_1 = 11.229$rad/s= 1.787Hz	$c_1 = 20.55$ N/(m/s)	$k_1 = 835$ N/m
$k_2 = 40970$ N/m^3	$\omega_2 = 5.6159$rad/s= 0.894 Hz	$c_2 = 0.076$ Nm/(rad/s)	$I_2 = 0.09363$ kgm^2
$k_3 = 4.15$ Nm/rad	$l = 0.155$m $\in [0.06, 0.5]$m	$l_b = 0.0683$ m	$F_0 = 15$ N

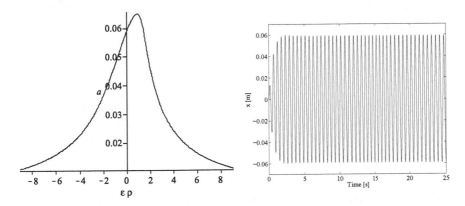

Fig. 10.2. Approximate frequency and transient responses for the primary Duffing system, without autoparametric interaction, when $F_0 = 15$ N and $\Omega = 1.787$ Hz.

responses (Fig. 10.4) are obtained from the nonlinear system (10.2)-(10.3), whose steady state amplitudes $a = 0.0197$ m and $b = 0.982$ rad are compatible with the approximate frequency responses in Fig. 10.3.

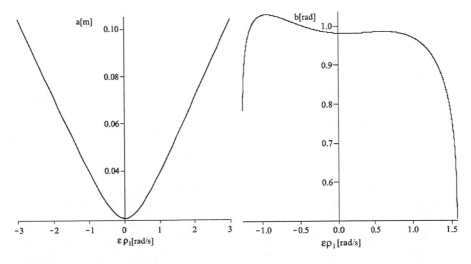

Fig. 10.3. Approximate frequency responses for the primary Duffing system and pendulum absorber with autoparametric interaction, when $F_0 = 15$ N.

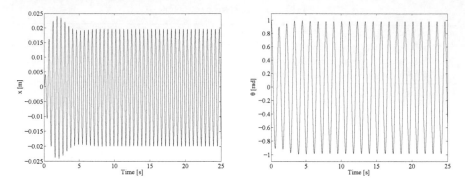

Fig. 10.4. Dynamic behavior of the primary Duffing system and pendulum absorber with autoparametric interaction, when $F_0 = 15$ N, $\Omega = 1.787$ Hz and $\varepsilon\rho_1 = 0$.

10.3 Passive/Active Autoparametric Pendulum Absorber

When the excitation frequency Ω in the perturbation force $F(t)$ is unknown or time varying, the passive vibration absorber may not be useful for vibration attenuation in the Duffing primary system. For time-varying excitation frequencies, satisfying the external resonance condition $\Omega = \omega_1$, the internal tuning condition $\omega_1 = 2\omega_2$ can be asymptotically achieved by the application of an active vibration control scheme. This can be achieved by means of a proper placement of the pendulum mass m_3. Thus, when $\Omega \neq \omega_1$ an active vibration absorber can still be used to automatically tune the pendulum absorber.

10.3.1 Motivation

When the frequency response function expressed by (10.18) (see also Fig. 10.3) is parameterized in terms of the pendulum length $s = l$, results the approximate frequency response described in Fig. 10.5. Here the nonlinear steady state amplitude (10.18) is shown in terms of $\varepsilon\rho_1$ and the pendulum length $s = l$, in such a way that, the internal resonance condition (10.7) can be accomplished to get the minimal attenuation gain. Note that, when $\varepsilon\rho_1 \equiv 0$, the external resonance condition (10.6) is also satisfied.

All the information contained in Fig. 10.5 will be used to asymptotically reach an optimal attenuation operation for the autoparametric pendulum absorber. In fact, there exists some region with minimal amplitudes, whose pendulum lengths can be computed to guarantee an optimal attenuation tuning for the passive/active pendulum absorber. Thus, the level curves and polynomial in Fig. 10.6 are associated to the approximate frequency response (Fig. 10.5), where a polynomial is obtained by applying curve fitting and least squares methods.

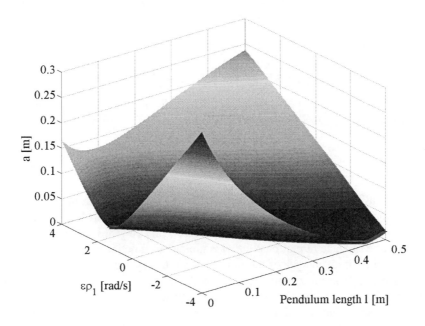

Fig. 10.5. Parameterized frequency response function in terms of the pendulum length when $F_0 = 15$ N.

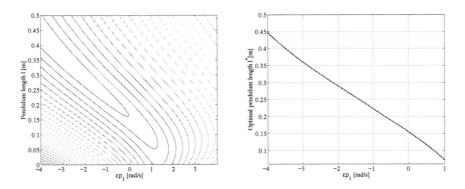

Fig. 10.6. Level curves for the approximate frequency response and polynomial fitted to the minimal gains in the primary Duffing system.

10.3.2 Synthesis of a Control Law Based on Partial Feedback Linearization Combined with Optimal Attenuation

The definition of the state variables

$$z_1 = x, z_2 = \theta, z_3 = s, z_4 = \dot{x}, z_5 = \dot{\theta}, z_6 = \dot{s} \qquad (10.22)$$

leads to the state space representation of the overall control system (10.1) in terms of the state vector $\mathbf{z} = [q, \dot{q}]^T$ as follows

$$\dot{\mathbf{z}}(t) = f(\mathbf{z}) + g(\mathbf{z})u(t) + p(\mathbf{z})F(t) \qquad (10.23)$$
$$y(t) = h(\mathbf{z}) = z_1 \qquad (10.24)$$

where $\mathbf{z} \in \mathbb{R}^6$, $u \in \mathbb{R}$, $F \in \mathbb{R}$ and $y \in \mathbb{R}$ denote the state vector, the force control, the harmonic perturbation force and the output to be controlled (displacement of the primary Duffing system), respectively. Here f, g and p are smooth vector fields (omitted due to space limitations) and h is the output function.

The main disadvantages of the nonlinear control system (10.23)-(10.24) is that this is strongly nonlinear, underactuated and the output to be controlled $y = z_1$ is not controllable from the control force u exactly at the equilibrium points of interest. These reasons suggest that the implementation of a direct control scheme is not recommended. There are, however, two variables $(s, \dot{s})$ which are always controllable from u, and, therefore, one can proceed to formulate an optimization problem to compute a suitable position $s = l$ leading to a minimal steady state amplitude a (*indirect control scheme*).

The passive/active control objective for the autoparametric pendulum absorber is formulated in two steps:

1. Given an excitation frequency Ω, compute the optimal attenuation position $s^* = z_3^*(\Omega) = l^*$ of the pendulum length, which minimizes the steady state amplitude of the primary Duffing system a for the passive vibration absorber, i.e.,

$$\min_{l_{min} \le l \le l_{max}} |a(\Omega, l)| \qquad (10.25)$$

where $a(\Omega, l)$ denotes the steady state amplitude in (10.18) parameterized in terms of Ω and $s = l$, for the closed interval $[l_{min}, l_{max}]$ associated to the total length of the pendulum. The solutions are computed numerically.

For practical purposes the optimal positions $s = l^*(\Omega)$ are computed and parameterized in terms of the excitation frequency Ω, as follows

$$l^*(\Omega) = s_N (\Omega - \omega_1)^N + s_{N-1} (\Omega - \omega_1)^{N-1} + ... + s_1 (\Omega - \omega_1) + s_0 \quad (10.26)$$

where ω_1 is the principal parametric frequency (linear) associated to the primary Duffing system, N is the degree of the polynomial and $s_i \in \mathbb{R}, i = 0, ..., N$ are its coefficients.

2. With the knowledge of the optimal attenuation position $s = l^*(\Omega)$ is synthesized a state feedback and feedforward control law $u(t) = \gamma(\mathbf{z}(t), \Omega)$ to get the asymptotic output tracking, such that

$$\lim_{t \to \infty} z_3(t) = l^*(\Omega) \tag{10.27}$$

Once the optimal position is achieved, the steady-state response of the passive/active control system converges to a fourth order perturbed zero dynamics. More precisely, the passive vibration control scheme (10.2)-(10.3), when the pendulum length is fixed at $z_3 = l^*(\Omega)$, which is locally asymptotically stable about the origin. Hence, the local aymptotic stability is guaranteed by construction (see Vazquez and Silva [9]).

In Fig. 10.7 is described the schematic diagram of the passive/active vibration control scheme for the primary Duffing system.

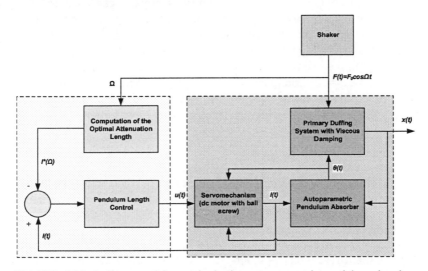

Fig. 10.7. A block diagram of the passive/active autoparametric pendulum absorber.

The control law is based on partial feedback linearization, combined with optimal attenuation, for the position error as an indirect output

$$y = h_2(\mathbf{z}) = \tilde{z}_3 = z_3 - l^*(\Omega) \tag{10.28}$$

where z_3 is the actual pendulum length and $l^*(\Omega)$ is the desired optimal length leading to the minimal amplitude in the primary Duffing system. This output is always controllable from the control force u and has relative degree 2 with respect to u and F.

The partial feedback linearization control is then expressed as

$$u = \gamma(\mathbf{z}, v, \Omega) = \frac{-L_f^2 \mathbf{h_2}(\mathbf{z}) + v}{L_g L_f \mathbf{h_2}(\mathbf{z})} \tag{10.29}$$

where v is a stabilizing control law, based on position and velocity feedback of the pendulum length z_3, which clearly corresponds to a conventional PD position control. Here $L_\Phi \lambda(\mathbf{z}) = \frac{\partial \lambda}{\partial z} \Phi(\mathbf{z})$ denotes the so-called Lie derivative of the scalar function $\lambda(\mathbf{z})$ with respect to the vector field $\Phi(\mathbf{z})$.

The PD control is synthesized as follows

$$v = k_{f1}(z_3 - l^*) + k_{f2}z_6 \tag{10.30}$$

where k_{f1} and k_{f2} are design parameters such that the closed-loop system dynamics be asymptotically stable and fast enough. In this case, the pole placement problem for the closed-loop dynamics of the servomechanism is performed in such a way that the asymptotic output tracking of the optimal length $l^*(\Omega)$ be fast enough, compared with the dynamics of the primary Duffing system; thus, the autoparametric resonance conditions (energy transfer) between both subsystems can be preserved.

Remark: The role of the above servomechanism is only to locate the mass of the pendulum absorber in order to achieve the optimal attenuation gain, where the pendulum works absorbing the external energy applied to the primary Duffing system. The control methodology employs feedback information of the overall system as well as the measurement of the excitation frequency Ω and, therefore, the closed-loop system results in a so-called hybrid (passive/active) absorber. Moreover, many of the existing control techniques can be used for the position control (e.g., sliding-mode, GPI, nonlinear output regulation), although the results are quite similar.

10.3.3 Simulation Results

In order to illustrate the dynamic performance of the passive (active pendulum vibration absorber, when the excitation frequency is changing to different constant values, we use the system parameters in Table 10.1.

The polynomial (10.26) is computed as follows:

$$l^*(\Omega) = -0.0016(\Omega - \omega_1)^3 - 0.0064(\Omega - \omega_1)^2 - 0.0736(\Omega - \omega_1) + 0.1549\,[\text{m}]$$

where $\omega_1 = 11.229$ rad/s= 1.787 Hz.

The initial conditions are set to $x(0) = 0$ m, $\theta(0) = 1.0$ rad, $s(0) = 0$ m, $\dot{x}(0) = 0$ m/s, $\dot{\theta}(0) = 0$ rad/s and $\dot{s}(0) = 0$ m/s. The harmonic force $F(t) = F_0 \cos(\Omega t)$ is started with $F_0 = 7.0711$N and an excitation frequency $\Omega = 1.787$ Hz (i.e., $\varepsilon\rho_1 = 0$ rad/s).

In Fig. 10.8 is described the dynamic behavior of the overall 3 degrees of freedom system (10.1) with the control laws (10.29) and (10.30), which includes the primary Duffing system, the passive/active pendulum absorber and the servomechanism controlling the pendulum length. Here, at $t = 30$ s the excitation frequency is decreased to $\Omega = 1.65$Hz ($\varepsilon\rho_1 = -0.8796$ rad/s).

Note how the passive/active autoparametric absorber is able to manipulate the overall system to yield the autoparametric interaction phenomenon, even in presence of variations in the excitation frequency Ω.

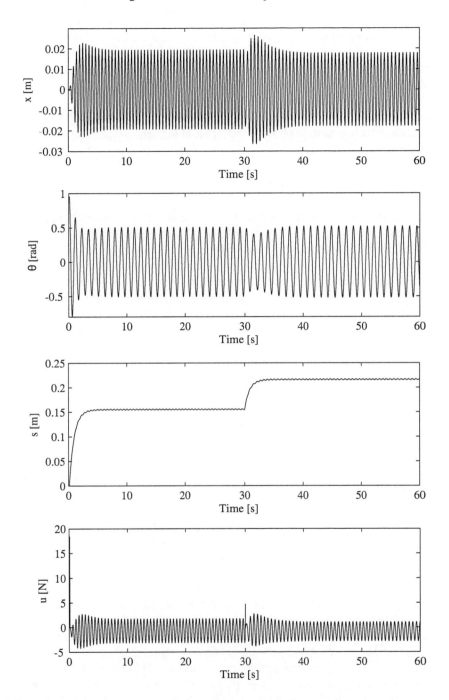

Fig. 10.8. Dynamic response of the Duffing system, with autoparametric interaction, using the passive/active pendulum absorber. The excitation frequency starts at $\Omega = 1.787$ Hz and this changes at $t = 30$ s to $\Omega = 1.65$ Hz ($\varepsilon \rho_1 = -0.8796$ rad/s), when $F = 7.0711$ N.

The primary Duffing system response is still maintained at small amplitudes, using small control efforts to move the pendulum mass m_3 from a length $l = 0.155$ m to $l = 0.216$ m, both leading to minimal attenuation gains.

Moreover, in Fig. 10.9 is illustrated the dynamic performance of the overall system when the excitation frequency is increased at $t = 30$ s from $\Omega = 1.787$Hz to $\Omega = 1.9$ Hz ($\varepsilon\rho_1 = 0.5027$ rad/s).

In general, the passive/active autoparametric absorber exhibits a robust dynamic performance, leading to the optimal attenuation of the primary Duffing system.

10.4 An Experimental Setup

To validate the above results an experimental platform has been designed and constructed, as shown in Figs. 10.10 and 10.11 (see Macias-Cundapi [6]).

This consists of a primary Duffing system, with a biconical spring, moving on a linear guide and excited at resonance by an electromagnetic shaker, which is also monitored and controlled with the PC by means of an accelerometer and force sensor. The pendulum absorber is mounted over the primary Duffing system, coupled to this by means of a torsional spring. The pendulum absorber has two disks pivoting and supporting a ball screw with a pendulum mass, which is actuated by a dc motor. The linear and angular displacements of the primary and secondary systems, respectively, are measured via optical incremental encoders. The passive/active autoparametric absorber is implemented on a Sensoray (Model 626) data acquisition card installed on a PC running on a Matlab/Simulink platform.

10.5 Conclusions

The application of a passive/active autoparametric pendulum vibration absorber to damped Duffing systems is addressed. The design of the active vibration control system is strongly based on the previous computation of a passive vibration absorber and the addition of a servomechanism (extra degree of freedom) to carry it to an optimal attenuation position. For a given excitation frequency the disturbance attenuation problem is indirectly solved in two steps: firstly, from an approximate frequency analysis are computed the optimal lengths for the pendulum absorber and, secondly, the application of a control scheme to place the absorber in the position leading to the optimal operation. The active vibration control scheme employs the measurement of the excitation frequency to tune the absorber according to the absorption conditions, guaranteeing robustness against variations on the excitation frequency. The position control is based on partial feedback linearization with a PD control providing good robustness against frequency and parameter variations. The overall dynamic performance proves the good robustness properties of the proposed control scheme for the attenuation in a highly nonlinear system, where the cubic stiffness is certainly transferred to the absorber.

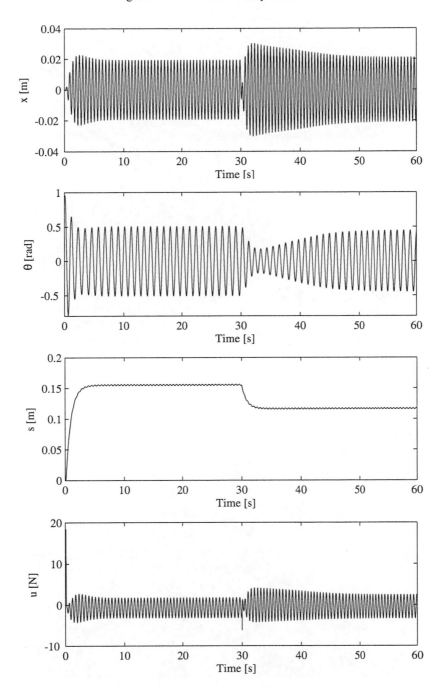

Fig. 10.9. Dynamic response of the Duffing system, with autoparametric interaction, using the passive/active pendulum absorber. The excitation frequency starts at $\Omega = 1.787$ Hz and this changes at $t = 30$ s to $\Omega = 1.9$ Hz ($\varepsilon\rho_1 = 0.5027$ rad/s), when $F = 7.0711$ N.

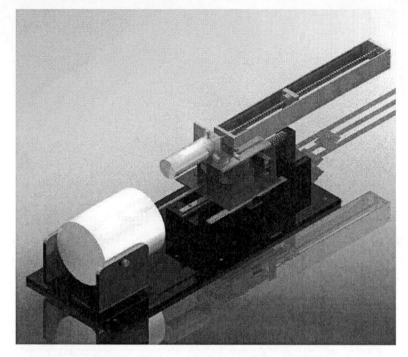

Fig. 10.10. CAD design of the prototype in *SolidWorks*.

Fig. 10.11. The experimental platform for the passive/active autoparametric pendulum absorber.

References

1. Cartmell, M.P., Lawson, J.: Performance enhancement of an autoparametric vibration absorber. Journal of Sound and Vibration, Vol. 177, No. 2, 173–195 (1994).
2. Carmell, M.P., On the need for control of nonlinear oscillations in machine system. Meccanica, 38, 185–212 (2003).
3. Cartmell, M.P.: Introduction to Linear, Parametric and Nonlinear Vibrations. Chapman and Hall, London (1990).
4. Haxton, R.S., Barr, A.D.S.: The autoparametric vibration absorber. Journal of Engineering for Industry, Vol. 94, No. 1, 119–124 (1972).
5. Korenev, B.G., Reznikov, L.M.: Dynamic Vibration Absorbers: Theory and Technical Applications. John Wiley & Sons, London (1993).
6. Macias-Cundapi, L.: Integration of an experimental platform with an autoparametric pendulum-type vibration absorber for Duffing mechanical systems. (In spanish). M.Sc. Thesis, Mechatronics Section, CINVESTAV-IPN, Mexico (August 2009).
7. Nayfeh, A.H., Mook, D.T.: Nonlinear Oscillations. John Wiley & Sons, NY (1979).
8. Tondl, A., Ruijgrok, T., Verhulst, F., Nabergoj, R.: Autoparametric Resonance in Mechanical Systems, Cambridge University Press, Cambridge (2000).
9. Vazquez-Gonzalez, B. and Silva-Navarro, G.: Evaluation of the autoparametric pendulum vibration absorber for a Duffing system. Shock and Vibration, Vol. 15, No. 3-4, 355–368 (2008).

Part II

Solid-State Materials and Electron Devices

11

Characterization of Nanocrystalline ZnO Grown on Silicon Substrates by DC Reactive Magnetron Sputtering

G. Juárez-Díaz[1], A. Esparza-García[3], M. Briseño-García[3], G. Romero-Paredes R.[1], J. Martínez-Juárez[2], and R. Peña-Sierra[1]

[1] Depto. de Ingeniería Eléctrica, Sección de Electrónica del Estado Sólido, Centro de Investigación y de Estudios Avanzados del I.P.N. A.P. 14-7 40,07000, México, D.F.
[2] Centro de Investigación en Dispositivos Semiconductores, BUAP, Puebla, Pue 14 Sur y Av. San Claudio C.U., C.P. 72570.
[3] Centro de Ciencias Aplicadas y Desarrollo de Tecnología-UNAM. Apdo. Postal. 04510, Ciudad Universitaria. México, D.F.

Summary. The aim of the work was to study the effect of postgrowth thermal annealing processes on the optical characteristics of the zinc oxide (ZnO) films grown on (001) silicon substrates by DC reactive magnetron sputtering. The growth temperature of the ZnO thin films was fixed at 230^oC and then the samples were annealed in dry air atmosphere at 800^oC from one hour up to seventeen hours. The surface of the ZnO samples was analyzed with a scanning electron microscope (SEM) and using an atomic force microscope (AFM). The structural properties were assessed by X-ray diffraction (XRD), Raman scattering, and Photoluminescence (PL) measurements. The XRD and Raman studies revealed the ZnO films crystallizes in the wurtzite structure with a certain amount of amorphous material in the as-grown films, after the thermal treatment a preferential orientation along the (002) direction was observed. The as grown ZnO films presented macrostrain and microstrain caused a shift of the line diffraction (002) and a broadening respectively. The films are constituted by crystallites of similar nanosize dimensions; however after some tens of hours annealing these strains disappear. The crystallites grain size of the ZnO films increased by increasing the annealing time. The as-deposited reactive sputtering ZnO films resulted semi-insulating with poor PL response. After high temperature annealing in air, the PL response considerably improved, but their semi-insulating property also increased. The PL spectra of the annealed samples showed well defined excitonic-like transitions and a wide visible deep-level band emission (430-640 nm) at room temperature. The PL response also shows a broad defect-related green band at 516 nm. Both bands are clearly linked to the nanostructure and the point defects content of the ZnO films. The main interest of this work was to enhance the PL response and to identify the origin of deep-level luminescence bands. The AFM, PL and XRD results indicated that the ZnO films annealing have potential applications in optoelectronic devices.

11.1 Introduction

Zinc oxide (ZnO) is an important material because it has a broad utility range. ZnO is a wide gap semiconductor (Eg=3.37 eV at room temperature) with applications in electronic displays, solar cells, electro-optical devices, and diverse sensors [1, 2, 3]. ZnO films have a lot of promising application in high frequency surface devices, optical devices, and UV detectors. The ZnO crystallizes in the wurtzite structure and possesses two formula units in the primitive cell, as is shown in the Fig. 11.1. As a wide band semiconductor with a large binding energy (about 60meV), ZnO is a promising candidate semiconductor material for the next generation of optoelectronic, light emission or high power and high frequency devices. The range of their technological usefulness expands when using ZnO nanostructured films for diverse applications such as cold electron nanoemitters [2]. ZnO is naturally n-type semiconductor, which can hardly be doped to become a p-type. This problem represents a challenge for optoelectronic applications of ZnO. Native defects, the non-stoichiometry, and hydrogen have been pointed out as the sources of n-type conductivity of ZnO. Low solubility of the dopants, and deep position of the impurity levels represent the main difficulties for achieving p-type material. Lately a growing number of reports have been claiming the growth of p-type ZnO using the diverse growing techniques. However, even any research group has proven to achieve reliable, reproducible device quality p-type ZnO. In spite of this, its high electrical conductivity and optical transmittance in the visible region make it useful for transparent conducting electrodes in flat panel displays or as optical windows [4, 5]. ZnO also find application in piezoelectric devices.

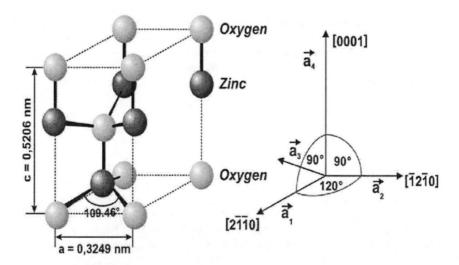

Fig. 11.1. The unit cell of the stable ZnO wurtzite structure.

ZnO thin films with excellent properties can be grown by numerous methods, including chemical-vapor deposition (CVD), molecular beam epitaxy (MBE), pulsed laser deposition (PLD), and the sputtering technique. The variants of sputtering are well developed methods and can be used to grow ZnO films with highly reproducible and controllable properties. The sputtering deposition technique offers the possibility to select the deposition rates in a wide range of values. When more complex alloys are required, the stoichiometry of the films can be modified by changing the substrate temperature, the pressure and the reactive atmosphere used during the deposition process. Furthermore post grown annealing processes can be applied to achieve the required film properties [5]. The nature of the annealing atmosphere and the temperature can also be chosen to control the material properties [6, 7]. In the present investigation, a study was made on the preparation of ZnO films by DC reactive magnetron sputtering deposited on silicon substrates. Most of ZnO films are grown on sapphire (Al_2O_3) substrates, but due to the large lattice mismatch (18%) with the ZnO films, other substrates such as crystalline silicon can also be used. Considering the possibility to produce an ideal interface between ZnO and Si, the resulting lattice mismatch between ZnO and (100) Si is very high $\approx$40%, however some special crystalline planes can be used to reduce the strains produced by the misfit between the crystalline lattices. For example, the lattice misfit on (111) Si substrates is of $\approx$11.5%. In real cases the existence of a native silicon oxide layer must be considered when depositing ZnO films. Native silicon oxide with thicknesses greater than 50 nm have been measured under well controlled deposition processes when using silicon wafers.

The growth temperature of the ZnO thin films was fixed at 230^oC and then the samples were annealed in dry air atmosphere at 800^oC for one hour. The ZnO films were deposited on silicon substrates and annealed in air atmosphere to study the correlation between optical and structural characteristics with respect to the initial properties. The main characteristics of the films and the effect of the post annealing process on the crystalline structure and their optical properties are reported. The goal of the study was to define a convenient route to growth ZnO films with the best quality for useful optoelectronic device applications.

The surface of the ZnO samples was observed with a scanning electron microscope (SEM) and using an atomic force microscope (AFM). The structural properties were assessed by X-ray diffraction (XRD), Raman scattering, and photoluminescence (PL) measurements.

The XRD and Raman studies revealed the ZnO films crystallize in the wurtzite structure with a certain amount of amorphous material in the as-grown films. After the ex-situ thermal treatment a preferential orientation along the c-axis is observed. The as-grown films are constituted by nanosized crystallites with an initial grain size, after the annealing process the grain size increases according to the processing time.

In a highly structured ZnO film the PL spectra is constituted by two bands; the near-band-edge-emission and the deep level band emission. The radiative transitions related to deep level emission associated to the structural defects, are even matter of discussion, especially when related to the growth conditions. Therefore studies to clarify the relation between the PL spectra with the sample preparation conditions are required. The most relevant result of this study was the excitonic structure of the room temperature PL of the near band edge band at 380 nm. The PL response also showed the broad green band at 516 nm, related to crystalline defects. The origin of the deep-level emission and the films stress condition of ZnO films grown by dc sputtering with post grown annealing processes in air atmosphere was investigated.

11.2 Experimental

The ZnO films were deposited onto p-type (100) silicon substrates of 5 Ω-cm by the DC reactive magnetron sputtering method. The complete details of the system have been published elsewhere [8]. The deposition system comprises a stainless steel vacuum chamber equipped with an array of turbomolecular and rotary vane pumps. The ultimate vacuum in the chamber is in the order of 10-7 mbar. The work pressure was $2\text{x}10^{-4}$ mbar, resulting from the Ar to O_2 flux ratio of 1:1 used in the runs. The oxygen pressure $p(O_2)$ was fixed at $1.5\text{x}10^{-3}$ mbar by controlling the flow of argon. A circular planar magnetron with 3" in diameter erosion zone was used as the cathode. The system is equipped with a DC US GUN II current supply with a maximum capacity of 1 Ampere. A constants DC power of 350W was used for all the runs. High purity Ar and O_2 (from LINDE) were used as the sputtering and reactive gases, respectively. The flow rates of both the Ar and O_2 gases were controlled separately by using MKS mass flow controllers. The target to substrate distance was maintained at 40 mm. ZnO sputtering targets (from LESKER), with a chemical purity of 99.999% were used routinely. The substrate temperature was fixed at 230°C.

Previous to the deposition process the silicon substrates were cleaned with the standard RCA procedure. After growth, the samples were cooled to room temperature while maintaining the growth chamber pressure. The growth rate increase with a linear behavior from 30 to 300 Å/min. For the used growth conditions the system provides uniform ZnO film surfaces with mirror like appearance. The ZnO thin films were annealed at 800°C in a quartz chamber in air atmosphere and at one atmosphere of pressure or oxygen-rich conditions. The annealing period were varied from some hours until several tens of hours.

The thickness of the films was measured with a Dektak IIA profilemeter. The samples used for this study has a thickness of 600 nm. The X-ray Diffractograms were measured with a BRUKER model D8 DISCOVER using the Cu Kα radiation (λ =1.5406 Å) with a voltage of 40kV and current of 40mA. The surface morphologies of the thin films were investigated with an atomic force microscope (AFM) of Park Scientific Autoprobe, model CP0175, and a Scanning Electron Microscope of JEOL. The micro-Raman spectra were recorded in a backscattering geometry with a dispersive DILOR Raman microprobe using an Ar-ion laser operating at the wavelength of 514 nm. The excitation source used in the PL measurements was a He-Cd laser operating at 325 nm with an output power of 50 mW. The emitting light from the sample was focused into the entrance slit of a double SPEX monochromator with a S1 type photomultiplier tube as detector. All of the measurements were performed in air at room temperature.

11.3 Results and Discussion

For a substrate temperature of 230°C the growth rate of the ZnO films was 29.6 nm/min. In the Fig. 11.2 is included the dependence of the growth rate for the ZnO films on (100) silicon by DC sputtering at 350 W with the substrates at 230 °C. The thickness of the studied films was fixed at 592 nm.

The AFM photomicrographs for the surface of the PD33 are shown in the Fig. 11.3. Fig. 11.3 a) is for the as grown ZnO sample and the Fig. 11.3 b) for the thermal annealed sample at 800 °C in dry air atmosphere.

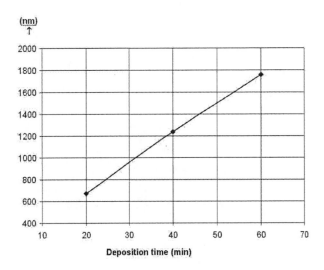

Fig. 11.2. Dependence of the growth rate of ZnO films on (100) silicon by DC sputtering at 350 W and the substrates at 230 °C

The surface of the as grown films (Fig. 11.3a) is composed of a uniform distribution of rounded particles with submicron sizes. The ex-situ annealing process revealed the rounded particles are composed of well defined smaller crystallites, as can be seen in the Fig. 11.3(b). The rms roughness of the as grown samples was of 61.4 nm, and decreased to 31.8 nm after the annealing process.

To discard the possible formation of stable compounds at the interface between the ZnO film and the substrate the annealed sampled was cleaved and observed by SEM. The Fig. 11.4 shows the SEM photomicrograph for the sample PD33-A ZnO, annealed at 800°C in dry air. The Fig. 11.4 a) show a detail of the cross section ZnO/silicon interface, the brilliant zone corresponds to the ZnO film and the right black zone, at the right, is the silicon substrate. Due to the submicrometric film thickness is difficult to identify the formation of a stable compound at the interface.

The Fig. 11.4 b) show a detail in plan view of the region near of the cleaved region. As the ZnO film can be ragged without great difficulty, it can be suggested that no stable compounds are formed at the interface for the used annealing conditions.

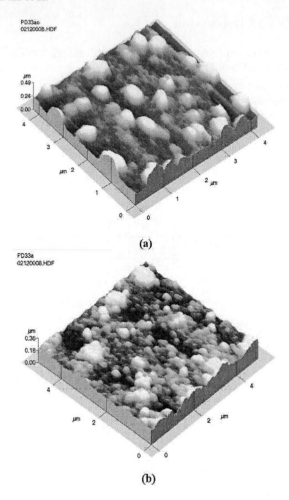

(a)

(b)

Fig. 11.3. Atomic force microscope (AFM) photomicrograph on the PD33 ZnO sample; a) as grown sample and b) thermal annealed sample at $800^{o}C$ in dry air atmosphere

The crystalline structure and orientation of the ZnO films were investigated by X-ray diffraction (XRD). Fig. 11.5 shows the θ-2θ diffractograms on the as grown ZnO films (a) and on the films after the annealing step (b).

The diffractograms show two dominant orientation of ZnO, at (002) and (101) of the hexagonal wurtzite crystal structure. The diffractograms exhibit a strong 2 peaks at 34.08^{o}, corresponding to the (002) peaks of ZnO. c-Axis (002) diffraction peaks prevail therefore it can be assumed that the c-axis oriented ZnO film is obtained.

It is important to point out that the peak intensity of ZnO (002) increased considerably after a short period of thermal annealing at 800 ^{o}C. In first instance the increase in the (002) peak intensity can be interpreted as the improvement on the

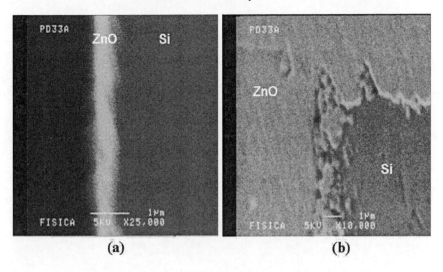

Fig. 11.4. SEM photomicrograph for the PD33-A ZnO sample annealed at 800^oC in dry air atmosphere. a) ZnO/silicon substrate interface, and b) plane view of the ZnO film and the silicon substrate.e

film quality with the annealing process (see grain size in the frame). The (002) peak position is shifted from its unstrained value of 34.48^o, this shift could be explained by the presence of compressive strain in the film.

To characterize the conditions of the as-grown ZnO films grown on the Si substrates, the films were annealed at $800°C$ by intervals of 1hours (h), 2h an 17 h in air ambient. Fig. 11.6 shows the θ-2 θ diffractograms for the annealed samples and in the Table 11.1 are included the details of the peak position, the full width at half maximum (FWHM) of the (002) orientation and the calculated grain size D (nm) on the ZnO samples.

The general behavior observed on the preference on the grain orientation on ZnO samples grown by sputtering has been reported by several authors in the literature. Since the XRD angle (2) on the bulk ZnO are 34.44^o and 0.5204 nm, respectively, we reveal that the compressive stress along the c-axis direction exists inside the ZnO thin deposited films.

The Raman backscattering measurements for the thin film ZnO samples are presented in Fig. 11.7. The structure of the wurtzite ZnO crystals belongs to the space group C_{6v} with two formula units in the primitive cell. The optical phonons at the Γ-point of the Brillouin zone can be described by the following irreducible representation [a, b, c]:

$$\Gamma_{opt} = 1A_1 + 2B_1 + 1E_1 + 2E_2 \qquad (11.1)$$

The modes A_1 (infrared active) and E_1 (infrared and Raman active) are polar and split into transverse (TO) and longitudinal optical (LO) phonons with different frequencies due to the electric fields associated with the LO phonons. The short-range

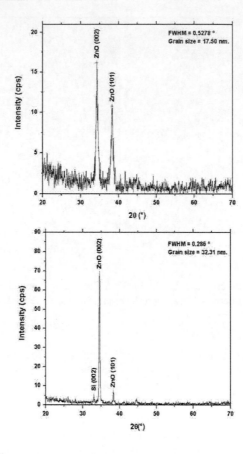

Fig. 11.5. X-ray diffractograms of ZnO thin films; a) as grown, b) after thermal annealing process. The values of FWHM and grain size of each sample are inside the frame.

Table 11.1. Peak Position, FWHM, crystallite size D on as grown and annealing ZnO thin films

Sample	Peakposition	FWHM	D(nm)
asgrown	34.08	0.8	14.02
annealed			
1h	34.42	0.22	114.65
2h	34.46	0.21	132.2
17h	34.48	0.17	308.4

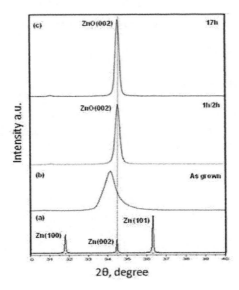

Fig. 11.6. X ray diffraction patterns of ZnO films: (a) Simulated ZnO pattern, (b) As grown ZnO, (c) 1h and 17h annealing ZnO samples for 800^oC in air, respectively

interatomic forces cause anisotropy, and A_1 and E_1 modes are produced at different frequencies. The two nonpolar E_2 modes [$E_2^{(1)}, E_2^{(2)}$] are Raman active but infrared inactive. The B_1 modes are either infrared or Raman inactive. The characteristic ZnO phonon modes of crystalline ZnO at low wave numbers with its usual grouping are in the Table 11.2. The as grown sample shows the characteristic ZnO phonon modes at low wave numbers. The detected modes on the ZnO films are the $E_2^{(2)}$ at 438 cm^{-1} and the E_1(LO) at 589 cm^{-1}. Other ZnO modes interfere with the strong silicon mode at 432 cm^{-1} [9].

Table 11.2. Phonon mode frequencies (in units of cm^{-1}) of the film and the bulk ZnO samples [[13], [14], [15]].

Mode	FTIR	Raman(this work)	Raman(Literature)
$v[E_1(TO)]$	409.1(0.9)		410(±3)
$v[E_1(LO)]$	588.3(0.7)	589	588(±3)
$v[A_1(TO)]$			380(±1)
$v[A_1(LO)]$	574.5(0.3)		378(±2)
$v[E_2^{(1)}]$			102(±1)
$v[E_2^{(2)}]$		438	437(±3)

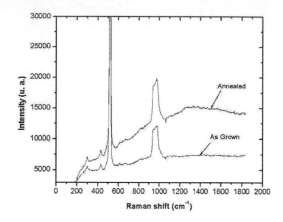

Fig. 11.7. Raman backscattering spectra on the PD33 ZnO sample; a) As grown sample and b) Thermal annealed sample at 800 ^{o}C in dry air atmosphere

The less intense lines at 332, 541, and 665 cm^{-1} have been mark described as possible multiphononic scattering processes. The other hand, the intense and wide signal centered at around ~1200 cm^{-1} is related to the photoluminescence response of the sample. The E2 mode is the characteristic band of the hexagonal wurtzite phase [9], which is the typical Raman active branch for the ZnO thin film. As can be seen from Fig. 11.7, the peak intensity of the E_2 mode of 438 cm^{-1} increased with the annealing process. This behavior correlates well with the XRD results and demonstrates the annealing process on the ZnO film improves the crystal quality ZnO films.

Photoluminescence (PL) spectroscopy is a powerful technique to characterize distinct kind of solids. The luminescence spectrum of ZnO consists of two main bands which are situated in the ultra-violet and visible region of the electromagnetic spectrum. The luminescence in ZnO is easily detectably at room temperature because of exciton binding energy of 60 meV, which also makes ZnO exceptionally attractive for optoelectronics. In the Fig. 11.8 the 10 K temperature PL spectrum taken from 350 to 650 nm for two quasi-epitaxial ZnO films grown by thermal oxidation of a monocrystalline Zn foil at two distinct temperatures clearly shows the cited bands. The wide band around 2.3 eV is produced by excitonic recombination trough lattice defects and the emission band or fundamental band around 3.0 eV comprise a variety of lines related to the free excitons and bound-exciton transitions.

The fundamental PL of the ZnO is constituted by the free excitonic, bound excitonic, and shallow donor-acceptor-pair (DAP) emission lines, among others The free-exciton transitions can be observed as three lines, produced by electrons from the conduction band and holes from the three degenerate valence bands, those lines are known as A (heavy holes), B (light hole), and C (produced by the crystal field splitting). Other paths for excitonic recombination are produced by discrete electronic energy levels in bandgap in ZnO are generated by the dopants or defects, influencing

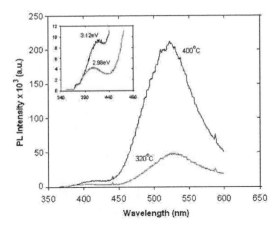

Fig. 11.8. 10K PL spectra on ZnO produced by thermal oxidation of a crystalline Zn foil at 320°C and 400°C

the optical absorption and emission processes. The type and band structure of the semiconductor material influences the electronic states of the bound exciton. Neutral or charged donors and acceptors can bind excitons, producing bounded excitions transition lines. The neutral shallow donor-bound exciton (BE) normally dominates the low-temperature PL spectrum of high-quality ZnO films, due to the presence of donor unintentional impurities or other shallow-level defects. The acceptor-bound exciton (ABE) lines are also sometimes observed in some ZnO films which contain substantial concentration of acceptors. The sharp lines in PL spectra generated by the recombination of bound excitons are commonly used to identify different defects or impurities source. Most sharp donor bound and acceptor-bound exciton lines are observed in the range from 3.34 to 3.38 eV in high quality ZnO films. Furthermore, shallow donor-acceptor-pair (DAP) emission exhibiting the main zero-phonon peak at about 3.22eV followed by at least two LO phonon replicas is important processes in the optical properties of ZnO according to the doping level.

Fig. 11.9 shows the spectra of PL of both samples at room temperature. A noteworthy improvement in PL intensity with the thermal annealed is observed. This result demonstrates that the thermal annealing process at 800°C enhances the structural nanocrystallites quality.

Three PL bands are observed; The tentative identification of the 3.29 eV band with the free exciton (FX) recombination is based on the consistency of this value with the reported result on ZnO single crystals [10], the 3.23 eV peak is assigned to the first order transverse optical TO phonon replica of the FX recombination, because the coincidence with the energy difference the FX peak and the TO phonon energy [11] The strong emission intensity of TO phonon peaks suggests a strong coupling of TO phonon and FX [12]. The peak at 3.18 eV is attributed to the FX emission accompanied by a TO and the first order longitudinal optical LO phonons 71.9 meV [11], since the energy difference between this line and the FX is close to the sum of

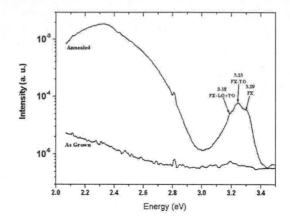

Fig. 11.9. Photoluminescence of the as-grown and thermal annealed ZnO thin films

the energies of a TO and a LO phonons. The main recombination process is due to free excitons with the LO phonons as in a bulk ZnO [10] because the well structured contact between the nonocrystallites. The possible influence of recombination paths through acceptors or donors are now under analysis and will be further reported.

11.4 Conclusions

ZnO thin films were deposited at $230^{o}C$ by dc reactive magnetron sputtering on silicon (100) substrates. The ZnO films were annealed ex-situ in dry air environment at 800 ^{o}C for a period of an hour. XRD studies of ZnO thin films showed that the annealing process improved the crystalline quality in comparison with the as grown samples. Raman and XRD results shown the annealed thin films acquired the c-axis preferential orientation towards the [001] crystal direction. The surface roughness determined by AFM shown the film roughness was notably reduced after the annealing process. The most striking effect of the ex-situ annealing process was the improvement of the PL response. In the annealed samples the PL spectra showed three excitonic related peaks at room temperature. As the excitonic transitions are only visible in high quality samples, the findings in the PL spectra confirm the XRD results. Therefore the ex-situ annealing process applied to the ZnO samples considerably improves the structural quality of the constituent nanosized crystallites. The set of experimental results demonstrate the ZnO films grown by dc reactive sputtering can be used in the realization of electronic devices.

References

[1] Ü. Özgür, Ya. I. Alivov, C. Liu, A. Teke, M. A. Reshchikov, S. Doğan, V. Avrutin, S.-J. Cho, and H. Morkoç, J. Appl. Phys. 98, 041301 (2005).

[2] M. Wei, Z-M Qi, M. Ichihara, I. Honma amd H. Zhou., Nanotechnology 18, 0956608 (2007).

[3] S.J. Pearton, D. P. Norton, K. Ip, Y.W. Heo, T. Steiner, Progress in Materials Science 50 293 (2005)

[4] X.-T. Hao, L.-W. Tan, K.-S. Ong and F. Zhu. J. Cryst. Growth 287, 44 (2006).

[5] S. J. Jiao, Z. Z. Zhang, Y. M. Lua, D. Z. Shen, B. Yao, J. Y. Zhang, B. H. Li, D. X. Zhao, X. W. Fan, and Z. K. Tang, Appl. Phys. Lett. 88, 031911 (2006).

[6] K. Ellmer, J. Phys.D: Appl. Phys. 33, R17 (2000).

[7] C. Guillén and J. Herrero, Thin Solid Films 515, 640 (2006)

[8] A Esparza, Síntesis y caracterización de películas delgadas de óxido de titanio y óxido de tungsteno, por la técnica de erosión catódica para su aplicación al sensado de gases. (Master of Science Thesis), México D. F.: Electric Engineer Department CINVESTAV-IPN (2004)

[9] T. S. Jeong. M. S. Han, C. J Youn. and Y. S. Park, J. Appl. Phys. 96, 175 (2004).

[10] R. L. Weiher and W. C. Tait, Phys. Rev. 166, 791 (1968).

[11] F. Decremps, J. Pellicer-Porres, A. M. Saitta, J. C. Cheervin, and A. Polian, Phys. Rev. B 65, 092102 (2002).

[12] T. Matsumoto, H. Kato, K. Miyamoto, M. Sano, E. A. Zhukov, and T. Yao, Appl. Phys. Lett. 81, 1231 (2002).

[13] C. A. Arguello, D. L. Rousseau, and S. P. S. Porto, Phys. Rev. 181, 1351(1969), "First-Order Raman Effect in Wurtzite-Type Crystals".

[14] J. M. Calleja and M. Cardona, Phys. Rev. B 16, 3753 (1977), "Temperature dependence of Raman scattering in ZnO".

[15] N. Ashkenov, B. N. Mbenkum, C. Bundesmann, V. Riede, M. Lorenz, D. Spemann, E. M. Kaidashev, A. Kasic, M. Schubert, M. Grundmann, G. Wagner, H. Neumann, V. Darakchieva, H. Arwin, and B. Monemar "Infrared dielectric functions and phonon modes of high-quality ZnO films", J. Appl. Phys. 93, 126 (2003)126-136

Thermodynamic Stability of $A^{III}B_x^V C_y^V D_{1-x-y}^V$ Semiconductor Alloys

Salvador F. D. Albarrán[1], Alicia G. G. Noguez[2], Patricia R. Peralta[1], and Vyacheslav A. Elyukhin[3]

[1] Department of Engineering in Computation, ESIME-IPN Culhuacan,
Av. Santa Ana 1000, México D. F., 04430
[2] Department of Radioactive Facilities CNSNS,
Dr. José Ma. Barragán 779, México D. F., 03020
[3] Department of Electrical Engineering, CINVESTAV-IPN,
Av. IPN 2508,México D. F., 07360
e-mail: felipondiaz@hotmail.com

Summary. Spinodal decomposition of the $GaSb_xN_yAs_{1-x-y}$ quaternary alloys lattice-matched to the GaAs as the result of the internal deformation and coherency strain energies is described. The alloys are represented as quaternary regular solutions. The internal deformation energy is represented by interaction parameters between the constituent compounds estimated within the framework of the valence force field model. Ranges of spinodal decomposition of the $GaSb_xN_yAs_{1-x-y}$ alloys up to $y \leq 0.035$ with and without coherency strain energy are demonstrated.

12.1 Introduction

Dilute nitride III-V alloys have attracted a lot of attention since the last decade. The earliest studies were initiated a long time ago on N-doped GaP, with nitrogen concentration in the range $10^{17} - 10^{18}$ cm^{-3} [1]. In this material, narrow photoluminescence lines were attributed to excitons bounded to isolated N centres or N-N pairs [1]. In the early 1990s, the development of wide-gap nitrides has considerably influenced the technology of N sources and precursors. As a consequence, new investigations on N-containing III-V were carried out from 1992, with modern epitaxial growth techniques. N concentrations around 1% and above could be easily achieved in GaAs or GaP. The introduction of nitrogen in GaAs also results in a smaller lattice parameter. Therefore, a dilute amount of nitrogen offers a unique feature of reducing simultaneously the band gap and the lattice parameter of a given III-V alloy [1].

It has been found that incorporating low concentrations of N has a profound effect on the electronic properties of the III-V alloy semiconductors composed of (B, Al, Ga, In) (N, P, As, Sb) [2]. A reduction on the band gap exceeding 0.1 eV per atomic percent of N content was observed in GaN_xAs_{1-x} for $x < 0.015$ [3]. Model calculations of the band structure of some of the group III-N-V alloys have shown that the reduction of the band gap is due to the highly localized nature of the perturbation introduced by N atoms [4]. The $GaSb_xN_yAs_{1-x-y}$ material system was recently proposed as promising candidate for GaAs-based optoelectronic devices. Compared to InGaAs, the $GaSb_xN_yAs_{1-x-y}$ material system has the advantage that for the same wavelength, GaAsSb has lower compressive strain compared to InGaAs.

The large difference between the atomic sizes of nitrogen, antimony and arsenic gives rise to the significant strain energy of such alloys. The internal deformation energy provides the tendency to disintegration that can lead to the appearance of the thermodynamically unstable states with respect to the phase separation [5]. The thermodynamically unstable states with respect to the descomposition may be realized as spinodal decomposition [5]. Spinodal decomposition results in formation of the macroscopic phases of different compositions decreasing the internal energy of the alloy [5]. At the same time, this decomposition leads to an occurrence of the coherency strain energy due to the stress between both formed regions with different compositions and these regions and other part an of alloy [6]. Thus, the internal deformation and coherency strain energies are two origins controlling spinodal decomposition in the $GaSb_xN_yAs_{1-x-y}$ semiconductor alloys. The aim of our chapter is the consideration of the spinodal decomposition region of $GaSb_xN_yAs_{1-x-y}$ layers grown on GaAs(001) substrates with all the described above origins. In the next section we briefly describe these solutions or alloys.

12.2 $GaSb_xN_yAs_{1-x-y}$ Quaternary Alloys

$GaSb_xN_yAs_{1-x-y}$ belong to $AB_xC_yD_{1-x-y}$ - type alloys where the anions (Sb,N and As) are surrounded by only one type cation (Ga), having one mixed sublattice. When we consider the chemical composition of such alloys the sum of the concentrations of the atoms in the mixed sublattice is supposed equal to unit. In the notation $GaSb_xN_yAs_{1-x-y}$, x is the concentration of Sb atoms and y is the concentration of N atoms. As a result, $GaSb_xN_yAs_{1-x-y}$ alloys have three types chemical bonds: Ga-Sb, Ga-N and Ga-As. Therefore, they are known as quaternary alloys of three binary compounds or quasiternary alloys since these alloys consist of three types of the chemical substances.

A special feature of these alloys is the one to one correspondence between the concentrations of the atoms and chemical bonds. This one to one correspondence can be written as

$$x_{GaSb} = x, \tag{12.1}$$

$$x_{GaN} = y, \tag{12.2}$$

$$x_{GaAs} = 1 - x - y . \tag{12.3}$$

We can also represent the crystal structure of these quaternary alloys as a structure consisting of molecules of the binary compounds. Concentrations of the bonds in the quaternary solutions of three binary compounds are also independent on the arrangement of the atoms in the mixed sublattice. In other words, the correlations in the arrangement of the atoms in the mixed sublattice do not affect the concentrations of the chemical bonds or the chemical composition of such solutions. These characteristics offer an advantage over other quaternary alloys represented as $A_xB_{1-x}C_yD_{1-y}$. Their crystal lattice consists of the mixed cation and anion sublattices, since two kinds of the atoms fill each of them. As a result, $A_xB_{1-x}C_yD_{1-y}$ alloys have four types of chemical bonds: A-C, A-D, B-C and B-D.

A special peculiarity of these alloys is the transformation of the A-C and B-D bonds in the A-D and B-C bonds or vice versa. This property is independent of the concentrations x and y. Therefore, the spinodal decomposition is accompanied by the transformation of the bonds and it modifies the chemical composition and should change the free energy of the alloy [7]. For example, spinodal decomposition range of the $In_xGa_{1-x}N_yAs_{1-y}$ alloys depends on the strain and coherency strain energies and transformation of the bonds. As it was shown in [8], the very extensive spinodal decomposition range of the $In_xGa_{1-x}N_yAs_{1-y}$ lattice mismatched to GaAs occurs due to the exchange of atoms. Then, there is one mixed sublattice in the crystal structure of the $GaSb_xN_yAs_{1-x-y}$ alloys in comparison to the $In_xGa_{1-x}N_yAs_{1-y}$ alloys. Mixing in one sublattice only should lead to the smaller internal deformation energy. We can say that from the spinodal decomposition standpoint, the $GaSb_xN_yAs_{1-x-y}$ alloys should be a more perspective material than the $In_xGa_{1-x}N_yAs_{1-y}$ alloys.

As mentioned above the internal deformation and coherency strain energies are two mechanisms controlling the decomposition in the $GaSb_xN_yAs_{1-x-y}$ quaternary alloys. The internal deformation energy is described by the valence force field approach. This approach is very important in our study on the spinodal decomposition, so we explain this model.

12.3 Valence Force Field Approach

The most useful phenomenological description of the short-range valence forces in the tetrahedrally coordinated crystals is the valence-force-field (VFF) approach, in which all interatomic forces are resolved into bond-stretching and bond-bending forces [9]. The are two primary virtues of the VFF model. First, because all distortions are described in terms of bond lengths and angles, the model is automatically rotationally invariant so that serious errors that may arise in the ordinary force-constant approach are avoided [10]. Second, in crystals in which atom pair bonds play an essential role, the VFF model is the most natural description of interatomic forces. Thus one expects the VFF model to involve the smallest possible number of parameters. In other words, the VFF model described the strained state of the elemental semiconductors or binary compounds by using two microscopic elastic constant. The deformation energy of the primitive or unit cell of the binary compounds with the zinc blende structure according to the VFF model is given as

$$u = \frac{1}{2}\alpha\left(\frac{3}{4R^2}\right)\sum_{i=1}^{4}\left[\Delta\left(\mathbf{r}_i^1\cdot\mathbf{r}_i^1\right)\right]^2 + \frac{1}{2}\sum_{s=1}^{2}\beta^s\left(\frac{3}{4R^2}\right)\sum_{i,j>i}\left[\Delta\left(\mathbf{r}_i^s\cdot\mathbf{r}_j^s\right)\right]^2 . \quad (12.4)$$

where α and β are the bond-stretching and bond-bending elastic constants, R is the bond length in undistorted crystal, $\Delta\left(\mathbf{r}_i^1 \cdot \mathbf{r}_i^1\right) = R_i^2 - r_i^2$ and $\Delta\left(\mathbf{r}_i^s \cdot \mathbf{r}_j^s\right) = R^2 \cos\varphi_0 - r_i^s r_j^s \cos\varphi$ are the scalar variations, $\mathbf{r}_{ij}^s$ are bond vectors about atom s, $\varphi_0 = 109.47^0$ and φ are the angles between the bonds in the unstrained and strained crystal, respectively [9].

The expressions for the elastic constants have the simple form [9]

$$C_{11} + 2C_{12} = \frac{\sqrt{3}}{4R}(3\alpha + \beta) - 0.355SC_0, \tag{12.5}$$

$$C_{11} - C_{12} = \frac{\sqrt{3}}{R}\beta + 0.53SC_0, \tag{12.6}$$

$$C_{44} = \frac{\sqrt{3}}{4R}(\alpha + \beta) - 0.136SC_0 - C\zeta^2, \tag{12.7}$$

where

$$C = \frac{\sqrt{3}}{4R}(\alpha + \beta) - 0.266SC_0, \tag{12.8}$$

and

$$\zeta = C^{-1}\left[\frac{\sqrt{3}}{4R}(\alpha - \beta) - 0.294SC_0\right]. \tag{12.9}$$

In particular (5), (6) and (9) may be used to derive a expression for the internal strain parameter

$$\zeta = \frac{2C_{12} - \mathscr{C}}{C_{11} + C_{12} - \mathscr{C}}, \tag{12.10}$$

where

$$\mathscr{C} = 0.314SC_0. \tag{12.11}$$

Thus, the equations (5)-(9) predict a relation among the elastic constants which may be checked experimentally. The relation may be written

$$\frac{2C_{44}(C_{11} + C_{12} - \mathscr{C})}{(C_{11} - C_{12})(C_{11} + 3C_{12} - 2\mathscr{C}) + 0.831\mathscr{C}(C_{11} + C_{12} - \mathscr{C})} = 1. \tag{12.12}$$

Now the internal deformation energy is represented by the interaction parameters between the constituent compounds of the $GaSb_xN_yAs_{1-x-y}$ alloys. The interaction parameters were obtained from the strain energies of the corresponding ternary alloys estimated by (4).

12.3.1 Strain Energy of the $A_xB_{1-x}C$ Ternary Alloys

The basic unit of the crystal lattice of the $A_xB_{1-x}C$ ternary alloys in our description is a tetrahedral cell with four atoms at the vertices and one atom inside. The atoms at the vertices are the atoms of the mixed sublattice and one atom inside such cell is an atom from another sublattice. There are five types of tetrehedral cells: 4A1C, 3A1B1C,

2A2B1C, 1A3B1C and 4B1C. It is supposed that the cells of any types have the same distances between the vertices because the mixed sublattice is slightly distorted. It is very important point for further description. Figure 1 shows the tetrahedral cells.

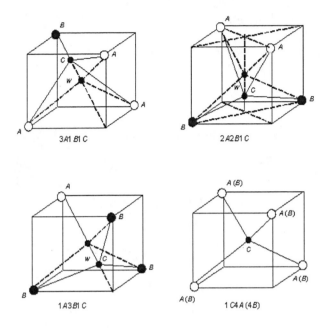

Fig. 12.1. 4A4(B)1C, 3A1B1C, 2A2B1C, 1A3B1C tetraherdral cells

The displacements of the central atom C are calculated by the minimum condition of the deformation energy of the cells. Average distances between the nearest atoms in the $A_xB_{1-x}C$ ternary alloy are written as

$$\overline{R_{AC}} = x^3 R_{AC}^{4A1C} + 3x^2(1-x)R_{AC}^{3A1B1C} + 3x(1-x)^2 R_{AC}^{2A2B1C} + \\ (1-x)^4 R_{AC}^{3A1B1C}, \tag{12.13}$$

$$\overline{R_{BC}} = x^3 R_{BC}^{3A1B1C} + 3x^2(1-x)R_{BC}^{2A2B1C} + 3x(1-x)^2 R_{BC}^{1A3B1C} + \\ (1-x)^4 R_{BC}^{4B1C}. \tag{12.14}$$

Where R_{AC}^{4A1C} is the distance between the A and C atoms in the 4A1C tetrahedral cell. The random distribution of the atoms in the mixed sublattice was taken account to obtain these formulas. The strain energies of the tetrahedral cells of the $A_xB_{1-x}C$ alloys are

$$u_{4A1C} = 2(1-x)^2 r^2 (3\alpha_{AC} + \beta_{AC}), \tag{12.15}$$

$$u_{3A1B1C} = \frac{3}{2}\left\{3\alpha_{AC}\left[(1-x)r - \frac{w_{3A1B}}{3}\right]^2 + \alpha_{BC}(xr - w_{3A1B})^2\right\} +$$
$$\frac{\beta_{AC} + \beta_{BC}}{8}\left[(1-2x)r - 2w_{3A1B}\right]^2 + \beta_{AC}\left[(1-x)r - w_{3A1B}\right]^2, \tag{12.16}$$

$$u_{2A2B1C} = 3\left\{\alpha_{AC}\left[(1-x)r - \frac{w_{2A2B}}{\sqrt{3}}\right]^2 + \alpha_{BC}\left(xr - \frac{w_{2A2B}}{\sqrt{3}}\right)^2\right\} +$$
$$\frac{\beta_{AC}}{3}\left[(1-x)r + \sqrt{3}w_{2A2B}\right]^2 + \frac{\beta_{AC}}{3}\left(xr + \sqrt{3}w_{2A2B}\right)^2 + \tag{12.17}$$
$$\frac{\beta_{AC} + \beta_{BC}}{6}(1-2x)r^2,$$

$$u_{1A3B1C} = \frac{3}{2}\left\{\alpha_{AC}\left[(1-x)r - w_{1A3B}\right]^2 + 3\alpha_{BC}\left(xr - \frac{w_{1A3B}}{3}\right)^2\right\} +$$
$$\frac{3}{2}\left\{\frac{\beta_{AC} + \beta_{BC}}{8}\left[(1-2x)r + 2w_{1A3B}\right]^2 + \beta_{BC}(xr + w_{1A3B})^2\right\}, \tag{12.18}$$

$$u_{4B1C} = 2x^2r^2(3\alpha_{BC} + \beta_{BC}), \tag{12.19}$$

$$u_{4A1C} = 0 . \tag{12.20}$$

and the displacements of the central C atoms

$$w_{3A1B1C} = \frac{3\left[\alpha_{AC} + (\alpha_{BC} - \alpha_{AC})x\right] + \frac{5\beta_{AC} + \beta_{BC}}{2} - (3\beta_{AC} + \beta_{BC})x}{\alpha_{AC} + 3\alpha_{BC} + 3\beta_{AC} + \beta_{BC}} r, \tag{12.21}$$

$$w_{2A2B1C} = \sqrt{3}\frac{\alpha_{AC}(1-x) + \alpha_{BC}x - \frac{\beta_{AC}}{3}(1-x) - \frac{\beta_{BC}}{3}x}{\alpha_{AC} + \alpha_{BC} + \beta_{AC} + \beta_{BC}} r, \tag{12.22}$$

$$w_{1A3B1C} = \frac{3\left[\alpha_{AC} + (\alpha_{BC} - \alpha_{AC})x\right] - \frac{\beta_{AC} + \beta_{BC}}{2} + (\beta_{AC} - \beta_{BC})x}{3\alpha_{AC} + 3\alpha_{BC} + \beta_{AC} + 3\beta_{BC}} r, \tag{12.23}$$

$$u_{4B1C} = 0 . \tag{12.24}$$

Where $r = R_{AC} - R_{BC}$, R_{AC} is the distance between A and C atoms in the AC unstrained compound, α_{AC} and β_{AC} are the bond length and bond-angle elastic constants of AC compound, respectively, w_{3A1B1C} is the displacement of central atom C in the 3A1B1C tetrahedral cell from the geometrical center of this cell. The supposition that the bond-angle elastic constant between the AC and BC bonds is equal arithmetical average of the bond-angle elastic constants of the AC and BC compounds

$$\beta_{AC-BC} = \frac{\beta_{AC-AC} + \beta_{BC-BC}}{2}, \tag{12.25}$$

is used in the formulas of the displacements of the central atoms and deformation energies.

After calculation the deformation energies of all types of the tetrahedral cells we can calculate the average deformation energy of a tetrahedral cell which is given as

$$\bar{u} = x^4 u_{4A1C} + 4x^3(1-x)u_{3A1B1C} + 6x^2(1-x)^2 u_{2A2B1C} +$$
$$(1-x)^3 u_{1A3B1C} + (1-x)^4 u_{4B1C} . \tag{12.26}$$

12.4 Coherency Strain Energy

Hookes Law states that

$$\varepsilon = s\sigma, \tag{12.27}$$

where s is a constant. s is called the elastic compliance constant or, shortly, the compliance. As an alternative we could write

$$\sigma = C\varepsilon, \tag{12.28}$$

with

$$C = \frac{1}{\sigma}, \tag{12.29}$$

where C is the elastic stiffness constant, or the stiffness. C is also Youngs Modulus [11]. The generalized form of Hookes Law may written as

$$\varepsilon_{ij} = s_{ijkl}\sigma_{kl}, \tag{12.30}$$

the s_{ijkl} are the compliances of the crystal. Equation (30) stands for nine equations, for example, $\varepsilon_{11} = s_{1111}\sigma_{11} + s_{1112}\sigma_{12} + s_{1113}\sigma_{13} + s_{1121}\sigma_{21} + s_{1122}\sigma_{22} + s_{1123}\sigma_{23} + s_{1131}\sigma_{31} + s_{1132}\sigma_{32} + s_{1133}\sigma_{33}$ and each with nine terms on the right-hand side. There are 81 s_{ijkl} coefficients.

As an alternative to (30) the stresses may be expressed in terms of the strains [11] by the equations

$$\sigma_{ij} = C_{ijkl}\varepsilon_{kl}, \tag{12.31}$$

where the C_{ijkl} are the 81 stiffness cosntants of the crystal. However, C_{ijkl} is a fourth-rank tensor that satisfies the equalities

$$C_{ijkl} = C_{jikl}, \tag{12.32}$$

and

$$C_{ijkl} = C_{ijlk} . \tag{12.33}$$

The equations (32) and (33) reduce the number of independent C_{ijkl} from 81 to 36. Therefore, (30) takes the shorter form

$$\sigma_i = C_{ij}\varepsilon_j \ (i,j = 1,2,\cdots,6), \tag{12.34}$$

it may be write in matrix notation as

$$\begin{pmatrix} \sigma_1 \\ \vdots \\ \sigma_6 \end{pmatrix} = \begin{pmatrix} C_{11} & C_{12} & \cdots & C_{16} \\ \vdots & \vdots & \ddots & \vdots \\ C_{61} & C_{62} & \cdots & C_{66} \end{pmatrix} \begin{pmatrix} \varepsilon_1 \\ \vdots \\ \varepsilon_6 \end{pmatrix} . \tag{12.35}$$

An important application of (35) is for cubic crystals [12]. In this case we have

$$
\begin{pmatrix} \sigma_x \\ \sigma_y \\ \sigma_z \\ \tau_{xy} \\ \tau_{yz} \\ \tau_{zx} \end{pmatrix} = \begin{pmatrix} C_{11} & C_{12} & C_{12} & 0 & 0 & 0 \\ C_{12} & C_{11} & C_{12} & 0 & 0 & 0 \\ C_{12} & C_{12} & C_{11} & 0 & 0 & 0 \\ 0 & 0 & 0 & C_{44} & 0 & 0 \\ 0 & 0 & 0 & 0 & C_{44} & 0 \\ 0 & 0 & 0 & 0 & 0 & C_{44} \end{pmatrix} \begin{pmatrix} \varepsilon_x \\ \varepsilon_y \\ \varepsilon_z \\ \gamma_{xy} \\ \gamma_{yz} \\ \gamma_{zx} \end{pmatrix}, \tag{12.36}
$$

where the ε_is and σ_is are the normal strain and stresses, respectively, and the γ_{ij}s and τ_{ij}s are the shear strains and stresses, respectively.

If the epitaxial film and its substrate are oriented along one of the (100) cubic symmetry directions, then (36) reduces to

$$
\begin{pmatrix} \sigma_\| \\ \sigma_\perp \end{pmatrix} = \begin{pmatrix} C_{11} + C_{12} & C_{12} \\ 2C_{12} & C_{11} \end{pmatrix} \begin{pmatrix} \varepsilon_\| \\ \varepsilon_\perp \end{pmatrix}, \tag{12.37}
$$

by [12], if the epitaxial film has a free surface, then

$$
\sigma_\perp = 0, \tag{12.38}
$$

and the perpendicular strain of the film is

$$
\varepsilon_\perp = \frac{-2C_{12}}{C_{11}} \varepsilon_\|, \tag{12.39}
$$

whereas the parallel componente is given as

$$
\sigma_\| = \frac{(C_{11} - C_{12})(C_{11} + 2C_{12})}{C_{11}} \varepsilon_\| . \tag{12.40}
$$

Therefore, the coherency strain energy of the $AB_x C_y D_{1-x-y}$ quaternary alloys may be written as the elastic energy of two epitaxial layers lattice-mismatched with the substrate that is given as [12]

$$
u^L = \sum_{i=1}^{2} \gamma_i v_i \frac{(C_{11}^i - C_{12}^i)(C_{11}^i + 2C_{12}^i)}{C_{11}^i} \left(\frac{a_i - a_{sub}}{a_{sub}} \right)^2 . \tag{12.41}
$$

Where γ_i, v_i and a_i are the portion, molar volume and the lattice-parameter of the ith phase of the alloy estimated by the Vegards Law, C_{11}^i and C_{12}^i are expressed as

$$
C_{11}^i = x C_{11}^{AB} + y C_{11}^{AC} + (1 - x - y) C_{11}^{AD}, \tag{12.42}
$$

$$
C_{12}^i = x C_{12}^{AB} + y C_{12}^{AC} + (1 - x - y) C_{12}^{AD} . \tag{12.43}
$$

The minimum value of the coherency energy u^L is achieved at the condition $\gamma_1 = \gamma_2$.

12.5 Spinodal Decomposition in the GaSb$_x$N$_y$As$_{1-x-y}$ Alloys

According to Gibbss classic treatment of phase stability, spinodal decomposition begins from the changes that are large in extent but small in degree [13] and develops when a negligibly small phase separation fluctuation decreases the free energy of an alloy. The initial stage of spinodal decomposition is accompanied by transfer of atoms on the distances of order of a lattice parameter. It was shown [14] that in cubic crystals the spinodal decomposition forms of the layers in a plane where their elastic energy is minimal. Accordingly, the transfer of atoms lead to an occurrence of thin two-layer region with negligibly small distinction in the composition. The composition of the formed layers at the initial stage of the decomposition can be considered as constant values due to their small thickness.

As the decomposition is developed the transfer of atoms and thickness of the layers become larger and composition of the layers varies with tickness. Afterwards, the difference in the mean concentrations of the phases is increased continuously [15]. The GaSb$_x$N$_y$As$_{1-x-y}$ quaternary alloys contain three types of the chemical bonds and the amounts of them are kept at the disintegration. Therefore, the Helmholtz free energy of mixing of the alloy is only varied at spinodal decomposition. In the cubic crystals minimal elastic energy corresponds the (100) planes if relation

$$2C_{44} - C_{11} + C_{12} > 0 \ . \tag{12.44}$$

Between the stiffness coefficients is fulfilled [5]. Thus, the initial stage of spinodal decomposition in the GaSb$_x$N$_y$As$_{1-x-y}$ alloys is considered as appearance thin two-layer objects oriented in the (100) plane.

The disintegration changes the value x, y or both of them in emerging two phases of the decomposed alloy. The variation of the Helmholtz free energy of mixing at the initial stage of spinodal decomposition can be represented as

$$\delta f \approx \frac{1}{2} \frac{\partial^2}{\partial x^2} \left[f^M(x,y) + u^L(x,y) \right] (\delta x)^2 +$$
$$\frac{\partial^2}{\partial x \partial y} \left[f^M(x,y) + u^L(x,y) \right] (\delta x)(\delta y) + \tag{12.45}$$
$$\frac{1}{2} \frac{\partial^2}{\partial y^2} \left[f^M(x,y) + u^L(x,y) \right] (\delta y)^2 \ .$$

An alloy reaches the spinodal decomposition range when the variation of its free energy becomes equal to zero [16]

$$\delta f = 0 \ . \tag{12.46}$$

This condition is fulfilled if one of two expressions

$$\frac{\partial^2 f}{\partial x^2}, \tag{12.47}$$

$$\frac{\partial^2 f}{\partial x^2} \times \frac{\partial^2 f}{\partial y^2} - \left(\frac{\partial^2 f}{\partial x \partial y}\right) . \tag{12.48}$$

Is equal to zero [17]. The Helmholtz free energy of the homogeneous alloys grown on crystaline substrates can be reprented as a sum

$$f = f^C + u^S + u^L - Ts, \tag{12.49}$$

where f^C, u^S, u^L are the free energy of the constituent compounds strain and lattice mismatch energies, respectively, s is the configurational entropy, and T is the absolute temperature. The negligibly small lattice mistmach between the alloy and substrate is introduced in order to include the coherency strain energy in our consideration.

The free energy of the constituent compounds of $GaSb_xN_yAs_{1-x-y}$ is given as

$$f^C = x\mu^0_{GaSb} + y\mu^0_{GaN} + (1 - x - y)\mu^0_{GaAs} . \tag{12.50}$$

Where μ^0_{GaSb} is the chemical potential of GaSb in the standard state. The internal deformation energy of the alloy is written as

$$u^C = xy\alpha_{GaSb-GaN} + x(1 - x - y)\alpha_{GaSb-GaAs} + y(1 - x - y)\alpha_{GaN-GaAs} . \tag{12.51}$$

Where $\alpha_{GaSb-GaN}$ is the interaction parameter between binary compounds GaSb and GaN. Thus, the internal deformation energy is represented by the interaction parameters between the constituent compounds of such alloys (section 2) [18].

The coherency strain energy of the decomposed alloy may be written as the elastic energy of two epitaxial layers lattice mismatched with the GaAs (001) substrate that is given by (41) with $a_{sub} = a_{GaAs}$. The configurational entropy of the alloy considered is obtained by the formula

$$s = k_B \ln g, \tag{12.52}$$

where g is the degeneracy factor. The expression for the factor g is

$$g = \frac{(N_{Sb} + N_N + N_{As})!}{N_{Sb}! N_N! N_{As}!}, \tag{12.53}$$

where N_{Sb}, N_N and N_{As} are the numbers of atoms Sb, N and As, respectively. Therefore, the configurational entropy of $GaSb_xN_yAs_{1-x-y}$ can be expressed as

$$s = -R[x \ln x + y \ln y + (1 - x - y) \ln(1 - x - y)] . \tag{12.54}$$

Equations (46) and (47) after taking into account the formulas (41), (48-50) and (52) are given, respectively, by

$$-2\alpha_{GaSn-GaAs} + RT \frac{1-y}{x(1-x-y)} + \frac{\partial^2 u^L}{\partial x^2} = 0, \tag{12.55}$$

and

$$\left[-2\alpha_{\text{GaSb}-\text{GaAs}} + RT\frac{1-y}{x(1-x-y)} + \frac{\partial^2 u^L}{\partial x^2}\right] \times$$

$$\left[-2\alpha_{\text{GaN}-\text{GaAs}} + RT\frac{1-x}{y(1-x-y)} + \frac{\partial^2 u^L}{\partial y^2}\right] - \quad (12.56)$$

$$\left[\alpha_{\text{GaSb}-\text{GaN}} - \alpha_{\text{GaSb}-\text{GaAs}} - \alpha_{\text{GaN}-\text{GaAs}} + \frac{RT}{1-x-y} + \frac{\partial^2 u^L}{\partial x \partial y}\right]^2 = 0 \ .$$

Where R is the universal gaseous constant.

12.6 Results

The spinodal decomposition ranges for the $GaSb_xN_yAs_{1-x-y}$ quaternary alloys lattice matched to GaAs with and without taking into account the coherency strain enegy are demonstrated in the Fig. 2. As can be see, the coherency strain energy emerging in the disintegration alloy substantially decreasing the temperature of spinodal decomposition. Therefore, the spinodal decomposition region of the alloy is narrower when we consider the coherency strain. $GaSb_{0.07}N_{0.023}As_{0.907}$ ($a = 5.655\,\text{Å}$, $\lambda = 1300\,\text{nm}$) alloys [19] is outside the spinodal decomposition range at its growth temperature. In the figure, this experimental value is represented by a circle.

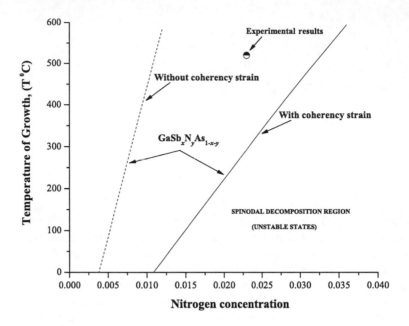

Fig. 12.2. Spinodal decomposition ranges of GaAs-enriched GaSb$_x$N$_y$As$_{1-x-y}$ quaternary alloys. Solid line indicates the range estimated considering the coherency strain. Dotted line demonstrated the range without the coherency strain.

The lattice, elastic and thermodynamic parameters: a(m), h(J/mol), c(J/molK), C_{11}(N/m^2), C_{12}(N/m^2), α(N/m) and β(N/m) for the GaSb$_x$N$_y$As$_{1-x-y}$ quaternary alloys are shown in the Table 1.

Table 12.1. Lattice, elastic and thermodynamic parameters of the GaSb$_x$N$_y$As$_{1-x-y}$ alloys

	a	h	c	C_{11}	C_{12}	α	β
GaSb	6.095×10^{-10}	-41868	49.379	0.883×10^{11}	0.403×10^{11}	33.16	7.22
GaN	4.50×10^{-10}	-109694	40.784	2.96×10^{11}	1.3×10^{11}	70.68	13.84
GaAs	5.65×10^{-10}	-81643	47.026	1.74×10^{11}	0.526×10^{11}	41.19	8.95

12.7 Discussion and Conclusions

As we are mentioned, the interaction parameters between the constituent compounds: $\alpha_{\text{GaSb–GaN}} = 409\,\text{kJ/mol}$, $\alpha_{\text{GaSb–GaAs}} = 22.920\,\text{kJ/mol}$ and $\alpha_{\text{GaN–GaAs}} =$

22.81 kJ/mol, were estimated from the internal deformation energies of the corresponding ternary alloys. The bond stretching and bond bending elastic constants of GaAs, GaSb and GaN were taken from [9, 20].

The lattice parameter of $GaSb_xN_yAs_{1-x-y}$ is given by Vegards Law

$$a = xa_{GaSb} + ya_{GaN} + (1 - x - y)a_{GaAs} \ . \tag{12.57}$$

In the estimations we supposed that the lattice parameter of the alloy and the substrate are closed to each other $a \approx a_{GaAs}$. Therefore, the concentration of Sb in these quaternary alloys is

$$x = \frac{a_{GaAs} - a_{GaN}}{a_{GaSb} - a_{GaAs}} \ . \tag{12.58}$$

Accordingly, the spinodal decomposition range of such semiconductors is function of one independent variable, which is nitrogen concentration. It is important to mention that the consideration of a wider temperature range is invalid since the GaAs-rich $GaSb_xN_yAs_{1-x-y}$ alloys grown up to 600 °C [21, 22]. Therefore, in spite of the significant strain energy, the $GaSb_xN_yAs_{1-x-y}$ alloys in the large composition regions are outside the spinodal decomposition range at their growth and lower temperatures.

In summary, we have described spinodal decomposition ranges in the $GaSb_xN_y$ As_{1-x-y} semiconductor alloys lattice matched to GaAs by taking into account the internal deformation and coherency strain energies. The internal deformation energy was estimated by the VFF model. The Helmholtz free energy of mixing controlling the disintegration is represented by using the strictly regular approximation.

References

1. J-C Harmand, et. al.: GaNAsSb:how does it compare with other dilute III-V nitride alloys?. Semicond. Sci. Tech. 17, 778-784 (2002)
2. W. Shan, et. al.: Band Anticrossing in GaInAsN Alloys. Phys. Rev. Lett. 82, 1221-1224 (1999)
3. M. Weyers, M. Sato, H. Ando.: Red Shift of Photoluminescence and Absorption in Dilute GaAsN Alloy Layers. Jpn. J. Appl. Phys. 31, L853-L855 (1992)
4. S. Sakai, Y. Ueta, Y. Terauchi.: Band Gap Energy and Band Lineup of III-V Alloys Semiconductors Incorporating Nitrogen and Boron. Jpn. J. Appl. Phys. 32, 4413-4417 (1993)
5. J. W. Cahn.:On spinodal decomposition. Acta Metall. 9, 795 (1961)
6. G. Stringfellow.: Organometallic Vapor-Phase Epitaxy:Theory and Practice. Academic Press, Boston, ch. 3, 94 (1989)
7. V. A. Elyukhin, M. K. Ebanoidze.: Spinodal decomposition in $A_x^3B_{1-x}^3C_y^5D_{1-y}^5$ solid solution. Russ. J. Phys. Chem. 61, 262 (1987)
8. R. Asomoza, V. A. Elyukhin, R. Peña-Sierra.: Spinodal decomposition range of $In_xGa_{1-x}N_yAs_{1-y}$ alloys. Appl. Phys. Lett. 81, 1785 (2002)
9. R. Martin.: Elastic Properties of ZnS Structure Semiconductors Phys. Rev. B 1, 4005-4011 (1970)

10. P. N. Keating.: Effect of Invariance Requeriments on the Elastic strain Energy of Crystals with Applications to the Diamond Structure Phys. Rev. 145, 637 (1966)

11. J. F. Nye.: Physical Properties of Crystals. Oxford University Press, London, ch. VIII, 131 (1979)

12. J. Y. Tsao.: Materials Fundamentals of Molecular Beam Epitaxy. Academic Press, MA, ch. 4, 107 (1993)

13. J. W. Cahn, J. E. Hilliard.: Free Energy of a Nonuniform System. III. Nucleation in a Two-Component Incompressible Fluid J. Chem. Phys. 31, 688-699 (1959)

14. J. W. Cahn.: The impurity-drag effect in grain boundary motion. Acta Metall. 10, 789-798 (1962)

15. R. Asomoza, V. A. Elyukhin, R. Peña-Sierra.: Spinodal decomposition in the $B_xGa_yIn_{1-x-y}As$ alloys. Appl. Phys. Lett. 78, 2494-2496 (2001)

16. P. Glansdorf and I. Progogine.: Thermodynamic Theory os Structure, Stability and Fluctuations. Wiley, New York, ch.3, 32 (1972)

17. G. A. Korn and T. M. Korn.: Mathematical Handbook for Scientists and Engineers. McGraw-Hill, New York, 420 (1968)

18. J. L. Martins, A. Zunger.: Bond lengths around isovalent impurities in semiconductors solid solutions. Phys. Rev. B 30, 6217-6220 (1984)

19. S. Wicaksono, S. F. Yoon, K. H. Tan, W. K. Loke.: Characterization of small-mismatch GaAsSbN on GaAs grown by solid source molecular beam epitaxy. J. Vac. Sci. Technol. 23, 1054-1059 (2005)

20. V. A. Elyukhin, S. A. Nikishin.: Internal Strain energy of $A_x^3B_{1-x}^3N$ ternary solid solution of cubic modifications. Semicond. Sci. Technol. 11, 917-920 (1996)

21. B. Kurnet, J. Koch, T. Torunski, K. Volz, W. Stolz.: MOVPE growth experiments of the novel (GaIn)(NP)/GaP material system. J. Cryst. Growth 272 , 753-759 (2004)

22. H. D. Sun, et. al.: Investigation of phase-separated electronic states in 1.5 μm GaInNAs/GaAs heterostructures by optical spectroscopy. J. Appl. Phys. 97, 033517/1-033517/5 (2005)

13

Electrical, Relaxation and Crystallization Properties of $Ge_2Sb_2Te_5$ Alloys

E. Morales-Sánchez[1], E. Prokhorov[2,3], G. Trapaga[2], M. A. Hernández-Landaverde[2], and J. González-Hernández[3]

[1] CICATA- IPN, Unidad Querétaro, Qro., México
[2] CINVESTAV del IPN, Unidad Querétaro, Qro., México
[3] CIMAV and Laboratorio Nacional de Nanotecnología, Chihuahua, Chih., México

Summary. In this work the properties of amorphous and fcc polycrystalline $Ge_2Sb_2Te_5$ phases have been investigated as well as the crystallization kinetics.

It has been found that the amorphous $Ge_2Sb_2Te_5$ phase shows two relaxation processes upon annealing. One in a low frequency range which has been associated with an α relaxation and the other a secondary β relaxation in a higher frequency range. One possible explanation of the structural relaxations process in the amorphous films can be associated with a cooperative and local rearrangement of GeTe and/or Sb_2Te_3 clusters or blocks.

In the crystallization process, there is an annealing temperature range for which two phases coexist: nuclei with $Ge_1Sb_4Te_7$ composition and fcc $Ge_2Sb_2Te_5$ phase. It is suggested that the appearance of the nuclei could be related to local fluctuations in the film composition. Electrical properties of the fcc polycrystalline phase strongly depend on the high resistivity at the grain boundaries which controls the current transport in these material. The various resistance states reached by annealing at different temperatures, suggest the possibility of using the $Ge_2Sb_2Te_5$ material as the active layer in a multi-state memory device.

13.1 Introduction

$Ge_2Sb_2Te_5$ chalcogenide semiconductor has been studied extensively due to its applications in optical and electrical memory devices. It is known that upon annealing, amorphous phase of this material first undergo the amorphous-to-crystalline (fcc) transition in the range of 140-180 °C and at higher temperatures (at about 230 °C) the crystal (fcc)-to-crystal (hex) transition.

In recent years, extensive experimental and theoretical studies have been conducted to understand the structure, electrical properties and crystallization phenomena in this material [1, 2, 3, 4, 5, 6, 7]. One of the main concerns in recent investigations is to establish the structure and electrical properties of amorphous phase and the nature of the incubation time (time necessary to form the critical nuclei previous to the crystallization of the amorphous phase). Last parameter is important because it limits the speed of the devices. Additional, from our knowledge, there are no reports in literature related with the thermal relaxation processes in amorphous $Ge_2Sb_2Te_5$

films. The study of the dielectric properties as a function of temperature in the amorphous phase provides information about structure and nature of dielectric loss, which is important for memory application. Phase change optical and electrical data storage is based on the reversible phase transformation in $Ge_2Sb_2Te_5$ films between amorphous and polycrystalline fcc states. The utilization of the two, amorphous (high resistance/low reflectivity) and the polycrystalline (low resistance/high reflectivity) states, allows only the implementation of binary memory devices.

In Refs. [8, 9, 10] have been proposed the concept of multi-state phase change memory. This means that the device can go from any arbitrary state (resistance or reflection) to any other arbitrary recording state with change of programming voltage [9] or optical pulse power [10]. The materials with multiple structural configurations, ranging from completely amorphous to completely crystalline, including a continuum of intermediate structures consisting of a mixture of amorphous and crystalline components have been proposed to explain multi-state electrical phase change memory [8, 9, 10]. So, multi-state memory operation depends on the electrical properties of polycrystalline fcc phase. But the transport and electrical properties of this phase are not yet well known.

The aim of this chapter is to investigate electrical, relaxation and crystallization properties of amorphous and fcc polycrystalline $Ge_2Sb_2Te_5$ phases.

13.2 Methodology

$Ge_2Sb_2Te_5$ ($Ge_22Sb_22Te_56$) thin films with thickness about 200 nm were prepared by vacuum evaporation of the bulk alloy onto unheated glass and Si substrates. The as-deposited films were confirmed to be amorphous by X-ray diffraction (using a Rigaku model Dmax 2100 diffractometer with a Cu tube). The film composition was determined by energy dispersive spectroscopy (Phillips brand Model XL 30 scanning electron microscope with EDS analysis). It was found that deviation from the target composition was not larger than 2 % for every element.

The electrical properties of the films deposited on the glass substrate were obtained from impedance measurements (Agilent Precision Impedance Analyzer 4294A) using two gold sputtering contacts in the frequency range of 40 Hz - 100 MHz and four collinear probe arrays in the frequency range of 40 Hz-300 kHz.

The 4-probe measurements in the low resistivity polycrystalline films allowed obtaining real resistivity values which do not depend on the properties of contacts [7].

DC 4-probe measurements were carried out on as-prepared films using a 485 Keithley amperometer, and the voltage was measured with a 34401A HP bench multimeter with input resistance > 10 GOhm.

The sample was heated using a resistive heater or a Peltier element. The temperature was controlled with a Watlow's Series 982, 1/8 DIN microprocessor-based, with a ramping controller. In isokinetics measurements the temperature controller was programmed to produce a constant heating rate of 3 °C/min. The measurements on the polycrystalline films were carried out on the samples annealed for 10 minutes at a temperature above the amorphous-fcc transition, but lower than the needed to induce the fcc-hexagonal transition.

DSC measurements in amorphous samples obtained by removing the as-deposited films from the glass substrates were carried out under nitrogen atmosphere using a TGA/SDTA 851^e Mettler Toledo apparatus.

13.3 Results and Discussion

13.3.1 Amorphous phase

Fig. 13.1 shows the dependence of conductivity obtained from the DC 4-probe measurements. In amorphous chalcogenide materials the DC conductivity in the high temperature range, when the dominating conduction mechanism is through extended states in the conduction band, can be described by:

$$\sigma = \sigma_0 \exp\left(-\Delta E/kT\right),$$

where the constant σ_0 for the chalcogenide semiconductors depends on composition [11], ΔE is the activation energy for conductance, k the Boltzman constant, and T is the absolute temperature.

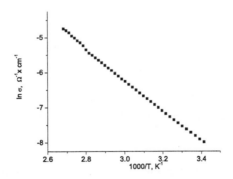

Fig. 13.1. Dependence of conductivity σ versus reciprocal temperature.

The obtained activation energy for conductance, 0.347±026 eV, doesn't depend on films thickness (in the range between 100 nm - 2 mkm) but differ from values reported in the literature [1] for sputtering films (0.45 eV). Such difference is probably related to the difference in the preparation technique.

Fig. 13.2 shows typical impedance spectra obtained in a $Ge_2Sb_2Te_5$ amorphous film (with 2 gold contacts) at a temperature of 80 °C; Fig. 13.3 shows the dependence of the loss coefficient (ε'') versus frequency calculated from impedance measurements (continuous line) for the same sample in Fig.13.2. Two critical issues in these measurements are: the relative high DC conductivity of amorphous phase and the influence of contacts (or polarization effects) in the low frequency range.

For semiconductor and semiinsulator materials complex permittivity contains the contribution of dielectric polarization and conductivity responses. So, DC conductivity strongly modifies the dielectric loss ε'' and all polarization relaxation effects are completely masked. For high conductivity materials, as a rule it is necessary to do the so named DC correction [13], in this case ε'' is written as:

$$\varepsilon'' = \varepsilon''_{exp} - \frac{\sigma_{dc}}{w\varepsilon_0},$$

where ε_{exp} is the experimental loss factor value, σ_{dc} ($\sigma_{dc} = B/R_{dc}$), B, R_{dc} and ε_0 are DC conductivity, geometrical factor, DC resistance and the permittivity of vacuum, respectively.

The value of DC resistance (R_{dc}) has been obtained from the extrapolation of the depressed semicircle in the impedance spectra to the Z' axis as shown in Fig. 13.2.

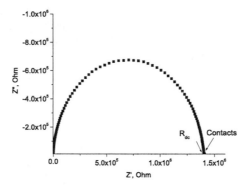

Fig. 13.2. Impedance spectra obtained on Ge$_2$Sb$_2$Te$_5$ amorphous film at a temperature of 80 °C.

Fig. 13.3 shows the dielectric loss vs frequency before and after dc correction (open circles). As can be seen, it clearly shows two relaxation processes separated in frequency. In addition, the contact/polarization effects become obvious in the range of 100-200 Hz, depending on sample and temperature; these effects are not included in the subsequent analysis.

Fig. 13.4 shows the experimental (points) dielectric loss as a function of frequency in amorphous Ge$_2$Sb$_2$Te$_5$ at temperature of 40, 60 and 80°C. The plots of dielectric loss versus frequency have two well resolved peaks in all investigated films: one at low (10^2-10^4 Hz) and the other at higher (10^5-10^7 Hz) frequency range. All peaks shift to higher frequencies with increasing temperature. These results allow fitting to separate the two relaxation processes.

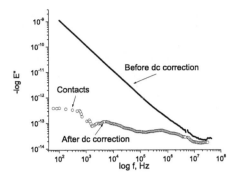

Fig. 13.3. Dependence of the loss coefficient (ε'') versus frequency calculated from impedance measurements.

Dielectric relaxation processes have been analyzed using the Havriliak-Negami (HN) empirical function for complex permittivity ε^* [12]:

$$\varepsilon^* = \varepsilon_\infty + \frac{\varepsilon_s - \varepsilon_\infty}{(1 + (iw\tau)^\alpha)^\beta} \tag{13.1}$$

where ε_s, ε_∞ and τ are dielectric constant at the low frequency limit, dielectric constant at the high frequency limit and relaxation time, respectively. The exponents α and β (or shape parameters) describe the symmetric and asymmetric broadening of the relaxation and $0 < \alpha, \beta < 1$. Results of fitting are shown on Fig. 13.4 as continuous lines.

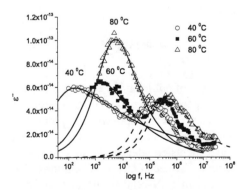

Fig. 13.4. Experimental (points) dielectric loss as a function of frequency in Ge$_2$Sb$_2$Te$_5$ obtained at the temperature 40, 60 and 80 °C. Continuous lines-results of fitting using HN model.

Fig. 13.5 shows the dependence of the relaxation time obtained fitting the model to the experimental data (black points) and results of fitting (continuous lines) using Vogel-Fulcher-Tammann and an Arrhenius equations.

The low frequency process observed in $Ge_2Sb_2Te_5$ amorphous films can be associated with an α relaxation process, the relaxation time follows Vogel-Fulcher-Tammann (VFT) equation, which for relaxation time τ is written as:

$$\tau = \tau_0 \exp\left(\frac{A}{T - T_0}\right),$$

where A is a constant and T_0 denotes the Vogel temperature.

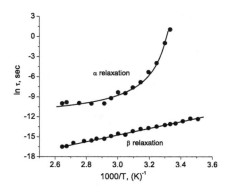

Fig. 13.5. Dependence of the relaxation time versus reciprocal temperature.

The secondary relaxation at high frequency follows an Arrhenius type relation corresponding to a so-called β relaxation:

$$\tau = \tau_0 \exp\left(\frac{E_a}{kT}\right),$$

where E_a is the activation energy.

Despite of the large number of publications about α and secondary β relaxations in different glass-formers materials the nature of these relaxations is not entirely understood. Many models have been proposed to explain these phenomena [13]. In these models an α relaxation is characterized by cooperative rearrangements of molecules, metastable islands, clusters, density fluctuations, etc.

A similar situation occurs for secondary β relaxations. This relaxation can be consider as precursor of the primary α-relaxation, rotational motion of molecules in metastable islands, reorientation motions of essentially all molecules with a nearly temperature-independent small angle, or a small number of molecules in fluctuations [13], etc.

The local amorphous structure of $Ge_2Sb_2Te_5$ are not precisely known yet, but according to extended X-ray absorption fine structure (EXAFS) studies, the amorphous phase is not a random mixture of atoms. EXAFS measurements [3], on as-deposited amorphous $Ge_2Sb_2Te_5$ films suggested that the structure of this material may be viewed as a random array of Ge_2Te_3 and Sb_2Te_3 structural units embedded in a tissue of a-Te, 17% of which is over coordinated. In [4] a structural model is proposed for the amorphous $(GeTe)_n$-$(Sb_2Te_3)_m$ material based on three-dimensional building blocks: a GeTe chain and a stibnite Sb2Te3.

Based on these arguments, one can propose that the relaxations process observed in amorphous $Ge_2Sb_2Te_5$ films can be associated with cooperative and local rearrangements of GeTe and/or Sb_2Te_3 clusters or blocks.

13.3.2 Crystallization

Fig. 13.6 shows results of DSC measurements taken on $Ge_2Sb_2Te_5$ amorphous material obtained by removing the as-deposited films from the glass substrates. Fig. 13.7 presents X-ray diffraction data obtained at room temperature on samples previously heated to the temperature indicated on graph. Upper pattern corresponds to a $Ge_1Sb_4Te_7$ sample annealed at 180 °C.

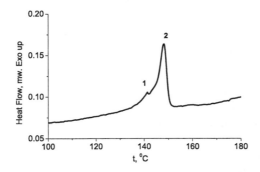

Fig. 13.6. DSC curve obtained for a $Ge_2Sb_2Te_5$ compound with a heating rate of 5 °C/min.

The DSC curve shown in Fig. 13.6 indicates that the crystallization process involves two exothermic reactions which, together with XRD measurements (Fig. 13.7), allow us to reach the following conclusions:

- at temperatures below 100 °C the sample shows only wide bands, which are characteristic of the amorphous material;
- the first exothermic peak in DSC measurements can be associated to the appearance of metastable nucleation centers embedded in the amorphous matrix. The X-ray data indicate that these nuclei have a nanometric size of about 12 nm and a fcc $Ge_1Sb_4Te_7$ composition (sampled annealed at 140 °C);
- the second (large) exothermic peak in the DSC measurements, observed at a higher temperature, corresponds to the process of transformation of these nucleation centers into an fcc stochiometric crystalline phase ($Ge_2Sb_2Te_5$), which

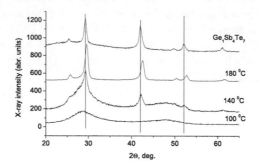

Fig. 13.7. X-ray diffraction patterns for $Ge_2Sb_2Te_5$ samples measured on films heated at temperatures indicated on graph. Upper pattern corresponds to a $Ge_1Sb_4Te_7$ sample annealed at 180 °C.

eventually covers the whole volume of the film (pattern at the sample annealed at 180 °C).

The electrical properties through impedance measurements of a system with more than two components (amorphous phase, nuclei of $Ge_1Sb_4Te_7$ and fcc $Ge_2Sb_2Te_5$ crystalline phase) can be described using the Bruggerman effective medium approximation [14, 15].

$$f_a\frac{\varepsilon_a-\varepsilon}{\varepsilon_a+2\varepsilon}+(1-f_a-f_n)\frac{\varepsilon_c-\varepsilon}{\varepsilon_c+2\varepsilon}+f_n\frac{\varepsilon_n-\varepsilon}{\varepsilon_n+2\varepsilon}=0$$

where ε_a, ε_c and ε_n are the dielectric constants of amorphous, crystalline and nucleation phases respectively; fa, $(1-f_a-f_n)$ and f_n are the corresponding volume fractions and ε is the effective dielectric constant calculated from impedance measurements.

Fig. 13.8 shows the total volume fractions of crystalline (curve 1, black circles) and of nuclei (curve 2, empty squares) as a function of temperature estimated from Bruggerman effective medium approximation [14, 15] through impedance measurements.

One can see that the volume fraction of nuclei measured in samples heated in the temperature range between about 100 °C and 140 °C first increases (Fig. 13.8), then for higher temperatures, when the volume fraction of the $Ge_2Sb_2Te_5$ fcc phase increases, the volume fraction of nuclei, determined from impedance measurements decreases, X-ray indicate that the $Ge_1Sb_4Te_7$ nuclei phase are gradually transformed into the main $Ge_2Sb_2Te_5$ material.

13.3.3 Polycrystalline fcc phase

Fig. 13.9 shows resistance as a function of temperature, obtained from 4- probe DC measurements on as-prepared $Ge_2Sb_2Te_5$ films (continuos line). The points represent

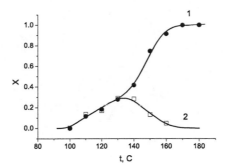

Fig. 13.8. Total volume fraction of crystalline (black circles) and volume fraction of nuclei as a function of temperature estimated from Bruggerman effective medium approximation (empty square).

the resistance obtained from 4-probe impedance measurements on the fcc polycrystalline sample previously annealed at 150 °C during 10 min (onset of crystallization temperature is of about 140 °C). The impedance measurements were taken during heating to a temperature of 170 °C (lower than needed to induce the fcc-hexagonal transition) and then cooled to room temperature.

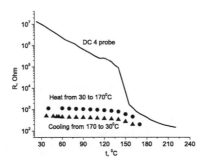

Fig. 13.9. Resistance versus temperature obtained from DC four probe measurements (continuous line). The points represent the resistance obtained from four probe impedance measurements on the fcc crystalline sample previously annealed to 150 °C during 10 min..

Fig. 13.10 shows typical impedance spectra for a $Ge_2Sb_2Te_5$ films measured using 4-probe impedance techniques in samples previously annealed at 170 °C during 10 min. The spectra were measured at the temperatures indicated in the graph and it consist of two well defined semicircles.

Insert in Fig. 13.10 shows the equivalent circuit (consisting of two RC circuit connected in series) used to fit the electrical response of the fcc polycrystalline films

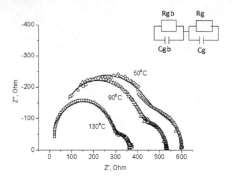

Fig. 13.10. Impedance spectra for a $Ge_2Sb_2Te_5$ film measured using four probe impedance techniques on a sample previously annealed to 170 °C during 10 min. Insert on Fig.13.2 shows the equivalent circuit used to fit the impedance measurements.

using the brick layer model. This model assumes cubic shaped grains separated by flat grain boundaries. The R_gC_g and $R_{gb}C_{gb}$ circuits are associated with the grains and the grain boundaries respectively. The frequency response using the proposed equivalent circuit has been fitted to the experimental 4-probe impedance spectra with help of the ZView program. The results of fitting are shown with the continuous lines on Fig. 13.10.

The brick layer model establishes the following relation between capacitances, dielectric constants of the grain ε_g, dielectric constant of grain boundaries ε_{gb} and the volume fraction of the grain boundaries x_{gb} [15]:

$$x_{gb} = 3 \frac{C_g}{C_{gb}} \bullet \frac{\varepsilon_{gb}}{\varepsilon_g}$$

These relations establish a correlation between microstructure and electrical properties. From electrical impedance measurements the volume fraction of grain boundaries are usually estimated by assuming $\varepsilon_g = \varepsilon_{gb}$ [13]. The volume fraction of grain boundaries x_{gb} calculated for the film is shown in Fig.13.11 (the symbols correspond to the same temperature cycles as in Fig.13.9). The graph shows, that an increase in the annealing temperature results in a reduction in the volume fraction of grain boundaries. The results obtained correlate well with TEM measurements which have shown that with increasing temperature the dimension of grains increases, therefore reducing the volume fraction of grain boundaries as shown in Fig.13.10 and 13.11.

The presence of high resistivity grain boundaries (as have been determined from 4-probe impedance measurements) has a strong influence on the current transport in the polycrystalline films. The grain boundaries formed by disordered atoms contain change trapping centers, which lead to an inter-grain band bending and potential barriers [17]. A conduction model, proposing a thermionic emission over the potential barriers formed at the grain boundaries have been applied to polycrystalline materials [17, 18, 19, 20].

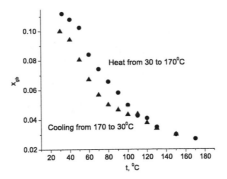

Fig. 13.11. The volume fraction of grain boundaries x_{gb} calculated for the film shown on Fig.9 (the symbols correspond to the same temperature cycles as in Fig.13.9).

According to ref. [20] for polycrystalline films, if the transport mechanism is dominated by the thermionic emission, the conductivity σ can be expressed as:

$$\sigma = Lq^2 p \left(\frac{1}{2\pi m^* kT}\right)^{\frac{1}{2}} \exp\left(-\frac{E_\sigma}{kT}\right),$$

where $L, q, p, m*, k$ and E_σ are, respectively, the average grain size, the electron charge, the carrier concentration, the effective mass, the Boltzmann constant and the potential barrier height at the grain boundaries.

According to these equations the plot $\ln(\sigma T^{\frac{1}{2}})$ versus T^{-1} for polycrystalline samples must show a linear dependence in which the slope determines the potential barrier height. Fig. 13.12 shows the corresponding experimental values of $\ln(\sigma T^{\frac{1}{2}})$ as a function of $1000/T$ for polycrystalline Ge$_2$Sb$_2$Te$_5$ film (previously annealed at 170 °C during 10 min.), results show a linear behavior in the measured temperature range with a potential barrier height at the grain boundaries equal 0.127 eV. It should be noted that, similar to Ge$_1$Sb$_2$Te$_4$ film [21] the value of the potential barrier height at the grain boundaries in Ge$_2$Sb$_2$Te$_5$ depends on the annealing temperature.

As can be observed, the room temperature resistivity values in Ge$_2$Sb$_2$Te$_5$ films presented in this work strongly depend on the thermal history of the film (Fig.13.8). For example, room temperature resistance in the film annealed to the temperature of 150 °C change from 1140 Ohm to the value 500 Ohm in the film annealed to 170°C.

The behavior of the room temperature resistance of annealed polycrystalline film at various temperatures increases the conductive dynamic range of the material, which has a potential use in a multi-state memory device. The idea proposed in [22] regarding the accumulation energy model to explain a multi-state transformation could be applied to polycrystalline structures with multiple resistivity states. According to the results obtained in this work the structure and electrical properties of the fcc phase are temperature dependent and any state could be "remembered" from the previous one, this opens the possibility of multi-stage transformations.

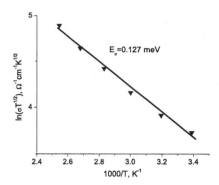

Fig. 13.12. The experimental dependencies of the $\ln(\sigma T^{\frac{1}{2}})$ as a function of $1000/T$ obtained on the polycrystalline $Ge_2Sb_2Te_5$ samples.

13.4 Conclusions

The present results show that the DC conductivity mechanism in the amorphous phase can be associated with conduction through extended states in the conduction band. Amorphous $Ge_2Sb_2Te_5$ phase shows two relaxation processes. One, in the low frequency range, having a non linear temperature dependence and therefore described by Vogel-Fulcher-Tammann equation associated with an α relaxation and a secondary β in a higher frequency range (10^5-10^7 Hz) which has an Arrhenius type dependence. One possible explanation of the structural relaxations process in the amorphous films can be associated with a cooperative and local rearrangement of GeTe and/or Sb_2Te_3 clusters or blocks.

In the crystallization process, there is an annealing temperature range for which two phases coexist: nuclei with $Ge_1Sb_4Te_7$ composition and fcc $Ge_2Sb_2Te_5$ phase. The appearance of the nuclei could be related to local fluctuations in the film composition, where small amorphous regions could have a composition closer to that of the $Ge_1Sb_4Te_7$ with a lower crystallization temperature than that of the $Ge_2Sb_2Te_5$ composition.

Properties of fcc phase strongly depend on the annealing temperature. The presences of high resistivity grain boundaries are responsible for the appearance of potential barriers and conduction can be described by a conduction model which includes thermionic emission over the potential barriers. Additional to the 3 orders of magnitude between amorphous and crystalline states it possible to increase the dynamic range of multi-state memory due to the variation in the resistivity in the crystalline state upon annealing, which makes this material a good candidate for this application.

13.5 Acknowledgements

This work was partially supported by CONACYT of Mexico. Authors are grateful to J.A. Muñoz-Salas for technical assistance in electrical measurements.

References

1. Kato, T., Tanaka K.: Electronic properties of amorphous and crystalline $Ge_2Sb_2Te_5$ films. Jpn. J. Appl. Phys. 44, 7340-7344 (2005)
2. Baily, S.A., Emin, D., Li H.: Hall mobility of amorphous $Ge_2Sb_2Te_5$. Sol. St. Com. 139, 161-164 (2006)
3. Paesler, M.A., Baker, D.A., Lucovsky, G., Edwards, A.E, Taylor, P.C.: EXAFS study of local order in the amorphous chalcogenide semiconductor $Ge_2Sb_2Te_5$. J. Phys Chem. Sol. 68, 873-877 (2007)
4. Im, J., Eom, J.H., Park, C., Park, K., Suh, D.S., Kim, K., Kang, Y.S., Kim, C., Lee, T.Y., Khang, Y., Yoon, Y.G., Ihm, J.: Hierarchical structure and phase transition of (GeTe)n(Sb2Te3)m used for phase-change memory. Phys. Rev. 78, 205205 (2008)
5. Claudio, D., Gonzalez-Hernandez, J., Licea, O., Laine, B., Prokhorov, E., Trapaga, G. : An analytical model to represent crystallization kinetics in materials with metastable phase formation. J. Non-Cryst. Sol. 352, 51-55 (2006)
6. Weidenhof, V., Friedrich, I., Ziegler, S., Wuttig, M.: Laser induced crystallization of amorphous $Ge_2Sb_2Te_5$ films. J. Appl. Phys. 89: 3168-3176 (2001)
7. Morales-Snchez, E., Prokhorov, E., Mendoza-Galvn, A., Gonzlez-Hernndez, J.: Electrical characterization of sputtered GeSbTe films using impedance measurements. Vacuum 68, 361-364 (2003)
8. Shi, L.P., Chong, T.C., Tan, P.K., Miao, M.S., Huang Y.M., Zhao, R.: Study of the partial crystallization properties of phase-change optical recording disks. Jpn. J. Appl. Phys. 38, 1645-1648 (1999)
9. Ovshinsky, S.R.: Optical Cognitive Information Processing - A New Field. Jpn. J. Appl. Phys. 43, 4695-4699 (2005).
10. Ovshinsky, S. R., Pashmakov, B.: Innovation providing new multiple functions in phase-change materials to achieve cognitive computing. Mat. Res. Soc. Symp. Proc. 803, HH1.1.1 (2004)
11. Mott, N. F., Davis, E. A.: Electronic processes in non-crystalline materials. Oxford: Clarendon Press (1979).
12. Raju, G.G.: Dielectrics in electrical fields. Marcel Dekker Inc (2003)
13. Ojovan, M. I.: Configurons: Thermodynamic parameters and symmetry changes at glass transition. Entropy 10, 334-364 (2008)
14. Landauer, R.: Electrical conductivity in inhomogeneous media. AIP Conf. Proc. 40, 2- 41 (1978)
15. Bergman, D.J.: Bulk effective moduli: their calculation and usage for describing physical properties of composite medium. MRS Symp. Proc. 195, 247-56 (1990)
16. MacDonald, J. R.: Impedance Spectroscopy: Emphasizing Solid Materials and Systems. John Wiley and Sons: New York (1987)
17. Orton, J. W., Powell, M. J.: The hall effect in polycrystalline and powdered semiconductors. Rep. Prog. Phys. 43, 1263-1307 (1980).
18. Petritz, R.: Theory of photoconductivity in semiconductor films. Phys. Rev. 104, 1508-1516 (1956).

19. Seto, J. Y. W.: The electrical properties of polycrystalline silicon films J. Appl. Phys. 46, 5247-5254 (1975)
20. Baccarani, G., Ricco B., Spadini, G.: Transport properties of polycrystalline silicon films. J Appl Phys. 49, 5565-5570 (1978).
21. Prokhorov,E., Trapaga, G., Gonzlez-Hernndez, J.: Structural and electrical properties of $Ge_1Sb_2Te_4$ face centered cubic phase. J. Appl. Phys. 104, 103712 (2008)
22. Wright, C. D., Blyuss K., Ashwin, P.: Master-equation approach to understanding multi-state phase-change memories and processors. Appl. Phys. Lett. 90, 063113 (2007)

Part III

Biomedical Engineering, Circuits and Communication Systems

Suitability of Alternative Methods of Time Delay Measurements for Ultrasonic Noninvasive Temperature Estimation in Oncology Hyperthermia

Arturo Vera[1], Lorenzo Leija[1], Abraham Tellez[1], Ivonne Bazán[1], and Antonio Ramos[2]

[1] Centro de Investigación y de Estudios Avanzados del Instituto Politécnico Nacional, Departamento de Ingeniería Eléctrica, México, D.F., 07300, México
[2] Instituto de Acústica, Departamento de Señales, Sistemas y Tecnologías Ultrasónicas. Consejo Superior de Investigaciones Científicas, Madrid, España

Summary. Noninvasive thermometry is one of the greatest challenges for hyperthermia treatment and ultrasound thermometry is one of the most attractive techniques used for this purpose. In this chapter, four time-delay estimation methods, which relate temperature changes inside tissue replica (phantom) with time shifts in ultrasound echo-signals, were tested and their results were comparatively evaluated. These methods were tested, for temperatures ranging from 30 °C to 50 °C and for both experimental and simulated echo-signals. Simulated signals were obtained from a numerical phantom which is proposed in this chapter, and experimental signals were acquired from a specially developed agar phantom with uniformly distributed scatterers. Regression coefficients obtained for each method and percentage errors are presented. In order to select the best method for delay estimation in hyperthermia context, a multi-index criterion for evaluation and comparison, which considers a) their suitability for simulated signals and for experimental signals, b) their associated processing time and c) their robustness under noisy conditions, is also proposed.

14.1 Introduction

14.1.1 Oncology hyperthermia and noninvasive temperature estimation: Generalities

Oncology hyperthermia has become a standard clinical procedure for cancer treatment. Its effects are obtained by increasing temperature in the tumor target up to 42-45 °C with the use of ultrasonic or electromagnetic radiation. It has been demonstrated extensively that cell heating induces conformational and metabolic changes. These changes lead to alterations of the microenvironment in tumors and have an impact on cellular death induced by heat. Hyperthermia therapy can act either as a cytotoxic agent on its own or as a sensitizing agent in combination with ionizing radiation or cytotoxic drugs (Dewhirst et al. [6]).

One of the advantages of hyperthermia is the noninvasive manner in which heat is delivered to tumor. This advantage is partially offset by the use of conventional temperature sensors employed for treatment feedback and control; most of the times, these sensors are invasive probes that are inserted in tumor. Only a limited number of measurements can be taken with conventional sensors; this may result in the acquisition of less data than needed in order to produce satisfactory temperature distribution for properly evaluating the thermal dosimetry. Furthermore, conventional temperature sensors are uncomfortable for patients and there is a risk of seeding tumor cells along the probe insertion track.

It is clear that an accurate and noninvasive temperature measurement inside tumors or their artificial replicas (phantoms) is important since this temperature measurement allows determining the success of the therapy and assessing the performance of new hyperthermia applicators. Therefore, one of the greatest challenges for hyperthermia is to find a way to measure internal temperatures noninvasively and within the whole tumor volume. Many researchers have explored the feasibility of using different techniques to achieve this goal.

Nuclear magnetic resonance (Bertsch et al. [2], de Poorter et al. [5]), microwaves (Leroy et al. [11], Hand et al. [8]), impedance tomography (Paulsen et al. [13]) and ultrasound are among the noninvasive temperature estimation methods that have been studied by many researchers and have yielded varying degrees of success. Ultrasound is one of the most interesting and promising alternatives for this purpose. Some of its advantages include real-time measurement and recording, significant penetration depth in tissues, potentially good temporal and spatial resolutions if a pulsed regime is applied, no appearance of undesirable interactions with the treatment, affordability and, finally, relative simplicity in signal processing requirements.

Ultrasound noninvasive temperature estimation exploits different techniques as (1) time domain analysis to estimate the echo shifts in ultrasonic signals due to tissue thermal expansion and changes in the speed of sound (Simon et al. [19], Maass-Moreno et al. [12], Seip et al. [17]), (2) frequency domain techniques that analyze the relationship between temperature and frequency components of the RF ultrasonic signal (Amini et al. [1]), and (3) changes in backscattered energy from tissue inhomogeneities (Seip and Ebbini [16]). The detection of echo shifts has received the most attention over the last decade.

This chapter reports a suitability study of four techniques by using the time delay estimation: the correlation, the phase related (phase shift and phase difference) and the demodulation methods. The temperature-dependent delay, δ, was estimated from the signals $y_1 = (z, t_1, T)$ and $y_2 = (z, t_2, T + \Delta T)$. These echo signals were acquired successively at times t1 and t2, and at temperatures T and T+ΔT.

14.1.2 Time delay temperature estimation methods

It is possible to determine the temperature inside tissues or phantoms by tracking local displacement of ultrasonic signals. In this approach, the local temperature of tissue can be estimated by tracking the velocity of the wave as it propagates through tissue; this calculation is based on the principle that the speed of sound propagation in a medium is directly related to the mechanical properties of that medium and to temperature changes. These time shift displacements are small, in the order of several nanoseconds.

Several methods for time delay estimation are described in literature; although they are not aimed directly at quantitative assessment of temperature measurement. These methods determine the velocity of the wave inside the media by using high frequency ultrasound. Wilson and Robinson ([23]) used a phase method to determine human tissue characteristics; Tristam et al. ([20]) developed a correlation method for determining kinetic characteristics of human tissues, and de Jong et al. ([3]) introduced a correlation interpolation of pulsed ultrasonic signal for determination of tissue motion velocity. In this research, these methods were applied and tested for the specific application of noninvasive estimation of temperature. Furthermore, the performance of four time delay estimation methods is discussed in this chapter. These methods were developed for specific frequency and resolutions. In this study, time delays up to 121 ns were analyzed at a frequency of 2.25 MHz (bandwidth @ -3dB=51.59 %, and with observation windows of 4.4 μs).

14.2 Time Delay Determination and Experimental Methodology

In this study, a set of R.F. signals was collected from the same region of interest in a phantom sample. Each R.F. pulsed signal consisted of multiple reflected ultrasonic echoes, and could thus be considered as a replica of the original pulse that was only shifted in echo depth and thus in time of flight. When the signal from phantom is in the initial position, $y_1(t)$ is collected. When the phantom is heated and the signal is a shifted replica of $y_1(t)$, i.e. $y_2(t) = y_1(t - \delta t)$, $y_2(t)$ is collected (Fig. 14.1).

14.2.1 Phase related procedures: phase-difference and phase-shift based

The instantaneous phase $\varphi(x,t)$ of a sinusoidal traveling wave $y(t) = a\sin(\omega t - kx)$ is simply defined as the value of $(\omega t - kx)$, expressed in module 2π. The unwrapped phase of a signal is an increasing value with the echo depth. So, when the phase signals of both echo signals (each one at a different temperature) are calculated, the time shift can be determined unambiguously. The phase signal can be calculated using the analytic signal (Wilson and Robinson [23]). The analytic signal, from a real band-limited signal, is defined as a complex signal in which the real part is the original signal and the Hilbert transform, or quadrature function, is the imaginary part. The analytic signal corresponding to $y(t)$ is:

$$s(t) = y(t) + iy_{Hi}(t), \tag{14.1}$$

where y_{Hi} denotes the Hilbert transform of $y(t)$, defined as:

$$y_{Hi}(t) = \frac{1}{\pi} \int_{-\infty}^{+\infty} \frac{y(\theta)}{t - \theta} d\theta. \tag{14.2}$$

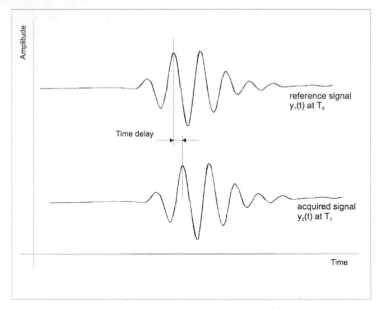

Fig. 14.1. RF signal $y_1(t)$ and its shifted replica, $y_2(t)$, at temperatures T_0 and T_1, respectively.

In practice, the analytic signal of $y(t)$ is computed by calculating the Fourier transform of $y(t)$, zeroing negative frequencies, and then the inverse Fourier transform is obtained. The absolute value of $|s(t)|$ the analytic signal represents the envelope of the original signal, and the phase of the original signal $y(t)$ is defined by:

$$\varphi(t) = \arctan(\frac{-y_{Hi}(t)}{y(t)}). \qquad (14.3)$$

The time shift between two signals can be determined using the two phase signals changes: phase difference and the phase shift methods, Fig. 14.2.

The phase signal is restricted to the range $-\pi$ to π radians, and must be unwrapped before further processing. Many phase-unwrapping algorithms have been developed, mainly to successfully unwrap phases in the vicinity of z-planes zeroes. It is important to mention that a displacement of the ultrasonic signal corresponding to a phase change of more than 2π will cause the temperature to be underestimated. To overcome this problem, for this study, a leap-phase detection algorithm was implemented.

Procedure for phase difference. A relative simple calculation can be performed when the phases of the echo signals are linear with the echo depth. In this case, the difference in phase represents the time shift between the echo signals. After the two echo signals have been filtered to increase the linearity between phase and echo depth (bandpass: 2.3 - 2.4 MHz; 5th order Butterworth filter), the time shift is determined

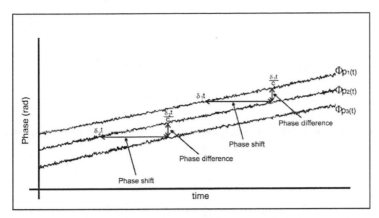

Fig. 14.2. Estimation of the time delay by the phase difference and the phase shift methods.

by multiplying the subtraction of the two signal phases by the coefficient r, which represents the relationship between phase difference and time shift:

$$\delta(t) = r(\varphi_1(t) - \varphi_2(t)). \tag{14.4}$$

Procedure for Phase shift. In this case, the time shift of two unwrapped phase signals can be unambiguously determined because the phase signal is an increasing signal in echo depth. In this method, the echo signals are filtered. The time shift is calculated by applying two-dimensional interpolation of the phase values and points in echo depth of the second phase signal to the phase values of the first phase signal. In this way, the points in echo depth of the second signal are changed, and the time shift between the two signals can be determined by subtracting the two echo depth values for equal phase values (Wilson and Robinson [23]).

14.2.2 Procedure based on Cross-correlation

The classical cross-correlation technique calculates the correlation coefficient between two signals while one is shifted with respect to the other. The shift with maximal value of the correlation coefficient represents the average time shift between the two signals. This method has, in the most basic approach, a time resolution of half the sampling period. A sub-sampling method can be used when a higher resolution is needed.

If the relationship between time shift and the correlation coefficient is known, the time, and thus temperature, can be estimated by calculating the correlation coefficient; this study takes advantage of this feature. In this method, the cross-correlation coefficient function R is defined as:

$$R_{y_1 y_2}(t, \delta t) = \frac{\sum_{t=1}^{N}[y_1(t) - \bar{y}_1(t)][y_2(t) - \bar{y}_2(t)]}{N\sigma_{y_1}(t)\sigma_{y_2}(t)}, \tag{14.5}$$

where, y_1 and y_2, i = 1 to N, are the sampled amplitudes of the ultrasonic signals consisting of N points; $\bar{y}_1(t), \bar{y}_2(t)$ and $\sigma_y(t)$ represent the mean and standard deviation (SD) of the y(t) ultrasonic signals (Tristam et al. [20]). The correlation can be taken and determined as a function that relates the correlation coefficient to the time delay.

Since R has a maximum normalized value of 1 when temperatures are identical and decreases monotonically as the rate of temperature increases, the monotonic dependence of the correlation coefficient on the rate of temperature suggests that R would be a useful tool to measure temperature.

14.2.3 Procedure based on Demodulation

The demodulation method is based on the classical Doppler technique, which estimates the phase of an R.F. continuous wave signal with respect to a reference signal by using demodulation and bandpass filtering (Hoeks et al [9]). The quadrature demodulation procedure was used in this study. When this option is used to estimate the time shift between two signals that are composed of a set of overlapping pulses, the results will be disturbed. To overcome this problem, a modified version of it, in which the demodulation terms were rearranged to calculate the time shift between the signals, was tested.

The objective of the quadrature demodulation is to preserve the real and imaginary component differences of the ultrasonic signals, and this is achieved by coherently demodulating the returning echo signal both with a master oscillator and with the signal derived from the master oscillator but shifted in phase, Fig. 14.3. The original and the shifted version of the ultrasonic signals are modeled, respectively, as:

$$p(t) = A(t)\cos(\omega_L t + \varphi), \tag{14.6}$$

$$p(t + \delta t) = A(t + \delta t)\cos(\omega_L(t + \delta t) + \varphi). \tag{14.7}$$

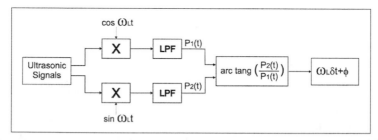

Fig. 14.3. Block diagram illustrating the quadrature demodulation of ultrasonic signals.

The emitted and the received ultrasonic signals are multiplied by a sine and cosine wave with angular frequency ω_L, and then the high frequency terms are rejected.

In practice, these terms are eliminated by filtering; for this study a 5th order Butterworth filter was used. Then, after a mathematical analysis, the phase of the demodulated signals can be determined with:

$$\arctan\left[\frac{\sin(\omega_L \delta t + \varphi)}{\cos(\omega_L \delta t + \varphi)}\right] = \omega_L \delta t + \varphi. \qquad (14.8)$$

When this operation is applied to the reference and shifted version signal, the time shift can be determined with:

$$\frac{1}{2}[(\varphi + \omega_L \delta t) - (\varphi - \omega_L \delta t)] = \omega_L \delta t. \qquad (14.9)$$

14.2.4 Phantom Numerical Simulation

The methods mentioned above were tested in simulated signals, generated from an algorithm implemented specifically for this thermometric purpose, and in experimental signals obtained from scattered phantoms. For the numerical phantom simulations performed for this chapter, the biological tissue was considered as a semi-regular lattice of scatterers separated by an average distance d, as suggested by other authors (Weng et al. [22], Kuc [10]). The echo signal coming from the tissue was the complex sum of echoes coming from every scatterer. A basic model of this phenomenon was taken for the analysis of time shift due to changes in temperature. It was considered that the echoes kept the same shape of the transmitted ultrasonic signals, but they changed their amplitudes according to depth. It is important to mention that in this model no important dispersion, associated to frequency in attenuation phenomenon, was considered. In addition, punctual reflectors and far-field conditions were considered so that diffraction effects due to the transducer or reflectors were neglected (Shankar[18]). The simulated signals were pulses constructed with a 2.25 MHz central frequency wave and a Gaussian envelope, which resulted in a 2.25 MHz pulse with a -6 dB bandwidth from 1.85 to 2.65 MHz:

$$p(t) = \frac{-t}{e^{4B^2 t^2}} \sin(2\pi f_c t), \qquad (14.10)$$

where f_c is the central frequency, $f_c = 2.25 MHz$, B is the bandwidth, and t is time.

According to the model proposed for this work, when the ultrasonic pulse $p(t)$ propagated inside a simulated tissue containing regularly spaced scatterers, the received signal, $r(t)$ in the time domain, could be considered as the complex sum of the echoes produced by each scatterer, then:

$$r(t) = \sum_{k=1}^{N} a_k p(t - (2x_k I c)), \qquad (14.11)$$

where a_k is the echo amplitude of the signal coming from the kth scatterer, t is time, x_k is the position of the scatterer, and c is the speed of ultrasound in the media. The simulated echo signals were shifted considering a temperature range from 30 °C to 50 °C, with increments of 2 °C, and a sampling rate of 2.5 GS/s.

14.2.5 Experimental setup

The experimental setup for acquiring all the measured data is depicted in Fig. 14.4. A custom-made pulser (Ramos et al [14], [15]) excited an unfocused piston transducer (Mod. 12C-0204-S, Harisonic, USA) having 6.35 mm in radius, 2.25 MHz of central frequency and 1.649 - 3.259 MHz of bandwidth @-6dB. A high voltage and short duration pulse, 300 V and 50 ns respectively, were selected for transducer driving in broadband conditions. The transducer was immersed in degassed and distilled water with a phantom placed in front of it. The ultrasound transducer was mounted on a firm support so that the distance between the face of the transducer and the phantom was constant during experiments. The measurements were made away of the irregular zone in the near-field, thus the transducer was placed at 24.2 mm from the surface of the phantom, water mediating. An aluminum reflector was placed at the bottom of the phantom in order to obtain differentiated echoes. The R.F. echo signals were received by the same transducer and were amplified (gain 40 dB) by a custom made amplifier. The R.F. signals were digitalized by a digital oscilloscope (Wave Runner 6000A, LeCroy, USA) at a sampling frequency of 2.5 GS/s. Each echo-signal trace consisted of 77784 data points. The R.F. data were stored in a computer for off-line processing.

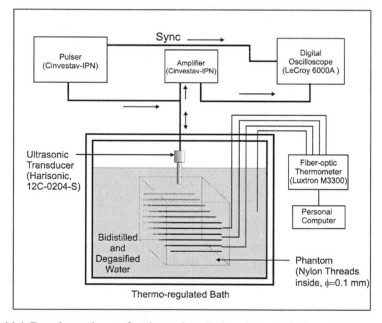

Fig. 14.4. Experimental setup for ultrasonic radiation, data acquisitions, and processing.

Echo-signal ultrasonic data were acquired, as a function of time, while the phantom was being heated. To ensure that the ultrasonic measurements were carried out for sufficiently accurate temperature values, the phantom was thermostatically stabilized to the different temperatures by using a proportional-integral-derivative (PID) controlled thermo-regulated bath (CD10-P5, Haake, Germany) with distilled and degassed water. The temperature inside the phantom was monitored with an optical fiber thermometer (Luxtron M3300, USA). This thermometer has a + 0.5 °C temperature accuracy, 0.01 °C in resolution, and a short-term drift ± 0.2°C/h. Its probes have a diameter of 0.5 mm. Four optic fibers were used, three inside the phantom (placed at 19.8 mm, 29.8 mm and 39.8 mm) and another one inside the water bath. Temperature measurements were taken every second. All experiments were performed under the same conditions. The interval of interest ranged between 30 °C and 50 °C with increments of 2 °C. The time allowed between two measurements was sufficiently long (20 min) to ensure the complete thermo-stabilization inside the phantom. Each experiment included the generation and acquisition of five successively echo-signal recordings in order to perform averaged measurements.

14.2.6 Phantom construction and characteristics

A soft-tissue equivalent phantom was used for obtaining experimental signals. It was based on a mixture of agar-agar (2.5 %), extra-pure glycerol (5%), and distilled degassed water (92.5 %). This phantom simulated the acoustical properties of soft tissue. The ultrasound velocity in the phantom was measured by the pulse-echo technique and was 1540 m/s. The procedure to prepare the phantom is described in Vazquez et al. ([21]). The phantom shape was a cube of 125 cm^3.

A regular scatterer lattice distribution was simulated by using 0.1 mm nylon threads. The threads were distributed in seven layers, uniformly spaced with a distance equal to 0.5 cm; each layer contained seven threads separated uniformly by 0.5 cm too. It is important to mention that the scatterer density of threads has only a minor influence on the estimation of the time shift due to temperature.

The time shifts between the two echo-signals here considered for analysis were achieved by changing the temperature in the phantom. The time shifts were obtained by acquiring the reflection signals from the nylon threads of the emitted ultrasonic pulse.

14.3 Results obtained for the four time-delay measurement methods

The performance of the measurement methods was tested on simulated R.F. signals and on R.F. data acquired from the phantom. After acquiring and simulating the ultrasonic signals, they were processed as mentioned in section 2 for the four methods: phase difference, phase shift, cross-correlation and demodulation. The experimental and simulated signals were divided in seven windows for their analyses. Each window represented a layer of the phantom which contains nylon threads, so it gave information about the temperature in that layer (scatterer depth). Each signal was taken at a precise temperature and the successive acquisitions were performed by incrementing the temperature approximately 2 °C in each step.

Concerning the phase-based methods, the signal processing consisted in computing magnitude and phase of the analytic signal of each window of the ultrasonic signal. The phase signal was restricted to the range $-\pi$ to π radians, and was unwrapped before further processing. For the phase difference method, the time shift was determined by multiplying the subtraction of the two-phase signals by a coefficient r according eq. 14.4. For the cross-correlation method, the coefficient R, as defined in eq. 14.5, was calculated between the windows, both consisting of 11112 samples and corresponding to temperatures with a $\Delta T= 2°C$.

Both, linear and quadratic regression coefficients for each method and for each window (scatterer depth) were calculated. The regression coefficients were calculated for the experimental and simulated signals. Tables 14.1-14.4 show the results for the phase-based methods: difference and shift. Tables 14.5-14.6 provide the coefficients for the cross-correlation method and, finally, Tables 14.7-14.8 give the results for the demodulation method. It is important to mention that the following parameters were considered in the simulated test data: scatterer density, time shift between the traces, and a step increase in time shift as function of echo depth.

Figures from 14.5 to 14.8 show the signal delays as a function of the measured temperature in the experimental data and for the four delay time methods in the first scatterer depth. A similar behavior was observed for the other depths. A reference time-delay was graphed so one can appreciate the differences between the reference and the calculated delays. Reference time delays were directly measured in the digitalized signals by using a very-high sampling resolution (0.4 ns). Fig. 14.9 shows the delays as a function of the measured temperature for the simulated signals and for the first scatterer depth. The delay differences among simulated, experimental and reference signals are shown in Table 14.9.

14.4 Discussion and Conclusions

In this chapter, time-delay based methods have been analyzed and applied to estimate temperature in a phantom, and the respective results were compared. These methods constitute a useful tool for the estimation of temperature increments inside tissue or phantoms during hyperthermia treatment. Each method has certain relative advantages and disadvantages that are discussed in this section.

Tables 14.1 to 14.8 show that the relation between time delay and temperature are almost linear in the range of temperature analyzed in this chapter (30 °C-50 °C). Delay increases appear to be proportional with depth; thus when using time delay techniques to estimate temperature with spatial resolution, it is necessary to introduce a correction factor to compensate accumulative effects on time delay due to the successive increments with depth. This aspect must be considered to quantify the spatial and temperature resolutions. Another consequence of this fact is that better resolutions could be obtained with ultrasonic echoes coming from the deepest layers of the phantom.

Table 14.1. Regression coefficients for data obtained from simulated signals using the phase-difference technique.

(a) Delay (δ) versus temperature (T), $\delta(T) = \delta_{T0} + \delta_{T1} \cdot T + \delta_{T2} \cdot T^2$				
	Scatterer depth (mm)	$\delta_{T0}(ns)$	$\delta_{T1}(ns/^\circ C)$	$\delta_{T2}(ns/^\circ C^2)$	R^2
Linear	28.5	1596.16	-54.183		0.99
	33.5	1633.64	-55.715		0.99
	38.5	1672.45	-57.316		0.99
	43.5	1710.80	-58.904		0.99
	48.5	1748.76	-60.479		0.99
	53.5	1786.36	-62.043		0.98
	58.5	1823.56	-63.593		0.98
Quadratic	28.5	2600.34	-105.68	0.64370910	0.99
	33.5	2925.40	-121.95	0.82805140	0.99
	38.5	3268.97	-139.18	1.02340900	0.99
	43.5	3611.26	-156.36	1.21824100	0.99
	48.5	3952.05	-173.46	1.41236600	0.99
	53.5	4290.78	-190.47	1.60539900	0.99
	58.5	4626.69	-207.34	1.79687700	0.99

(a) Temperature (T) versus delay (δ), $T(\delta) = T_{\delta 0} + T_{\delta 1} \cdot \delta + T_{\delta 2} \cdot \delta^2$				
	Scatterer depth (mm)	$T_{\delta 0}(ns)$	$T_{\delta 1}(ns/^\circ C)$	$T_{\delta 2}(ns/^\circ C^2)$	R^2
Linear	28.5	29.50	-0.01837		0.99
	33.5	29.39	-0.01781		0.99
	38.5	29.29	-0.01726		0.99
	43.5	29.19	-0.01673		0.99
	48.5	29.11	-0.01624		0.99
	53.5	29.03	-0.01576		0.98
	58.5	28.96	-0.01531		0.98
Quadratic	28.5	30.21	-0.01400	0.000004024	0.99
	33.5	30.27	-0.01252	0.000004748	0.99
	38.5	30.33	-0.01110	0.000005367	0.99
	43.5	30.39	-0.00982	0.000005853	0.99
	48.5	30.45	-0.00867	0.000006230	0.99
	53.5	30.50	-0.00764	0.000006513	0.99
	58.5	30.56	-0.00671	0.000006719	0.99

Table 14.2. Regression coefficients for data obtained from experimental signals using the phase-difference technique.

(a) Delay (δ) versus temperature (T), $\delta(T) = \delta_{T0} + \delta_{T1} \cdot T + \delta_{T2} \cdot T^2$

	Scatterer depth (mm)	$\delta_{T0}(ns)$	$\delta_{T1}(ns/°C)$	$\delta_{T2}(ns/°C^2)$	R^2
Linear	28.5	1496.96	-50.0399		0.99
	33.5	1646.64	-56.0147		0.99
	38.5	1892.13	-64.3041		0.99
	43.5	2019.74	-68.8852		0.99
	48.5	2073.77	-71.2346		0.99
	53.5	2250.59	-77.1172		0.99
	58.5	2350.65	-80.4225		0.99
Quadratic	28.5	2126.81	-81.7442	0.38905510	0.99
	33.5	3230.10	-136.441	0.99653010	0.99
	38.5	3349.64	-138.591	0.92340400	0.99
	43.5	3590.27	-149.200	1.00139300	0.99
	48.5	4428.69	-191.512	1.49754400	0.99
	53.5	4365.83	-184.923	1.34090200	0.99
	58.5	4483.71	-189.269	1.35375700	0.99

(a) Temperature (T) versus delay (δ), $T(\delta) = T_{\delta0} + T_{\delta1} \cdot \delta + T_{\delta2} \cdot \delta^2$

	Scatterer depth (mm)	$T_{\delta0}(ns)$	$T_{\delta1}(ns/°C)$	$T_{\delta2}(ns/°C^2)$	R^2
Linear	28.5	29.96	-0.01989		0.99
	33.5	29.51	-0.01765		0.99
	38.5	29.50	-0.01543		0.99
	43.5	29.40	-0.01440		0.99
	48.5	29.27	-0.01382		0.99
	53.5	29.30	-0.01282		0.99
	58.5	29.33	-0.01231		0.99
Quadratic	28.5	30.52	-0.01641	0.000003317	0.99
	33.5	30.61	-0.01135	0.000005526	0.99
	38.5	29.50	-0.01089	0.000003477	0.99
	43.5	30.30	-0.01013	0.000003062	0.99
	48.5	30.48	-0.00824	0.000003855	0.99
	53.5	30.37	-0.00833	0.000002846	0.99
	58.5	30.37	-0.00812	0.000002556	0.99

Table 14.3. Regression coefficients for data obtained from simulated signals using the phase-shift technique.

(a) Delay (δ) versus temperature (T), $\delta(T) = \delta_{T0} + \delta_{T1} \cdot T + \delta_{T2} \cdot T^2$

	Scatterer depth (mm)	$\delta_{T0}(ns)$	$\delta_{T1}(ns/^\circ C)$	$\delta_{T2}(ns/^\circ C^2)$	R^2
Linear	28.5	1631.56	-55.175		0.99
	33.5	1675.52	-56.890		0.99
	38.5	1720.87	-58.680		0.99
	43.5	1766.54	-60.475		0.99
	48.5	1812.25	-62.275		0.99
	53.5	1857.38	-64.059		0.99
	58.5	1902.94	-65.851		0.99
Quadratic	28.5	2377.29	-93.417	0.47803030	0.99
	33.5	2674.16	-108.10	0.64015150	0.99
	38.5	2986.96	-123.60	0.81159670	0.99
	43.5	3303.00	-139.26	0.98490680	0.99
	48.5	3618.89	-154.92	1.15810000	0.99
	53.5	3934.38	-170.57	1.33141000	0.99
	58.5	4248.30	-186.12	1.50343800	0.99

(a) Temperature (T) versus delay (δ), $T(\delta) = T_{\delta0} + T_{\delta1} \cdot \delta + T_{\delta2} \cdot \delta^2$

	Scatterer depth (mm)	$T_{\delta0}(ns)$	$T_{\delta1}(ns/^\circ C)$	$T_{\delta2}(ns/^\circ C^2)$	R^2
Linear	28.5	29.59	-0.01807		0.99
	33.5	29.49	-0.01750		0.99
	38.5	29.39	-0.01693		0.99
	43.5	29.30	-0.01638		0.99
	48.5	29.22	-0.01587		0.99
	53.5	29.15	-0.01538		0.99
	58.5	29.09	-0.01491		0.99
Quadratic	28.5	30.11	-0.01494	0.000002834	0.99
	33.5	30.16	-0.01356	0.000003456	0.99
	38.5	30.21	-0.01224	0.000003979	0.99
	43.5	30.26	-0.01104	0.000004397	0.99
	48.5	30.31	-0.00997	0.000004716	0.99
	53.5	30.36	-0.00899	0.000004960	0.99
	58.5	30.41	-0.00812	0.000005132	0.99

Table 14.4. Regression coefficients for data obtained from experimental signals using the phase-shift technique.

(a) Delay (δ) versus temperature (T), $\delta(T) = \delta_{T0} + \delta_{T1} \cdot T + \delta_{T2} \cdot T^2$

	Scatterer depth (mm)	$\delta_{T0}(ns)$	$\delta_{T1}(ns/°C)$	$\delta_{T2}(ns/°C^2)$	R^2
Linear	28.5	1507.99	-50.528		0.99
	33.5	1665.39	-56.469		0.99
	38.5	1799.35	-61.433		0.99
	43.5	1901.72	-65.355		0.99
	48.5	2017.54	-69.481		0.99
	53.5	2109.27	-72.828		0.99
	58.5	2205.84	-76.200		0.99
Quadratic	28.5	2363.00	-93.566	0.52813620	0.99
	33.5	3014.43	-124.98	0.84899730	0.99
	38.5	3428.17	-144.45	1.03193400	0.99
	43.5	3778.67	-161.34	1.19677400	0.99
	48.5	4214.94	-181.71	1.39737700	0.99
	53.5	4646.93	-202.24	1.60867800	0.99
	58.5	4887.52	-213.04	1.70193800	0.99

(a) Temperature (T) versus delay (δ), $T(\delta) = T_{\delta0} + T_{\delta1} \cdot \delta + T_{\delta2} \cdot \delta^2$

	Scatterer depth (mm)	$T_{\delta0}(ns)$	$T_{\delta1}(ns/°C)$	$T_{\delta2}(ns/°C^2)$	R^2
Linear	28.5	29.90	-0.0196		0.99
	33.5	29.58	-0.0175		0.99
	38.5	29.39	-0.0161		0.99
	43.5	29.22	-0.0151		0.99
	48.5	29.18	-0.0142		0.99
	53.5	29.13	-0.0135		0.99
	58.5	29.12	-0.0129		0.99
Quadratic	28.5	30.63	-0.0152	0.000004249	0.99
	33.5	30.52	-0.0122	0.000004648	0.99
	38.5	30.44	-0.0106	0.000004391	0.99
	43.5	30.35	-0.0095	0.000004218	0.99
	48.5	30.42	-0.0084	0.000004080	0.99
	53.5	30.49	-0.0074	0.000004045	0.99
	58.5	30.49	-0.0070	0.000003728	0.99

Table 14.5. Regression coefficients for data obtained from simulated signals using the cross-correlation technique.

(a) Delay (δ) versus temperature (T), $\delta(T) = \delta_{T0} + \delta_{T1} \cdot T + \delta_{T2} \cdot T^2$				
Scatterer depth (mm)	$\delta_{T0}(ns)$	$\delta_{T1}(ns/°C)$	$\delta_{T2}(ns/°C^2)$	R^2
Linear				
28.5	1632.00	-55.185		0.99
33.5	1677.09	-56.970		0.99
38.5	1722.65	-58.763		0.99
43.5	1768.43	-60.561		0.99
48.5	1813.74	-62.350		0.99
53.5	1859.38	-64.147		0.99
58.5	1904.65	-65.934		0.99
Quadratic				
28.5	2378.90	-93.488	0.47878790	0.99
33.5	2695.27	-109.18	0.65268070	0.99
38.5	3009.20	-124.74	0.82470860	0.99
43.5	3324.98	-140.38	0.99778550	0.99
48.5	3637.20	-155.86	1.16888100	0.99
53.5	3954.10	-171.56	1.34277400	0.99
58.5	4266.83	-187.07	1.51421900	0.99

(a) Temperature (T) versus delay (δ), $T(\delta) = T_{\delta0} + T_{\delta1} \cdot \delta + T_{\delta2} \cdot \delta^2$				
Scatterer depth (mm)	$T_{\delta0}(ns)$	$T_{\delta1}(ns/°C)$	$T_{\delta2}(ns/°C^2)$	R^2
Linear				
28.5	29.59	-0.01807		0.99
33.5	29.48	-0.01747		0.99
38.5	29.38	-0.01690		0.99
43.5	29.29	-0.01636		0.99
48.5	29.21	-0.01584		0.99
53.5	29.14	-0.01535		0.99
58.5	29.08	-0.01489		0.99
Quadratic				
28.5	30.11	-0.01493	0.000002838	0.99
33.5	30.16	-0.01346	0.000003506	0.99
38.5	30.21	-0.01215	0.000004026	0.99
43.5	30.27	-0.01096	0.000004435	0.99
48.5	30.31	-0.00990	0.000004742	0.99
53.5	30.36	-0.00893	0.000004981	0.99
58.5	30.41	-0.00806	0.000005147	0.99

Table 14.6. Regression coefficients for data obtained from experimental signals using the cross-correlation technique.

(a) Delay (δ) versus temperature (T), $\delta(T) = \delta_{T0} + \delta_{T1} \cdot T + \delta_{T2} \cdot T^2$

	Scatterer depth (mm)	$\delta_{T0}(ns)$	$\delta_{T1}(ns/°C)$	$\delta_{T2}(ns/°C^2)$	R^2
Linear	28.5	1539.60	-51.530		0.99
	33.5	1693.38	-57.346		0.99
	38.5	1836.25	-62.640		0.99
	43.5	1952.78	-67.009		0.99
	48.5	2052.81	-70.630		0.99
	53.5	2136.75	-73.718		0.99
	58.5	2234.20	-72.130		0.99
Quadratic	28.5	2378.26	-93.745	0.51803560	0.99
	33.5	3012.42	-124.34	0.83012270	0.99
	38.5	3463.00	-145.55	1.03062100	0.99
	43.5	3820.53	-162.52	1.19091100	0.99
	48.5	4247.60	-182.72	1.39571500	0.99
	53.5	4662.40	-202.52	1.60106000	0.99
	58.5	4900.92	-213.21	1.69244800	0.99

(a) Temperature (T) versus delay (δ), $T(\delta) = T_{\delta 0} + T_{\delta 1} \cdot \delta + T_{\delta 2} \cdot \delta^2$

	Scatterer depth (mm)	$T_{\delta 0}(ns)$	$T_{\delta 1}(ns/°C)$	$T_{\delta 2}(ns/°C^2)$	R^2
Linear	28.5	29.93	-0.01929		0.99
	33.5	29.61	-0.01729		0.99
	38.5	29.41	-0.01581		0.99
	43.5	29.26	-0.01476		0.99
	48.5	29.20	-0.01397		0.99
	53.5	29.15	-0.01336		0.99
	58.5	29.13	-0.01276		0.99
Quadratic	28.5	30.63	-0.01508	0.000003931	0.99
	33.5	30.52	-0.01223	0.000004341	0.99
	38.5	30.44	-0.01054	0.000004134	0.99
	43.5	30.35	-0.00945	0.000003896	0.99
	48.5	30.42	-0.00838	0.000003880	0.99
	53.5	30.49	-0.00748	0.000003883	0.99
	58.5	30.48	-0.00710	0.000003579	0.99

Table 14.7. Regression coefficients for data obtained from simulated signals using the demodulation technique.

(a) Delay (δ) versus temperature (T), $\delta(T) = \delta_{T0} + \delta_{T1} \cdot T + \delta_{T2} \cdot T^2$				
	Scatterer depth (mm)	$\delta_{T0}(ns)$	$\delta_{T1}(ns/°C)$	$\delta_{T2}(ns/°C^2)$	R^2
Linear	28.5	1507.45	-50.976		0.99
	33.5	1549.57	-52.635		0.99
	38.5	1591.56	-54.291		0.99
	43.5	1633.54	-55.945		0.99
	48.5	1675.72	-57.603		0.99
	53.5	1717.43	-59.253		0.99
	58.5	1759.67	-60.914		0.99
Quadratic	28.5	2197.15	-86.346	0.44211490	0.99
	33.5	2489.38	-100.83	0.60244270	0.99
	38.5	2778.27	-115.14	0.76071370	0.99
	43.5	3068.90	-129.55	0.92009880	0.99
	48.5	3361.29	-144.04	1.08049600	0.99
	53.5	3652.70	-158.49	1.24055800	0.99
	58.5	3943.70	-172.91	1.40002000	0.99

(a) Temperature (T) versus delay (δ), $T(\delta) = T_{\delta0} + T_{\delta1} \cdot \delta + T_{\delta2} \cdot \delta^2$				
	Scatterer depth (mm)	$T_{\delta0}(ns)$	$T_{\delta1}(ns/°C)$	$T_{\delta2}(ns/°C^2)$	R^2
Linear	28.5	29.59	-0.01956		0.99
	33.5	29.48	-0.01891		0.99
	38.5	29.38	-0.01829		0.99
	43.5	29.29	-0.01771		0.99
	48.5	29.21	-0.01715		0.99
	53.5	29.14	-0.01662		0.99
	58.5	29.08	-0.01612		0.99
Quadratic	28.5	30.11	-0.01616	0.000003324	0.99
	33.5	30.16	-0.01458	0.000004105	0.99
	38.5	30.21	-0.01316	0.000004709	0.99
	43.5	30.26	-0.01188	0.000005187	0.99
	48.5	30.32	-0.01072	0.000005560	0.99
	53.5	30.36	-0.00967	0.000005838	0.99
	58.5	30.41	-0.00872	0.000006037	0.99

Table 14.8. Regression coefficients for data obtained from experimental signals using the demodulation technique.

(a) Delay (δ) versus temperature (T), $\delta(T) = \delta_{T0} + \delta_{T1} \cdot T + \delta_{T2} \cdot T^2$

	Scatterer depth (mm)	$\delta_{T0}(ns)$	$\delta_{T1}(ns/{}^\circ C)$	$\delta_{T2}(ns/{}^\circ C^2)$	R^2
Linear	28.5	1265.78	-42.237		0.99
	33.5	1480.95	-49.503		0.99
	38.5	1543.49	-52.603		0.99
	43.5	1435.58	-49.197		0.99
	48.5	1638.78	-55.573		0.99
	53.5	1644.24	-56.473		0.99
	58.5	1713.20	-58.827		0.99
Quadratic	28.5	1874.60	-72.883	0.37606200	0.99
	33.5	2103.56	-81.127	0.39183540	0.99
	38.5	3136.09	-133.77	0.01008985	0.99
	43.5	2683.70	-113.02	0.79582400	0.99
	48.5	2487.28	-98.910	0.53957790	0.99
	53.5	3624.27	-157.45	1.25518600	0.99
	58.5	3684.42	-159.41	1.25104100	0.99

(a) Temperature (T) versus delay (δ), $T(\delta) = T_{\delta 0} + T_{\delta 1} \cdot \delta + T_{\delta 2} \cdot \delta^2$

	Scatterer depth (mm)	$T_{\delta 0}(ns)$	$T_{\delta 1}(ns/{}^\circ C)$	$T_{\delta 2}(ns/{}^\circ C^2)$	R^2
Linear	28.5	30.02	-0.02355		0.99
	33.5	29.95	-0.02012		0.99
	38.5	29.49	-0.01874		0.99
	43.5	29.28	-0.02012		0.99
	48.5	29.57	-0.01785		0.99
	53.5	29.31	-0.01738		0.99
	58.5	29.30	-0.01671		0.99
Quadratic	28.5	30.63	-0.01899	0.000005208	0.99
	33.5	30.45	-0.01683	0.000003289	0.99
	38.5	30.60	-0.01170	0.000006651	0.99
	43.5	30.30	-0.01344	0.000006672	0.99
	48.5	30.08	-0.01488	0.000002611	0.99
	53.5	30.61	-0.00967	0.000006744	0.99
	58.5	30.55	-0.00963	0.000005944	0.99

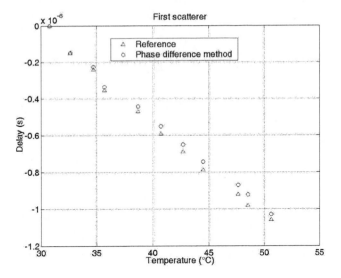

Fig. 14.5. Time delay vs. temperature curve for the first depth (first scatterer) using the phase-difference method for the experimental data.

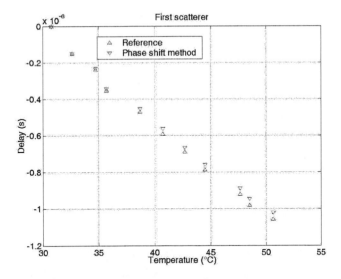

Fig. 14.6. Time delay vs. temperature curve for the first depth (first scatterer) using the phase-shift method for the experimental data.

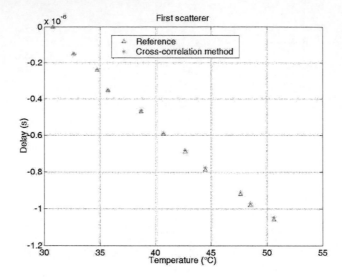

Fig. 14.7. Time delay vs. temperature curve for the first depth (first scatterer) using the cross-correlation method for the experimental data.

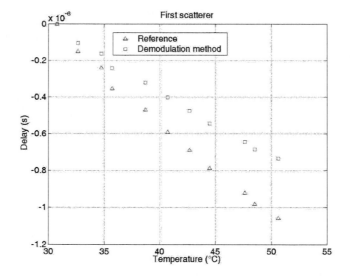

Fig. 14.8. Time delay vs. temperature for the first depth (first scatterer) using the demodulation method for the experimental data.

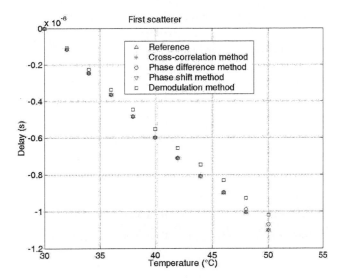

Fig. 14.9. Time delay vs. temperature for the first depth (first scatterer) using all methods for the simulated data.

Table 14.9. Maximal variation of time for the experimental and simulation data in the first depth (first scatterer).

Method	Maximum difference (ns)	
	Experimental signals	Simulated signals
Phase difference	59.8	32.1
Phase shift	38.2	0.4
Cross correlation	13.2	0.6
Demodulation	32.4	84.4

Concerning the comparative performance of the four methods for estimating time delay, the first parameter of comparison was the time delay difference with respect to a reference measurement (direct estimation) in numerical phantom. The delay for each simulated signal was established according to the proposed model thus it was consequently well-known. Delays obtained from the four methods for simulated signals were contrasted with these well-known time delays. Percentage errors for the four time-delay methods are shown in Table 14.10. Cross-Correlation method reported the lowest comparative error estimation in the numerical phantom, with no noise added to the ultrasonic signal. The highest error, 3.95 %, was observed for the demodulation method.

Another parameter taken into account was the suitability of the methods for experimental signals, for which there are non-ideal conditions, as in the simulated signals. In this experimental case, the time delay results were compared to the direct

Table 14.10. Suitability of methods for numerical phantom expressed in error percentage.

Depths (Echo signal positions)	Error for TDE Methods (%)			
	CC	PD	PS	DM
1st depth	0.04	0.28	0.03	3.99
2nd depth	0.03	0.45	0.19	4.00
3rd depth	0.05	0.43	0.14	3.90
4th depth	0.03	0.47	0.13	3.90
5to depth	0.04	0.52	0.12	3.93
6th depth	0.04	0.60	0.14	3.98
7th depth	0.03	0.66	0.14	3.98
Total Error	**0.16**	**1.65**	**0.50**	**15.81**
Average Error	**0.04**	**0.41**	**0.12**	**3.95**

measurements of time delays obtained from the time-signal by using the highest resolution in the oscilloscope (0.4-ns of resolution) included in the experimental setup. Table 14.11 shows the percentage differences obtained from this comparison. In this case, again, the Cross-Correlation method had the lowest error percentage average (0.4 %) and the demodulation method, the highest (18.77 %).

Table 14.11. Suitability of methods for experimental signals expressed in error percentage.

Depths (Echo signal positions)	Error for TDE Methods (%)			
	CC	PD	PS	DM
1st depth	0.59	2.70	1.84	18.42
2nd depth	0.47	1.20	1.30	17.45
3rd depth	0.31	0.35	1.34	16.53
4th depth	0.25	0.62	1.37	22.67
5to depth	0.27	0.63	1.04	21.62
6th depth	0.31	0.73	0.88	21.42
7th depth	0.15	0.75	0.80	21.61
Total Error	**1.63**	**4.88**	**5.87**	**15.81**
Average Error	**0.40**	**1.22**	**1.46**	**3.95**

A third complementary parameter used for comparison among methods was the time necessary for signal processing. For this particular comparison, it was considered that the processing time was related to two ultrasonic signals contiguous in temperature (for increments of 2 °C), and that the processing time included the consideration of all the seven windows (77784 data points). The computing system was a 2GB RAM-memory PC, with a 1.6 GHz processor and Windows XP platform. The times of signal processing (with and without pre-processing) are shown in Table 14.12. Under these conditions, the fastest method was the phase shift method.

Table 14.12. Processing times for the four time-delay estimation methods.

Time Delay Estimation Methods(TDE)	Number of essays	Processing Time (s)	
		Method	Method + pre-processing
Phase Difference (PD)	1000	0.0840	1.19390
Phase Shift (PS)	1000	0.0228	0.42275
Cross-Correlation (CC)	1000	0.1524	0.63072
Demodulation (QPD)	1000	0.3966	1.17370

Considering the results obtained a criterion for the selection of the best method for time delay measurement applied to the estimation of temperature in oncology hyperthermia was proposed. This criterion was based on three indexes that consider the parameters already explained above: suitability of the methods for numerical phantom, suitability of the methods for experimental signals, and processing time. Considering the lowest average values for each parameter, and dividing these values by the average values obtained for each method, three numerical indexes normalized to the unity, which could be summed in a global index normalized to three, was obtained. The results are shown in Table 14.13. Notice that, for this table, the higher the value, the better performance of the method. From this table, one can conclude that the best method, regarding only the suitability for numerical phantom and experimental signals, is clearly the cross-correlation method, and even considering the global index (2,67/3) this method seems to have the best performance. Time of processing can be considered as a secondary parameter as far as the time recording in oncology hyperthermia treatment, according to the guidelines given by the ESHO, is every 6 s (Hand et al. [7]). In fact, all the methods considered here had processing times under this value.

Another important issue considered in this research was the robustness of the methods under noisy conditions. Basic signals from the numerical phantom, which were contaminated with additional noise levels at - 3 dB and - 6 dB, were employed for this analysis. The SNR based on average power of the simulated signals and noise was used in order to calculate the amplitude of the added noise to the simulated ultrasonic signals. Then, these artificial noisy signals were tested for the four methods. It was observed that the cross-correlation and the demodulation methods had the same linear behavior as presented in section 3, but it was noticed that the phase-based methods were very sensitive to the presence of noise. Time delays were under or over estimated by the phase-based methods, with large variations of up to

Table 14.13. Processing times for the four time-delay estimation methods.

Time Delay Estimation Methods	Suitability of Method for Numerical Phantom	Suitability of Method for Experimental Signals	Processing Time M	Processing Time M+P
PD	0.09	0.32	0.27	0.35
PS	0.33	0.27	**1.00**	**1.00**
CC	**1.00**	**1.00**	0.14	0.67
QPD	0.01	0.02	0.05	0.36

2.5 μs when - 6-dB noise was added. Due to this fact, the phase-based methods were tested with lower levels of noise ranging from - 60 dB to - 90 dB in increments of 2 dB. This analysis was carried out for the seventh window because it contains the ultrasonic signal with the lowest amplitude and for this reason it could be easily affected by noise. According to the ESHO protocol(Hand et all [7]) for hyperthermia treatments, the accuracy of the thermometer should be 0.2 °C. This accuracy would correspond to 13 ns for the seventh window and 11 ns for the first window, under the conditions presented in section 2. For the phase difference method, the average error for the seventh window was 12.7 ns when the signal is noise-free, and 25.9 ns for the signals with - 80-dB added noise. For the phase shift method, the average error for the seventh windows was 1.7 ns when the signal is noise-free, and 11.8 ns for the signals with - 82-dB added noise.

From all these analyses, it can be concluded that the method with the highest robustness against the noise is the cross-correlation method, which has small errors, 0.78 ns, with high noise levels (- 3dB). Demodulation method is stable when noise is added but time delay estimation is poor. Finally, one of the disadvantages of the phase-based methods is that they are very sensitive to noise, and small amplitudes in noise signals, at the beginning and the end of the experimental data, easily disturb the results, thus consequently producing spurious phase changes on the signal of interest. These undesirable phase changes are produced by the different frequency content of the noise oscillations. Another disadvantage of the phase-based methods is that it is necessary to have the same voltage level at the beginning of the signals for all the temperatures. In this chapter, these inconveniences were overcome, for experimental signals, by removing the noise oscillations contained in the shifted signal. These noise oscillations were removed by using a thresholding determined as a percentage of the peak value in the reference signal. This percentage depends on the SNR of the ultrasonic signal: for a high SNR the use of a low threshold is sufficient; whereas for low SNR signals, the application of a higher threshold is required. According to the obtained results, correlation method has the best performance for detection of small time delays and thus for estimating temperature in tissues by using this time option; however the other time methods analyzed can be used when less accuracy is needed or when a robust pre-processing is performed before the signals are fed to the processing method.

Acknowledgement

This work was supported by: European Project ALFA-Contract Number AML/B7-311/97/066/ii-0343-FA-FCD-FI; IberoAmerican Research Project Pulsets (CYTED); Spanish R&D National Plan of the Ministry of Science & Innovation (Project DPI2005-00124), and the Mexican projects Conacyt 45041 and 60903.

The authors thank M. in C. Hugo Zepeda for his invaluable technical assistance.

References

1. Amini, A. N., Ebbini, E. S. and Georgiou, T. T.: Noninvasive estimation of tissue temperature via high-resolution spectral analysis techniques. IEEE Trans. Biomed. Eng. 52, 221-228 (2005)

2. Bertsch, F., Mattner, J., Stehling, M. K., Mller-Lisse, U., Peller, M., Loeffler, R., Weber, J. Memer, K., Wilmanns, W., Issels, R. and Reiser, M.: Non-invasive temperature mapping using MRI: comparison of two methos based on chemical shift and T1-relaxation. Magn. Res. Imaging. 16, 393-403 (1998)

3. de Jong, P. G. M., Arts, T. G. E., Hoeks, A. P. G. and Reneman, R. S.: Determination of tissue motion velocity by correlation interpolation of pulsed ultrasonic echo signal. Ultrason. Imaging. 12, 84-98 (1990)

4. de Korte, C. L., van der Steen, A. F. W., J Dijkman, B. H. and Lance, C. T.: Performance of time delay estimation methods for small time shifts in ultrasonic signals. Ultrasonics. 35, 263-274 (1997)

5. de Poorter, J., de Wageter, C., de Deene, Y., Thompson, C., Stahlberg, F. and Achten, E.: Noninvasive MRI thermometry with the proton resonance frequency (PRF) method: in vivo results in human muscle. Magn. Res. Med. 33, 74-81 (1995)

6. Dewhirst, M. W., Viglianti, B. L., Lora-Michiels, M., Hanson, M. and Hoopes, P. J.: Basic principles of thermal dosimetry and thermal thresholds for tissue damage from hyperthermia. Int. J. Hyperthermia. 19, 267-294 (2003)

7. Hand, J. W., Lagendijk, J. J. W., Andersen, J. B. and Bolomey, J. C.: Quality assurance guidelines for ESHO protocols. Int. J. Hyperthermia. 5, 421-428 (1989)

8. Hand, J. W., Van Leeuwen, G. M., Mizushina, S., Van de Kamer, J. B., Maruyama, K., Sugiera, T., Azzopardi, D. V. and Edwards, A. D.: Monitoring of deep brain temperature in infants using multi-frequency microwave radiometry and thermal modeling. Phys. Med. Biol. 46, 1885-1903 (2001)

9. Hoeks, A. P. G., Hennerici, M. and Reneman, R. S.: Spectral composition of Doppler singnals. Ultrasound Med. Biol. 17, 751-760 (1991)

10. Kuc, R.: Estimating acoustic attenuation from reflected ultrasound signals: Comparison of spectral-shift and spectral-difference approaches. IEEE Trans. Acoust. Speech Signal Process. 1, 1-6 (1984)

11. Leroy, Y., Bocquet B. and Mamouni, A.: Non-Invasive microwave radiometry thermometry. Physiol. Meas. 19, 127-148 (1998)

12. Maass-Moreno, R., Damianou, C. A. and Sanghvi, N. T.: Noninvasive temperature estimation in tissue via ultrasound echo-shift. Part II. In vitro study. J. Acoust. Soc. Am. 100, 2522-2530 (1996)

13. Paulsen, K., Moskowitz, M., Ryan, T., Mitchell, S. and Hoopes, P.: Initial in vivo experience with EIT as a thermal estimator during hyperthermia. Int. J. Hyperthermia. 12, 573-591 (1996)

14. Ramos, A., San Emeterio, J. L., Sanz, P. T.: Depedence of pulser driving responses on electrical and motional characteristics of NDE ultrasonic probes. Ultrasonics. 38, 553-558 (2000)

15. Ramos, A., San Emeterio, J. L., Sanz, P. T.: PSpice circuital modelling of ultrasonic imaging transcievers including frequency-dependent acoustic losses and signal distrontions in electronic stages. Ultrasonics. 44, 995-1000 (2006)

16. Seip, R. and Ebbini, E. S.: Noninvasive estimation of tissue temperature response to heating fields using diagnostic ultrasound. IEEE Trans. Biomed. Eng. 42, 828-839 (1995)

17. Seip, R., Van Baren, P., Cain, C. A. and Ebbini, E. S.: Noninvasive real-time multipoint temperature control for ultrasound phased array treatment. IEEE Trans. Ultrason. Ferroelctr. Freq. Control. 43, 1063-1073 (1996)

18. Shankar, P. M.: A model for ultrasonic scattering from tissues based on the K distribution. Phys. Med. Biol. 40, 1633-1649 (1995)

19. Simon, C., Van Baren, P. and Ebbini, E. S.: Two-dimensional temperature estimation using diagnostic ultrasound. IEEE Trans. Ultrason. Ferroelectr. Freq. Control. 45, 1088-1099 (1998)

20. Tristam, M., Barbosa, D. C., Cosgrove, D. O., Nassiri, D. K., Bamber, J. C. and Hill, C. R.: Ultrasonic study of in vivo kinetic characteristics of human tissues. Ultrasound Med. Biol. 12, 927-937 (1986)

21. Vazquez, M., Ramos, A., Leija, L. and Vera, A.: Noninvasive temperature estimation in oncology hyperthermia using phase changes in pulse-echo ultrasonic signals. Jpn. J. Appl. Phys. 45, 7991-7998 (2006)

22. Weng, L., Reid, J. M., Shankar, P. M., Soetanto, K. and Lu, X-M.: Nonuniform phase distribution in ultrasound speckle analysis. II. Parametric expression and a frequency sweeping technique to measure mean scatterer spacing. IEEE Trans. Ultrason. Ferroelectr. Freq. Control. 39, 360-365 (1992)

23. Wilson, L. S. and Robinson, D. E.: Ultrasonic measurement of small displacement and deformations of tissue. Ultrason. Imaging. 4, 71-82 (1982)

Assessing the Statistical Nature of the MAI in a WSN based on IR-UWB for Cognitive-like Operation

Fernando Ramírez-Mireles[1] and Angel Almada[2]

[1] Instituto Tecnológico Autónomo de México (ITAM)
  México City, D.F. C.P. 01000, México
  `ramirezm@ieee.org`
[2] Georgia Institute of technology
  Atlanta, Georgia, 30332, USA
  `saalm56@gmail.com`

Summary. We propose a method for assessing the statistical nature of the multiple-access interference(MAI) in wireless sensor networks (WSN). More specifically, the scenario contemplates uncoordinated nodes communicating using Impulse-Radio (IR) Ultra Wideband (UWB) communications (IR-UWB). The statistical nature of the MAI at the output of a single-correlator receiver depends, among other things, on the number N_u of interfering nodes and the number of frames per symbol N_s. The interfering nodes form part of the environment in which the node of interest operate, and determining how the statistical nature of the MAI changes with N_u and N_s can be useful environmental information that can be used by a cognitive radio to perform calculations or adaptations in the operation, for example, for adaptation of the power control to different conditions, or for calculation of the maximum throughput of the system. We propose a low-complexity method to establish "Gaussianity regions" in a two dimensional plane consisting of pairs (N_u, N_s). We determine the boundary of these "Gaussianity regions" for three different conditions: An ideal propagation channel with perfect and imperfect power control, as well as a multipath channel with "perfect average" power control.
Keywords: UWB, impulse radio, ad hoc networks, sensor networks, environmental sensing.

15.1 Introduction

Wireless sensor networks [1] are flexible networks composed of nodes with sensing capability for which there is no need of a central coordinator and for which the numbers of nodes and the topology of the network are not predetermined. The WSNs have a large number of nodes and are deployed in close proximity to the phenomena under study, and the topology of the network can change constantly due, for example, to nodes prone to fail (usually nodes have limited power, computational capabilities and memory).

UWB is an indoor communication technique currently under intense research activity [2] due to many attributes, including its robustness against multipath conditions, its high capacity in a multiple access environment, the capability to achieve high transmission rates using a low amount of power, and, for pulse based UWB, the possibility of operating using a carrier-less modulation. According to [3] a signal is considered of UWB nature if the 10 dB bandwidth of the signal is at least 20% of its center frequency, or if this 10 dB bandwidth is at least 500 MHz.

In this chapter we work with the pulse-based UWB spread spectrum multiple access (SSMA) technique called impulse radio proposed by Scholtz [4]. This technique uses binary PPM for data modulation and TH for code modulation. Multiple access is achieved using different TH codes for different users. To decode the signal of the desired user we consider a correlator receiver. The IR-UWB has been proposed for wireless ad-hoc networks [5]. The many desirable characteristics of IR-UWB can be used in WSN for simultaneous communication, ranging and positioning [6].

In a WSN the nodes mainly use a broadcast communication paradigm that can suffer severe degradation due to the presence of MAI. The statistical nature of the MAI produced by N_u interfering nodes $\{S_i\}$, $i = 1, 2, \ldots, N_u$, operating in the vicinity of a designated node S_o depends on the signal format (e.g. the number of frames per symbol N_s), the propagation conditions, as well as the number N_u of nodes [7]. The $\{S_i\}$ form part of the environment in which S_o operate, and determining how the statistical nature of the MAI changes with N_u can be useful environmental information that can be used by a cognitive radio [8] to perform calculation or adaptations in the operation of S_o, for example, for adaptation of the power control to different conditions [9], or for calculation of the maximum throughput of the system [10].

We are interested in studying how the statistical nature of the MAI (e.g. Gaussian or not) change for different conditions [11] [12]. Previous analysis (e.g. [4] [13] [14]) assume that the MAI at the output of a correlation receiver is a Gaussian random variable (r.v.). This Gaussian assumption is based on the central limit theorem [15], and it leads to a simplified analysis. This assumption is reasonable for SSMA systems with both low per-user data rate and a large number of users, however, it has been shown that for high data rate or with a limited number of users this is a poor approximation [7] [16] - [22], and some work has been done considering approaches to calculate performance without the Gaussian assumption [17] - [20].

The works in [7] [16]-[22] deal with the asymptotic conditions to determine the domain of validity for the Gaussian assumption, but they don't address the issue of finding for which specific sets of parameters the Gaussian assumption is valid. In our method we use a low-complexity Gaussianity test based on entropy to construct a Gaussianity index used to establish "Gaussianity regions" in a two dimensional plane consisting of pairs (N_u, N_s). We determine the domain of validity for the Gaussian assumption for 3 different scenarios: An ideal propagation channel with perfect and imperfect power control, as well as a multipath channel with "perfect average" power control.

In our calculations we left fix certain signal design parameters (e.g. pulse duration T_p, pulse position and frame time T_f), and we vary N_u and N_s to find for which pairs (N_s, N_u) the Gaussianity indexes is above a certain predetermined threshold. We then set different scenarios to study tradeoffs between N_s and/or N_u for changing T_f and for fixed T_f. Asymmetry on the range of values for N_s and N_u needed to reach Gaussianity in different conditions is explored and explained.

15.2 System Model

We consider a system formed by a designated node S_o operating in the presence of N_u interfering nodes $\{S_i\}$, $i = 1, 2, \ldots, N_u$. The nodes communicate with each-other asynchronously, in a full-duplex mode uisng code division duplexing. We also assume that all of them operate in the same propagation environment. Next we provide various model details.

15.2.1 Scenarios

Fig. 15.1 depicts the power profiles for the three scenarios considered: 1) Ideal propagation with perfect power control, where the power of each component in the MAI is equal to a constant 2) Ideal propagation with imperfect power control, where the power of each component in the MAI is a r.v., and 3) A multipath channel with "perfect average" power control. For the random powers we consider two power profiles: Line of sight (LOS) and non-line of sight (NLOS). We use $D = 3, 6, 9$ m for LOS and $D = 1, 2, 3$ m for NLOS. We produce 49 profiles for each transmitter-receiver distance D and average over the 3 distances in each case.

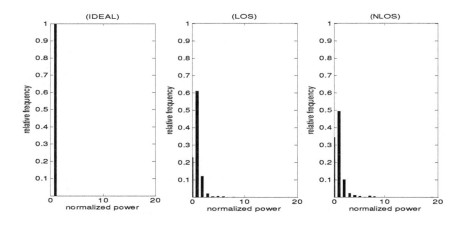

Fig. 15.1. Power profiles considered in this work.

15.2.2 Signal format

The binary TH-PPM signals received in multipath are[3]

---

[3] The received signal $x^{(v)}(t)$ includes both the channel and antenna effects.

$$x^{(v)}(t) = \sum_{m=0}^{N_s-1} w(\xi^{(v)}, t - mT_f - c_m^{(v)} T_c - T_d d_{\lfloor m/N_s \rfloor}^{(v)}), \tag{15.1}$$

where the superscript (v), $1 \leq v \leq N_u + 1$, denotes quantities related to user v (there are one desired users and N_u other users). The T_f is the frame time, T_d is the data time shift, and T_c is the code time shift. The $c_m^{(v)}$ is the pseudorandom time-hopping sequence for user v with range $0 \leq c_m^{(v)} < N_h$ and with sequence period N_p. The $d_{\lfloor m/N_s \rfloor}^{(v)}$ is the data of user v that could be 0 or 1 and changes every N_s hops. The $\xi^{(v)}$ denotes a random index used to indicate the random multipath trajectory from the transmitters of the interferer users ($1 < v <= N_u + 1$) to the receiver of the desired user (user with $v = 1$).

Note that the data symbol changes every N_s frames, therefore the symbol duration is $N_s T_f$ and the bit rate is $R = \frac{1}{N_s T_f}$.

The $w(\xi^{(v)}, t)$ is the basic UWB signal used to convey information. It has average duration T_a (i.e. mean delay spread $T_a < T_f$), energy

$$E_w(\xi^{(v)}) = \int_{-\infty}^{\infty} \left| w(\xi^{(v)}, t) \right|^2 dt, \tag{15.2}$$

and normalized signal correlation

$$\gamma(\xi^{(v)}, \tau) = \frac{1}{E_w(\xi^{(v)})} \int_{-\infty}^{\infty} w(\xi^{(v)}, t) w(\xi^{(v)}, t - \tau) dt. \tag{15.3}$$

For a channel with ideal propagation conditions the random index $\xi^{(v)}$ is dropped and the pulse $w(t)$ has duration T_w (instead of T_a), with $T_w \ll T_f$.

15.2.3 Multipath channel Model

We consider three types of channels: one with ideal propagation conditions, one multipath channel with LOS, and one with NLOS. We use the autoregressive model for an indoor residential environment proposed in [23] [24]. This model has an average delay spread of $Ta = 160$ ns and includes a path loss model that allow to calculate multipath profiles for different transmitter-receiver distances D.

We assume that all the users experience the same multipath effects with the same probability and independently from the other users, i.e., $\xi^{(v)}$ are independent and identically uniformly distributed.

15.2.4 Multiple-access interference

We consider an asynchronous multiple access system. In the system model under consideration all the users utilize the same type of binary time hopping PPM signals in (15.1) to convey information, the difference being the TH code used for each user. All the users experience the same multipath environment, although each one has its own multipath trajectory $\xi^{(v)}$. When N_u asynchronous transmitters are active, the received signal at user one's receiver position is

$$r(t) = \sum_{v=1}^{N_u+1} A^{(v)} x^{(v)}(t - \tau^{(1)}) + n(t)$$

$$\stackrel{\triangle}{=} x^{(1)}(t - \tau^{(1)}) + n_{\text{MUI}}(t) + n(t) \qquad (15.4)$$

where $\tau^{(v)}$ represent time asynchronisms between the clock of user v's transmitter and user one's receiver, $(A^{(v)})^2$ is the ratio of average power used by user v's transmitter with respect to the average power used by user one's transmitter (with $|A^{(1)}|^2 = 1$), $n_{\text{MUI}}(t)$ represents MAI, and $n(t)$ represents non-MAI interference modeled as AWGN with two-sided power spectrum density (PSD) $N_o/2$.

We consider three cases. 1) Free space propagation conditions with perfect power control. In this case the pulse $w(t)$ suffers no distortion due to multipath, and $(A^{(v)})^2 = (A^{(1)})^2$ for $v = 2, 3, \ldots, N_u + 1$. 2) Free space propagation conditions with imperfect power control. As before, the pulse $w(t)$ suffers no distortion due to multipath, but now $(A^{(v)})^2$ is a r.v. that depends on $v = 2, 3, \ldots, N_u + 1$. In particular, we chose this r.v. $(A^{(v)})^2$ to follow the statistics of the power variations in a multipath channel. Notice, however, that in this case there is no variation in the received energy of the desired user. 3) Multipath propagation conditions with "perfect average" power control. In this cases all the signals $x^{(v)}(t)$, $v = 1, 3, \ldots, N_u + 1$, are received with random energy and random correlation values, and the random variations in energy are normalized to have mean one (that's what we mean for "perfect average").

15.2.5 Processing at the receiver

We assume that the user one's receiver is perfectly matched and perfectly synchronized with its own transmitter (that is $\tau_1 = 0$). We use the classical filter matched to the signal $v(\xi^{(1)}, t) = w(\xi^{(1)}, t) - w(\xi^{(1)}, t - T_d)$, so the decision statistic is

$$\beta = \sum_{m=0}^{N_s-1} \int_{mT_f}^{(m+1)T_f} r(t) v(\xi^{(1)}, t - mT_f) dt. \qquad (15.5)$$

A 0 is decoded when $\beta \geq 0$ (no time shift was added to the signal) and 1 when $\beta < 0$ (a time shift of T_d ns. was added to the signal).

15.2.6 Model Parameters.

We are assuming the following points for the rest of the chapter and to program the simulations.

Pulse shape.

We consider two forms of pulse shape. The first is the second derivative of a Gaussian pulse [4]

$$w(t) = \left[1 - 4\pi\left[\frac{t}{t_n}\right]^2\right] \exp\left(-2\pi\left[\frac{t}{t_n}\right]^2\right), \tag{15.6}$$

for $-T_w/2 \le t \le T_w/2$, where $t_n = 0.2877$ ns is a parameter that determine the pulse duration $T_w \simeq 0.7$ ns, with energy $E_w = 3t_n/8$. The signal correlation function is

$$\gamma(\tau) = \left[1 - 4\pi\left[\frac{\tau}{t_n}\right]^2 + \frac{4\pi^2}{3}\left[\frac{\tau}{t_n}\right]^4\right] \times \exp\left(-\pi\left[\frac{\tau}{t_n}\right]^2\right), \tag{15.7}$$

for $-T_w \le \tau \le T_w$, with a minimum value $\gamma_{\min} \overset{\triangle}{=} \gamma(\tau_{\min}) \simeq -0.6181$ at $\tau_{\min} \simeq 0.156$ ns. The second pulse shape is based on a gated sine wave [25]

$$w(t) = \sin\left(2\pi\frac{Q}{T_w}t\right), \tag{15.8}$$

for $-T_w/2 \le t \le T_w/2$, $T_w = 2.0$ ns, where $Q = 10$ is a positive integer, resulting in a signal spectrum centered at $\frac{Q}{T_w} = 5$ GHz. This pulse have energy $E_w = \left(\frac{T_w}{2}\right)$ and signal correlation function

$$\gamma(\tau) = \frac{1}{E_w}\frac{T_w - |\tau|}{T_w}\cos\left(2\pi\frac{Q}{T_w}\tau\right), \tag{15.9}$$

for $-T_w \le \tau \le T_w$, with a minimum value $\gamma_{\min} \overset{\triangle}{=} \gamma(\tau_{\min}) \simeq -0.9501$ at $\tau_{\min} \simeq 0.0995$ ns for $Q = 10$.

Table 15.1. Values for TH-PPM time parameters.

Parameters	Gaussian Pulse	Gated sine wave
T_w	0.7 ns	2.0 ns
T_f	70, 150, 250 ns	200 ns
T_d	0.156 ns	0.0995 ns
T_c	0.9 ns	0.1 ns
N_h	8	200

Time parameters

Values of the parameters are shown in Table 15.1. General assumptions are described in detail in [14] [10]. Under these assumptions, the MAI at the output of the correlator is

$$\alpha = \sum_{v=2}^{N_u+1} \alpha^{(v)} \tag{15.10}$$

$$\alpha^{(v)} = \sum_{m=0}^{N_s-1} A^{(v)} [R_w(\lambda_m^{(v)}) - R_w(\lambda_m^{(v)} - T_d)] \tag{15.11}$$

$$\lambda_m^{(v)} = [c_{m-\Phi}^{(v)} - c_m^{(1)}] T_c + d_{\lfloor m/N_s \rfloor}^{(v)} T_d + \tau_v, \tag{15.12}$$

where $\lambda_m^{(v)}$ is a weighted sum of uniformly distributed r.v.s $c_j^{(v)}$, $d_{\lfloor m/N_s \rfloor}^{(v)}$ and τ_v, the $R_w(\cdot)$ are cross correlation terms, the τ_v are random transmission delays, and Φ is a r.v. that depends on the transmission delays. We notice that the r.v.'s $[c_{m-\Phi}^{(v)} - c_m^{(1)}]$ for distinct values of m are conditionally independent, given the value of the time shift parameter Φ. We also notice that the r.v.'s $\alpha^{(v)}$ for distinct values of v are conditionally independent, given the code $\{c_m^{(1)}\}$ of user one.

For a multipath channel we add the random index $\xi^{(v)}$ to the previous expressions, e.g. $\alpha^{(v)}(\xi^{(v)})$.

15.3 Gaussianity Test

In this section we describe a procedure to determine the Gaussianity of α in (15.10). More specifically, we collect a sample of α and calculate an entropy function which is then compared with the entropy of a Gaussian r.v. with the same mean and variance.

Let's consider the following hypothesis testing

$$\mathcal{H}_0 : \alpha \sim N(0, \sigma^2)$$
$$\mathcal{H}_1 : \alpha \nsim N(0, \sigma^2), \tag{15.13}$$

where α is assumed to be a sample with zero-mean[4] and with variance σ^2, and $N(\cdot)$ is the normal distribution. $\mathcal{H}_1$ say's that the distribution of α is not Gaussian. By exploiting the fundamental fact that a r.v. has maximum differential entropy if and only if its Gaussian distributed [26], the following equivalent hypothesis testing problem can be established:

$$\mathcal{H}_0 : \text{entropy}(\alpha) = \tfrac{1}{2} \ln(2\pi e \sigma^2)$$
$$\mathcal{H}_1 : \text{entropy}(\alpha) < \tfrac{1}{2} \ln(2\pi e \sigma^2) \tag{15.14}$$

where e is the base of $\ln(\cdot)$.

The (differential) entropy of a random variable X is [26]

$$H(f) = \int_{-\infty}^{\infty} f_X(x) \log f_X(x) dx$$
$$= \int_0^1 \log \left\{ \frac{d}{du} F_X^{-1}(u) \right\} du, \tag{15.15}$$

---

[4] The zero-mean can be verified using the model parameters in section 15.2.6.

where f_X and F_X are the probability density function (p.d.f.) and the cumulative distribution function (c.d.f.) [27] of X, respectively. The second equality follows from a change of variable and the properties of the function F_X [27]. Approximating the derivative in (15.15) by the formula in [28, chapter 9, exercise 22.] and writing the integral as a Riemann sum we get the following approximation:

$$H_{nm} = \frac{1}{n} \sum_{i=1}^{n} \log \left\{ \frac{n}{2m} \left(x_{(i+m)} - x_{(i-m)} \right) \right\}, \tag{15.16}$$

where n is the number of samples taken, $0 < m < n/2$ is a natural number and $x_{(i+m)}$ are order statistics: $x_{(1)} \leq x_{(2)} \leq \dots. \leq x_{(n)}$. The quantity $x_{(i+m)} - x_{(i-m)}$ should be replaced by $x_{(i+m)} - x_{(1)}$ or $x_{(n)} - x_{(i-m)}$ when $i - m < 1$ or $i + m > n$, respectively.

The main result in [29] is that the point estimator in (15.16) satisfies that

$$H_{nm} \xrightarrow{P} H, : n \to \infty, : m \to \infty, : \frac{m}{n} \to 0, \tag{15.17}$$

where $\xrightarrow{P}$ denotes convergence in probability [27]. By defining a normalized normality index as

$$K_{nm} \triangleq \frac{e^{H_{mn}}}{\overline{\sigma}}, \tag{15.18}$$

where $\overline{\sigma}^2 = \frac{1}{n}\sum_{i=1}^{n}(x_i - \bar{x})^2$ is the sample variance and $\bar{x} = \frac{1}{n}\sum_{i=1}^{n} x_i$ is the sample mean. we get a further elaboration of the hypothesis testing in (15.14)

$$\begin{aligned} \mathcal{H}_0 &: K_{nm} \xrightarrow{P} \sqrt{2\pi e} = 4.1327\dots \\ \mathcal{H}_1 &: K_{nm} < \sqrt{2\pi e} \end{aligned}, \tag{15.19}$$

where K_{nm} is calculated as [29]

$$K_{nm} = \frac{n}{2m\overline{\sigma}} \left\{ \prod_{i=1}^{n} \left(\alpha_{(i+m)} - \alpha_{(i-m)} \right) \right\}^{1/n}. \tag{15.20}$$

To establish the Gaussianity test, a critical value K^* is proposed so that when $K_{nm} \geq K^*$ then, with certain probability, we accept that α is a Gaussian r.v. For this purpose we use $K^* = 3.35$ that has a 95% confidence level for $n = 50$ and $m = 5$, and was originally proposed in [29]. With these assumptions we can rewrite the final hypothesis testing problem as

$$\begin{aligned} \mathcal{H}_0 &: K_{nm} \geq K^* \\ \mathcal{H}_1 &: K_{nm} < K^*. \end{aligned} \tag{15.21}$$

For a given $p(t)$ and T_f, we can use this test to define the *Gaussian regime* as the set of all pair of points (N_s, N_u) such that the normality index $K_{n,m} \geq K^*$.

15.4 Simulation

Samples of the r.v. α in (15.10) were generated using Matlab . The α can be seen as a double sum of independent and identically distribute (i.i.d.) r.v.'s containing a bunch of r.v.s $R_w(\lambda_m^{(v)})$.

Using the values on Table 15.1 we simulate 1000 samples of the r.v. α. Fig. 15.2 shows the histograms for the cases $(N_u, N_s) = (2,2)$ and $(N_u, N_s) = (16,16)$, respectively, as well as the Gaussian p.d.f. that "fit" the data (i.e., normal distribution with same mean and variance). Clearly, the case $(N_u, N_s) = (16,16)$ results in a "better fit" than the case $(N_u, N_s) = (2,2)$.

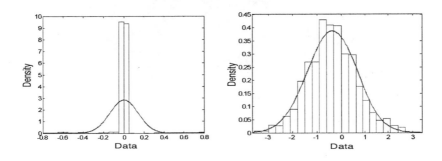

Fig. 15.2. The empirical p.d.f. corresponding to 1000 samples of α and the corresponding Gaussian fit for $(N_u, N_s) = (2,2)$ (Left figure) and $(N_u, N_s) = (16,16)$ (Right figure).

Fig. 15.3 shows that normality index $K_{n,m}$ in (15.20) is consistent, i.e., for more terms in the r.v. α the index is closest to the objective value K^*. In this figure we can also verify that the pace of "convergence speed" is different for different values of T_f. Intuitively, $T_f = 50$ ns results in more pulse collisions than $T_f = 150$ ns, hence when $T_f = 50$ Gaussianity is achieved at lower values of N_s and/or N_u.

Next we use the normality index $K_{n,m}$ in (15.20) to obtain the regions where α can be considered a Gaussian r.v. The boundary of the Gaussianity region is determined by those (N_s, N_u) where the entropy test $K_{50,5} \geq 3.35$ is satisfied. Fig. 15.4 illustrates the boundary between regions.

We simulate several cases considering a large amount of permutations of (N_s, N_u). Since the entropy estimator is itself a r.v., the "random boundary" is actually a region determined by many random realizations. To study boundary changes for different T_f we smooth the plots by averaging over 10 of such realizations (cf. fig. 15.5). To

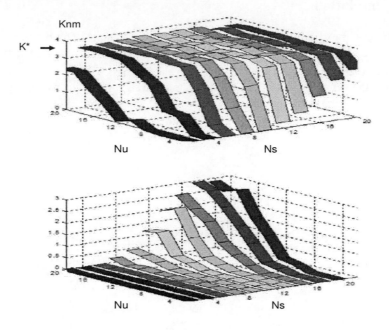

Fig. 15.3. Index K_{nm} with $n = 50$ and $m = 5$. Upper figure: $T_f = 50$ ns. Lower figure: $T_f = 150$ ns.

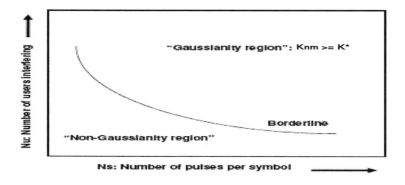

Fig. 15.4. Gaussian Regions as a function of K_{nm} for different N_u and N_s.

study the boundary with a fixed T_f value we generate 100 of such realizations (cf. figs. 15.6 and 15.7).

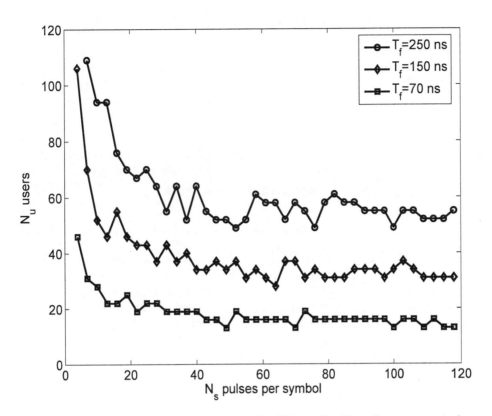

Fig. 15.5. Boundaries of the Gaussianity region for different T_f with perfect power control.

15.4.1 Ideal propagation with perfect power control

Fig. 15.5 show how the boundary changes for different T_f using the Gaussian pulse in (15.6). Comparing our results with previous works, we found a good match. For example, fig. 2(a) in [22] shows that for $N_s = 8$ and $T_f = 150$ ns about $N_u = 50$ users are needed to be in the Gaussian region, a number that agrees with results in fig. 15.5. As another example, propositions 6 and 7 in [7] shows that for a fixed N_s the larger T_f is, the larger N_u should be to reach Gaussianity, a situation that agrees with results in fig. 15.5.[5] Intuitively, a shorter T_f results in more pulse collisions than a longer T_f, hence with a shorter T_f Gaussianity is achieved with lower values of N_s and/or N_u.

15.4.2 Ideal propagation with imperfect power control

Fig. 15.6 show the range of pairs (N_s, N_u) defining the "random boundary" using the pulsed sinusoid in (15.8). We notice that there is an asymmetry on the range of values for N_s and N_u needed to reach Gaussianity. On one hand, it is observed that for low N_s we need a large N_u to reach Gaussianity. This can be explained by recalling that $\alpha^{(v)}$ in (15.11) for $v = 1, 2, \ldots, N_u + 1$ are dependant r.v.'s trough user one's code $\{c_m^{(1)}\}$. On the other hand, it is observed that N_u reaches a sort of minimum floor as N_s is increased. This can be explained recalling that $\lambda_m^{(v)}$ in (15.12) for $m = 0, 2, \ldots, N_s - 1$ are dependant r.v.'s trough Φ in (15.12), and that τ_v, being uniformly distributed over $\left[-\frac{T_f}{2}, \frac{T_f}{2}\right]$, is the largest component in $\lambda_m^{(v)}$ compared to

$$[c_{m-\Phi}^{(v)} - c_m^{(1)}]T_c + d_{\lfloor m/N_s \rfloor}^{(v)} T_d. \tag{15.22}$$

Fig. 15.6 also shows that the boundary with imperfect power control have a higher spread of (N_s, N_u) values than the boundaries with perfect power control This is expected since the weak components of the MAI are masked by the strong ones, hence it takes more r.v. to get to the Gaussianity region. We also point out that boundaries for NLOS have a higher spread than the boundaries for LOS. This is explained observing that the NLOS power profile in fig. 15.1 have a larger energy variance than the LOS power profile.

15.4.3 Multipath channel with perfect average power control

Fig. 15.7 show the random boundary using the pulsed sinusoid. We notice that in this case there is also an asymmetry on the range of values for N_s and N_u needed to reach Gaussianity, i.e., the range of N_u needed to reach Gaussianity is approximately constant for all values of N_s considered, and this range contains just a few interferer users, hence, in a dense multipath Gaussianity can be reached with few values of both N_s and N_u. These results are consistent with previous studies where a single-user UWB signal in multipath can be modeled as a non-stationary Gaussian random process [30]. Notice, however, that fig. 7 in [7] shows that a higher N_u is needed to reach Gaussianity. But results in [7] are for a multipath channel with a limited number of discrete paths, while in this work we are using a dense multipath channel with a continuous impulse response.

---

[5]Notice that [7] uses a signal format and parameters in which N_s, N_h and T_f are coupled, i.e., they define $T_f = N_h T_c$ and a processing gain $N = N_s N_h$, hence $T_f = N T_c / N_s$. By keeping N fixed, a low value of N_s implies a large value of T_f. Notice that in our case we consider PPM with $T_f/2 > N_h T_c + 2T_d$, hence N_h and N_s can be changed without necessarily affecting T_f.

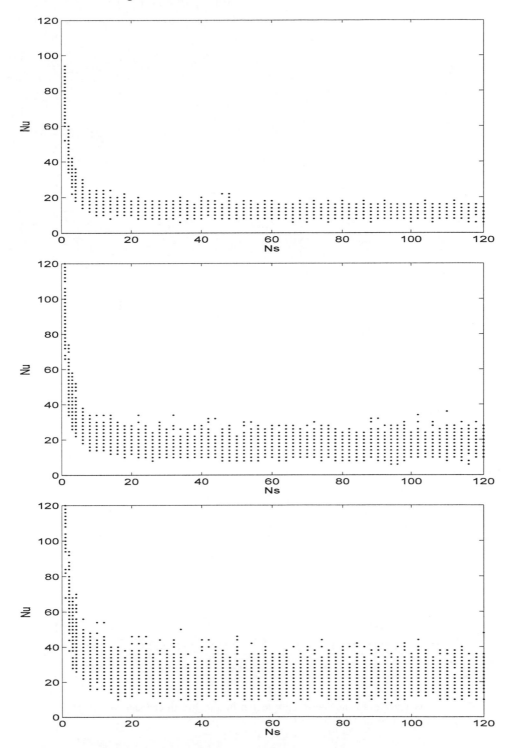

Fig. 15.6. Boundaries of the Gaussianity region with imperfect power control. Top: Ideal power profile. Middle: LOS power profile. Bottom: NLOS power profile.

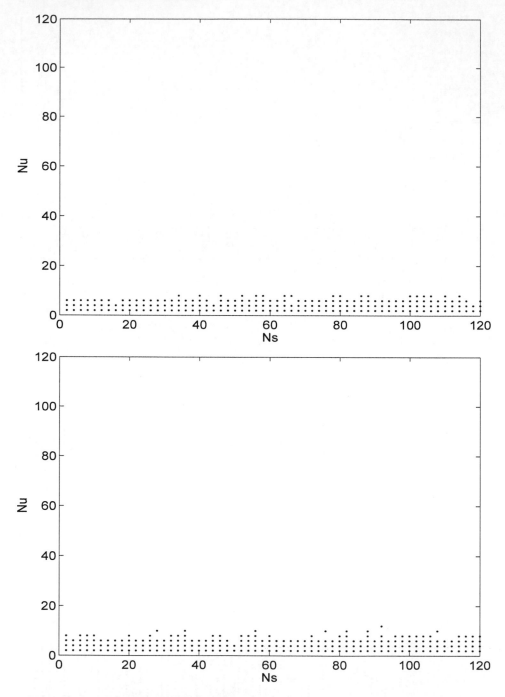

Fig. 15.7. Boundaries of the Gaussianity region in a multipath channel with average power control. Top for LOS and bottom for NLOS.

15.5 Conclusion.

In this paper we assessed the statistical nature of IR-UWB MAI produced by N_u interfering nodes $\{S_i\}$, operating in the vicinity of a designated node S_o.

We established a statistical test to determine when the MAI at the output of a correlator can be considered a random variable with a Gaussian distribution. Using an entropy point estimator we determine the minimum number of users N_u and the minimum number of pulses per symbol N_s necessary to model the MAI component as a Gaussian random variables. Asymmetry on the range of values for N_s and N_u needed to reach Gaussianity in different conditions is explored and explained.

Results for perfect power control and for multipath cases agrees with previous results. Results for imperfect power control suggest that even in this case the Gaussian approximation can be used if we consider a large enough number of N_u and/or N_s, e.g., for low data rate systems with large number of users. Comparison with previous results also shows the dependence of the Gaussian assumption on the signal structure, time parameters, and statistics of the multipath channel.

The statistical nature of the MAI is useful environmental information that can be used by a cognitive radio to perform calculation or adaptations in the operation of the radio. As an example, let's consider the plot at the top in fig. 15.6 that shows that Gaussianity is reached at $(N_u, N_s) = (20, 20)$ with perfect power control. Using this information, [9] shows that we could possibly reduce the transmission power by several dB and still maintain the target bit error rate (BER) value. Also using this information, [10] shows that we could calculate the maximum number of users to support a target BER, and use this number, for example, for admission control.

Future work include studies considering using other pulse shapes, modulations, channel models, Gaussianity tests, and calculation of α with mis-matched filters.

References

1. I. F. Akyildiz, W. Su, Y. Sankarasubramaniam, and E. Cayirci, A survey on sensor networks, in IEEE Commun. Mag., vol. 40 (2002), pp. 102- 114.
2. R. C. Qiu, H. Liu, X. Shen, Ultra-Wideband for Multiple Access Communications, in IEEE Commun. Magazine, Vol. 43 (2005), pp. 2-8.
3. U.S. Federal Communications Commission, First Report and Order for UWB Technology, U.S. Federal Communications Commission, (2002), February.
4. R. A. Sholtz, Multiple-access with time-hopping impulse modulation, in Proc. Military Communications Conf. (1993), pp. 447-450.
5. L. De Nardis, P. Baldi, and M.-G. Di Benedetto, UWB Ad-Hoc networks, in Proc. IEEE UWBST Conf.(2002), pp. 219 - 223.
6. L. De Nardis, and M.-G. Di Benedetto, Joint communication, ranging, and positioning in low data-rate UWB networks, in Proc. 2nd. Workshop on positioning, navigation and communication and 1st UWB expert talk (2005), pp. 191-200.

7. J. Fiorina and W. Hachem , On the Asymptotic Distribution of the Correlation Receiver Output for Time-Hopped UWB Signals, in *IEEE Trans. on Signal Processing*, vol. 54 (2006), pp. 2529 - 2545.

8. H. Celebi and H. Arslan, Enabling location and environment awareness in cognitive radios, Elsevier Computer Communications, vol. 31 (2008), pp. 1114-1125.

9. F. Ramírez-Mireles, Transmission Power Management for IR-UWB WSN Based on Node Population Density, in Wireless Sensor and Actor Networks II, IFIP International Federation for Information Processing Book Series, Springer Boston (2008), pp. 37-48.

10. F. Ramírez-Mireles, Quantifying the Degradation of Combined MUI and Multipath Effects in Impulse-Radio UWB, IEEE Trans. Wireless Commun., Vol. 6 (2007), pp. 2831-2836.

11. F. Ramírez-Mireles and A. Almada, Testing the Gaussianity of UWB TH-PPM MUI with Perfect and Imperfect Power Control, in Proc. IEEE ROCC Conf. (2007), pp. 1-6.

12. F. Ramírez-Mireles and A. Almada, Testing the Gaussianity of UWB TH-PPM MUI with Imperfect Power Control and Multipath, in Proc. IEEE 5th Intl Conf. CCE (2008) pp. 242-247.

13. M.Z. Win and R.A. Sholtz, Ultra-Wide Bandwidth Time-Hopping Spread-Spectrum Impulse Radio for Wireless Multiple-Access Communications, IEEE Trans. Commun. (2000), pp. 679-691.

14. F. Ramírez-Mireles, Performance of ultrawideband SSMA using time hopping and M-ary PPM, in IEEE J. Select. Areas Commun., vol. 19 (2001), pp. 1186-1196.

15. Kallenberg Olav, Modern Foundations of Probability, Springer-Verlag: New York (2002).

16. G. Durisi and G. Romano, On the validity of Gaussian approximation to characterize multiuser capacity of UWB TH-PPM, in Proc. IEEE Conf. Ultra Wideband Systems and Technologies (2002)157-161.

17. B.Hu and N.C. Beaulieu, Accurate evaluation of multiple-access performance in TH-PPM and TH-BPSK UWB systems, IEEE Trans. Commun., vol. 52 (2004).

18. B. Hu and N.C. Beaulieu, Exact bit-error analysis of TH-PPM UWB systems in the presence of multiple-access interference, IEEE Comm. Letters, vol. 7 (2003), pp. 572-574.

19. G. Durisi and S. Benedetto, Performance evaluation of TH-PPM UWB systems in the presence of multiuser interference, in IEEE Commun. Lett. vol. 7 (2003), pp. 224-226.

20. A.R. Forouzan, M. Nasiri-Kenari, and J.A.Salehi, Performance analysis of time hopping spread-spectrum multiple-access systems: Uncoded and Coded schemes, in *IEEE Trans. Wireless Commun.* , vol. 1 (2002), pp. 671-681.

21. G. Giancola, L. De Nardis and M.-G. Di Benedetto, Multi User Interference in Power-Unbalanced Ultra Wide Band systems: Analysis and Verification, in Proc. UWBST Conf. (2003), pp. 325-329.

22. Y. Dhibi and T. Kaiser, On the impulsiveness of multiuser interferences in TH-PPM-UWB systems", in IEEE Trans. on Signal Processing, vol. 54 (2006), pp. 2853 - 2857.

23. W. Turin, R. Jana, S. Ghassemzadeh, C. Rice, and V. Tarokh, Autoregressive Modeling of an Indoor UWB Channel, in Proc. UWBST Conf. (2002), pp. 71-74.

24. S. Ghassemzadeh, R. Jana, C. Rice, W. Turin, and V. Tarokh, A Statistical Path Loss Model for In-Home UWB Channel," in Proc. UWBST Conf. (2002) Conf., pp. 59-64.

25. F. Ramírez-Mireles, Performance of UWB N-Orthogonal PPM in AWGN and Multipath Channels, in IEEE Trans. on Vehic. Technol., vol. 56 (2007), pp. 1272-1285.

26. Robert B. Ash, Information Theory, Dover Publications, New York, (1990).

27. P. Billingsley, Probability and Measure, New York: Wiley, (1986).

28. M. Spivak, Calculus, Addison-Wesley Higher Education (1967).

29. Vasicek Oldrich, A Test for Normality Based on sample Entropy, in Journal of the Royal Statistical Society Series B, vol. 38 (1976), pp. 54-59, 1976.

30. Q. T. Zhang and S. H. Song, Parsimonious Correlated Nonstationary Models for Real Baseband UWB Data, in IEEE Trans. on Vehic. Technol., vol. 54 (2005), pp. 447-455.

Hybrid Space-Time Codes for MIMO Wireless Communications*

Joaquin Cortez[1], Alberto Sanchez[2], Miguel Bazdresch[3], Omar Longoria[2], and Ramon Parra[2]

[1] Instituto Tecnológico de Sonora, Cd. Obregón, México
[2] CINVESTAV-GDL, Zapopan, México
[3] ITESO University, Periférico Sur 8585, Tlaquepaque, México

Summary. Hybrid MIMO space-time codes combine spatial multiplexing with diversity gain to achieve both high spectral efficiency and link reliability. In this chapter, we present in detail a family of hybrid codes, known as *LD STBC-VBLAST* codes, along with a receiver architecture suitable for low-complexity hardware implementation. We also present a correlated MIMO channel model and explore the impact of correlation on code performance. We show that *LD STBC-VBLAST* codes exhibit higher performance than other recently-proposed hybrid codes.

16.1 Introduction

A multiple-input, multiple-output (MIMO) communications system uses n_T antennas to transmit and n_R antennas to receive. Compared to a single-input, single-output (SISO) system, where one antenna is used at each end, a MIMO system can potentially achieve a higher data rate with lower probability of error, without increasing bandwidth or power. In this chapter, we will present and analyze transmission schemes that are able to extract this potential extra performance.

Let **s** be a vector of symbols to be transmitted. Each symbol x_i may be transmitted at certain time intervals, over certain antennas. A *space-time block code* (STBC) is a mapping of symbols to combinations of time intervals and antennas. This mapping may be represented by a table or a matrix, where each row corresponds to a transmit antenna and each column to a time interval. As an example, consider a system with $n_T = 2$; the vector to be transmitted is $\mathbf{s} = [s_1 \, s_2]$. The STBC given by

$$S = \begin{bmatrix} s_1 & -s_2^* \\ s_2 & s_1^* \end{bmatrix} \tag{16.1}$$

*This work was supported by Intel research grant INTEL-CERMIMO2008 and by CONACYT grants 84559-Y and 51332-Y, and PROMEP.

specifies that, at the first symbol interval, antenna 1 transmits s_1 and antenna 2 transmits s_2. During the second interval, antenna 1 transmits $-s_2^*$ and antenna 2 transmits s_1^*.

Compared to a SISO system, there are two main ways an STBC may exploit the extra space resources offered by MIMO: diversity and spatial multiplexing. Diversity increases the reliability of wireless communications by transmitting each symbol over several different, uncorrelated channels. The probability that all channels are in a deep fade decreases as the number of channels increases. Intuitively, an STBC may increase diversity by transmitting each symbol s_i over a large number of antennas and time intervals. The maximum diversity of a MIMO channel is given by $n_T n_R$.

Spatial multiplexing, on the other hand, means that a MIMO system is equivalent to a system with $n_{min} = \min(n_R, n_T)$ parallel, independent channels. Each of these channels may be used to transmit a different data stream, thereby increasing the data rate by a factor n_{min} compared to the SISO system. In other words, the received signal has n_{min} complex dimensions (or n_{min} degrees of freedom).

These two types of gains may appear to be mutually exclusive and, indeed, the first STBCs offered either maximum diversity gain or maximum data rate gain. In order to obtain maximum diversity, each symbol s_i should be repeated over space and time as many times as possible, which precludes any rate gain. On the other hand, maximum data rate gain means repeating symbols as few times as possible, limiting the diversity gain.

In fact, it has been found [1] that MIMO systems have a *diversity-multiplexing tradeoff*. It is possible, in theory, to design STBCs where the data rate R scales as

$$R = r \log_2(SNR) \tag{16.2}$$

and the error probability p_e scales as

$$p_e \approx SNR^{-d(r)}. \tag{16.3}$$

Here, r is the multiplexing gain and $d(r)$ is the diversity gain, which is given by

$$d(r) = (n_R - r)(n_T - r), \tag{16.4}$$

where r takes values between 0 and n_{min}, and $d(r)$ takes values between 0 and $n_T n_R$. In practice, a given code may achieve gains that are lower than these optimal values. Such a code is said to achieve a suboptimal tradeoff. The most representative example of a code that achieves maximum data rate gain is V-BLAST [2]; the Alamouti scheme [3] is one of the best well-known maximum diversity gain codes.

How to design space-time codes that provide the desired gains while still being decodable with reasonable complexity? This is an area of ongoing research. We may divide the current efforts in two camps. On one hand, there are results based on information theory that concentrate on designing optimal codes with maximum $d(r)$ for all possible values of r. Usually, joint, maximum-likelihood decoding at the receiver is assumed (for an overview of work in this area, see [4]). One of the stronger results obtained so far are the so-called Perfect space-time codes, which meet a large number of design criteria that theoretically guarantee optimum performance. However, it has been proved that Perfect STBCs exist only for a very limited range of n_T and n_R [5].

On the other hand, a more empirical technique known as *hybrid coding* has been proposed. A hybrid space-time code tries to achieve desired gains r and $d(r)$ by simply mixing, in the transmitter, two or more STBCs, some of which provide diversity gain and the rest provide data rate gain. More specifically, the n_T transmit antennas are partitioned, and each partition is used to transmit a different data stream using a specific STBC. This technique offers the advantages of being simple to design, and offering high flexibility over data rate and performance.

The question is, how can such a mixture of codes be efficiently decoded? Several methods have been proposed; see [6, 7, 8] for some examples. In this chapter we focus on a method based on Linear Dispersion Codes (LDCs) [9]. LDCs are a very general family of STBCs, which allow for a large amount of flexibility in the way each symbol is spread over time and space. In fact, codes as diverse as V-BLAST, Alamouti and Perfect STBCs are particular cases of LDCs. One key feature of LDCs is that they may be decoded using a technique known as Ordered Successive Interference Cancellation (OSIC), first introduced as a decoder for V-BLAST. This technique, while suboptimal, is well understood and feasible to implement. There is a link between hybrid codes and LDCs, explored in [10], which allows hybrid codes to be expressed as equivalent LDCs and, as a consequence, allowing them to be decoded with OSIC.

In section 16.2 of this chapter, we will present a MIMO channel model (that may include correlation) that justifies the system model and assumptions made later. In section 16.3, we introduce a hybrid STBC and present the transmitter and receiver architectures. In section 16.4, we present in detail a low-complexity detection algorithm based on OSIC, whose performance (both on uncorrelated and correlated channels) is obtained and analyzed in section 16.5. We also compare its performance with other hybrid coding schemes, and study the implications of implementing it using the CORDIC algorithm.

16.2 Channel Model

16.2.1 MIMO Channel System

A MIMO system with n_R receive antennas and n_T transmit antennas ($n_R \times n_T$) system can be seen as multiple data streams that travel in a fading channel, and are received by a collection of antennas, in such a way that the rich scattering can be exploited. The mathematical base-band model of the communication mechanism is given by

$$\mathbf{r} = \mathbf{Hs} + \mathbf{n}, \tag{16.5}$$

where $\mathbf{s} \in \mathbb{C}^{n_T \times 1}$ represents the data stream from the n_T antennas; $\mathbf{n}$ an additive white circularly symmetric complex Gaussian noise (AWGN) vector with mean given by $E\{\mathbf{n}\} = 0$ and covariance matrix by $E\{\mathbf{nn}^H\} = N_0 \mathbf{I}_{n_R}$, where $\mathbf{I}_{n_R}$ is the

$n_R \times n_R$ identity matrix, $(\bullet)^H$ denotes conjugate transpose, and $E\{\bullet\}$ represents the expectation operator; $\mathbf{r}$ is the $n_R \times 1$ received symbols vector. Then the $n_R \times n_T$ channel matrix

$$\mathbf{H} = \begin{bmatrix} h_{1,1} & \cdots & h_{1,n_T} \\ \vdots & \ddots & \vdots \\ h_{n_R,1} & \cdots & h_{n_R,n_T} \end{bmatrix} \tag{16.6}$$

includes complex fading gains h_{ji} expressing the tap gains between transmit antenna i and receive antenna j.

In this way, the Spatial Correlation Function (SCF) sampled at the antennas' position results in the Spatial Correlation Tensor (SCT) or full correlation matrix given by

$$\mathbf{R}_H = E\{vec(\mathbf{H})vec(\mathbf{H})^H\}, \tag{16.7}$$

where $vec(\bullet)$ stands for an operator that stacks as a vector the matrix argument column-wise.

Initial investigations [2] around this topic considers that the channel matrix elements are independent and identically distributed (i.i.d.) complex Gaussian random variables with zero means and unit variances, so that the SCF is $\mathbf{R}_H = \mathbf{I}_{n_R n_T}$. Under this assumption, a MIMO system has large capacity [11]. However, in real propagation environments, the fades are not independent due to insufficient spacing of antennas and relative displacement between transmitter and receiver. This phenomenon is typically called Spatial Selectivity (SS), i.e., the channel is under Spatial Correlation (SC) conditions.

It has been observed [12] that when the fades are correlated, the channel capacity is significantly smaller than when fades are i.i.d. This section intends to describe the phenomenology behind the fading in correlated scenarios specifically considering SS, but not time nor frequency selectivity.

When independence (separability condition) between transmitter and receiver is assumed, the channel model can be characterized by partial information from the SCF via the Kronecker Model [13]:

$$\mathbf{H} = \left(\mathbf{R}_H^{R_x}\right)^{\frac{1}{2}} \mathbf{G} \left(\mathbf{R}_H^{T_x}\right)^{\frac{H}{2}}, \tag{16.8}$$

where

$$\mathbf{R}_H^{T_x(R_x)} = \left(\mathbf{R}_H^{T_x(R_x)}\right)^{\frac{1}{2}} \left(\mathbf{R}_H^{T_x(R_x)}\right)^{\frac{H}{2}}, \tag{16.9}$$

$\mathbf{G}$ is a i.i.d. random complex matrix whose elements have zero mean normal distribution and variance equal to one, and $\mathbf{R}_H^{T_x(R_x)}$ stands for the transmitted (received) Spatial Correlation Matrix (SCM). The SCT and the SCM of each transceiver are related via the Kronecker product $\mathbf{R}_H = \mathbf{R}_H^{T_x} \otimes \mathbf{R}_H^{R_x}$, where $\otimes$ denotes the Kronecker product.

16.2.2 Power Azimuth Spectrum

In order to evaluate the effects of SS on the performance of MIMO systems, it is necessary to develop a channel realization mechanism capable of accomplishing (16.7), for a given SCF (see [17] and references therein).

This chapter is oriented only to Gaussian processes, which are completely described by their second order statistics, that is, by their SCF. For this reason, it is sufficient to have the SCF to obtain suitable channel realizations. These processes' power distribution function may be understood intuitively, since it depends on the spatial reference angles. It is a useful starting point for channel simulation. In section 16.2.3 we will explain how to calculate the SCF starting from the distribution function.

Let us first describe the typical MIMO communication system scenario where the transmitter and receiver sides are implemented with antenna arrays with a given topology. Here we focus on the Uniform Linear Array (ULA) [17]; an example of this system is shown in Fig. 16.1. As its name indicates, the antennas are arranged uniformly over a straight line. Spherical coordinates in Fig. 16.1 are used to indicate the position of each antenna, where ϑ and φ are the elevation and azimuthal angles respectively, and ρ is the position vector given by

$$\rho = [\cos(\varphi)\cos(\vartheta)\hat{x}, \sin(\varphi)\cos(\vartheta)\hat{y}, \sin(\vartheta)\hat{z}]^T , \tag{16.10}$$

where $(\bullet)^T$ is the transposition operator.

The vector **k** represents the wave vector of the departing/incident wavefront in a far field scenario, and is expressed as

$$\mathbf{k} = k_0 [\cos(\varphi)\cos(\vartheta)\hat{x}, \sin(\varphi)\cos(\vartheta)\hat{y}, \sin(\vartheta)\hat{z}]^T , \tag{16.11}$$

where $k_0 = \frac{2\pi}{\lambda}$ is the free-space wave number of a single wavefront, where λ is the wavelength of the carrier frequency. Hereafter, we assume the array orientation with respect to y-axis (as depicted in Fig. 16.1).

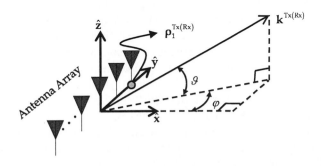

Fig. 16.1. Representation of wave fronts as vectors in spherical coordinates.

A physical consequence of a SS scenario on signal behavior is that transmission may be done at different Angles of Departure (AoD) in elevation and azimuthal

Table 16.1. Most commonly used PAS functions.

PAS	Formula	Parameters
Uniform	$p(\varphi) = \frac{1}{\varphi_{max}-\varphi_{min}}, -\pi \leq \varphi_{min} \leq \varphi \leq \varphi_{max} \leq \pi$	Interval $(\varphi_{min}, \varphi_{max})$
Laplacian	$p(\varphi) = \frac{e^{-\frac{1}{k}|\varphi|}}{2k\left(1-e^{\frac{-\pi}{k}}\right)}, -\pi \leq \varphi \leq \pi$	Decay Parameter k
Gaussian	$p(\varphi) = \frac{e^{-\frac{1}{2\sigma^2}\varphi^2}}{2\pi\sigma^2 \mathrm{Erf}\left(\frac{\pi}{\sqrt{2}\sigma}\right)}, -\pi \leq \varphi \leq \pi$	Variance σ^2

coordinates, and even the same signal can impact the receiver array at different Angles of Arrival (AoA). For practical purposes, only the azimuthal channel model is considered; that is, the direction of each wavefront is parallel to a given plane (*xy* in Fig. 16.1); therefore, the elevation angle is zero valued.

Under SS conditions, the power of the signal is treated as a random variable that depends on AoA (φ^{Rx}) and AoD (φ^{Tx}), with Probability Density Function (PDF) performing as a power distribution. For this reason it is necessary to introduce the cross-Power Azimuth Spectrum (xPAS), denoted by $p\left(\varphi^{Rx}, \varphi^{Tx}\right)$. This describes the power distribution of the transmitted and received signals as a function dependent of φ^{Rx} and φ^{Tx}.

Under separability conditions it is useful to have a function that describes separately the power distribution at the transmitter and receiver; such a function is called Power Azimuth Spectrum (PAS) and is denoted by $p\left(\varphi^{Tx}\right)$ as the PAS in the transmitter, and by $p\left(\varphi^{Rx}\right)$ as the PAS in the receiver. These functions satisfy $p\left(\varphi^{Rx}, \varphi^{Tx}\right) = p\left(\varphi^{Rx}\right)p\left(\varphi^{Tx}\right)$.

Several standards and measurement campaigns have been developed to estimate the PAS at each side in different environments, such as the IEEE 802.15 standard proposed for wireless Personal Area Networks (PAN) at 60 GHz; the METRA project [14], in macrocell and microcell environments; the COST 259 standard [15], for macrocell, urban and suburban environments. An important selection of parameters for the angle aperture is gathered in [16]. Table 16.1 shows the corresponding mathematical expressions for typical PAS considered in MIMO communication systems. All of them are centered on zero degrees.

An example of each cited PAS is depicted in Fig. 16.2. An uniform PDF with maximum aperture angle of 30° is shown in Fig. 2(a); Fig. 2(b) shows a truncated Gaussian PDF with maximum aperture angle of 90°.

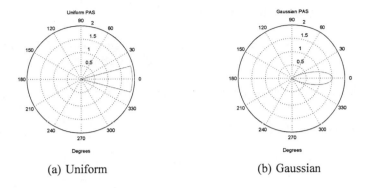

Fig. 16.2. Polar coordinate plots of power azimuth spectra.

16.2.3 PAS and the SCF

As explained above, the main goal of channel characterization is to be capable of replicating the model established in (16.8); for this purpose we require to know the SCM. However, most measurement campaigns provide second order statistics information only for the PAS. A practical mathematical tool that relates $p\left(\varphi^{Tx(Rx)}\right)$ to $\mathbf{R}_H^{Tx(Rx)}$ at the transmitter (or receiver) is introduced in [17]. It uses the well known Array Manifold Vector (AMV), that is described by

$$\mathbf{V}^{Tx(Rx)}\left(\varphi^{Tx(Rx)}\right) = \left[e^{-\mathbf{k}^{Tx(Rx)}\left(\varphi^{Tx(Rx)}\right)\cdot\rho_1^{Tx(Rx)}} \cdots e^{-\mathbf{k}^{Tx(Rx)}\left(\varphi^{Tx(Rx)}\right)\cdot\rho_{n_T n_R}^{Tx(Rx)}}\right], \quad (16.12)$$

where $\mathbf{k}^{Tx(Rx)}$ is the wave vector at the transmitter (or receiver); $\varphi^{Tx(Rx)}$ is the AoD (AoA); $\rho_{m(n)}^{Tx(Rx)}$ is the position vector of the mth (nth) antenna (given a designated order) at the transmitter (or receiver); being $m = 1,2,\cdots,n_T$ ($n = 1,2,\cdots,n_R$), and n_T (n_R) is the number of antennas at the transmitter (or receiver).

For ULA, the AMV is reduced to

$$\mathbf{V}^{Tx(Rx)}\left(\varphi^{Tx(Rx)}\right) = \left[e^{\frac{-2\pi j}{\lambda}d_1^{Tx(Rx)}\sin(\varphi^{Tx(Rx)})} \cdots e^{\frac{-2\pi j}{\lambda}d_{n_T(n_R)}^{Tx(Rx)}\sin(\varphi^{Tx(Rx)})}\right] \quad (16.13)$$

where $d_k^{Tx(Rx)}$ represents the linear position of the kth antenna at the transmitter (or receiver), and λ is the wavelength of the signal. In (16.13) the array is oriented as in Fig. 16.1.

The direct formula to calculate a SCM is

$$\mathbf{R}_H^{Tx(Rx)} = \int_{\varphi^{Tx(Rx)}} p\left(\varphi^{Tx(Rx)}\right)\mathbf{V}^{Tx(Rx)}\left(\varphi^{Tx(Rx)}\right)\mathbf{V}^{Tx(Rx)}\left(\varphi^{Tx(Rx)}\right)^H d\varphi^{Tx(Rx)}.$$
$$(16.14)$$

Once the SCM of the transmitter or receiver are known it is possible to develop the channel realizations necessary for the desired performance test.

16.3 Hybrid Code Description

We now present a hybrid code based on Linear Dispersion Codes [9] that may be used to transmit over a MIMO channel such as that described in the previous section. We call this code *LD STBC-VBLAST* [18]. We first present the space-time mapping in the transmitter, and then we present a receiver architecture.

16.3.1 Space-Time Mapping

This hybrid code partitions its transmit antennas into $n_L = n_S + n_B$ sets, called *spatial layers*. Layers one to n_S operate as V-BLAST, transmitting one symbol per period without repetition. Each of the n_B layers is a 2-antenna Alamouti encoder. A single data stream is converted to n_L parallel streams, each of which is mapped to a given constellation and then transmitted on one of the spatial layers. The total number of transmit antennas is $n_T = n_S + 2n_B$. A system block diagram is shown in Fig. 16.3.

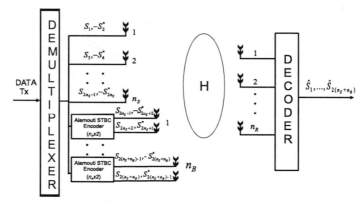

Fig. 16.3. *LD STBC-VBLAST* Transceiver Architecture. H represents the channel among each pair of transmit-receive antennas.

More specifically, a block of symbols $s_1, s_2, \cdots, s_{n_{sym}}$, where $n_{sym} = 2(n_S + n_B)$, is mapped to each antenna as specified in Table 16.2; on average, $(2n_S + n_B)$ symbols are transmitted per channel use. The code rate, then, is $\frac{2n_S + n_B}{2n_T}$. Note that the rate may be varied by changing the number and spatial layering distribution of the transmit antennas; also, in general, the rate will be higher than that obtained by the Alamouti scheme, but less than that obtained with V-BLAST. The power allocated to the V-BLAST layers is given by:

$$P_S = \frac{2}{n_{sym}} \ ,$$

while the power allocated to the STBC encoder is given by:

$$P_A = \frac{P_S}{2} \ .$$

This allocation results in all symbols being transmitted with the same energy, which is the optimal strategy in the absence of Channel State Information (CSI) at the transmitter.

Table 16.2. *LD STBC-VBLAST* Symbol to Antenna Mapping with $k = B - 1 + n_S$

Time	Spatial Antennas	STBC Blocks $B = 1, \cdots, n_B$	
	Antenna $V = 1, 2, \cdots, n_S$	Antenna 1	Antenna 2
t	s_V	s_{2k+1}	s_{2k+2}
$t+T$	$-s_{V+1}^*$	$-s_{2k+2}^*$	s_{2k+1}^*

16.3.2 Receiver Architecture

We assume that $n_R \geq n_S + n_B$ and perfect CSI at the receiver. The transmitted signal may be written as $S = [S_{spa} S_A]^t$, where S_{spa} corresponds to the symbols transmitted by V-BLAST layers and the matrix S_A to the symbols transmitted by the Alamouti encoders, so that

$$S_{spa} = \begin{bmatrix} s_1 & -s_2^* \\ s_3 & -s_4^* \\ \vdots & \vdots \\ s_{2n_S-1} & -s_{2n_S}^* \end{bmatrix} = \begin{bmatrix} s_1^{(1)} & s_1^{(2)} \\ s_2^{(1)} & s_2^{(2)} \\ \vdots & \vdots \\ s_{n_S}^{(1)} & s_{n_S}^{(2)} \end{bmatrix}, \qquad (16.15)$$

and

$$S_A = \begin{bmatrix} S_1^A \\ \vdots \\ S_{n_B}^A \end{bmatrix} = \begin{bmatrix} S_1^{(1)A} & S_1^{(2)A} \\ \vdots & \vdots \\ S_{n_B}^{(1)A} & S_{n_B}^{(2)A} \end{bmatrix}, \qquad (16.16)$$

where each element of (16.16) is given by:

$$\begin{bmatrix} S_B^{(1)A} & S_B^{(2)A} \end{bmatrix} = \begin{bmatrix} s_{2k+1} & -s_{2k+2}^* \\ s_{2k+2} & s_{2k+1}^* \end{bmatrix}$$

with $B = 1, 2, \cdots, n_B$ and $k = B - 1 + n_S$. Note that we have performed the mapping described in Table (16.2). In this notation, the super-index indicates time interval and the sub-index indicates spatial layer number. An A in the super-index indicates an Alamouti layer.

It is assumed that the channel remains constant during the transmission of S, so that the received signal may be written as

$$
\begin{bmatrix} y_1^{(1)} & y_1^{(2)} \\ y_2^{(1)} & y_2^{(2)} \\ \vdots & \vdots \\ y_{n_R}^{(1)} & y_{n_R}^{(2)} \end{bmatrix} = \begin{bmatrix} h_{1,1} & \cdots & h_{1,n_T} \\ h_{2,1} & \cdots & h_{2,n_T} \\ \vdots & \ddots & \vdots \\ h_{n_R,1} & \cdots & h_{n_R,n_T} \end{bmatrix} \begin{bmatrix} S_{spa} \\ S_A \end{bmatrix} + \begin{bmatrix} n_1^{(1)} & n_1^{(2)} \\ n_2^{(1)} & n_2^{(2)} \\ \vdots & \vdots \\ n_{n_R}^{(1)} & n_{n_R}^{(2)} \end{bmatrix} , \qquad (16.17)
$$

or, in matrix form,

$$
Y = HS + N . \qquad (16.18)
$$

Matrix $Y \in \mathbb{C}^{n_R \times 2}$ represents the symbols received in a block. Matrix $H \in \mathbb{C}^{n_R \times n_T}$ is the channel matrix. Matrix $N \in \mathbb{C}^{n_R \times 2}$ represents the Gaussian noise added to each received symbol.

Reformulating the system equation (16.17) as a linear dispersion code [10], we have:

$$
\begin{bmatrix} y_1^{(1)} \\ y_1^{(2)*} \\ \vdots \\ y_{n_R}^{(1)} \\ y_{n_R}^{(2)*} \end{bmatrix} = \begin{bmatrix} H_{spa} & H_A \end{bmatrix} S_{LD} + \begin{bmatrix} n_1^{(1)} \\ n_1^{(2)*} \\ \vdots \\ n_{n_R}^{(1)} \\ n_{n_R}^{(2)*} \end{bmatrix} ; \qquad (16.19)
$$

which can be expressed as a matrix equation:

$$
Y_{LD} = H_{LD} S_{LD} + N_{LD} , \qquad (16.20)
$$

where H_{LD} is composed of two blocks and is called a Linear Dispersion Matrix. One block of H_{LD} corresponds to the V-BLAST layers, and the other to the Alamouti layers. The V-BLAST block H_{spa} is given by:

$$
H_{spa} = \begin{bmatrix} H_{1,1}^{spa} & H_{1,2}^{spa} & \cdots & H_{1,n_S}^{spa} \\ H_{2,1}^{spa} & H_{2,2}^{spa} & \cdots & H_{2,n_S}^{spa} \\ \vdots & \vdots & \ddots & \vdots \\ H_{n_R,1}^{spa} & H_{n_R,2}^{spa} & \cdots & H_{n_R,n_S}^{spa} \end{bmatrix} , \qquad (16.21)
$$

where

$$
H_{ij}^{spa} = \begin{bmatrix} h_{i,j} & 0 \\ 0 & -h_{i,j}^* \end{bmatrix} , \qquad (16.22)
$$

for $i = 1,2,\cdots,n_R$ and $j = 1,2,\cdots,n_S$. The Alamouti block H_A is itself a block matrix; it is given by:

$$
H_A = \begin{bmatrix} H_{1,1}^A & H_{1,2}^A & \cdots & H_{1,n_B}^A \\ H_{2,1}^A & H_{2,2}^A & \cdots & H_{2,n_B}^A \\ \vdots & \vdots & \ddots & \vdots \\ H_{n_R,1}^A & H_{n_R,2}^A & \cdots & H_{n_R,n_B}^A \end{bmatrix} , \qquad (16.23)
$$

where each element of H_A is given by:

$$H_{kB}^A = \begin{bmatrix} h_{k,2B+n_S-1} & h_{k,2B+n_S} \\ h_{k,2B+n_S}^* & -h_{k,2B+n_S-1}^* \end{bmatrix} , \tag{16.24}$$

for $k = 1, 2, \cdots, n_R$ and $B = 1, 2, \cdots, n_B$. The matrix H_{ij}^{spa} is the portion of H_{LD} that links the jth spatial antenna with the ith receiver antenna. Likewise, H_{kB}^A links the Bth Alamouti block to the kth receiver antenna. To complete the reformulation of system equation (16.18), we need to rearrange matrix S. We define S_{LD} as:

$$S_{LD} = \begin{bmatrix} S_{LD}^{spa} \\ S_{LD}^A \end{bmatrix} , \tag{16.25}$$

where

$$S_{LD}^{spa} = \begin{bmatrix} s_1^{(1)} s_1^{(2)} \cdots s_{n_S}^{(1)} s_{n_S}^{(2)} \end{bmatrix}^t , \tag{16.26}$$

and

$$S_{LD}^A = \begin{bmatrix} S_1^{(1)A} \\ \vdots \\ S_{n_B}^{(1)A} \end{bmatrix} . \tag{16.27}$$

The reformulation of equation (16.17) as equation (16.19) allows us to consider the hybrid MIMO system as a simpler, equivalent, purely spatial system with $N_T = 2(n_S + n_B)$ transmit antennas and without distinction between the Alamouti and V-BLAST layers. This simpler system is shown in Fig. 16.4. We are now ready to propose a receiver algorithm, based on the *sorted-QR decomposition* (SQRD) [19] and OSIC linear detection, that takes advantage of the structure of the linear dispersion matrices to achieve low complexity and high performance.

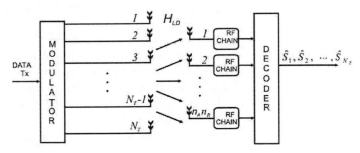

Fig. 16.4. *LD STBC-VBLAST* hybrid code expressed as a purely spatial linear dispersion code.

16.4 Detection Algorithm

16.4.1 OSIC Detection for *LD STBC-VBLAST* Hybrid Code

We first find $Q_{LD}R_{LD} = H_{LD}$, where Q_{LD} is unitary and R_{LD} is upper triangular. By multiplying the received signal Y_{LD} by Q_{LD}^H, we obtain a new received vector

$$\tilde{Y}_{LD} = Q_{LD}^H Y_{LD} = R_{LD} S_{LD} + \tilde{N}_{LD} .$$

Note that the statistical properties of the noise term $\tilde{N}_{LD}$ are the same as N. Due to the upper triangular structure of R_{LD}, the jth element of $\tilde{Y}_{LD}$ is given by:

$$\tilde{y}_j = r_{j,j} s_j + \sum_{i=j+1}^{N_T} r_{j,i} \cdot s_i + \tilde{n}_j .$$

The symbols are estimated in sequence, from lower stream to higher stream, using OSIC; assuming that all previous decisions are correct, the interference can be perfectly canceled at each step except for the noise. The estimated symbol $\hat{s}_j$ is given by:

$$\hat{s}_j = \mathbf{D} \left[\frac{\tilde{y}_j - \sum_{i=j+1}^{N_T} r_{j,i} \cdot \hat{s}_i}{r_{j,j}} \right] ,$$

where $\hat{s}_j$ is the estimate of s_j and $\mathbf{D}[\cdot]$ is a decision device that maps its argument to the closest constellation point. The SQRD may be obtained using Givens rotations using an algorithm we call *Hybrid Coding (HC) Sorted QR Decomposition* or HC-SQRD.

16.4.2 *HC Sorted QR* Decomposition

Note that matrix $H_{LD} \in \mathbb{C}^{2n_R \times N_T}$. A direct application of Givens rotations to calculate QR decomposition on it would result in very high complexity. This complexity may be reduced substantially by taking advantage of the matrix's structure. At the same time, we will implement Givens rotations using CORDIC, to allow an eventual hardware implementation.

From (16.21), (16.22), (16.23) and (16.24) we can see that many of the elements of H_{LD} are equal, and their locations in each matrix are fixed and can be calculated in advance. The method proposed involves obtaining the QR decomposition of H_{LD} in two stages: first we obtain the QR decomposition corresponding to the spatial layers of the hybrid system; in the second stage we calculate the QR decomposition for the diversity layers. A block diagram of the process is shown in Fig. 16.5.

In the first step, we calculate the Sorted QR decomposition of matrix H_m defined as:

$$H_m = \begin{bmatrix} h_{1,1} & h_{1,2} & \cdots & h_{1,n_T} & y_{1,1} & y_{1,2} \\ h_{2,1} & h_{2,2} & \cdots & h_{2,n_T} & y_{2,1} & y_{2,2} \\ \vdots & \vdots & \ddots & \vdots & \vdots & \vdots \\ h_{n_R,1} & h_{n_R,2} & \cdots & h_{n_R,n_T} & y_{n_R,1} & y_{n_R,2} \end{bmatrix} . \tag{16.28}$$

We apply a series of orthogonal transformations Θ using Givens rotations on H_m to obtain matrix $\overline{H}_m$:

$$
\overline{H}_m = \begin{bmatrix}
r_{1,1} & r_{1,2} & \cdots & r_{1,n_S} & r_{1,k} & \cdots & r_{1,n_T} & \tilde{y}_{1,1} & \tilde{y}_{1,2} \\
0 & r_{2,2} & \cdots & r_{2,n_S} & r_{2,k} & \cdots & r_{2,n_T} & \tilde{y}_{2,1} & \tilde{y}_{2,2} \\
\vdots & \vdots & \ddots & \vdots & \vdots & \vdots & \vdots & \vdots & \vdots \\
0 & 0 & \cdots & r_{n_S,n_S} & r_{n_S,k} & \cdots & r_{n_S,n_T} & \tilde{y}_{n_S,1} & \tilde{y}_{n_S,2} \\
0 & 0 & \cdots & 0 & \overline{h}_{k,k} & \cdots & \overline{h}_{k,n_T} & \overline{y}_{k,1} & \overline{y}_{k,2} \\
0 & 0 & \cdots & 0 & \overline{h}_{k+1,k} & \cdots & \overline{h}_{k+1,n_T} & \overline{y}_{k+1,1} & \overline{y}_{k+1,2} \\
\vdots & \vdots & \ddots & \vdots & \vdots & \ddots & \vdots & \vdots & \vdots \\
0 & 0 & \cdots & 0 & \overline{h}_{n_R,k} & \cdots & \overline{h}_{n_R,n_T} & \overline{y}_{n_R,1} & \overline{y}_{n_R,2}
\end{bmatrix}, \tag{16.29}
$$

where $k = n_S + 1$. In this process, we also produce a vector *order* which specifies the detection order of the spatial layers. The operations performed on H_m are detailed in algorithm (1), where CORDIC operations are denoted as follows. Let $Z = a + jb$ a complex number. The CORDIC vectorial operation is denoted $[X_r\, \theta] = \mathbf{V}(a,b)$, and the CORDIC rotation is denoted $[X_r X_i] = \mathbf{R}(a,b,\theta)$.

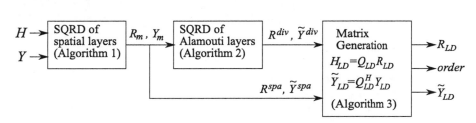

Fig. 16.5. *HC Sorted QR* process using Givens rotations.

Next, we select the first n_S rows of $\overline{H}_m$ to build the matrices R^{spa} and $\tilde{Y}^{spa}$:

$$
R^{spa} = \begin{bmatrix}
r_{1,1} & r_{1,2} & \cdots & r_{1,n_T} \\
0 & r_{2,2} & \cdots & r_{2,n_T} \\
\vdots & \vdots & \ddots & \vdots \\
0 & 0 & \cdots & r_{n_S,n_T}
\end{bmatrix}, \tag{16.30}
$$

$$
Y^{spa} = \begin{bmatrix}
\tilde{y}_{1,1} & \tilde{y}_{1,2} \\
\tilde{y}_{2,1} & \tilde{y}_{2,2} \\
\vdots & \vdots \\
\tilde{y}_{n_S,1} & \tilde{y}_{n_S,2}
\end{bmatrix}. \tag{16.31}
$$

The matrices R^{spa} and $\tilde{Y}^{spa}$ represent the contribution of the spatial layers in the hybrid scheme. The next step is building matrix H_{div}^{ala} for the Alamouti layers, from elements $\overline{h}_{i,j}$ in (16.29), as follows (assuming $k = n_S + 1$):

Algorithm 1 Spatial Decomposition of *HC Sorted QR* using Givens Rotations

1: INPUT: $H_m^{n_R \times n_T + L}$, L, N_T, n_S.
2: OUTPUT: $\overline{H}_m$ and vector *order*
3: $\overline{H}_m = H_m$, $order = [1 : 1 : N_T]$.
4: **for** $i = 1$ to n_S **do**
5:      $k = \text{argmin}_{j=i:n_S} \|(\overline{H}_m(:,j))\|_2$
6:      exchange columns i and k of $\overline{H}_m$.
7:      exchange columns $Li : -1 : Li - (L-1)$ and $Lk : -1 : Lk - (k-1)$ of *order*
8:      **for** $l = i + 1$ to n_R **do**
9:          **for** $m = n_T + 2 : 2$ to $n_T + L$ **do**
10:             $\overline{H}_m(l,m) = -\overline{H}_m(1,m)^*$
11:         **end for**
12:     **end for**
13:     **for** $l = i$ to n_R **do**
14:         $[x_r\, \theta] = \mathbf{V}\left(\mathbf{Re}(\overline{H}_m(l,i)), \mathbf{Im}(\overline{H}_m(l,i))\right)$
15:         $\overline{H}_m(l,i) = x_r + j * 0$
16:         **for** $m = i + 1$ to $n_T + L$ **do**
17:             $[x_r\, x_i] = \mathbf{R}\left(\mathbf{Re}(\overline{H}_m(l,m)), \mathbf{Im}(\overline{H}_m(l,m)), \theta\right)$
18:             $\overline{H}_m(l,m) = x_r + j * x_i$
19:         **end for**
20:     **end for**
21:     **for** $l = i + 1$ to n_R **do**
22:         **for** $m = n_T + 2 : 2$ to $n_T + L$ **do**
23:             $\overline{H}_m(l,m) = -\overline{H}_m(1,m)^*$
24:         **end for**
25:     **end for**
26:     **for** $l = n_R : -1$ to $i + 1$ **do**
27:         $[x_r\, \phi] = \mathbf{V}\left(\mathbf{Re}(\overline{H}_m(l-1,i)), \mathbf{Re}(\overline{H}_m(l-1,i))\right)$
28:         $\overline{H}_m(l-1,i) = x_r + j * 0$
29:         $\overline{H}_m(l,i) = 0 + j * 0$
30:         **for** $m = i + 1$ to $n_T + L$ **do**
31:             $[x_1\, x_2] = \mathbf{R}\left(\mathbf{Re}(\overline{H}_m(l-1,m)), \mathbf{Re}(\overline{H}_m(l,m)), \phi\right)$
32:             $[x_3\, x_4] = \mathbf{R}\left(\mathbf{Im}(\overline{H}_m(l-1,m)), \mathbf{Im}(\overline{H}_m(l,m)), \phi\right)$
33:             $\overline{H}_m(l-1,m) = x_{r1} + j * x_{i1}$
34:             $\overline{H}_m(l,m) = x_{r2} + j * x_{i2}$
35:         **end for**
36:     **end for**
37: **end for**

$$H_{div}^{ala} = \begin{bmatrix} \overline{h}_{k,k} & \overline{h}_{k,k+1}^* & \cdots & \overline{h}_{k,n_T-1} & \overline{h}_{k,n_T}^* & \tilde{y}_{k,1} & \tilde{y}_{k,2} \\ \overline{h}_{k+1,k} & \overline{h}_{k+1,k+1}^* & \cdots & \overline{h}_{k+1,n_T-1} & \overline{h}_{k+1,n_T}^* & \tilde{y}_{k+1,1} & \tilde{y}_{k+1,2} \\ \vdots & \vdots & \ddots & \vdots & \vdots & \vdots & \vdots \\ \overline{h}_{n_R,k} & \overline{h}_{n_R,k+1}^* & \cdots & \overline{h}_{n_R,n_T-1} & \overline{h}_{n_R,n_T}^* & \tilde{y}_{n_R,1} & \tilde{y}_{n_R,2} \end{bmatrix}. \tag{16.32}$$

The SQRD of this matrix may be carried out using algorithm 2.

Finally, the matrices R_{LD} and $\tilde{Y}_{LD}$ are generated from the matrices R^{spa}, R^{div}, $\tilde{Y}^{spa}$ and $\tilde{Y}^{div}$. The construction process is described in algorithm 3. Once the matrices R_{LD} and $\tilde{Y}_{LD}$ are generated the detection of the received symbols may be carried out according to the procedure described in section 16.4.1.

16.5 Analysis of Code Performance

In this section we compare the BER versus average signal-to-noise ratio (SNR) performance of *LD STBC-VBLAST*, STBC-VBLAST [7], QR Group Receiver [8], and Precoded STBC-VBLAST [20]. In all cases we assume 16-QAM modulation and a block length $L = 2$. Simulations are run until 400 block errors are found. The code rate has been fixed to 3 symbols (12 bits) per channel use.

In Fig. 6(a), we show the BER performance comparison (using double-precision floating point arithmetic and Modified Gram-Schmidt to calculate *HC Sorted QR*) between QR Group Receiver 6×6, STBC-VBLAST 6×6 (2,2,3), and *LD STBC-VBLAST* with $n_R = 6$, $n_T = 6$ and $n_B = 3$. With $n_T = 6$, $n_R = 6$ and $n_B = 3$, at BER= 10^{-3}, *LD STBC-VBLAST* performs substantially better (by several dB).

In Fig. 6(b), we compare *LD STBC-VBLAST* with Precoded STBC-VBLAST. This hybrid code requires CSI at the transmitter, which is used to optimally allocate antenna power. 3 symbols per channel use requires a precoded system of size 6×6. Without CSI and the extra cost of precoding, *LD STBC-VBLAST* with 6 antennas per side outperforms the precoded system by 2dB for a BER= 10^{-4}.

As previously mentioned, the *HC Sorted QR* matrix may be calculated with help from the CORDIC algorithm. In Fig. 16.7 we compare MGS with CORDIC with 5, 6 and 7 iterations for a 6×6 system with 3 Alamouti encoders in the transmitter. As can be seen in the figure, 7 iterations suffice to achieve BER performance just a fraction of a dB worse than MGS with double precision.

16.5.1 Performance on Correlated Channels

So far it has been assumed that the channel matrix has independent elements. A more realistic situation is that there is some degree of correlation present, which results in a loss of diversity and performance. To illustrate this loss, we introduce channel correlation in the following way: we assume a 90 degree transmit angle, 30 degree receive angle, and three different correlation distributions: Uniform, Laplacian, and Gaussian. In Fig. 8(a) the performance loss is clearly visible, with Laplacian correlation the worst case (a loss of nearly 8dB at BER= 10^{-3}).

Algorithm 2 Diversity Decomposition of *HC Sorted QR* using Givens Rotations

1: INPUT: H_{div}^{ala}, n_S, n_B, L, *order*.
2: OUTPUT: R^{div}, $\tilde{Y}^{div}$ and vector *order*
3: $R^{div} = H_{div}^{ala}$, $m = Ln_S$,*indice* $= 1$.
4: **for** $i = 1 : i = i + 2$ to $2n_B$ **do**
5:      $k = \text{argmin}_{j=i:2:2n_B} \|R^{div}(:,j)\|_2$
6:      Exchange columns k and $k+1$ for i and $i+1$ of R^{div}.
7:      Exchange columns $m + Li$ and $m + Li - 1$ for $m + Lk$ and $m + Lk - 1$ of *order*
8:      **for** $l = indice$ to n_R **do**
9:          $[x_1 \; \theta_1] = \mathbf{V}\left(\mathbf{Re}(R^{div}(l,i)), \mathbf{Im}(R^{div}(l,i))\right)$
10:         $[x_2 \; \theta_2] = \mathbf{V}\left(\mathbf{Re}(R^{div}(l,l+1)), \mathbf{Im}(R^{div}(l,i+1))\right)$
11:         $R^{div}(l,i) = x_1 + j*0$, $R^{div}(l,i+1) = x_2 + j*0$
12:         $theta1(l) = -\theta_1$, $theta2(l) = -\theta_2$
13:         **for** $m = i+2 : 2$ to $n_T + L$ **do**
14:             $[x_r \; x_i] = \mathbf{R}\left(\mathbf{Re}(R^{div}(l,m)), \mathbf{Im}(R^{div}(l,m)), \theta_1\right)$
15:             $R^{div}(l,m) = x_r + j*x_i$
16:         **end for**
17:         **for** $m = i+3 : 2$ to $n_T + L$ **do**
18:             $[x_r \; x_i] = \mathbf{R}\left(\mathbf{Re}(R^{div}(l,m)), \mathbf{Im}(R^{div}(l,m)), \theta_2\right)$
19:             $R^{div}(l,m) = x_r + j*x_i$
20:         **end for**
21:     **end for**
22:     **for** $l = indice$ to n_R **do**
23:         $[x_r \; \phi] = \mathbf{V}\left(\mathbf{Re}(R^{div}(l,i)), \mathbf{Re}(R^{div}(l,i+1))\right)$
24:         $R^{div}(l,i) = x_r + j*0$, $R^{div}(l,i+1) = 0 + j*0$
25:         **for** $m = i+2 : 2$ to $n_T + L$ **do**
26:             $[x_{r1} \; x_{r2}] = \mathbf{R}\left(\mathbf{Re}(R^{div}(l,m)), \mathbf{Re}(R^{div}(l,m+1)), \phi\right)$
27:             $[x_{i1} \; x_{i2}] = \mathbf{R}\left(\mathbf{Im}(R^{div}(l,m)), \mathbf{Im}(R^{div}(l,m+1)), \phi\right)$
28:             $R^{div}(l,m) = x_{r1} + j*x_{i1}$, $R^{div}(l,m+1) = x_{r2} + j*x_{i2}$
29:         **end for**
30:     **end for**
31:     **for** $l = indice$ to n_R **do**
32:         **for** $m = i+3 : 2$ to $n_T + L$ **do**
33:             $[x_{r1} \; x_{i1}] = \mathbf{R}\left(\mathbf{Re}(R^{div}(l,m)), \mathbf{Im}(R^{div}(l,m)), theta_1(l)\right)$
34:             $[x_{r2} \; x_{i2}] = \mathbf{R}\left(\mathbf{Re}(R^{div}(l,m)), \mathbf{Im}(R^{div}(l,m)), theta_2(l)\right)$
35:             $R^{div}(l,m) = x_{r1} + j*x_{i1}$, $R^{div}(l,m) = x_{r2} + j*x_{i2}$
36:         **end for**
37:     **end for**
38:     **for** $l = indice : -1$ to $indice + 1$ **do**
39:         $[x_r \; \phi] = \mathbf{V}\left(\mathbf{Re}(R^{div}(l-1,i)), \mathbf{Re}(R^{div}(l,i))\right)$
40:         $R^{div}(l-1,i) = x_r + j*0$, $R^{div}(l,i) = 0 + j*0$
41:         **for** $m = i+2$ to $n_T + L$ **do**
42:             $[x_{r1} \; x_{r2}] = \mathbf{R}\left(\mathbf{Re}(R^{div}(l-1,m)), \mathbf{Re}(R^{div}(l,m)), \phi\right)$
43:             $[x_{i1} \; x_{i2}] = \mathbf{R}\left(\mathbf{Im}(R^{div}(l-1,m)), \mathbf{Im}(R^{div}(l,m)), \phi\right)$
44:             $R^{div}(l-1,m) = x_{r1} + j*x_{i1}$, $R^{div}(l,m) = x_{r2} + j*x_{i2}$
45:         **end for**
46:     **end for**
47:     $indice = indice + 1$
48: **end for**
49: $\tilde{Y}^{div} = \begin{bmatrix} R^{div}(1,n_T+1) & R^{div}(1,n_T+2) \\ R^{div}(2,n_T+1) & R^{div}(2,n_T+2) \\ \vdots & \vdots \\ R^{div}(n_B,n_T+1) & R^{div}(n_B,n_T+2) \end{bmatrix}$

Algorithm 3 Obtain R_{LD} and $\tilde{Y}_{LD}$ from R^{spa}, R^{div}, $\tilde{Y}^{spa}$ and $\tilde{Y}^{div}$

1: INPUT: R^{spa}, R^{div}, $\tilde{Y}^{spa}$, $\tilde{Y}^{div}$, n_S, n_B.
2: OUTPUT: R_{LD}, $\tilde{Y}_{LD}$.
3: Let $R_{LD}^{2n_R \times 2(n_S + nB)} = 0$.
4: Built R_{LD}^{spa} and R_{LD}^{A}
5: $row = 1$
6: **for** $k = 1$ to n_S **do**
7:     $R_{LD}^{spa}(row, 1 : 2 : 2n_S - 1) = R^{spa}(k, 1 : n_S)$
8:     $R_{LD}^{spa}(row + 1, 2 : 2 : 2n_S) = -R^{spa}(k, 1 : n_S)^*$
9:     $row = row + 2$
10: **end for**
11: $row = 1$, $col = 2n_S + 1$
12: **for** $i = 1$ to $n_S + n_B$ **do**
13:     **for** $j = 1$ to n_B **do**
14:         $R_{LD}^{spa}(row, col) = R^{spa}(i, 2j - 1 + n_S)$
15:         $R_{LD}^{spa}(row, col + 1) = R^{spa}(i, 2j + n_S)$
16:         $R_{LD}^{spa}(row + 1, col) = R^{spa}(i, 2j + n_S)^*$
17:         $R_{LD}^{spa}(row + 1, col + 1) = -R^{spa}(i, 2j - 1 + n_S)^*$
18:         $col = col + 2$
19:     **end for**
20:     $col = 2n_S + 1$
21:     $row = row + 2$
22: **end for**
23: $R_{LD}^{A}(2 * n_S + 1 : 2 * (n_S + n_B), 2 * n_S + 1 : 2 * (n_S + n_B)) = R^{div}$
24: $R_{LD} = \begin{bmatrix} R_{LD}^{spa} \\ R_{LD}^{A} \end{bmatrix}$,
25: Built vector $\tilde{Y}_{LD}$
26: $rows = 1$
27: **for** $k = 1$ to n_S **do**
28:     $\tilde{Y}_{LD}^{spa}(rows, 1) = \tilde{Y}^{spa}(k, 1)$
29:     $\tilde{Y}_{LD}^{spa}(rows + 1, 1) = \tilde{Y}^{spa}(k, 2)$
30:     $rows = rows + 2$
31: **end for**
32: $rows = 1$
33: **for** $k = 1$ to $2n_B$ **do**
34:     $\tilde{Y}_{LD}^{div}(rows, 1) = \tilde{Y}^{div}(k, 1)$
35:     $\tilde{Y}_{LD}^{div}(rows + 1, 1) = \tilde{Y}^{div}(k, 2)$
36:     $rows = rows + 2$
37: **end for**
38: $\tilde{Y}_{LD} = \begin{bmatrix} \tilde{Y}_{LD}^{spa} \\ \tilde{Y}_{LD}^{div} \end{bmatrix}$

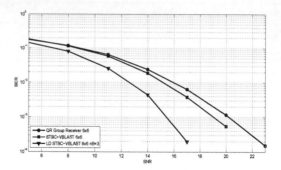

(a) BER vs. average SNR of *LD STBC-VBLAST* , STBC-VBLAST 6×6 $(2, 2, 3)$, and QR Group Receiver 6×6.

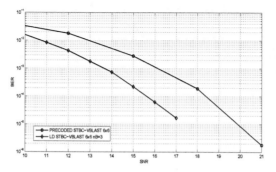

(b) BER vs. average SNR of *LD STBC-VBLAST* and Precoded STBC-VBLAST.

Fig. 16.6. BER vs. average SNR comparison.

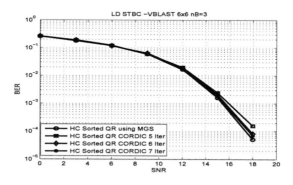

Fig. 16.7. BER vs. SNR of *LD STBC-VBLAST* 6×6, $n_B = 3$ using CORDIC algorithm

The results presented above assume the *HC Sorted QR* matrix has been calculated with MGS with double precision arithmetic. It is interesting to see the effect that correlation has on a more feasible practical implementation using the CORDIC algorithm. As seen in Fig. 8(b), the number of CORDIC iterations required to get results comparable to MGS has increased to 9. Correlation not only negatively impacts the receiver's BER performance, but it results in increased complexity as well. These results correspond to Laplacian distribution with the same angles as before.

The reason for increased receiver complexity is explained by the magnitude of the diagonal elements of matrix R_{LD}. Correlation results in smaller values, a good approximation of which requires more CORDIC iterations.

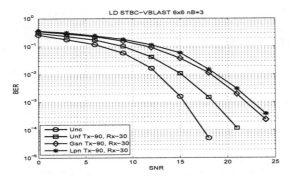

(a) BER vs. SNR of 6×6 *LD STBC-VBLAST* with $n_B = 3$ on a correlated channel with 90 degree transmit angle and 30 degree receive angle. *Unc* is uncorrelated channel (included for reference), *Unf* is uniformly distributed correlation, *Gsn* is Gaussian, and *Lpn* is Laplacian.

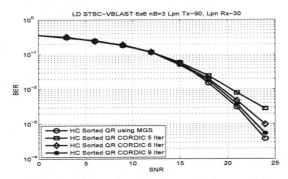

(b) BER vs. SNR of 6×6 *LD STBC-VBLAST* , $n_B = 3$, using CORDIC algorithm and correlated (Laplacian) channel.

Fig. 16.8. Performance of *LD STBC-VBLAST* on correlated channels.

References

1. L. Zheng and D. Tse, "Diversity and multiplexing: fundamental tradeoff in multiple antenna channels", *IEEE Transactions on Information Theory*, vol. 49, pp. 1073-96, May 2003.
2. G. J. Foschini, "Layered Space-Time Architecture for Wireless Communication in a Fading Environment When Using Multi-Element Antennas", Bell Lab. Tech. J., Vol. 1, No. 2, 1996, pp. 41-59.
3. S. M. Alamouti, "A Simple Transmit Diversity Technique For Wireless Communications", IEEE J. Selected Areas in Communications, Vol. 16, No. 8, Oct. 1998, pp. 1451-1458.
4. Oggier F., Belfiore J.-C. and Viterbo E. (2007). Cyclic Division Algebras: A Tool for Space-Time Coding. Foundations and Trends in Communications and Information Theory, Vol. 4, No. 1, pp. 1-95, 2007, ISBN: 978-1-60198-050-2
5. Berhuy, G. and Oggier, F., "On the Existence of Perfect SpaceTime Codes," Information Theory, IEEE Transactions on, vol.55, no.5, pp.2078-2082, May 2009
6. H. Kim, H. Park, T. Kim and I. Eo: Performance Analysis of a DSTTD System with Decision-Feedback Detection, in Proceedings of IEEE International Conference on Acoustics, Speech and Signal Processing, Vol. 4, (2006), 749–752.
7. T. Mao and M. Motani: STBC-VBLAST for MIMO wireless communication systems, in Proceedings of International Conference on Communications, Vol. 4, (2005), 2266–2270.
8. L. Zhao and V. K. Dubey: Detection Schemes for Space-time Block Code and Spatial Multiplexing Combined System, IEEE Communication Letters, Vol. 9, No. 1, (2005), 49–51.
9. B. Hassibi and B. M. Hochwald: High-rate codes that are linear in space and time, IEEE Transactions on Information Theory, Vol. 48, No. 7, (2002), 1804–1824.
10. O. Longoria-Gandara et al: Linear Dispersion Codes Generation from Hybrid STBC-VBLAST Architectures, in Proceedings of 4th International Conference on Electrical and Electronics Engineering, (2007).
11. I. E. Telatar, "Capacity of multiple-antenna Gaussian channels," European Trans. Telecomm., vol. 10, no. 6, pp. 585-595, 1999.
12. Gesbert, D.; Bolcskei, H.; Gore, D.A.; Paulraj, .J, "Outdoor MIMO wireless channels: models and performance prediction." IEEE Trans. Commun., vol. 50, no. 12, Dec. 2002, pp. 1926-1934.
13. D–S Shiu, G. J. Foshini, M. J. Gans & J. M. Kahn, "Fading Correlation and Its Effect on the Capacity of Multielement Antenna systems," IEEE Transactions on Communications, Vol. 48, No. 3, pp 502–512, March 2000.
14. J.R. Fonollosa, R. Gaspa, X. Mestre, A. Pages, M. Heikkila, J.P. Kermoal, L. Schumacher, A. Pollard & J. Ylitalo. "The IST METRA Project". IEEE Communications Magazine, Vol. 40, No. 7, pp. 78-86. July 2002.
15. A. F. Molisch, H. Asplund, R. Heddergott, M. Steinbauer & T. Zwick. "The COST259 Directional Channel Model–Part I: Overview and Methodology". IEEE Transactions on Wireless Communications, Vol. 5, No. 12, pp. 3421–3433. December 2006.
16. S. Sesia, I. Toufik, M. Baker. "LTE The UMTS Long Term Evolution, from Theory to Practice". John Wiley & Sons Ltd, West Sussex, UK, 2009.
17. R. Parra-Michel, A. Alcocer-Ochoa, A. Sanchez-Hernandez, & Valeri Kontorovich. "Recent Advances in Signal Processing". Chapter "MIMO Channel Modeling and Simulation". ISBN 978-953-7619-41-1. I-Tech Education and Publishing KG.
18. J. Cortez, M. Bazdresch and A. Garcia, "Low Complexity Detector for STBC-VBLAST Architecture Based on SQRD Decomposition", 20th IEEE Personal, Indoor and Mobile Radio Communications Symposium, Tokyo, Japan, September 2009.

19. D. Wubben et al: Efficient algorithm for decoding layered space-time codes, Electronics Letters, Vol. 37, No. 22, (2001), 1348–150.
20. C. Meng and J. Tuqan: Precoded STBC-VBLAST for MIMO Wireless Communication Systems, in Proceedings of IEEE International Conference on Acoustics, Speech and Signal Processing (ICASSP 2007), Vol. 3 (2007), 337–340.

17

Calculation of the Radiation Pattern of On-Board Antennas Placed over Complex Structures using Physical Optics Including the Contribution of the Shadowed Areas

Lorena Lozano, Francisco Saez de Adana, and Manuel Felipe Cátedra

Dprt. of Computer Science, Universidad de Alcalá, Alcalá de Henares (Madrid), Spain
kiko.saez@uah.es

Summary. A numerical integration considering shadowed and illuminated currents in Physical Optics (PO) is used to calculate the radiation pattern of antennas on complex structures modeled by NURBS. This method is used when the antenna is placed at a distance less than one wavelength of the structure. The contribution of the illuminated area is obtained using the classical PO formulation. A creeping wave approach is used to determine the induced current in the shadowed areas. The contribution of shadow currents improves the results compared to measurements with other PO techniques, such as the Stationary Phase Method (SPM) or numerical integration considering only illuminated currents.

17.1 Introduction

The application of numerical techniques to the analysis of on-board antennas over arbitrary shaped structures is a problem that has been extensively treated in the literature in the last decades [1, 2, 3, 4, 5, 6]. Two elements must be considered when this problem is proposed. The first one is purely related to electromagnetic field analysis and consists of selecting the proper numerical technique to solve the problem under study. The second element is more related to geometric design and requires finding an accurate and suitable representation of the body to improve the quality of the predictions without increasing the computational load associated with the geometric information.

Regarding the first element (the numerical technique), high frequency or asymptotic techniques have been widely used when the structure under analysis is electrically large (several wavelengths). The most used asymptotic techniques for this kind of problem are the Uniform Theory of Diffraction (UTD)[7, 8] and Physical Optics (PO) [9, 10]. The first is a ray-based technique, and the second is based on calculating the surface induced current over the body under analysis. There are several examples in the open literature of the application of both techniques to the analysis of on-board antennas [4, 5, 6].

The PO approach has revealed itself as one of the more appropriate techniques for the analysis of high frequency problems [11, 12]. The main concern with the application of this technique is the computational cost associated with the calculation of the PO integral. Several techniques have been presented to reduce this computational cost, especially if the structure is very large in terms of the wavelength. Recently, one of the most popular methods is the Stationary Phase Method (SPM) [13, 14].

With respect to geometric representation, in the world of the Computer Aided Graphic Design (CAGD), the representation of complex bodies with arbitrary shape by means of Non Uniform Rational B-Splines (NURBS) surfaces is becoming one of the most popular representations because it combines precision in the representation of the body and low amount of required data. Thus, this representation, which initially was born in the aeronautic industry, has been adopted extensively throughout the world of electromagnetic analysis [6, 11, 15]. For instance, a combination of the PO with a NURBS representation has been proposed [6].

However, the application of the PO to the analysis of complex structures modeled by NURBS presents a problem directly related with the PO formulation. The classical PO formulation states that the induced current is zero outside the illuminated area of the structure (the area directly visible from the antenna). This approach does not cause any problem when the antenna is far enough from the structure and, therefore, most of the structure is illuminated. However, if the antenna is very close (less than one lambda) to the structure, the boundary between the illuminated and the shadow areas (the shadow boundary) is very close to the antenna. In this case, considering only the contribution of the illuminated area produces considerable error in the radiation pattern prediction. This error is especially sensitive in the case of the application of the SPM, where some areas can appear without the contribution of any stationary phase point.

This problem has been resolved in the past using several approaches. One of the most popular is to consider the contribution of the shadow boundary as a diffraction that introduces some diffraction coefficients in the PO formulation [16, 17]. This solution, which is very efficient from an electromagnetic point of view, causes problems when the structure is complex because the exact shadow boundary must be determined. In this case, for instance, determining the shadow boundary is not an easy task when the NURBS representation is used and can be very expensive from a computational point of view.

An alternative to avoid this problem is presented here. The idea is to solve the PO integral numerically, but obtain the induced currents in both the illuminated and the shadowed areas. These currents will be calculated at some sample points over the NURBS surface and the contribution of all these points will be considered when the PO integral is computed. The induced currents at the illuminated points are obtained using the classical PO approach, and in the shadowed areas, the induced current is obtained making use of a ray-tracing approach. The field at the shadow boundary is considered to be propagated along the shadow area following a creeping wave. Then, the magnetic field (and thus, the induced current) can be obtained at any point using the creeping wave's expressions of the UTD.

The chapter is organized as follows: Section 17.2 presents a brief summary of a representation using NURBS surfaces. Section 17.3 shows the formulation used to obtain the radiated field by the structure for both the illuminated and the shadowed areas. Some examples that illustrate the effect of the inclusion of the shadowed currents are shown in Section 17.4, and finally, Section 17.5 concludes the chapter.

17.2 Description of the geometry using NURBS surfaces

The representation of a given geometry using NURBS surfaces is a technique that is widely used in the aeronautic, automobile, ship, and other industries because it provides many advantages when representing complex objects. The NURBS scheme is able to manipulate both free-form surfaces and primitive quadric surfaces (e.g., cylinders, spheres, or cones) with a low number of patches, and therefore with a small amount of information. For instance, a complex body, such as a complete aircraft, can be described with nearly all its details by only a few hundred NURBS patches (Fig. 17.1). Today, most of the available computer-aided geometric design (CAGD) tools provide descriptions of the designed objects in terms of NURBS curves and surfaces.

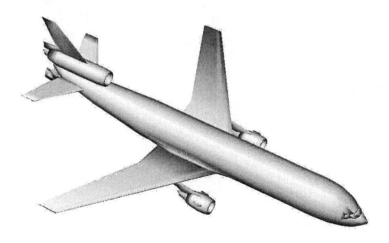

Fig. 17.1. Geometric model of an aircraft using 58 NURBS surfaces

Given these advantages, the use of NURBS patches has been extended in the last years to geometrically represent complex bodies for the numerical analysis of electromagnetic problems, such as the RCS computation or the calculation of the radiation of on-board antennas [6, 11, 15]. The application of this representation for electromagnetic problems has several advantages with respect to other representations including, for instance, the use of flat facets. The following are some of the advantages:

1. Artificial edges are not introduced.
2. The number of patches is much lower.
3. A better fit to the object geometry is provided.
4. A flat facet is a particular case of a NURBS surface.
5. Geometric parameters of the body surface (e.g., normal vectors, curvatures, and principal directions) are easy to obtain from the surface description.

A NURBS surface is a rational piecewise polynomial parametric surface. The coefficients of the polynomials depend on just a few control points (with associated weights) and two knot vectors (Fig. 17.2). NURBS surfaces are quite useful in geometric design for two reasons: they are invariant under affine transformations (e.g., rotation, scaling, and translation) of the control points and the "local control" property [18], which means that altering the position of a single data point only causes changes in part of the surface.

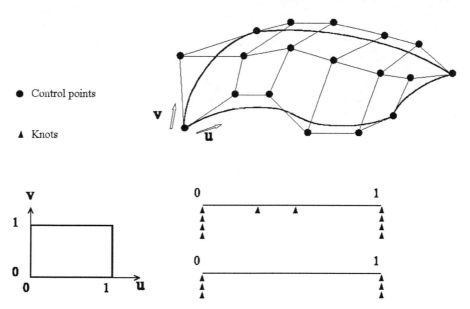

Fig. 17.2. Control points and knot vector used for the NURBS surface description

NURBS surfaces must be written like a combination of rational Bezier patches [18, 19] for application to numerical electromagnetic problems because they are more suitable for numerical computation of parameters associated with its geometry (curvatures, derivatives, integrations). This is a result of the fact that a rational Bezier patch is, as a NURBS patch, a parametric surface, but in this case, it is defined in terms of a linear combination of Bernstein polynomials [18, 19]. The use of the Bernstein basis provides the necessary numerical stability to the Bezier patches to calculate the required parameters for the electromagnetic problems under study

The transformation of NURBS into rational Bezier surfaces is straightforward by applying the Cox-De Boor algorithm [19]. Therefore, for electromagnetic problems, the NURBS format is used for the design and storage of the body's shape and the Bezier format for the surface interrogation of parameters, such as point coordinates and parametric derivatives, as in other applications that use the NURBS representation.

The application of the Cox-De Boor algorithm to a NURBS surface provides a set of rational Bezier patches. The number of Bezier patches obtained from a NURBS depends on its knot vectors. The union of these patches makes up the primitive surface. The continuity between adjacent Bezier patches is determined by the knot vectors of the primitive NURBS surface [18, 19]. The mathematical treatment of Bezier surfaces is simple. They are polynomial parametric surfaces normalized with a weight function. A Bezier patch is defined by two degrees (one for each parametric coordinate), a mesh of control points and a set of associated weights. The coordinates of the surface points and their parametric derivative functions are given in terms of linear combinations of Berstein polynomials that can be easily computed and are numerically stable [18].

The surface points of a rational Bezier surface are given by

$$
r(u,v) = \frac{\sum_{i=0}^{m}\sum_{j=0}^{n} w_{ij} b_{ij} B_i^m(u) B_j^n(v)}{\sum_{i=0}^{m}\sum_{j=0}^{n} w_{ij} B_i^m(u) B_j^n(v)}
\tag{17.1}
$$

where b_{ij} are the control points, w_{ij} are the control point weights, m and n are the surface degrees, and $B_i^m(u)$ and $B_j^n(v)$ are the Berstein polynomials:

$$
B_i^n(t) = \frac{n!}{i!(n-i)!} t^i (1-t)^{n-i}
\tag{17.2}
$$

The other parameters necessary for the numerical electromagnetic treatment, such as parametric derivatives and normal vectors, can be derived from expression (17.1) [18].

17.3 Numerical computation of the PO integral considering the contribution of the illuminated and the shadowed areas

The PO integral to obtain the radiated field when the on-board antenna generates the incident wave is found after the Stratton-Chu integral equations are formulated [20]. This integral is given by the following expression:

$$E_s(r) = -\frac{jw\mu}{2\pi} \frac{e^{-jkr}}{r} \int_S \{\hat{k}_s \times [\hat{n} \times H_0(r')] \times \hat{k}_s\} \, e^{jk(\hat{k}_s \cdot r' - d)} dS' \qquad (17.3)$$

where S is the surface over which the integral is computed, $\hat{n}$ is the unit normal vector to the surface at the integration point, k_s is the observation direction, w is the angular frequency, H_0 is the incident magnetic field and E_s is the electric field radiated by the induced currents at point r.

One of the premises of the PO formulation is that the PO solution is confined to the illuminated area of the body under analysis. However, in certain situations, this premise does not provide accurate results enough because there are geometric areas that, despite being in the shadowed area, contribute to the radiation due to the propagation by surface waves. Therefore, in this case, the integration surface must be divided in two regions: one illuminated and the other shadowed (Fig. 17.3). The total field radiated by the antenna will be the sum of both contributions.

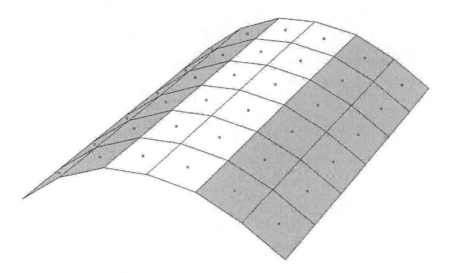

Fig. 17.3. Representation of the sample of points and the illuminated area (in white) and the shadowed area (in gray)

The first step to obtain both contributions is to determine the shadow boundary. A sampling of the surface is accomplished to achieve this goal. Six samples per wavelength are taken, and for every sample, a test is performed to determine whether or not it is illuminated by the source. The criterion to see if a point is illuminated is the normal vector criterion, where the illumination of the point is stated if the scalar product between the incident wave and the normal vector to the surface at this point is positive or zero and shadowed if it is negative. This criterion can be written as follows:

$$\hat{k}_i \cdot \hat{n} \geq 0 \quad \text{illuminated}$$
$$\hat{k}_i \cdot \hat{n} < 0 \quad \text{shadowed}$$

(17.4)

Once the illuminated and shadowed regions are determined, the next step is to compute the contribution of each region.

17.3.1 Contribution of the illuminated area

The contribution of the illuminated area is obtained using the classical PO formulation. Therefore, the incident magnetic field at each point of the visible zone $H_0(r')$ is computed, and after that, the current induced by this magnetic field is computed considering $J(r') = 2\hat{n} \times H_0(r')$.

Once the induced current at each of those points has been obtained, the radiated field is evaluated by calculating the numerical integral using a finite summation:

$$V(i,j) = -\frac{jw\mu}{4\pi} \left(\hat{k}_s \times J_{ij} \times \hat{k}_s \right) e^{jk\hat{k}_s \cdot r'_{ij}}$$

(17.5)

$$E_{\text{Silluminated}} = \sum_{i=2}^{NDu} \sum_{j=2}^{NDv} \Delta_u \Delta_v V(i,j)$$

(17.6)

where NDu is the number of divisions in the parametric coordinate u of the surface, NDv is the number of divisions in the other coordinate v, and Δ_u and Δ_v are the increments in u and v (lambda/6 is as mentioned above).

If the entire area of the body is illuminated by the antenna, the radiated field is only the classical PO integral and the results are accurate enough for high frequency applications, which has been widely proved in the literature [6, 15, 11].

17.3.2 Contribution of the shadowed area

If there is an important part of the body that is shadowed, the formulation of the previous section is not enough to provide a good prediction of the radiated field; the classical PO formulation states that the induced current in the shadowed area is zero. However, using an analogy with ray-tracing (which, it must not be forgotten, is also a high frequency technique), if the surface under analysis is smooth and convex, a new electromagnetic phenomenon appears: diffraction by creeping waves. This phenomenon starts at the boundary between the illuminated area and the shadowed area of the surfaces and produces creeping wave propagation along the shadowed area.

Thus, it can be stated that in the points at the shadow boundary, rays that follow geodesic trajectories along the surface are propagated, which generate the creeping wave. Therefore, the propagation of these rays produces a magnetic field over the points of the shadowed area that were determined previously. Further, this magnetic field will induce a current over these points that must be taken into account in the PO formulation. Therefore, the procedure only consists of determining the geodesic path that crosses the sample points of the surface that are in the shadowed area. These geodesic paths correspond to a creeping wave that originates a magnetic field $H(r')$ on that point. Then, it is easy to obtain the induced current over the point by using the well-known formula $J(r') = 2\hat{n} \times H(r')$.

Therefore, the only thing remaining is to obtain the magnetic field mentioned above. However, it can be easily obtained from the electric field produced by the creeping wave at that given point. This electrical field can be obtained from the Uniform Theory of Diffraction (UTD) formulation for creeping waves [21]. The expression for the electric field is

$$E^d(r) = E^d(Q_2)e^{-jk(r_{Q_2} \cdot \hat{k}_s)} \tag{17.7}$$

where Q_2 is the point over the geodesic trajectory, r_{Q_2} is the position vector of that point, $\hat{k}_s$ is the direction of propagation and $E^d(Q_2)$ is the diffracted field at the point Q_2.

The diffracted field at the point Q_2 is obtained using the following expression:

$$E^d(Q_2) = E^i(Q_1)\overline{\overline{T}} \tag{17.8}$$

where $E^1(Q_1)$ is the incident field at the point Q_1 (field of the incident at the starting point of the trajectory, Fig. 17.4) and $\overline{\overline{T}}$ is the matrix of the creeping wave diffraction coefficients given by

$$T_{s,h} = -\left[\sqrt{m(Q_1)m(Q_2)}\sqrt{\frac{2}{k}}\left(\frac{e^{-j(\pi/4)}}{2\sqrt{\pi}\zeta^d}\left[1 - F(X^d)\right] + \hat{P}_{s,h}(\zeta^d)\right)\right]\sqrt{\frac{s^i}{s^i + t}}e^{-jkt} \tag{17.9}$$

where $m(Q_i)$ is a parameter that depends on the surface curvature at the point Q_i as follows:

$$m(Q_i) = \left(\frac{k\rho_g(Q_i)}{2}\right)^{1/3} \tag{17.10}$$

where $\rho_g(Q_i)$ are the radii of curvature at the point Q_i; s^i is the distance between the antenna and Q_1 and $F(X^d)$ is the Fresnel transition function [8]. The argument X^d is calculated as follows:

$$X^d = \frac{kL^d(\zeta^d)^2}{2m(Q_1)m(Q_2)} \tag{17.11}$$

where L^d is the distance parameter, which is calculated as follows:

$$L^d = \frac{s^d s^i}{s^d + s^i} \tag{17.12}$$

The Fock parameter, ζ^d, is

$$\zeta^d = \int_{Q_1}^{Q_2} dt' \frac{m(t')}{\rho_g(t')} \tag{17.13}$$

where $t = \int_{Q_1}^{Q_2} dt'$ is the arc length covered by the ray over the surface, which is obtained as the product between the number of points in the ray trajectory and the distance between them ($\Delta\sigma$).

The previous integral (17.13) is evaluated as the product of the sum of the value of $\frac{m(t')}{\rho_g(t')}$ at each point of the trajectory with $\Delta\sigma$.

The function $\hat{P}_{s,h}$ is the Pekeris function and can be defined from the Fock functions $p*$ and $q*$ [8] as follows [21]:

$$\hat{P}_{s,h}(\zeta^d) = \left\{\begin{array}{c} p*(\zeta^d) \\ q*(\zeta^d) \end{array}\right\} e^{-j(\pi/4)} - \frac{e^{-j(\pi/4)}}{2\sqrt{\pi}\zeta^d} \tag{17.14}$$

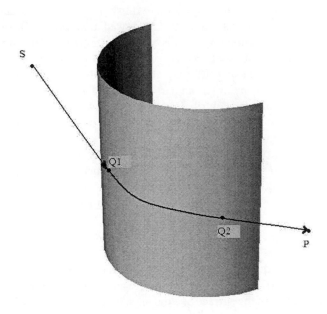

Fig. 17.4. Trajectory of the creeping wave

It is obvious that several geodesic trajectories can exist in the shadowed area of a surface. Therefore, a current is induced on all of these trajectories, and all these currents contribute to the radiated field. This means that the procedure mentioned above must be followed for all the geodesic trajectories over the sampling points. Once the induced current over all the shadowed area is known, the radiated field is obtained using the following expression:

$$V(i,j) = -\frac{jw\mu}{4\pi} \left(\hat{k}_s \times J_{ij} \times \hat{k}_s\right) e^{jk\hat{k}_s \cdot r'_{ij}} \tag{17.15}$$

$$E_{\text{Sshadowed}} = \sum_{i=1}^{Ngeo} \Delta_u \Delta_v V(i,j) \tag{17.16}$$

where *Ngeo* is the number of geodesic points obtained as the sum of the points over all the geodesic curves in the shadowed area, and Δ_u and Δ_v are the steps in u and v, where u and v are the parametric coordinates of the NURBS surface.

Obviously, the radiation pattern of the on-board antenna will be obtained as the sum of the contribution of the radiated field from the illuminated and the shadowed areas.

17.4 Some examples

Some examples are shown in this section to illustrate the improvement obtained with the inclusion of the contribution of the shadowed area to the PO formulation. The results obtained are compared not only with measurements, but also with the case in which only the illuminated area is considered to see the difference. The results considering integration using SPM are also included to demonstrate the inaccuracy of this technique when an important part of the body is shadowed. The results could be improved by adding the contribution of the shadow boundary as mentioned in the introduction, but the associated computational cost makes the procedure unsuitable for complex problems.

In this section, two cases are shown. The first is very simple and consists of a section of a cylinder modeled by only one NURBS surface. Although is not a realistic case, it is interesting because it perfectly illustrates the effect of the inclusion of the shadowed currents. The second case is an aircraft modeled by 58 NURBS surfaces.

17.4.1 Cylindrical section

This geometry is a quarter of a cylinder that is 7.166λ long with a radius of 3.58λ at a frequency of 1075 MHZ (Fig. 17.5). The antenna (two infinitesimal dipoles) is placed in the center of the surface at a distance of $\lambda/8$ from the surface. The analysis of the radiation pattern was performed for a cut with $\phi=90$ as θ varied from 0 to 180. Fig. 17.6 shows the comparison between measurements performed for the θ component of the radiated field with the numerical computation with and without shadowed currents. It is evident that the behavior in the shadowed area is only obtained numerically when the shadowed currents are considered. The computation using SPM is also included, which proves the statement given below about the inaccuracy of the techniques for antennas that are very close to the surface.

17.4.2 Aircraft

Fig. 17.7 shows the geometric model of a realistic aircraft. The source is an electric dipole placed at a distance $\lambda/8$ over the aircraft cabin. The frequency is 12900 MHz. The radiation pattern was obtained for a cut with $\theta=90$ as ϕ varied from 0 to 180.

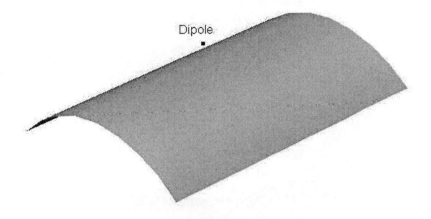

Fig. 17.5. Geometric model and dipole positions

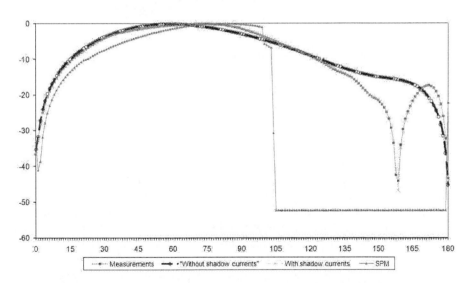

Fig. 17.6. Theta component of the radiation pattern of the cylindrical section

Figs. 17.8 and 17.9 show the comparison between the contribution of both areas for the θ and ϕ components of the radiated field. The difference between the results with and without shadowed currents is illustrated, as well as the inaccuracy of the SPM for this kind of analysis.

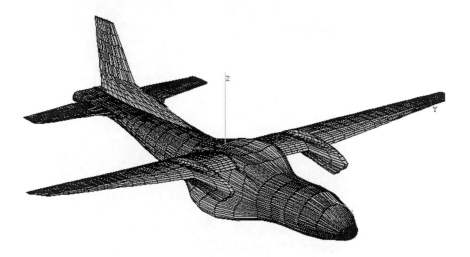

Fig. 17.7. Geometric model of the aircraft and dipole position

17.5 Conclusions

The analysis of the radiation of on-board antennas using PO presents the problem of the inaccuracy of the technique when part of the analyzed body is shadowed from the source. The induced current in that case is not zero, and the classical PO approach is not good enough to accurately predict the radiation pattern of the antenna. In particular, this happens when the antenna is very close to the structure. A method to obtain the induced currents at the shadowed area was presented. These currents are obtained considering that the field at the shadow boundary is propagated along the shadow area following a creeping wave. This ray approach allows the magnetic field at the shadow boundary to be obtained using the UTD formulation for creeping waves. As a result, the induced currents are obtained using the same expression as that for the illuminated areas. Some examples were shown to illustrate the improvement obtained by adding the shadow currents distribution and the advantages with respect to other techniques that compute the PO integral as the SPM.

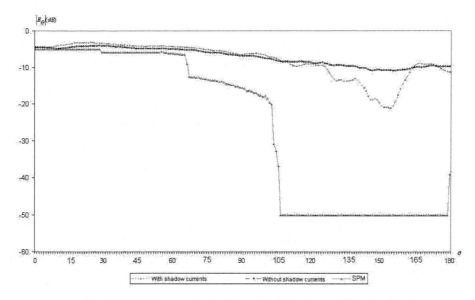

Fig. 17.8. Theta component of the radiation pattern of the aircraft

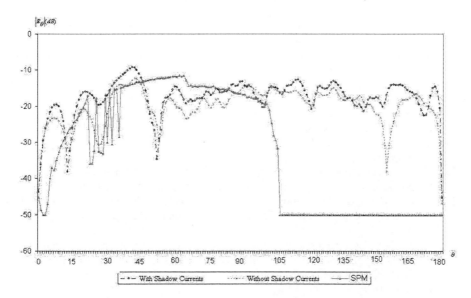

Fig. 17.9. Phi component of the radiation pattern of the aircraft

References

1. Burnside, W.D., Marhefka, R.J., Yu, C.L.: Roll-Plane Analysis of Aircraft Antennas. IEEE Transactions on Antennas and Propagation 21, 780–786, (1973).
2. Burnside, W.D., Gilreath, M.C., Marhefka, R.J., Yu, C.L.: A Study of KC-135 Aircraft Antenna Patterns. IEEE Transactions on Antennas and Propagation 23, 309–316, (1975).
3. Balanis, C.A., Cheng, Y.B.: Antenna Radiation Modeling for Microwave Landing System. IEEE Transactions on Antennas and Propagation 22, 490–497, (1976).
4. Yu, C.L., Burnside, W.D., Gilreath, M.C.: Volumetric Pattern Analysis of Airborne Antennas. IEEE Transactions on Atennas and Propagation, 26, 636–641, (1978).
5. Perez, J., Saiz, J.A., Conde, O.M., Torres, R.P., Catedra, M.F.: Analysis of Antennas on Board Arbitrary Structures Modeled by NURBS Surfaces. IEEE Transactions Antennas and Propagation 45, 1045–1053, (1997).
6. Conde, O.M., Perez, J., Catedra, M.F.: Stationary Phase Method Application for the Analysis of Radiation of Complex 3D Conducting Structures. IEEE Transactions on Antenas and Propagation 49, 724–731 (2001)
7. McNamara, D.A., Pistorius C.W.I., Malherbe J.A.G. Introduction to the Uniform Geometrical Theory of Diffraction. Artech House (1988)
8. Kouyoumjian, R.C., Pathak, P.H.: A Geometrical Theory of Diffraction for an Edge in a Perfectly Conducting Surface. Proceedings of IEEE 62, 1448–1461, (1974).
9. Harrington, R.F.: Time-Harmonic Electromagnetic Fields. New York, McGraw-Hill, (1961).
10. Balanis, C.A.: Advanced Engineering Electromagnetics. John Wiley and Sons, (1989).
11. Perez, J., Catedra, M.F.: Application of Physical Optics to the RCS Computation of Bodies Modeled with NURBS Surfaces. IEEE Transactions on Antennas and Propagation 42, 1404–1411, (1994).
12. Shifflett, J.A.: CADDRAD: A Physical Optics Radar/Radome Analysis code for arbitrary 3D geometries. IEEE Transactions on Antennas and Propagation 39, 73–79, (1997).
13. Ludwig, A.C.: Computation of Radiation Patterns Involving Numerical Double Integration. IEEE Transactions on Antennas and Propagation 16, 767–769, (1968).
14. Jones, D.S., Kline, M.: Asymptotic Expansion of Multiple Integrals and the Method of Stationary Phase. Journal Math. Phys. 37, 1–28, (1958).
15. Rius, J.M., Ferrando, M., Jofre, L.: GRECO: graphical electromagnetic computing for RCS prediction in real time. IEEE Antennas and Propagation Magazine 35, 7–17, (1993).
16. Chuang, C.W.: An Asymptotic Solution for Currents in the Penumbra Region with Discontinuity in Curvature. IEEE Transactions on Antennas and Propagation 34, 728–732, (1986).
17. Michaeli, A.: Elimination of infinities in equivalent edge currents, part I: Fringe current components. IEEE Transactions on Antennas and Propagation 34, 912–918, (1986).
18. Farin, G.: Curves and Surfaces for Computer Aided Geometric Design. San Diego, Academic (1988)
19. Boehm, W.: Generating the Bezier Points of B-Spline Curves and Surfaces. Computer Aided Design 13, 365–366 (1981)
20. Ruck, G.T.: Radar Cross Section Handbook. New York: Plenum Press, (1970).
21. Pathak, P.H. An Asymptotic Analysis of the Scattering of Plane Waves by a Smooth Convex Cylinder. Radio Science 14, 419–435 (1979).

Analisys of both a Voltage-to-Current Converter and a Memory Cell Implemented using the Multiple-Input Floating Gate MOSFET Transistor

A. Medina-Vázquez[1], J. Moreno-Cadenas[1], F. Gómez-Castañeda[1], Luis Martín-Flores[1], and M. E. Meda-Campaña[2]

[1] Department of Electrical Engineering, CINVESTAV-IPN, México City, Mexico
[2] CUCEA, Guadalajara University, Zapopan, Jalisco, México

Summary. Differences and advantages that the Multi-Input Floating Gate MOS (MIFGMOS) transistor has versus the conventional CMOS transistor are shown. In order to do this, the design and implementation of both a Voltage to Current Converter (VIC) cell and a Memory Current Cell (MIC) using MIFGMOS transistors is presented. The development is based in mathematical and simulation analysis as well as in experimental results. Both cells present good performance and linearity according to theoretical analysis and presents low voltage operation and low power consumption, despite the long channel technology. These characteristics are very important in analog and mixed signal applications, like mobile communications systems. The cells presented here can be part of a low-voltage sample and hold circuit but applications are not restricted. Additionally, some comparison between simulation and experimental results obtained when testing five 3-input MIFGMOS transistor are included in order to show the properties and behavior of this transistor.

18.1 Introduction: Multiple-Input Floating Gate MOSFET Transistor

The floating gate MOSFET (FGMOS) transistor is a simple MOSFET transistor with its gate completely insolated, that means, without metal contact to the outside. The gate is immersed in an oxide layer and the electric charge into the floating gate does not flow to either side. For this reason, the floating gate transistor was used in implementing EPROM or EEPROM memories some decades ago [1], because the electric charge remain in the floating gate for a long time. However, currently it is possible to implement a great variety of analog and digital circuits based in FGMOS transistors [2] and [3], respectively. These kinds of circuits can be applied in communications systems as shown in [4], neural network [5], biomedical systems, etc. In general, in all these applications, low supply voltage and low power consumption are desired.

Because the gate of the MOSFET in a FGMOS is floating, it is necessary implement some mechanism to manipulate the electric charge or voltage on the

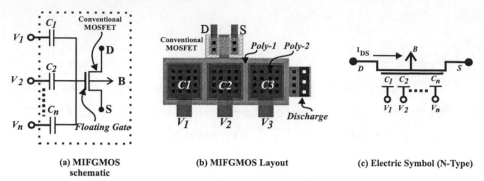

Fig. 18.1. Basic MIFGMOS transistor structure: (a) schematic, (b) 3-input MIFGMOS layout, (c) electric symbol

floating gate. The most popular techniques are: using ultraviolet light [6], Fowler-Nordheim tunneling [7], hot electron injection [8], quasi-floating gate [9], and capacitive coupling [10]. The approach presented here is focused to using the capacitive coupling in order to control the voltage on the floating gate. It is necessary to note that using this technique, multiple inputs are possible.

As shown in Fig. 18.1(a), Multi-input Floating Gate MOS (MIFGMOS) transistor has many input gates (V_1, V_2,..., V_n) coupled to the gate of a conventional MOSFET trough capacitors ($C_1, C_2,..., C_n$). Fig. 18.1(b) shows the layout of a three-input MIFGMOS and part (c) shows the electrical symbol.

The coupling capacitors are formed by the poly1 layer (the floating gate) and poly2 layer. It is recommendable to use poly1-poly2 capacitors because the gate is always built with poly1 layer and furthermore, there is more capacitance per unit area than using metal1-metal2 capacitors. Hence, input voltages (V_1, V_2,..., V_n) are coupled through the capacitors to the floating gate to modify the drain-source current [10]. In this manner, the input voltages can be coupled and added arithmetically to the floating gate. This is because the floating gate voltage is a function of the input voltages V_i and the coupling capacitors C_i. In general, the input voltages V_i are called control inputs. Thus, the floating gate voltage can be approximated by [11]:

$$V_{FG} = \sum_{i=1}^{n} \frac{C_i}{C_T} V_i + \frac{C_{FGS}}{C_T} V_S + \frac{C_{FGD}}{C_T} V_D + \frac{Q_{FG}}{C_T} \qquad (18.1)$$

where n is the number of inputs coupled capacitively, C_i are the coupling capacitances, V_i are the input voltages, V_S and V_D are the source and drain voltages, respectively, Q_{FG} is the initial charge in the floating gate, and C_T is given by:

$$C_T = \sum_{i=1}^{n} C_i + C_{FGD} + C_{FGS} \qquad (18.2)$$

In (18.2), C_{FGD} and C_{FGS} are the floating-gate-drain and floating-gate-source parasitic capacitances, respectively.

In this way, the first approximation equation to determine the drain-source current in sub-threshold regime, triode and saturation region are, respectively:

$$I_{DS} = I_{DO}\frac{W}{L}\exp\left(\frac{V_{FGS} - V_{th}}{\eta V_{ther}}\right) \text{ if } V_{FGS} < V_{th} \text{ and } V_{DS} < 4V_{ther} \qquad (18.3)$$

where V_{ther} is the thermal voltage approximated by $\frac{kT}{q}$ (k is the Bolztman constant, T is temperature and q the electron electric charge). I_{DO} is a drain-source current when $V_{FGS} = V_{th}$ and η is a constant, usually 1, and V_{th} is the MOSFET threshold voltage.

$$I_{DS} = K'\frac{W}{L}\left((V_{FGS} - V_{th}) - \frac{V_{DS}}{2}\right)V_{DS} \text{ if } 0 < V_{DS} < (V_{FGS} - V_{th}) \qquad (18.4)$$

where $K' = \mu_o C_{ox}$ and W/L is the MOSFET geometry

$$I_{DS} = K'\frac{W}{L}(V_{FGS} - V_{th})^2 \text{ if } 0 < (V_{FGS} - V_{th}) < V_{DS} \qquad (18.5)$$

In (18.3)-(18.5) V_{FGS} is approximated by (18.1).

On the other hand, an important consideration is the unpredictable charge stored in the floating gate when the MIFGMOS transistor is fabricated. This charge remain on the floating gate for a long time, so it can changes the circuit performance. To solve this problem, the technique presented in [12] is used here, where the undesirable electrical charge is removed during the manufacturing process using switchs formed by metal contacts.Other methods to discharge the initial charge in the floating gate (do not used here) are: ulraviolest light, Fowler-Nordheim technique and quasi-floating gate technique, all referenced above.

18.1.1 Simulation and d.c. results

To get the current-voltage simulation of the MIFGMOS transistor and compare it with the experimental results, some simulations models published in the literature were considered [13], [14], [15]. All this models have positive and negative characteristics and the search for better models continues. However, the model presented in [15] was considered in this work, because it calculates the floating gate voltage based in the parameters provided for the manufacturer and its SPICE implementation is simpler. Then, to compare the d.c. simulation results with the experimental results, five 3-input MIFGMOS coupled inputs were fabricated in 1.2μm technology. Table 18.1 shows the values of the coupling capacitances and ratio aspect as well other important parameters of each MIFGMOS transistors after they were fabricated. Additionally, Table 18.2 shows the capacitive parameters of the 1.2μm process.

Thus, the three inputs $V_{G1} = V_{G2} = V_{G2}$ of the MIFGMOS represented in Fig. 18.2 are connected toghether and they has been sweep while V_{DD} is kept constant. In Fig. 18.3, the experimental average results (measuring the experimental median results of five transistors) are ploted when $V_{DD} = 1.0V$ and $V_{DD} = 3.0V$, respectively. It is possible to see simulation results are according to experimental results.

Table 18.1. Coupling capacitances and ratio aspect of the 3-input MIFGMOS transistor

Parameter	Value	Description
C_1	50.1984fF	Coupling capacitor 1
C_2	40.9953	Coupling capacitor 2
C_3	96.6319	Coupling capacitor 3
W_n	6.0μm	Channel wide
L_n	3.6μm	Channel lenght
λ	1.2μm	CMOS process technology
μ_o	647.91 cm^2/Vs	Low-field Mobility (n-type)
$K'(\mu_o C_{ox})$	21.1$\mu A/V^2$	Transconductance parameter
V_{th}	0.59V	n-type MOSFET sub-threshold voltage
t_{ox}	316 Å	Gate-oxide thickness

Table 18.2. Capacitance parameters in the 1.2μm CMOS process

Parameter	N+	P+	Poly1	Poly2	Metal1	Metal2	Units
Area(sustrate)	290	304	37	37	24	16	aF/μm^2
Area(N+ active)	-	-	1094	696	52	27	aF/μm^2
Area(P+ active)	-	-	1079	690	-	-	aF/μm^2
Area(Poly1)	-	-	-	581	46	23	aF/μm^2
Area(Poly2)	-	-	-	-	47	23	aF/μm^2
Area(metal1)	-	-	-	-	-	38	aF/μm^2
Fringe(sustrate)	73	157	-	-	30	26	aF/μ m
Fringe(poly1)	-	-	-	-	60	43	aF/μ m
Fringe(metal1)	-	-	-	-	-	55	aF/μ m
Overlap(N+ active)	-	-	256	-	-	-	aF/μ m
Overlap(P+ active)	256	-	-	-	-	-	aF/μ m

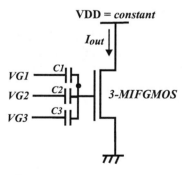

Fig. 18.2. Configuration to analyze d.c. curves of the 3-input-MIFGMOS

18.1.2 Programmable threshold voltage

As it was mentioned early, the MIFGMOS transistor has a tuning property because it can increase or decrease its relative threshold voltage dynamically using a single

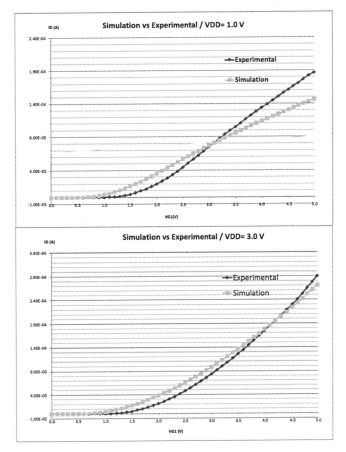

Fig. 18.3. Comparation between experimental and simulation I_{DS} vs. V_{G1} curves

coupled input gate. This can be interpreted as a programmable apparent threshold voltage. To show this, an experimental test is done setting the inputs of the transistor in Fig. 18.2 as follows: two inputs (V_{G2} and V_{G3}) are connected together to apply a sweep input voltage, and the input V_{G1} will be used to control the relative threshold voltage. The V_{G1} voltages take values of -5.0, -3.0, -1.0, 0.0, +1.0, +3.0 and +5.0 volts. In this case, the supply voltage V_{DD} is maintained constant with a value of 5.0V. Fig. 18.4 shows the I_{DS} vs. V_{G1} curves, where I_{DS} is the drain current of the transistor. We can note a particular case when V_{G1}=+5.0V, there is a small current flowing in the transistor even if the input voltage V_{G1} is zero. This could be a disadvantage in a circuit based on MIFGMOS transistor because is necessary to cut this current when the transistor is not required, using additional transistors as switches.

However, modifying the relative MIFGMOS threshold voltage, we can get positive consequence since the supply voltage and operation voltage can be reduced, e.g. in cascode configurations in a CMOS design, where to maintain the transistor working in the saturation region it is needed that each transistor satisfy $V_{DS} =$

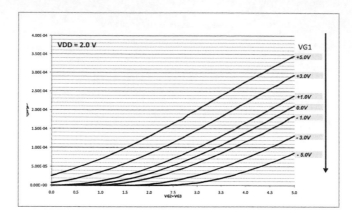

Fig. 18.4. Drain-source current of MIFGMOS transistor when the apparent threshold voltage is varied (experimental results).

$V_{GS} - V_{th}$, thus, if the threshold voltage V_{th} or the gate-source voltage V_{GS} can be manipulated, V_{DS} can be reduced, [16] and [17] and, consequently, we can reduce the supply voltage V_{DD}. Moreover, decreasing the supply voltage, the total power dissipation of the circuit is decreased.

18.1.3 Other properties of the MIFGMOS

The MIFGMOS transistor has another propierty, that is, the possibility of a feedback loop to improve the linearity in a particular desing and this will be shown later. More even, using MIFGMOS transistors the circuit complexity is reduced, signal processing is simplified and the quiscient point in a circuit can be shifted [13]. All advantages mentioned above are very important both in analog and in mixed signal circuits design, particularly in applications which require low supply voltage and low power dissipation, as used in [18] and [19].

However, the MIFGMOS has some disadvantages when is compared with the conventional MOSFET. The most important disadvantages are: lower effective transconductance, lower output resistance, and lower frequency response. If we define the effective MIFGMOS transconductance (in saturation region) respect to a single input V_i as

$$g_{mi} = \frac{\delta I_{DS}}{\delta V_i} \tag{18.6}$$

from (18.5) and (18.6) we get the expression:

$$g_{mi} = \frac{\delta I_{DS}}{\delta V_i} = \frac{\delta}{\delta V_i} \left(\frac{1}{2} \mu_o C_{ox} \frac{W}{L} \left(\frac{C_i}{C_T} V_i + \sum_{x=2}^{n} \frac{C_x}{C_T} V_x - V_{th} \right) \right) \tag{18.7}$$

and

$$g_{mi} = \frac{C_i}{C_T} g_m \qquad (18.8)$$

where g_m is the transconductance of the conventional MOSFET transistor. Note than the transconductance of the MIFGMOS respect to a single input is lower than the conventional MOSFET ($\frac{C_i}{C_T} < 1$).

18.2 Voltage to current converter based on MIFGMOS

The Voltage-to-current converter (VIC) cell based on the MIFGMOS transistor is presented and analyzed here to show the properties of the MIFGMOS such as: low-power operation, good linearity at low voltage operation and configurable threshold voltage. The advantages mentioned above are very important both in analog and in mixed signal circuits design, particularly in applications which require low voltage operation and low power dissipation.

The main function of the VIC is to convert an input voltage to an output current. We are interested in a VIC designed with MIFGMOS transistors to take advantages about its properties mentioned above.

This cell consists of a linear transconductor and a p-type current mirror as a load, both implemented with MIFGMOS transistors, as shown in Fig. 18.5. In this cell M1 is the transconductor with M6 as a triode load. The p-type current mirror (M8-M11) operates as a load in order to reduce the channel-length modulation and improve the circuit gain. M1 transistor has a feedback loop between its drain and the floating gate through C_f capacitance to improve the linearity of the cell. However, if the cell has a V_{DDFG} voltage applied, it will be a voltage V_f coupled to the floating gate through C_f, thus, we have a current IM6, even if V_{in} is null. This could be an undesired effect. To cut this undesired current when the cell is not needed, we use the transistors M3, M5 and M7 as switches.

In the next subsection each part is analyzed and it is showed how the VIC cell proposed here requires less supply voltage maintaining good linearity than conventional MOSFET version.

18.2.1 Transconductor stage

Now, it is showed how the transconductor stage in the VIC in Fig. 18.5, proposed here, has better linearity than conventional MOSFET version.

Analyzing M1 transistor and based in (18.1), the floating gate voltage is calculated by:

$$V_{FG1} \approx \frac{C_{in} V_{in}}{C_T} \qquad (18.9)$$

ignoring C_{GD}, C_{GS}, C_{GB} and C_{ox} effects in order to facilitate hand calculations without losing generality.

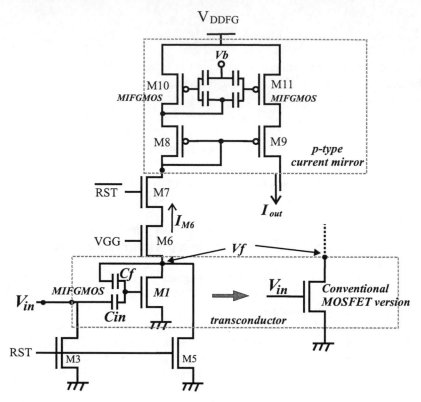

Fig. 18.5. Voltage to current converter implemented with MIFGMOS transistors. M1 MIFGMOS is compared with a conventional MOSFET

Taking M6 triode load, M7 switch, and p-type current mirror together shown in Fig. 18.5 as a load called R_D, the drain voltage V_f of M1 in saturation region is:

$$V_f = V_{DDFG} - R_D \frac{\mu_o C_{ox}}{2} \frac{W_1}{L_1} \left(\frac{C_{in}}{C_T} V_{in} + \frac{C_f}{C_T} V_f - V_{th} \right)^2 \qquad (18.10)$$

where μ_o is the mobility of charge carriers, C_{ox} is the MOSFET gate oxide capacitance by unit area, V_{th} is the threshold voltage for n-type MOSFET, and L and W are the large and width of the MOSFET transistor, respectively.

Developing the squared term in (10), we get:

$$V_f = V_{DDFG} - R_D \frac{\mu_o C_{ox}}{2} \frac{W_1}{L_1} (V_{th}^2 - \frac{2C_{in}}{C_T} V_{in} V_{th} - \frac{2C_f}{C_T} V_f V_{th} \qquad (18.11)$$

$$+ \frac{2C_{in} C_f}{C_T^2} V_f V_{in} + \frac{C_{in}^2}{C_T^2} V_{in}^2 + \frac{C_f^2}{C_T^2} V_f^2)$$

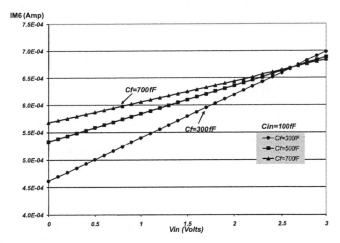

Fig. 18.6. I_{out} vs. V_{in} in MIFGMOS transconductor (simulation)

In (18.11) there are three squared terms. In this case, V_{th} is constant ($\approx$ 0.7V, according to the 1.2μm technology), but we can have control in the two last terms: $\frac{C_{in}^2}{C_T^2}V_{in}^2$ and $\frac{C_f^2}{C_T^2}V_f^2$.

So, linearity improves increasing C_T since $(C_T = C_{in} + C_f)$, it means, it is necessary to increase C_{in} or C_f. We opted arbitrary to increase the value of C_f to increase C_T. It should be noted this property is not possible using conventional MOSFET transistor.

Fig. 18.6 shows the IM6 current (drain of M6 transistor), when the input voltage of the VIC (V_{in}) is swept, also, C_f is varied. Note the good linearity with 2.0V voltage supply (V_{DDFG}) using a resistor as a load ($R_D = 10k\Omega$) . Also, note that the output current IM6 is different to zero when $V_{in} = 0V$. This is because the feedback voltage V_f is present when V_{DDFG} is applied. This current means static power consumption but it can be cut using the switches mentioned above. Here, $C_{in} = 100fF$.

Fig. 18.7 shows a simulated comparison between the drain current flowing through M1 transistor working as the transconductor shown in Fig. 18.5, both MOSFET and MIFGMOS version. In these cases, the supply voltage is 3.0V. We can see better linearity in MIFGMOS version but better dynamic range (RD) in MOSFET case.

18.2.2 Low-supply voltage p-type current mirror

Now, it is showed how the p-type current mirror proposed and implemented with MIFGMOS transistors (Fig. 18.8(a)) will require less supply voltage than conventional MOSFET version (Fig. 18.8(b)). So, we demonstrate that:

$$V_{DDC} > V_{DDFG} \tag{18.12}$$

Each version is analyzed and compared.

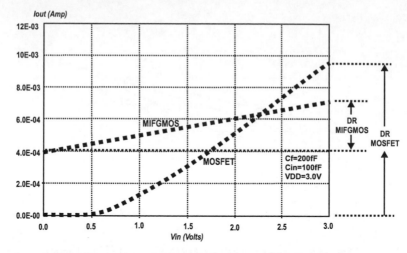

Fig. 18.7. Comparison between MIFGMOS and conventional MOSFET transconductor (simulation)

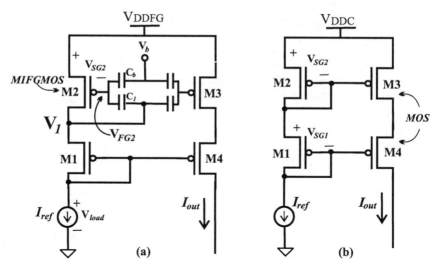

Fig. 18.8. (a) P-type current mirror implemented with MIFGMOS transistor and (b) implemented with MOSFET transistors only

a. Current mirror implemented using conventional MOSFET transistors only

Analyzing the circuit showed in Fig. 18.8(b), the supply voltage V_{DDC} is:

$$V_{DDC} = V_{GS2} + V_{GS1} + V_{load} \qquad (18.13)$$

where V_{load} is the voltage provided for the current source I_{ref}. In this case, we consider all the transistors working in the saturation region, hence the source to gate voltage of M1 or M2 is:

$$V_{SG1} = \sqrt{\frac{2I_{ref}}{\mu_o C_{ox}} \cdot \frac{L_1}{W_1}} + V_{thp} \qquad (18.14)$$

where V_{thp} is the threshold voltage for p-type MOSFET transistor.

Now, combining (13) and (14) we obtain:

$$V_{DDC} = 2V_{thp} + \sqrt{\frac{2I_{ref}}{\mu_o C_{ox}}} \left(\sqrt{\frac{L_2}{W_2}} + \sqrt{\frac{L_1}{W_1}} \right) + V_{load} \qquad (18.15)$$

The result obtained in (18.15) is very important because it shows the way the supply voltage depends of V_{load} and both the aspect ratio and V_{thp}. Also, we will use (18.15) to demonstrate (18.12).

b. Current mirror implemented using the conventional MOSFET transistor

.

Similar analysis is realized for the current mirror MIFGMOS version. In Fig. 8(a), the current through M2 transistor in saturation region is:

$$I_{ref} = \frac{\mu_o C_{ox}}{2} \frac{W_2}{L_2} \left(V_{DDFG} - V_{FG2} - V_{thp} \right)^2 \qquad (18.16)$$

where the floating gate voltage of M2 (V_{FG2}) is given by:

$$V_{FG2} = \frac{C_1 V_1 + C_b V_b}{C_T} \qquad (18.17)$$

In (18.17), V_b is a small bias voltage applied to the p-type MIFGMOS transistors in order to decrease the relative threshold voltage of M2 and consequently to reduce the supply voltage (V_{DDFG}). C_1 and C_b are coupling capacitors. Later, ignoring the parasitic capacitances C_{GD}, C_{GS}, and C_{GB} and combining (18.16) and (18.17), the supply voltage is:

$$V_{DDFG} = \sqrt{\frac{2I_{ref}}{\mu_o C_{ox}} \frac{L_2}{W_2}} + \frac{C_b}{C_T} V_b + \frac{C_1}{C_T} V_1 + V_{thp} \qquad (18.18)$$

where the voltage in node V_1 is given by:

$$V_1 = V_{SG1} + V_{load} \qquad (18.19)$$

Finally, combining (18.14), (18.18), (18.19):

$$V_{DDFG} = V_{thp} + \frac{C_1}{C_T} V_{thp} + \frac{C_b}{C_T} V_{load} + \sqrt{\frac{2I_{ref}}{\mu_o C_{ox}}} \left(\sqrt{\frac{L_2}{W_2}} + \frac{C_1}{C_T} \sqrt{\frac{L_1}{W_1}} \right) \qquad (18.20)$$

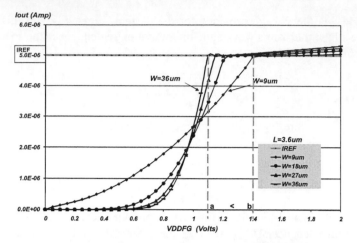

Fig. 18.9. I_{out} vs. V_{DDFG} for different geometries (simulation)

Equation (18.20) calculates the minimum supply voltage needed for the current mirror transistors working in the saturation region. Fig. 18.9 shows the curves of V_{DDFG} voltage versus I_{out}. In this case, a current of $5\mu A$ is applied as a target and $C_1 = C_b = 100fF$.

It is possible to watch that modifying the ratio aspect of the transistor the supply voltages V_{DDFG} ca be reduced. This is shown in the simulation results in Fig. 18.9 for different values of W, maintaining $L = 3.6\mu m$. When W is increased, less supply voltage V_{DDFG} is required to copy the target current maintaining the transistors in saturation region (see points a and b). However, a tradeoff exists between low supply voltage and linearity and this must be taken into account in the design process.

In order to analyze the effect of the bias voltage V_b in the current mirror in Fig. 18.8(a), Fig. 18.10 shows the current mirror operation varying V_b. We can see that applying a very small or negative voltage on V_b, the supply voltage V_{DDFG} is reduced maintaining the transistor in the saturation region (see points c and d). In this case, $I_{ref} = 5\mu A$ and $C_1 = C_b = 100fF$.

Now, it is demonstrated than (18.12) is true. First, note that V_{DDC} in (18.15) and V_{DDFG} in (18.20) are always positive. Then subtracting (18.20) and (18.15), we have:

$$V_{DDC} - V_{DDFG} = V_{load}\left(1 - \frac{C_1}{C_T}\right) + V_T\left(1 - \frac{C_1}{C_T}\right) \tag{18.21}$$

$$+\sqrt{\frac{2I_{ref}}{\mu_o C_{ox}}\frac{L_1}{W_1}\left(1 - \frac{C_1}{C_T}\right)} + \frac{C_b}{C_T}V_b$$

Since $0 < C_1 < C_T$:

$$0 < (1 - \frac{C_1}{C_T}) < 1 \tag{18.22}$$

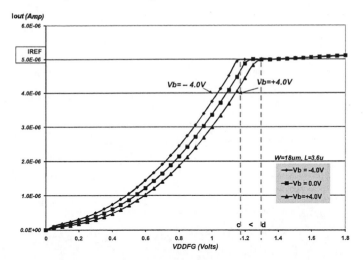

Fig. 18.10. V_b impact in the p-type curent mirror. I_{out} vs. V_{DDFG} changing V_b (simulation)

Hence, all terms in (18.21) are always positives (V_b could be zero), so we conclude that $V_{DDC} > V_{DDFG}$.

18.3 VIC implementation

The VIC cell was fabricated in a $1.2\mu m$ CMOS technology, double poly, double metal. Fig. 18.11 shows the experimental results when we measured the linearity of the VIC. To do this, an input voltage is swept in V_{in} input and the output current I_{out} is measured. Note the good linearity of this cell with a voltage supply of 1.7V.

Fig. 18.12 (scope image) shows the operation of the VIC cell when is used as a code detector in a signal processing. In this case, V_{in} is a binary sequence "10011100", with amplitude of $0V$ to $500mV$ at $500kHz$ and a supply voltage V_{DDFG} of $1.7V$. For reason of easy, the output voltage was measure using a $1.0k\Omega$ resistive load between the drain of M9 transistor and ground, which it has a maximum value of $8.0mV$. Now, the maximum output current of the VIC will be $8mV/1.0K\Omega = 8\mu A$. Meanwhile, if this cell operates with a higher supply voltage of $3.0V$, a speed operation of $2.0MHz$ is possible; but, it must take into account the load (discrete resistor of $1.0k\Omega$) in the drain of M9 used in the experimental measures because it will be a limiting factor. As we can see, this cell has a limited speed operation, thus, it is necessary to consider this characteristic in the design process.

18.4 Memory current cell (MIC)

In this section, a current memory cell implemented using MIFGMOS transistor is introduced. This cell is formed by n-type transistors current mirror. In previous

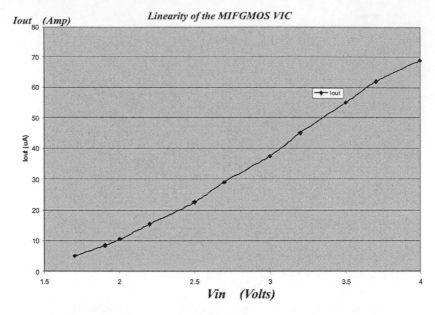

Fig. 18.11. linearity of the MIFGMOS VIC (experimental results)

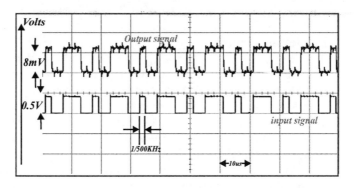

Fig. 18.12. Operation of the VIC at 1.7 V supply voltage (experimental result).

section the low-voltage operation of the current mirror implemented with MIFGMOS transistor was shown. Thus, the memory current cell shown in Fig. 18.13 stores an analog current as a charge and, connecting both the VIC and MIC cells in a cascade array, a low-voltage sample and hold circuit in current mode can be implemented.

The MIC implemented with MOSFET transistor is shown in Fig. 18.13(a) and the MIFGMOS version proposed here is shown in Fig. 18.13(b). The MIFGMOS MIC is a current mirror (M2, M4 and M5) used as a memory, where M1, M3 and M6 operate as switches.

This cell operates as follows: when M1 and M3 switches are closed and the M6 switch is opened, I_{in} generates a voltage in the gate of M4. Later, cutting M1

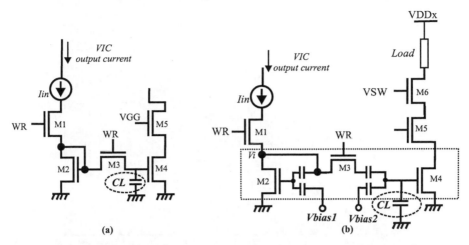

Fig. 18.13. Memory Current Cell (a) conventional MOS, (b) MIFGMOS version

and M3, a charge is stored in the gate-source parasitic capacitance (C_L) inherent to the MIFGMOS transistor M4. This stored charge will be proportional to the I_{in} amplitude, representing the memory. When the M6 is closed, the voltage stored in C_L will generate a current through M4, M5 and M6. In this case, the bias voltages V_{bias1} and V_{bias2} decrease the relative threshold voltage to get a minimal supply voltage in cascode configuration.

The MIC implemented with MIFGMOS transistors will have a better linearity with a minimal supply voltage V_{DDx} because transistors M4-M6 are working in saturation region with less voltage thanks to the applied bias voltages to the floating gate. Thus, we take advantages from the properties of the MIFGMOS transistor to build a low-voltage MIC with a good linearity.

18.4.1 MIC implementation

The MIC shown in Fig. 18.12(b) using MIFGMOS transistors was implemented in a$1.2\mu m$ CMOS technology, double poly, double metal. In order to see the linearity of this cell, Fig. 18.15 shows a swept input current I_{in} versus the output current through the transistors M4-M6 of the MIC. Here, the current is reduced intentionally with a rate of 3:1. The output current was measured between the drain of M6 and V_{DDX} voltage.

Finally, Fig. 18.17 shows the operation of the MIC with a supply voltage V_{DDx} of 1.7V (scope image). In this case, a low frequency sinusoidal signal of 500Hz is sampled using a 500 kHz sampling pulsed signal (0-3V). For reason of easy, the output is read in voltage mode using a R_D load of $220K\Omega$ to get an observable signal. Again, the speed of the cell is limited for high frequency applications because the high value of the resistor and the RC constant involved in all the measuring process. In order to enhance the bandwidth of the cell, a more elaborate I/V converter

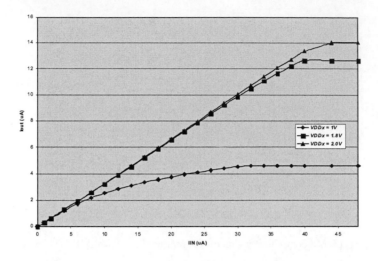

Fig. 18.14. Linearity of the MIFGMOS MIC. Iin vs. Iout. (experimental results)

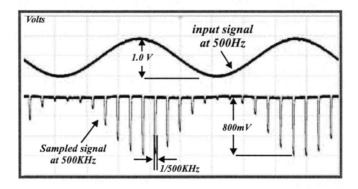

Fig. 18.15. Operation of the MIC with a voltage supply of 1.7V (experimental result).

is recommended. However, both cells are recommended for current mode circuit design.

18.5 Conclusions

This document shows than MIFGMOS transistor is an important alternative to implement analog circuits with low voltage operation and good linearity based in the advantages of this transistor. To demonstrate this, the analysis and implementation of two cells using MIFGMOS transistors were presented: a voltage to current converter cell and a memory current cell. Two important properties of MIFGMOS transistors

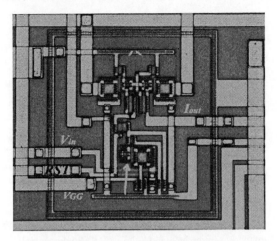

Fig. 18.16. MIcrofotography of the VIC

were shown: variable threshold voltage and feedback possibility. Combining these properties, a voltage to current converter with good linearity at low supply voltage was implemented. We compare the performance of the VIC in a MOSFET version versus a MIFGMOS version and we demonstrated with analytical and experimental results than MIFGMOS has a better linearity applying a low supply voltage, but with a low frequency operation. This must be taken into account in the process design.

On the other hand, a MIC using the MIFGMOS transistor was implemented in order to obtain good linearity with low supply voltage. We got satisfactory results despite the long-channel technology.

Both cells were implemented in a $1.2\mu m$ CMOS technology, double poly, double metal. The VIC operates satisfactory with 1.7V supply voltage at 0.5MHz with a power consumption of $20\mu W$. Also, the MIC can operate with a supply voltage of 1.7V with $12\mu W$ power consumption at 0.5MHz.

Finally, these cells can be used to implement a sample and hold circuit operating in current mode, but theirs applications are not limited. Fig. 18.16 and 18.17 show the micropotography of the VICand the MIC, respectively.

Acknowledgment

We appreciate the support of CONACyT, Mexico

References

[1] Asquith, J., Chang-Ling S., Hung Bun C. "A two-dimensional numerical model of a floating-gate EEPROM transistor", Int. J. Electronics, Vol. 85, No. 6, pp. 697 - 712 (1998)
[2] Topor-Kamiski, L., and Holajn, P. "Multiple-input floating-gate MOS transistor in analogue electronics circuit". Bulletin of the Polish Academy of Sciences, Vol. 52, No. 3. pp. 251 - 256. (2004)

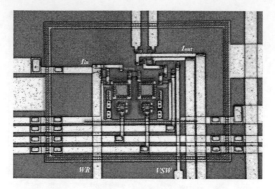

Fig. 18.17. MIcrofotography of the MIC

[3] Kotani, K., Shibata, T., Imai, M., and Ohmi, T. "Clock-controlled neuron-MOS logic gates" IEEE Transactions on Circuits and Systems II: Analog and Digital Signal Processing, Vol. 45, No. 4, pp. 518 - 522. (1998)

[4] Yamasaki, T., Nakayarna, T., and Shibata, T. "A Low-Power Switched-Current CDMA Matched Filter With On-Chip V-I and I-V Converters". Symposium on VLSl Circuits Digest of Technical Papers. pp. 214 - 217.(2004)

[5] Suresh, V., Katharine, R., and Baiju, M. "Adaptive Neuron Activation Function with FGMOS Based Operational Transconductance Amplifier", IEEE Computer Society Annual Symposium on VLSI, pp. 353 - 356.(2008)

[6] Yngvar Berg, ivind Nss, Mats E. Hvin and Henning Gundersen. "Ultra Low-Voltage Floating-Gate (FGUVMOS) Amplifiers" Analog Integrated Circuits and Signal Processing. Springer Netherlands. Volume 26, No. 1 (2001)

[7] Lenzlinger, M. and Snow, E. H., "Fowler-Nordheim tunneling in thermally grown SiO2," Journal of Applied Physics, vol. 40, p. 278 (1969)

[8] Hasler, P. "Continuous-Time Feedback in Floating-Gate MOS Circuits", IEEE Transactions on Circuits and Systems-II: Analog and Digital Signal Processing, Vol. 48, No. 1, pp. 56 - 64. (2001)

[9] Ramrez-Angulo, J., Urquidi, C., Gonzlez-Carvajal, R., Torralba, A., and Lpez-Martn, A. "A new family of low-voltage analog circuits based on quasi-floating-gate transistors". IEEE Transactions on Circuits and Systems II: Analog and Digital Signal Processing, Vol. 50, No. 5, pp. 214 - 220. (2003)

[10] Shibata, T. and Ohmi, T. "A Functional MOS Transistor Featuring Gate-Level Weighted Sum and Threshold Operations". IEEE Transactions on Electron Devices, Vol. 39, Issue 6, pp. 1444-1455. (1992)

[11] Antonio J. Lopez Martin, Alfonso Carlosena, Jaime Ramrez Angulo. "D/A Conversion Based on Multiple-Input Floating-Gate MOST" 42nd Midwest Symposium on Circuits and Systems. Volume: 1, On page(s): 149-152 vol. 1. (1999)

[12] Rodriguez-Villegas, E. "Solution to trapped Charge in FGMOS Transistors", Electronics Letters, Volume 39, Issue 19, pp. 1416 - 1417. (2003)

[13] Rodriguez-Villegas, E. "Low Power and Low-Voltage Circuit Design with the FGMOS Transistor", London, United Kingdom. The Institution of Engineering and Technology. pp. 6 - 7, and 22 - 28. (2006)

[14] Ochiai, T. and Hatano, H. "A Proposition on Floating Gate Neuron MOS Macromodeling for Device Fabrications". IEICE Trans. Fundamentals, Vol. E82-A, No. 11, pp. 2485 - 2491. (1999)

[15] Guan, H. and Tang, Y. "Accurate and efficient models for the simulation of neuron MOS integrated circuits". Int J. Electronics, Vol. 87, No. 5, pp. 557-568. (2000)

[16] Rajput, S., and Jamuar, S. "Low-Voltage Analog Circuit Design Techniques". Circuits and Systems Magazine, IEEE. Vol. 2, Issue: 1, pp. 24-42.(2002)

[17] Drakaki, M., Fikos, G., and Siskos, S. (2005), "Analog Signal Processing Circuits Using Floating Gate MOS Transistor", 5th International Conference on Technology and Automation ICTA'05. http://icta05.teithe.gr/papers/59.pdf. (2005)

[18] Liou, M. and Chiueh, T. "A low-power digital matched filter for direct-sequence spread-spectrum signal acquisition". IEEE J. Solid-State Circuits, vol. 36, no. 6, pp. 933-943. (2001)

[19] Yamasaki, T., Nakayama, Y., and Shibata, T. "A Low-Power and Compact CDMA Matched Filter Based on Switched-Current Technology". IEEE Journal of Solid-State Circuits, Vol. 40, No. 4, pp. 926 - 932. (2005)

A CMOS Pipelined Multiplier Based on a New High-Speed Low-Power Full-Adder Cell

Mariano Aguirre-Hernandez[1] and Monico Linares-Aranda[2]

[1] Intel Corporation, Communications Research Center, Mexico
   mariano.aguirre@intel.com
[2] National Institute of Astrophysics, Optics and Electronics, Mexico
   mlinares@inaoep.mx

Summary. This work presents an 8×8-bits CMOS pipelined multiplier built using a full-adder cell with a new internal logic structure and a pass-transistor logic style that allow to get reduced delay and power consumption. For achieving a full-adder cell with a reduced power-delay product (PDP), we analyzed and compared different XOR-XNOR gates and full-adders that have been reported as good candidates for low-power designs, and developed a new proposal that presents improved performance in terms of speed, area and power consumption. Post-layout simulations showed that the designed multiplier is able to operate up to 1.2GHz when it is supplied with 3.3V. The power savings obtained when it is compared to similar pipelined multipliers are up to: 20% when operating with transitioning input data, 25% with non-transitioning input data, and 80% with the clock signal disabled. A test chip containing the multiplier was fabricated using a $0.35\mu m$ CMOS technology. The experimental measurements confirm its operation at 1.2GHz with a power consumption of 180mW, for a supply voltage of 3.3V.

19.1 Introduction

With the availability of high-speed VLSI CMOS digital technologies, multimedia applications are being used increasingly in many embedded and portable systems such as mobile phones, PDAs, and laptops, which require both low-power operation and high-throughput capabilities.

The processing of multimedia information has high throughput requirements because of the huge amount of data produced with the direct digitalization of the audio and video signals. Compression techniques are used to store and to process this kind of data, so the throughput of today's high-end microprocessors become sufficient to implement multimedia applications [1].

Besides, due to the limited power supply capability of current battery technologies, designers are required to use low-power design techniques, in order to extend the working time of portable multimedia systems [2]. Thus, the increasing demand for the high fidelity of multimedia on portable devices has emphasized the development of low-power and high-performance systems.

The broad spectrum of digital signal processing applications implemented in current multimedia systems include algorithms such as filtering, coding, transformation, etc. These algorithms require optimized arithmetic modules that allow efficient realizations in terms of power consumption and delay. It is known that a multiplier module usually falls on the critical path of several algorithms performed within these applications. Thus, this module must be deeply optimized in terms of speed and power consumption, in order to attain a high-throughput and low-power system.

This work shows the design of an 8×8-bits CMOS pipelined multiplier, built using a full-adder cell implemented with a new proposed internal logic structure and a swing-restored complementary pass-transistor logic style, exhibiting high-speed and low-power performance.

This report is organized as follows: Section 19.2 presents the design of the main constituent cell for the multiplier, including an analysis carried out to obtain the best implementation of a XOR-XNOR gate in terms of the PDP, and the proposed alternative internal logic structure used to build new energy-efficient full-adders. Section 19.3 presents the design of the 8×8-bits pipelined multiplier using the proposed full-adder, and establishes a comparison with three pipelined multipliers published previously. Section 19.4 presents simulations and experimental results for the fabricated chip. Finally, Section 19.5 concludes this work.

19.2 High-Speed Low-Power Full-Adder Design

19.2.1 Analysis of XOR-XNOR Gates

Many papers have presented new or modified 1-bit full-adder cells, where the XOR-XNOR gate is shown to be the critical constituent component [3, 4, 5, 6, 7, 8, 9, 10, 11, 12, 13, 14, 15, 16, 17, 18, 19, 20, 21, 22, 23, 24, 25, 26, 27]. Other works have shown new isolated XOR-XNOR structures arguing its wide application in larger circuits such as comparators and compressors [4]. It can be seen from those works that the overall performance of these larger arithmetic cells is affected by the individual performance of the XOR-XNOR gates that are used to build them. Therefore, it is always desirable to have an optimized implementation for these gates in terms of area, delay and power consumption.

Fig. 19.1 shows most of the XOR-XNOR structures published in previous papers. A common subject in these works has been looking for a low-power XOR-XNOR structure dealing with each component of the power consumption, namely dynamic, short circuit, static and leakage. The present work includes XOR-XNOR gates designed with CPL and DPL logic styles (d, e, f and g) that were not taken into account in the analysis performed in [5, 6], and three recently published schemes designated as k [7], l [8] and m [9].

In general, the structures showing lowest power consumption are those exhibiting an incomplete voltage swing in internal or output nodes. Obviously, that feature results from the fact that less energy from the power supply is required to charge the nodes having an incomplete voltage swing. This behavior is certain as long as the

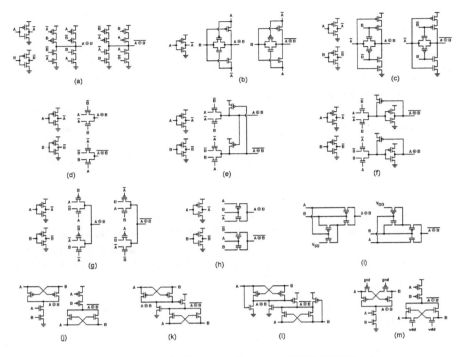

Fig. 19.1. Schematic diagrams of compared XOR-XNOR gates.

XOR-XNOR gate is simulated alone, i.e. with no other gates connected to the inputs and outputs.

However, in a real scenario, when there are circuits (a static inverter, as a generalization) being controlled by the node with an incomplete voltage swing, larger power consumption can result because of the chance of enabling a static DC path due to some partially turned-on transistors. Furthermore, the operation speed is degraded when the XOR-XNOR gate is used as part of a larger circuit where its output signals pass through or control a multiplexer, and the new output signal drives an inverter, due to the reduction of the current-driving capability of the node having an incomplete voltage swing.

Thus, a comparison of XOR-XNOR gates was carried out paying special attention in the performance of the schemes with full swing because of the reasons above explained.

The simulation environment used for comparing the XOR-XNOR gates performance is the same as suggested in [5, 6, 11]. The input signals of the XOR-XNOR gates are fed through inverters and the output signals are sent to other inverters. This test-bed represents a generalization of static CMOS gates driving and been driven by the XOR-XNOR cell under test. Hence, the main advantage of using this simulation environment is that the following power components are taken into account, in addition to the dynamic one:

- The short-circuit consumption of the inverters connected to the device under test (DUT) inputs. This power consumption varies according to the capacitive load that the DUT offers at its inputs. Furthermore, the energy required to charge and discharge the DUT internal nodes, when the module has no direct power supply connections(as for the case of pass-transistor logic styles), comes through these inverters connected at the DUT inputs.
- The short-circuit consumption of the DUT, as it is receiving signals with finite slopes coming from the buffers connected at the inputs, instead of ideal ones coming from voltage sources.
- The short-circuit and static consumption of the inverters connected to the DUT outputs, due to the finite slopes and possible degraded voltage swing of the DUT output signals.

Table 19.1 shows the simulation results for the XOR-XNOR gates performance comparison regarding power consumption and delay. It can be seen than the XOR-XNOR cell with the best PDP performance is the one implemented with the SR-CPL logic style shown in Fig. 19.1(e). Refer to [10], for a detailed analysis of this comparison.

Table 19.1. Simulation results of XOR-XNOR gates (power in μW and delay in ns.)

scheme		avg power	pwr supply	dynamic	static	short-circuit	% wasted/total	% xor/top	delay	pwr * delay
a	top	849.4	849.4	731.0	-	118.5	13.9	37.7	0.432	366.9
	xor	320.4	320.8	291.4	-	29.0	9.1			
b	top	573.9	573.9	458.7	-	115.2	20.0	31.6	0.281	161.3
	xor	181.2	54.5	166.7	-	14.5	8.1			
c	top	693.3	693.3	577.5	-	115.8	16.7	34.5	0.478	331.4
	xor	238.9	238.9	218.5	-	20.5	8.6			
d	top	739.9	739.9	418.1	17.2	304.3	43.4	15.5	0.219	162.0
	xor	114.5	73.3	106.9	-	7.6	6.7			
e	top	583.1	583.1	444.2	-	138.6	23.8	30.1	0.226	131.8
	xor	175.2	133.3	137.9	-	37.3	21.3			
f	top	716.8	716.8	535.6	-	181.2	25.3	42.5	0.414	296.8
	xor	304.9	261.0	200.6	-	104.3	34.2			
g	top	586.7	586.7	479.5	-	107.3	18.3	37.9	0.236	138.5
	xor	222.1	106.3	196.4	-	25.7	11.6			
h	top	771.2	771.2	440.8	16.8	313.5	42.8	22.1	0.247	190.5
	xor	170.6	78.5	152.8	-	17.8	10.4			
i	top	804.2	804.2	418.1	13.0	372.9	47.9	16.7	0.368	295.9
	xor	134.0	27.7	129.0	-	5.0	3.6			
j	top	689.4	689.4	474.2	5.0	215.2	31.2	27.6	0.300	206.8
	xor	190.1	41.9	182.5	-	7.6	4.0			
k	top	1008.5	1008.5	523.7	-	484.8	48.0	53.8	0.527	531.5
	xor	542.5	258.1	214.2	-	328.4	60.5			
l	top	1049.7	1049.7	484.4	112.9	451.8	53.8	58.3	0.397	416.7
	xor	612.2	196.7	184.1	112.9	315.2	69.9			
m	top	758.3	758.3	489.4	-	268.6	35.4	35.1	0.491	372.3
	xor	266.3	52.5	255.1	-	11.2	4.3			

19.2.2 High-Speed Low-Power Full-Adder Cell

Several papers have been published regarding the design of low-power full-adders, dealing with the logic style (Standard CMOS [15], Differential Cascode Voltage Switch (DCVS) [16], Complementary Pass-Transistor Logic (CPL) [17], Double Pass-Transistor Logic (DPL) [18], and Swing Restored CPL (SR-CPL) [14], and also with the logic structure used to build the adder module [3, 19].

The internal logic structure shown in Fig. 19.2(a) has been adopted as the standard configuration in most of the enhancements developed for the 1-bit full-adder. This structure was built based on the transmission function theory [20]. The full-adder is formed by three main logical blocks: a XOR-XNOR gate to obtain $A \otimes B$ and $\overline{A \otimes B}$ (Block 1), and XOR blocks or multiplexers to obtain the SUM (So) and CARRY (Co) outputs (Blocks 2 and 3).

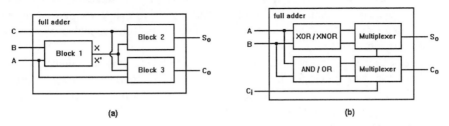

Fig. 19.2. (a) standard and (b) proposed full-adder internal logic structure.

Since the proposal presented in [20], several papers have introduced new full-adder cells proposing different realizations for the three logic blocks of Fig. 19.2(a). Chronologically, some of them are: 14TA [21], 14TB [22], wu_ng [23], 16T [24], 10TA [25], 10TB [13], full_rest [7], mux_based [26], and wey_chow [27].

In [10], it was shown that the most efficient realization for the block 1 is the SR-CPL logic style. Furthermore, in the same paper [10], another important conclusion was pointed out: the major problem regarding the propagation delay for a full-adder built with the logic structure shown in Fig. 19.2(a), is that it is necessary to obtain intermediate $A \otimes B$ and $\overline{A \otimes B}$ signals, which are then used to drive other blocks in order to generate the final outputs. Thus, the overall propagation delay and, in most of the cases, the power consumption of the full-adder, depend on the delay and voltage swing of the $A \otimes B$ and $\overline{A \otimes B}$ signals generated within the cell.

Therefore, to increase the operational speed of the full-adder, it is necessary to develop a new logic structure that does not require the generation of intermediate signals to control the selection or transmission of other signals located on the critical path.

A new internal logic structure for a full-adder.

Examining the full-adder's true-table, it can be seen that the So output is equal to the $A \otimes B$ value when $C = 0$, and it is equal to $\overline{A \otimes B}$ when $C = 1$. Thus, a multiplexer can be used to obtain the respective value taking the C input as the selection signal. Following the same criteria, the Co output is equal to the $A \bullet B$ value when $C = 0$, and it is equal to $A + B$ value when $C = 1$. Again, C can be used to select the respective value for the required condition, driving a multiplexer.

Hence, an alternative logic scheme to design a full-adder cell, can be formed by a logic block to obtain the $A \otimes B$ and $\overline{A \otimes B}$ signals, another block to obtain the $A \bullet B$ and $A + B$ signals, and two multiplexers being driven by the C input to generate the So and Co outputs, as shown in Fig. 19.2(b). The features and advantages of this alternative logic structure are:

- There are no internally generated signals controlling the selection of the output multiplexers. Instead, the C input signal, exhibiting a full voltage swing and no extra delay, is used to drive the multiplexers, thus reducing the overall propagation delay.
- The capacitive load for the C input has been reduced, as it is connected only to some transistor gates and no longer to some drain or source terminals, where the diffusion capacitance is becoming very large for sub-micrometer technologies. Therefore, the overall delay for larger modules where the C signal falls on the critical path can be reduced.
- The propagation delay for the So and Co outputs can be tuned up individually by adjusting the XOR/XNOR and the AND/OR gates; this feature is advantageous for applications where the skew between arriving signals is critical for a proper operation (e.g. wave-pipelining).
- The inclusion of buffers at the full-adder outputs can be implemented by inter-changing the XOR/XNOR signals, and the AND/OR gates to NAND/NOR gates at the input of the multiplexers, improving so the performance for load sensitive applications.
- There is no need to wait for the computation of one output signal (e.g. carry signal) to obtain the other one, as it just selects between the outputs of simple logic functions, leading to a faster operation.

Comparison of full-adders' performance.

Based on [10] and the results presented above, a new full-adder has been designed using the SR-CPL logic style and the new logic structure presented in Fig. 19.2(b). Fig. 19.3 shows the schematic for the proposed adder. The AND/OR gates of this adder have been built using a powerless/groundless pass-transistor configuration, and also a pass-transistor based multiplexer to get the Co output.

Several full-adders were compared with regards to power consumption and delay. They were named: cmos_26 [15], cmos_28 [15], cpl [17], sr_cpl [14], dcvs [16], bay_10a [25], bay_10b [13], bay_14a [21], bay_14b [22], bay_16 [24], full_rest [7],

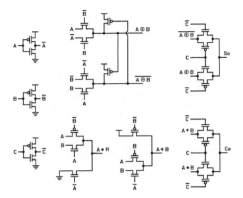

Fig. 19.3. Full adder designed with the proposed logic structure and a SR-CPL logic style.

mux_based [26], tran_funct [20], wey_chow [27], wu_ng [23], and the proposal shown in Fig. 19.3 (Ours).

The full-adders were designed using an AMS 0.35 μm CMOS technology, simulated using the BSIM3v3 model (level 49), and supplied with 3.3 V. Simulations were carried out using Nanosim [28] to determine the power consumption features of the designed full-adders, and Hspice [29] to measure the propagation delay for the output signals (the longest time required for one output signal to reach the 50% of its voltage swing measured from the moment when one of the input signals reached the 50% of its voltage swing).

Table 19.2 shows the simulation results from the full-adder's performance comparison regarding power consumption and delay. It is worth to observe that in some cases, the power consumed from the power-supply for the full-adders is smaller than the total average power. This is because of, for some logic styles (e.g. pass-transistor style), some current is taken from the inputs of the full-adder and is used to charge the internal nodes. As mentioned above, it is the importance of considering the power consumption of the input buffers in the test-bed. The following paragraphs about power consumption and delay can be extracted from these results:

The full-adders designed with pass-transistor logic styles (cmos_26, cpl, cpl_sr, Ours) exhibit a shorter delay than the other ones; this can be expected because of the fact of reduced internal parasitic capacitances, as stated in [7] for these logic styles.

On the other hand, the full-adders designed using the logic structure shown in Fig. 19.2(a) (bay_10a, bay_10b, bay_14a, bay_14b, bay_16, full_rest, tran_funct, wey_chow) have larger propagation delays (around or exceeding 1ns) as expected, due to the internal XOR/XNOR gates that generate intermediate signals having an extra delay, used to control the output blocks.

The full-adders presenting an incomplete voltage swing (bay_14a, bay_14b, bay_16) present lower power consumption than other ones (cmos_26, tran_funct), but only when the surrounding circuitry consumption is neglected (row "add"). If

Table 19.2. Simulation results of the compared full-adders (power in μW, delay in ns, width in μm, and Vdd in V).

scheme		avg power	pwr supply	dynamic	static	short-circuit	% wasted/total	% add/top	delay	pwr * delay	Σ width	Vdd min
cmos_26	top	1286.3	1286.3	875.2	4.3	406.9	31.9	81.7	0.703	904.3	67.8	1.3
	add	1051.1	811.5	683.8	4.3	363.0	34.9					
cmos_28	top	1736.8	1736.8	1420.7	-	315.8	18.2	59.0	0.984	1709.0	184.8	1.3
	add	1024.7	1025.0	746.8	-	277.9	27.1					
cpl	top	2975.9	2975.9	984.4	1991.6	37.6	66.9	89.1	0.781	2324.2	113.2	1.8
	add	2650.2	2504.4	702.6	37.6	1910.0	73.4					
cpl_sr	top	2264.1	2264.1	1097.9	-	1165.9	51.5	85.6	0.812	1838.4	116.4	1.4
	add	1937.8	1804.8	810.2	-	1127.6	58.1					
cpl_uye	top	2179.7	2179.7	1190.3	-	989.3	45.3	89.3	0.853	1859.3	116.4	1.4
	add	1946.0	1688.6	982.7	-	963.3	49.5					
dcvs	top	2965.4	2965.4	1579.7	-	1385.3	46.7	84.8	1.107	3282.7	182.4	1.3
	add	2515.3	2515.6	1179.1	-	1336.2	53.1					
bay_10a	top	1576.4	1576.4	632.0	55.1	889.4	60.0	83.8	1.955	3081.9	51.4	2.8
	add	1321.7	1237.5	436.9	55.1	829.6	67.0					
bay_10b	top	1565.5	1565.5	960.6	2.0	602.6	38.6	85.4	1.157	1811.3	84.8	2.4
	add	1336.2	953.7	773.2	2.0	561.0	42.1					
bay_14a	top	1221.0	1221.0	848.8	0.7	371.6	30.5	81.1	1.220	1489.6	68.7	2.4
	add	989.7	684.8	658.0	0.7	331.0	33.5					
bay_14b	top	1290.6	1290.6	975.5	0.7	314.2	24.4	82.3	1.366	1763.0	84.0	2.4
	add	1061.6	663.3	785.7	0.7	275.2	26.0					
bay_16	top	1343.8	1343.8	919.4	0.3	423.7	31.5	83.2	1.688	2268.3	80.4	2.8
	add	1118.0	699.6	724.4	0.3	393.4	35.2					
full_rest	top	1795.5	1795.5	894.3	266.3	634.6	50.1	87.4	1.022	1835.0	72.7	1.8
	add	1569.2	1027.0	687.7	266.3	615.1	56.2					
mux_based	top	1560.2	1560.2	750.8	4.6	804.5	51.8	85.2	1.362	2125.0	65.6	2.4
	add	1329.2	1100.2	566.6	4.6	758.0	57.3					
tran_funct	top	1225.0	1225.0	796.6	-	428.0	34.9	81.0	0.932	1141.7	63.0	1.3
	add	992.0	808.8	612.2	-	379.8	38.3					
wey_chow	top	1346.4	1346.4	942.2	-	404.3	30.0	83.1	1.024	1378.7	75.8	1.7
	add	1119.4	739.9	747.1	-	372.2	33.2					
wu_ng	top	1161.6	1161.6	959.6	-	202.0	17.4	78.4	1.067	1239.4	79.4	1.7
	add	910.5	728.6	741.8	-	168.6	18.5					
Ours	top	835.6	835.6	710.8	-	124.7	15.0	66.6	0.734	613.3	50.4	1.5
	add	556.4	364.7	466.6	-	89.8	16.1					

the whole test-bed consumption (row "top") is considered, then those proposals have no longer better performance than the other ones.

Besides, the propagation delay for those adders is longer, due to the current-driving capability degradation of the multiplexers being controlled by the nodes exhibiting an incomplete voltage swing, making the power-delay product even worse than the value exhibited for the other adders.

Furthermore, the minimum voltage supply (column "Vdd min") that maintains the right operation for those circuits is higher than the supply required for the ones having internal nodes with a full voltage swing.

Regarding the proposals in this work, it can be clearly seen the advantage of the alternative logic structure derived above, since the realization designed using this scheme (Ours) exhibits the lowest power consumption, delay and power-delay product.

In addition, since this realization has neither static consumption, nor internal direct paths from Vdd to Gnd (except for the inverters at the inputs), it is a good candidate for battery-operated applications where low consumption modules with stand-by modes are required. Even more, the power consumption can be further reduced for this circuit, as it can operate properly with voltage supplies as low as 1.5V.

The importance of the simulation setup and the inclusion of the power consumption components for the surrounding circuitry are now evident, as some realizations reported previously as low-power cells have been shown to perform worse than other ones when considering the whole test-bed consumption.

Now, addressing the required implementation area (column "$\sum width$"), it can be noticed that the pass-transistor based circuits occupy less area than the static ones. In particular, the proposed full-adder require the smallest area, which can also be considered as one of the factors for presenting lower delay and power consumption, as it implies smaller parasitic capacitances being driven.

19.3 8×8-bits Pipelined Multipliers

To verify the operation of the proposed full-adder in a more complex application, four 8×8-bits pipelined multipliers were designed in order to compare their performance regarding speed and power consumption. This architecture was selected since the registers placed between the rows of the array allow the maximum operating frequency to be determined by the longest delay of either an AND gate or the full-adder cell being used. Fig. 19.4 shows the block diagram for these pipelined multipliers.

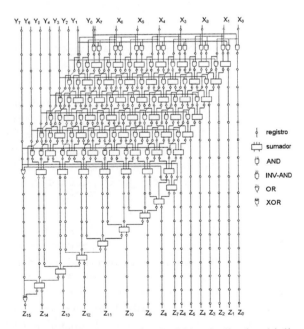

Fig. 19.4. Block diagram of an 8×8 bits pipelined multiplier.

Three of these multipliers use the full-adders shown in Fig. 19.5, and the last one was built upon the SR-CPL full-adder cell proposed in Sect. 19.2 (Fig. 19.3), with the addition of the C^2MOS latches required for the pipelined scheme. The full-adder cells in Fig. 19.5 were used in previous works to desing pipelined multipliers: (a) quasi-domino [31], (b) static & dynamic [32], and (c) static & pass-tx [33].

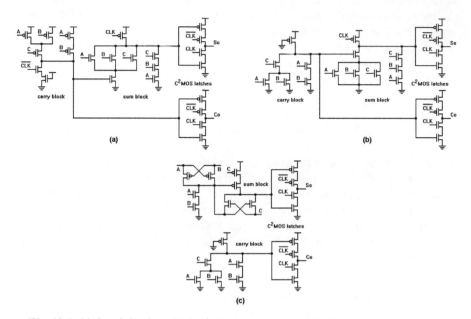

Fig. 19.5. (a) Quasi-domino, (b) Static & dynamic, and (c) Static & pass-tx full-adders.

In [34], there is a detailed analysis of the modifications that those proposals suggested to improve the full-adders performance. It was also shown that the suggested enhancements that were done trying to increase the full-adder's speed have introduced new power consumption components or worsen the existing ones. The justification for the usefulness of those proposals was that at high-frequency operation, the static consumption can be neglected, as the circuit was transitioning most of the time.

Nevertheless, this scenario has changed recently, since the clock-gating and stand-by operational modes have become very popular, and even mandatory in most of the current high-performance VLSI designs. So, the static power consumption for the modules used as building blocks must be considered and optimized wherever it is possible.

All multipliers were designed using an AMS 0.35μm CMOS technology, and were simulated using the BSIM3v3 model (level 49) and supplied with 3.3V. Several Hspice and Nanosim simulations were carried out to obtain the performance of the multipliers for the following scenarios:

- With transitioning data at the inputs, and the clock signals (CLK and $\overline{CLK}$) activated, to simulate a normal operation.
- With stable data at the inputs, and the clock signals activated, to simulate a module isolation technique applied to the multiplier and its surrounding circuitry.
- With stable data at the inputs, and the clock signals stopped, to simulate a clock gating technique applied to the multiplier.

19.4 Multipliers Performance Evaluation

Table 19.3 summarizes the Nanosim simulation results. It is worth to mention that the highest operational frequency obtained for the multiplier using the proposed full-adder was 1.2GHz, with a power consumption of 195mW, while the other multipliers worked properly up to 1GHz.

Table 19.3. Power consumption results for compared multipliers (power in mW @ 1GHz).

multiplier	operation mode	avg power	pwr supply	dynamic	static	short-circuit	% wast/total
quasi-domino	normal	187	187	155	0	32	17
	stable inputs	162	162	131	0	31	19
	clock gated	116	116	11	72	34	91
static & dynamic	normal	201	201	149	21	31	26
	stable inputs	175	175	122	27	26	30
	clock gated	125	125	13	79	33	90
static & pass-tx	normal	181	181	141	23	17	22
	stable inputs	152	152	114	24	14	25
	clock gated	64	64	12	43	9	81
Our proposal	normal	161	161	151	0	10	6
	stable inputs	130	130	121	0	9	7
	clock gated	26	26	14	6	6	46

The following paragraphs can be stated from the results in Table 19.3:

Both dynamic multipliers, quasi-domino and static & dynamic, exhibit a large short-circuit consumption, due to the elimination of the clock-driven transistors in the dynamic stages.

When operating under normal conditions (with transitioning input data and the clock signals enabled), the multiplier with the highest power consumption is the one designed with the static & dynamic logic style. This is because of the short-circuit power component, as well to the increase of static consumption due to the pseudo-NMOS stage.

The multiplier designed with static & pass-transistor logic style reduces the dynamic power consumption by eliminating the pre-charging behavior of the dynamic stage, but it holds the static consumption drawback from the pseudo-NMOS stage.

It can be seen that these 3 multipliers exhibit a large static consumption when the clock gating scheme is applied, due to the existence of direct paths from Vdd to Gnd, when the clock-driven transistors are stopped at one state.

In normal operation mode, our proposal improves the power consumption by 10% with respect to the static & pass-transistor logic, and more than 20% with respect to the other ones. This is because there is neither static consumption, nor short circuit consumption, except for the consumption of the inverters at the inputs.

In the scenario of having stable input data and the clock signals running free, the power consumption for all the multipliers is reduced, since the dynamic component is decreased because there are less transitions within the multiplier array. Our proposal shows a saving of 15% against the static & pass-transistor logic multiplier, and 25% against the other ones.

For the clock signals being deactivated, our proposal exhibits a saving of 60% with respect to the static & pass-transistor logic, and up to 80% with respect to the other multipliers. The main advantage of the proposed multiplier is the large reduction of dynamic consumption as the latches are stopped, but also the very small amount of wasted power from static and short-circuit components due to the elimination of direct paths from Vdd to Gnd within the full-adder cell built with the SR-CPL logic style and the powerless/groundless AND/OR gates.

Thus, the advantages with regards to speed and power consumption of the proposed pipelined multiplier have been confirmed with these simulation results, showing that this module is well suited for high-speed low-power arithmetic modules.

The pipelined multiplier proposed in this work was fabricated using a $0.35\mu m$ CMOS technology from Austria Micro Systems through the EUROPRACTICE mini-ASIC prototyping service. Fig. 19.6(a) shows a photograph of the fabricated chip, and Fig. 19.6(b) presents the layout of the core logic, which includes the following components: (1) 8×8-bits array multiplier, (2) parallel-to-serial converter, (3) Voltage-Controlled Oscillator (VCO), (4) clock distribution network, (5) clock divider, (6) clock signal multiplexer, and (6) 1-bit full-adder cell. The whole chip size including the I/O pads is $1.2mm^2$, while the core area is $0.6mm^2$.

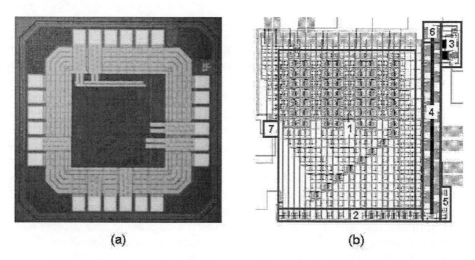

(a)   (b)

Fig. 19.6. (a) Photograph of the fabricated chip and (b) layout of the core logic.

Due to the small number of I/O pads that resulted from the chip's layout, the Vdd pin for the multiplier array was also used to supply the serial-to-parallel converter. Thus, in order to obtain the multiplier's power consumption, the power consumption value of the serial-to-parallel converter obtained from simulations was subtracted from the value measured when the chip was operating.

Fig. 19.7 presents experimental waveforms of the multiplier's function. The operation -65×91 was performed with the multiplier array running at 1.2GHz, and the result was stored in the shift-register by activating its load signal. Then, the clock multiplexer was switched to a 50MHz clock signal applied at the *CLKext* pin, in order to get the result in a serial stream at the *Zserial* pin. The *CLKdiv* wave is the clock signal divided by 16, which marks the boundaries of the 16-bits word for the result (-5915).

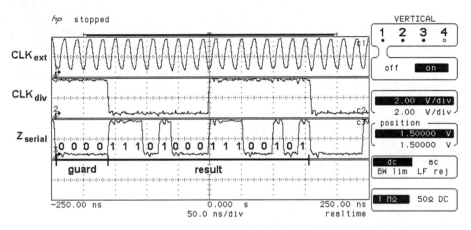

Fig. 19.7. Measured waveforms for operation of -65×91 = -5915 $(1110100011100101)_2$.

19.5 Conclusions

An 8×8-bits pipelined multiplier has been designed using a full-adder cell built with a proposed alternative internal logic structure. A SR-CPL logic style was used to implement the full-adder, in order to reduce the power consumption, due to short-circuit and static consumption components.

Hspice and Nanosim simulations showed that this multiplier is able to work up to 1.2GHz, with a power consumption around 195mW when supplied with 3.3V; this also exhibits savings up to 20%, 25% and 80% when operating at 1GHz for normal, stable data input, and clock-gated modes, respectively, when compared to other three pipelined multipliers designed using full-adder cells reported previously. The fabricated chip showed to be functional when it was clocked at 1.2GHz, and exhibited a power consumption of 180mW. The measured power when operating in normal conditions at 1GHz was around 153mW.

References

1. T. Nishitani, "Low-Power architectures for Programable Multimedia Processors", IEICE Transactions Fundamentals, Vol. E82-A, No. 2, February 1999, pp. 184-195.
2. C. Him, H. Kim and S. Ha, "Dynamic Voltage Scheduling Technique for Low-Power Multimedia Applications Using Buffers", ISLPED'01, August 2001, pp.34-39.
3. S. Agarwal, V.K. Pavankumar and R. Yokesh, "Energy-efficient High Performance Circuits for Arithmetic Units", 21st International Conference on VLSI Design '08, January 2008, pp. 371-376.
4. H. T. Bui, A. K. Al-Sheraida and Y. Wang, "New 4 Transistor XOR and XNOR designs", Proceedings of the Second IEEE Asia Pacific Conference on ASICs 2000, pp. 25-28.
5. H. T. Bui, Y. Wang and Y. Jian, "Design an Analysis of low-Power10-Transistor Full-Adders Using Novel XOR-XNOR Gates", IEEE Transactions on Circuits and Systems, Vol. 49, N0. 1, January 2002, pp. 25-30.
6. A. M. Shams, T. K. Darwish and M. Bayoumi, "Performance Analysis of Low-Power1-Bit CMOS Full-Adder Cells", IEEE Trans. on VLSI Systems, Vol. 10, No. 1, February 2002, pp. 20-29.
7. D. Radhakrishnan, "Low-Voltage Low-Power CMOS Full-Adder", IEE Proceedings on Circuits Devices and Systems, Vol. 148, No. 1, February 2001, pp. 19-24.
8. M. Elgamel, S. Goel and M. Bayoumi, "Noise Tolerant Low-Voltage XOR-XNOR for Fast Arithmetic", GLVLSI2003, April 2003, USA, pp. 285-288.
9. M. Aguirre-Hernandez and M. Linares-Aranda, "A Low-Power Bootstrapped CMOS Full-Adder", 2nd International Conference on Electrical and Electronics Engineering, September 2005, pp. 243-246.
10. M. Aguirre and M. Linares, "An Alternative Logic Approach to Implement High-Speed Low-Power Full-Adder Cells", SBCCI'05, Florianopolis, Brazil, September 2005, pp. 166-171.
11. A. M. Shams and M. Bayoumi, "A Structured Approach for Designing Low Power Adders", ASILOMAR Conference on Signals, Systems and Computers, Vol. 1, 1998, pp. 757-761.
12. A. Fayed and M. Bayoumi, "A Low-Power 10-Transistor Full-Adder Cell for Embedded Architectures", IEEE International Symposium on Circuits and Systems, 2001, pp. 226-229.
13. A. M. Shams and M. Bayoumi, "A Low Power 10-Transistor Full-Adder Cell for Embedded Architectures", IEEE International Symposium on Circuits and Systems, 2001, pp. 226-229.
14. R. Zimmerman and W. Fichtner, "Low-power logic styles: CMOS versus pass-transistor logic", IEEE JSSC, Vol. 32, No. 7, July 1997, pp. 1079-1090
15. N. Weste and K. Eshraghian, "Principles of CMOS design, A system perspective", Addison-Wesley, 1988
16. K. M. Chu and D. Pulfrey, "A comparison of CMOS circuit techniques: differential cascode voltage switch logic versus conventional logic", IEEE JSSC, Vol. sc-22, No. 4, August 1987, pp. 528-532
17. K. Yano, et al, "A 3.8ns CMOS 16 16-b multiplier using complementary pass-transistor logic", IEEE JSSC, Vol. 25, April 1990, pp. 388-395.

18. M. Suzuki, et al, "A 1.5ns 32-b CMOS ALU in double pass-transistor logic", IEEE JSSC, Vol. 28, No. 11, November 1993, pp. 1145-1150.
19. D. Patel, P. G. Parate, P. S. Patil and S. Subbaraman, "ASIC Implementation of 1-bit Full-Adder", 1st International Conference on Emerging Trends in Engineering and Technology, July 2008, pp. 463-467.
20. N. Zhuang and H. Wu, "A new design of the CMOS full-adder", IEEE JSSC, Vol. 27, No. 5, May 1992, pp. 840-844.
21. E. Abu-Shama and M. Bayoumi, "A new cell for low power adders", IEEE ISCAS '96, Vol. 4, May 1996, pp. 49-52.
22. E. Abu-Shama, A. Elchouemi, S. Sayed and M. Bayoumi, "An efficient low power basic cell for adders", Proceedings of the 38th Midwest Symposium on Circuits and Systems, Vol. 1, August 1995, pp. 306-309.
23. A. Wu and C. K. Ng, "High performance low power low voltage adder", IEE Electronic Letters, Vol. 33, No. 8, April 1997, pp. 681-682.
24. A. M. Shams and M. Bayoumi, "A novel low-power building block CMOS cell for adders", Proceedings of IEEE ISCAS '98, Vol. 2, June 1998, pp. 153-156.
25. A. M. Shams and M. Bayoumi, "A 10-transistor low-power high-speed full-adder cell", Proceedings of IEEE ISCAS '99, Vol. 1, June 1999, pp. 43-46.
26. B. Alhalabi and A. Al-Sheraidah, "A novel low-power multiplexer-based full-adder cell", IEEE ICECS 2001, Vol. 3 , September 2001, pp. 1433-1436.
27. I. Wey, C. Huang and H. Chow, "A new low-voltage CMOS 1-bit full-adder for high performance applications", Proceedings of the IEEE Asia-Pacific Conference on ASICs '02, August 2002, pp. 21-24.
28. "NanosimUser Guide" (A-2008.03), Synopsys, March 2008.
29. "HSPICE User Guide" (A-2007.12), Synopsys, December 2007.
30. T. G. Noll et al, "A pipelined 330-MHz multiplier", IEEE JSSC, Vol. SC-21, pp. 411-416, June 1986.
31. F. Lu and H. Samueli, "A 200-MHz CMOS pipelined multiplier-accumulator using a quasi-domino dynamic full-adder cell design", IEEE JSSC, Vol. 28, No. 2, pp. 123-132, February 1993.
32. S. Jou, C. Chen, E. Yang and C. Su, "A pipelined multiplier-accumulator using a high-speed low-power static and dynamic full-adder design", IEEE JSSC, Vol. 32, No. 1, pp. 114-118, January 1997.
33. A. M. Shams, W. M. Badawy and M. A. Bayoumi, "An enhanced low-power computational kernel for a pipelined multiplier-accumulator unit", ICM'98, Monastir, Tunisia, December 1998, pp. 33-36.
34. M. Aguirre and M. Linares, "Energy-Efficient High-Speed CMOS Pipelined Multiplier", CCE '08, Mexico, D.F., Mexico, November, 2008, pp. 460-464

A Reaction-Diffusion Model for the Production of Autowaves and its Application to Navigation Control of Mobile Robots

José Antonio Medina Hernández[1,2], Felipe Gómez Castañeda[1], and José Antonio Moreno Cadenas[1]

[1] Department of Electrical Engineering. CINVESTAV-IPN, Av. Instituto Politécnico Nacional 2508, 07360, México D.F., México
[2] Department of Mathematics and Physics, Aguascalientes Autonomous University, Av. Universidad 940, 20100, Aguascalientes, México jmedina@cinvestav.mx

Summary. There are many examples of natural processes in which a specimen travels along a trajectory avoiding obstacles until it reaches a goal using a mechanism of orientation and navigation. Such mechanism can be implemented using attraction by autowaves. Necessary analytic conditions based on linear theory are established in such a way that a reaction-diffusion system has an oscillating behavior in the initial phase. A new model for the production of autowaves is described. Some results of simulation are shown, and an algorithm for navigation control of a mobile robot is described.

Keywords: Reaction-Diffusion Cellular Neural Network, Autowaves, Navigation Control, Discrete Fourier Transform

20.1 Introduction

There are many natural processes in which a specimen should travel avoiding obstacles to reach an objective, using a mechanism for orientation and navegation. An example are the enzymes in the living beings. These species move and stay in some part of the body for inducing physical development. Another example is the mechanism used by birds and fish for arriving to some place [1]. A possible way for implementing such mechanisms is using attractive waves as drivers [11]. A procedure for the generation of such waves is the use of a reaction-diffusion model [1]-[12].

The analysis of the dynamics in a reaction-diffusion model has been an important issue of study in the last five decades, and particularly, the analysis of autowaves [1],[2],[4],[5],[7]-[10] and the generation of Turing patterns [1],[3]-[6],[9]. These models have the form

$$\frac{\partial U(\mathbf{x},t)}{\partial t} = f(U,V) + D_U \left(\frac{\partial^2 U}{\partial x^2} + \frac{\partial^2 U}{\partial y^2} \right) \tag{20.1}$$

$$\frac{\partial V(\mathbf{x},t)}{\partial t} = g(U,V) + D_V \left(\frac{\partial^2 V}{\partial x^2} + \frac{\partial^2 V}{\partial y^2} \right) \tag{20.2}$$

where $f(U,V)$ and $g(U,V)$ are reaction functions, being D_U and D_V the diffusion coefficients for the species U and V, respectively. In the simplest case, D_U and D_V are constant values. When the functions f and g depend only on U and V, but not on the spatial coordinates (x,y) neither on the time t, the system is named autonomous [5]. A nonautonomous system or a system with variable difussion coefficients can have a much more complicated dynamics. The linear theoretical analysis is possible when we have an autonomous system with constant coefficients D_U and D_V [1],[3],[6],[9].

Reaction-diffusion models explain a great variety of static and oscillating patterns present in many scientific disciplines, and they can be used for resolving some engineering problems [11]-[12]. Two applications for autowaves are the establishment of an algorithm to guide a mobile robot [11], and the synchronization of the moving parts in the case of a walking robot [12]. In this chapter we show a model for the generation of autowaves and several properties of its behavior are analyzed.

20.2 The Reaction-Diffusion Model

The behavior of reaction-diffusion systems can be approximated using a mesh with finite points in a bounded rectangular domain of the $x - y$ plane. If we put M equidistant points in the horizontal direction and N equidistant points in the vertical direction, we can construct a mesh with MN points. We are interesed in the values of the U and V variables on the points of the mesh, so the continuous model is reduced to a system of MN differential equations, named Reaction-Diffusion Cellular Neural Network (RDCNN) [5]. Every spatial point (i,j) is the center of a cell having two functions of activation U_{ij} and V_{ij}, so we say the RDCNN is a second order system, and it means that the RDCNN is costructed with two layers. For obtaining information about the behavior of the system we use a linear approximation, described in Section 20.2.

The reaction-diffusion model proposed in this chapter for the generation of autowaves has the reaction functions

$$f(U,V) = \alpha_u U + \beta_u U^2 + \gamma_u U^3 + \frac{\delta_u}{1 + e^{\lambda_u V}} + a \tag{20.3}$$

$$g(U,V) = \frac{\delta_v}{1 + e^{\lambda_v U}} + \alpha_v V + \beta_v V^2 + \gamma_v V^3 + b \tag{20.4}$$

so the associated RDCNN has the model

$$\frac{\partial U_{ij}}{\partial t} = \alpha_u U_{ij} + \beta_u U_{ij}^2 + \gamma_u U_{ij}^3 + \frac{\delta_u}{1 + e^{\lambda_u V_{ij}}} + D_U \left(\frac{\partial^2 U_{ij}}{\partial x^2} + \frac{\partial^2 U_{ij}}{\partial y^2} \right) + a \tag{20.5}$$

$$\frac{\partial V_{ij}}{\partial t} = \frac{\delta_v}{1 + e^{\lambda_v U_{ij}}} + \alpha_v V_{ij} + \beta_v V_{ij}^2 + \gamma_v V_{ij}^3 + D_V \left(\frac{\partial^2 V_{ij}}{\partial x^2} + \frac{\partial^2 V_{ij}}{\partial y^2} \right) + b. \tag{20.6}$$

In Section 20.4 we assign values to the parameters and some simulations are shown.

20.3 Linear Analysis of Reaction-Diffusion Systems

In this section we review the linear theory of the reaction-diffusion systems [1],[3],[6]. Using adequate parameters and initial conditions a reaction-diffusion system can have an emergent behavior, generating static or dynamic patterns. For generating a pattern, the system must be no stable with respect to small perturbations around an equilibrium point. An equilibrium point (U_o, V_o) is a constant solution of the reaction-diffusion system, so we must have $f(U_o, V_o) = 0$ and $g(U_o, V_o) = 0$. The behavior of the system around an equilibrium point is studied using the associated linear system, which is obtained approximating the reaction functions $f(U,V)$ and $g(U,V)$ by the Taylor series around the equilibrium point (U_o, V_o). We suppose that at the starting time every point of the mesh has U and V values very near to the equilibrium point (U_o, V_o). We define the variables $u = U - U_o$ and $v = V - V_o$, taken as small perturbations around (U_o, V_o). The Taylor expansions of first order for f and g around (U_o, U_o) are given by

$$f(U,V) = f(U_o, V_o) + (U - U_o)\frac{\partial f(U_o, V_o)}{\partial U} + (V - V_o)\frac{\partial f(U_o, V_o)}{\partial U} \quad (20.7)$$

$$g(U,V) = g(U_o, V_o) + (U - U_o)\frac{\partial g(U_o, V_o)}{\partial U} + (V - V_o)\frac{\partial g(U_o, V_o)}{\partial U}. \quad (20.8)$$

As long as (U_o, V_o) is an equilibrium point for the system, then we have $f(U_o, V_o) = 0$ and $g(U_o, V_o) = 0$, so the reaction functions can be written as

$$f(U,V) = (U - U_o)\frac{\partial f(U_o, V_o)}{\partial U} + (V - V_o)\frac{\partial f(U_o, V_o)}{\partial V} \quad (20.9)$$

$$g(U,V) = (U - U_o)\frac{\partial g(U_o, V_o)}{\partial U} + (V - V_o)\frac{\partial g(U_o, V_o)}{\partial V} \quad (20.10)$$

which are equivalent to

$$f(U,V) = f_u u + f_v v \quad (20.11)$$

$$g(U,V) = g_u u + g_v v \quad (20.12)$$

where

$$f_u = \frac{\partial f(U_o, V_o)}{\partial U} \quad f_v = \frac{\partial f(U_o, V_o)}{\partial V} \quad g_u = \frac{\partial g(U_o, V_o)}{\partial U} \quad g_v = \frac{\partial g(U_o, V_o)}{\partial V} \quad (20.13)$$

are the partial derivatives of $f(U,V)$ and $g(U,V)$ evaluated at the equilibrium point (U_o, V_o). We also have

$$\frac{\partial U}{\partial t} = \frac{\partial (U_o + u)}{\partial t} = \frac{\partial u}{\partial t} \quad \frac{\partial V}{\partial t} = \frac{\partial (V_o + v)}{\partial t} = \frac{\partial v}{\partial t}. \quad (20.14)$$

The previous identities produce the associated reaction-diffusion linear system

$$\frac{\partial u}{\partial t} = f_u u + f_v v + D_u \nabla^2 u \tag{20.15}$$

$$\frac{\partial v}{\partial t} = g_u u + g_v v + D_v \nabla^2 v \tag{20.16}$$

where $D_u = D_U$ and $D_v = D_V$.

As we have indicated in Section 20.2, for studying the behavior of the reaction-diffusion systems and of their linear associated systems, a discretization of the spatial domain is used, constructing a rectangular mesh with MN points, which are equidistant in the axes. In such a case, the resulting model is named reaction-diffusion cellular neural network [5]. In every point (i,j) of the mesh we must find the values $U_{ij}(t)$ and $V_{ij}(t)$, obtaining the discrete system

$$\frac{\partial U_{ij}}{\partial t} = f(U_{ij}, V_{ij}) + D_U \nabla^2 U_{ij} \tag{20.17}$$

$$\frac{\partial V_{ij}}{\partial t} = g(U_{ij}, V_{ij}) + D_V \nabla^2 V_{ij} \tag{20.18}$$

and their associated linear system

$$\frac{\partial u_{ij}}{\partial t} = f_u u_{ij} + f_v v_{ij} + D_u \nabla^2 u_{ij} \tag{20.19}$$

$$\frac{\partial v_{ij}}{\partial t} = g_u u_{ij} + g_v v_{ij} + D_v \nabla^2 v_{ij} \tag{20.20}$$

where $i = 0, 1, 2, ..., M-1$ and $j = 0, 1, 2, ..., N-1$. The variables U_{ij}, V_{ij}, u_{ij} and v_{ij} depend only on the time. The symbols $\nabla^2 u_{ij}$ and $\nabla^2 v_{ij}$ in the linear system are the discrete Laplace operators applied to u_{ij} and v_{ij}. The discrete Laplace operator is defined as [3], [6]

$$\nabla^2 \phi(i,j) = \phi(i-1,j) + \phi(i+1,j) + \phi(i,j-1) + \phi(i,j+1) - 4\phi(i,j). \tag{20.21}$$

In the next, we describe the procedure used for the analysis of the associated linear systems. We consider MN eigenfunctions

$$\phi_{mn}(i,j) = exp\left[\varsigma\left(\frac{2\pi mi}{M} + \frac{2\pi nj}{N}\right)\right], \quad \varsigma^2 = -1 \tag{20.22}$$

which are orthogonal respect to the wave numbers [6], satisfying the property

$$\sum_{i=1}^{M-1} \sum_{j=1}^{N-1} \overline{\phi}_{mn}(i,j)\phi_{pq}(i,j) = \begin{cases} MN & \text{if } (m,n) = (p,q) \\ 0 & \text{if } (m,n) \neq (p,q) \end{cases} \tag{20.23}$$

There are MN eigenvalues

$$k_{nm}^2 = 4\left[sin^2\left(\frac{\pi m}{M}\right) + sin^2\left(\frac{\pi n}{N}\right)\right] \tag{20.24}$$

satisfying the relations

$$\nabla^2 \phi_{mn}(i,j) = -k_{mn}^2 \phi_{mn}(i,j).$$
(20.25)

Now, solutions of the form

$$u_{ij}(t) = \sum_{m=0}^{M-1} \sum_{n=0}^{N-1} \phi_{mn}(i,j)\widehat{\mu}_{mn}(t), \quad v_{ij}(t) = \sum_{m=0}^{M-1} \sum_{n=0}^{N-1} \phi_{mn}(i,j)\widehat{v}_{mn}(t)$$
(20.26)

are supposed, where $i = 0, 1, 2, ..., M-1$ and $j = 0, 1, 2, ..., N-1$. Substituting these expressions in the system (19)-(20), summarizing over all the points (i,j) in the mesh, and using the orthogonal property, we have

$$\begin{pmatrix} \dot{\widehat{\mu}}_{mn}(t) \\ \dot{\widehat{v}}_{mn}(t) \end{pmatrix} = \begin{pmatrix} f_u - k_{mn}^2 D_u & f_v \\ g_u & g_v - k_{mn}^2 D_v \end{pmatrix} \begin{pmatrix} \widehat{\mu}_{mn}(t) \\ \widehat{v}_{mn}(t) \end{pmatrix}.$$
(20.27)

This 2×2 system of differential equations has constant coefficients, so we suppose the solution

$$\begin{pmatrix} \widehat{\mu}_{mn}(t) \\ \widehat{v}_{mn}(t) \end{pmatrix} = e^{\lambda_{mn}t} \begin{pmatrix} v_{mn}^1 \\ v_{mn}^2 \end{pmatrix}$$
(20.28)

where $(v_{mn}^1, v_{mn}^2)^T$ is a constant vector and λ_{mn} is a constant value. Substituting such a solution in the previous system, we have

$$\lambda_{mn} \begin{pmatrix} v_{mn}^1 \\ v_{mn}^2 \end{pmatrix} = \begin{pmatrix} f_u - k_{mn}^2 D_u & f_v \\ g_u & g_v - k_{mn}^2 D_v \end{pmatrix} \begin{pmatrix} v_{mn}^1 \\ v_{mn}^2 \end{pmatrix}$$
(20.29)

so λ_{mn} must be an eigenvalue for the matrix

$$A_{mn} = \begin{pmatrix} f_u - k_{mn}^2 D_u & f_v \\ g_u & g_v - k_{mn}^2 D_v \end{pmatrix} = J - k_{mn}^2 \begin{pmatrix} D_u & 0 \\ 0 & D_v \end{pmatrix}$$
(20.30)

associated to the eigenvector

$$\begin{pmatrix} v_{mn}^1 \\ v_{mn}^2 \end{pmatrix}$$
(20.31)

where

$$J = \begin{pmatrix} f_u & f_v \\ g_u & g_v \end{pmatrix}.$$
(20.32)

If the system has an eigenvalue different from zero, the determinant

$$\begin{vmatrix} f_u - k_{mn}^2 D_u - \lambda_{mn} & f_v \\ g_u & g_v - k_{mn}^2 D_v - \lambda_{mn} \end{vmatrix}$$
(20.33)

is zero and so we have the equation

$$\lambda_{mn}^2 - tr(A_{mn})\lambda_{mn} + det(A_{mn}) = 0$$
(20.34)

where

$$tr(A_{mn}) = f_u + g_v - k_{mn}^2(D_u + D_v) \tag{20.35}$$

$$det(A_{mn}) = D_u D_v (k_{mn}^2)^2 - (f_u D_v + g_v D_u)k_{mn}^2 + f_u g_v - f_v g_u. \tag{20.36}$$

These expressions are the trace and the determinant of the matrix A_{mn}, respectively. This algebraic equation of second degree has two solutions, namely λ_{mn1} and λ_{mn2} given by

$$\lambda_{mn1} = \frac{tr(A_{mn}) + \sqrt{tr(A_{mn})^2 - 4 \cdot det(A_{mn})}}{2} \tag{20.37}$$

$$\lambda_{mn2} = \frac{tr(A_{mn}) - \sqrt{tr(A_{mn})^2 - 4 \cdot det(A_{mn})}}{2}. \tag{20.38}$$

As long as the system is linear its solution is given by

$$\begin{pmatrix} \widehat{\mu}_{mn}(t) \\ \widehat{v}_{mn}(t) \end{pmatrix} = c_{mn}^1 e^{\lambda_{mn1}t} \begin{pmatrix} v_{mn}^{11} \\ v_{mn}^{21} \end{pmatrix} + c_{mn}^2 e^{\lambda_{mn2}t} \begin{pmatrix} v_{mn}^{12} \\ v_{mn}^{22} \end{pmatrix} \tag{20.39}$$

where

$$\begin{pmatrix} v_{mn}^{11} \\ v_{mn}^{21} \end{pmatrix}, \qquad \begin{pmatrix} v_{mn}^{12} \\ v_{mn}^{22} \end{pmatrix} \tag{20.40}$$

are the unitary eigenvectors of A associated to λ_{mn1} and λ_{mn2}, being c_{mn}^1 and c_{mn}^2 arbitrary constants that depend on the initial conditions.

The general solution of the system (19)-(20), obtained from (39), is given by [1],[3],[6]

$$u_{ij}(t) = \sum_{m=0}^{M-1} \sum_{n=0}^{N-1} (c_{mn}^1 v_{mn}^{11} e^{\lambda_{mn1}t} + c_{mn}^2 v_{mn}^{12} e^{\lambda_{mn2}t}) \phi_{mn}(i,j) \tag{20.41}$$

$$v_{ij}(t) = \sum_{m=0}^{M-1} \sum_{n=0}^{N-1} (c_{mn}^1 v_{mn}^{21} e^{\lambda_{mn1}t} + c_{mn}^2 v_{mn}^{22} e^{\lambda_{mn2}t}) \phi_{mn}(i,j). \tag{20.42}$$

These expressions indicate that the variables $u_{ij}(t)$ and $v_{ij}(t)$ have a raising oscillatory behavior when the eigenvalues λ_{mn1} and λ_{mn1} are complex numbers with positive real part. It is equivalent to satisfying the next two inequalities:

$$0 < tr(A_{mn}) \tag{20.43}$$

$$0 > tr(A_{mn})^2 - 4 \cdot det(A_{mn}). \tag{20.44}$$

These inequalities give a criterion for determining when an associated linear system has an oscillatory behavior. This criterion is equivalent to verify the validation of the chain of inequalities

$$0 < tr(A_{mn}) < 2 \cdot \sqrt{det(A_{mn})}. \tag{20.45}$$

Observe that if $m << M$ and $n << N$, then $k_{mn}^2 \approx 0$ and $A_{mn} \approx J$, so the eigenvalues of A_{mn} are approximated by the eigenvalues of J. If D_U and D_V are small, we also have $A_{mn} \approx J$.

20.4 Simulation Results

In this section we show some results obtained with the reaction functions

$$f(U,V) = 0.972\,U - 0.3\,U^3 - \frac{2}{1+e^{-2V}} + 0.7 \tag{20.46}$$

$$g(U,V) = \frac{2}{1+e^{-2U}} + 0.972\,V - 0.3\,V^3 - 0.7 \tag{20.47}$$

when the parameters take the values: $\alpha_u = \alpha_v = 0.972$, $\beta_u = \beta_v = 0$, $\gamma_u = \gamma_v = -0.3$ $\delta_u = -\delta_v = -2$, $\lambda_u = \lambda_v = -2$, $a = -b = 0.7$. In the simulations we also assume $D_U = D_V = 0.1$ [10].

The equillibrium point is obtained solving the algebraic non-linear system $f(U,V) = g(U,V) = 0$, having the solution $U_0 = -0.0041, V_0 = -0.3139$. The Jacobian matrix evaluated at this point is

$$J = \begin{pmatrix} 0.9720 & -0.9076 \\ 1.0000 & 0.8833 \end{pmatrix} \tag{20.48}$$

with determinant $D = 1.7661$, trace $T = 1.8553$ and eigenvalues $\lambda_1 = 0.9276 + 0.9516i$, $\lambda_2 = 0.9276 + 0.9516i$. If k_{mn}^2 is small, then $A_{mn} \approx J$. The value $2 \cdot \sqrt{D} = 2.6579$ indicates that the inequality $0 < T < 2 \cdot \sqrt{D}$ is satisfied. So, the linear analysis predicts an oscillatory behavior with low and medium spatial frequencies for this model, around the equilibrium point $(U_0, V_0) = (-0.0041, -0.3139)$.

From Figs. 20.1 to 20.4 it is observed how it can be produced an oscillatory behavior using adequate initial conditions [4],[7],[10] and periodic boundary conditions. The squares in the left side corresponds to the U variable, and the squares on the right side correspond to the V variable. The white color correspons to a high value and the black color correspons to a small value.

In Fig. 20.1 is shown the result of assigning large values to a central column in the layer U and low values to a small square in the layer V: a source of concentric autowaves is produced. The vertical column in the layer U rapidly acquires a bend form in the extremes, producing two 'hooks'. They grow until their previous parts in the branch are annihilated by collision. Later, we have again the formation of the initial pattern and the propagation of an additional circular front.

In Fig. 20.2 is shown the behavior of the system when the rectangle of the layer V in Fig. 20.1 is diminished. A unique front is produced and then it expands outside to all the domain. The front repeats again, and the system obtains a periodic behavior.

In Fig. 20.3 is shown the way to produce a diagonal array of autowave sources, using initial conditions very similar to the used in Fig. 20.1. Again, the front waves are propagated across the periodic boundaries.

In Fig. 20.4 we used random initial conditions and the system still generates autowaves. It is seen how are formed autowave sources of different sizes progressively along the time. The production of fronts becomes periodic, inducing a periodic oscillatory behavior in all the domain, because the system acquires a periodic production and annihilation of fronts.

20.5 Navigation using Autowaves

A fundamental problem in dealing with mobile robots is to set an algorithm capable to trace a trajectory avoiding obstacles from an origen to a goal.

20.5.1 Basic Algorithm for Navigation using Autowaves

In [11] the next algorithm to guide a mobile robot is proposed using autowaves and cellular neural network for front detection [12]:

Basic algorithm for navigation using autowaves

1)Use two coupled Reaction-Diffusion Cellular Neural Networks to generate two sets of fronts. The first set of fronts has its source at the goal and it is attractive. The second set of fronts comes from the obstacles and it is repulsive. The attractive and repulsive fronts are generated in the layers U of the first and second RDCNN nets, respectively.

2)For every direction north, south, east and west, apply steps 3)-4) over first type RDCNN.

3)Use a first order CNN to detect attractive moving fronts in the actual direction [11].

4)If the attractive front detected in the position of robot has a nonzero intensity in the d_1 current direction, then the robot should move along $-d_1$ direction.

5)For every direction north, south, east and west, apply steps 6)-7) over repulsive fronts.

6)Use a first order CNN to detect repulsive moving fronts in the actual direction.

7)If the repulsive front detected in the position of robot has a nonzero intensity in the d_2 current direction, then push robot to d_2 direction.

8)Apply steps 2)-7) until reaching the goal.

For some situations this algorithm can produce successful results. However, exhaustive experiments show drawbacks such as standstill, close contact with obstacles, a dominant front over other, deficient detection of fronts and asynchrony in reception of fronts. The following algorithm was designed to overcome these problems.

20.5.2 PROPOSED ALGORITHM FOR NAVIGATION USING AUTOWAVES

In the next, we describe a navigation control algorithm, extending the mechanism described in Section 20.5.1. Some reaction-diffusion systems, such as FitzHug-Nagumo, are activator-inhibitor type, where one of the substances activates the system [1]. This idea can be adapted to navigation systems. A navigation mechanism is named *driven by activation* if and only if a movement on the mobile robot is induced only when it is *touched by an attractive front*. In the next, such a mechanism is described:

Navigation Mechanism driven by activation

The points 1) and 2) in the following were also used in Section 20.5.1:

1)Two RDNN are used to generate two sets of fronts. The first set of fronts has its source in the goal and it is attractive. The second set of fronts has its source in the obstacles and it is repulsive. The attractive and repulsive fronts are generated in the layers U of the first and second RDCNN nets, respectively.

2)Use four first order motion detection CNNs [11] to detect the moving of the fronts in the north, south, east and west directions. For every sample, calculate a total attractive direction d_1, obtained summing all the *inverted directions* of the attractive fronts touching the robot. Similarly, calculate the total repulsive direction d_2, obtained summing all the repulsive fronts touching the robot, corresponding to the north, south, east and west directions.

3)If the dot product between d_1 and d_2 is positive, take $d_2 = 0$. This is done to neutralize the repulsive front emitted from the background obstacles. If this consideration is omitted, the robot could receive an impulse towards the objective larger than necessary, producing an impact with a near obstacle .

4)The detected movements of the attractive and repulsive fronts are combined to determine which is the direction that the robot should move on. If the attractive front is in the nonzero direction $-d_1$ and the repulsive fronts are in the d_2 direction (it could be zero), then the direction of the robot is given by a vector in the set

$$\{(1,0),(1,1),(0,1),(-1,1),(-1,0),(-1,1),(0,-1),-(1,1)\}$$

which approaches to $d_1 + d_2$.

The points 3) and 4) above, which were not used in Section 20.5.1, and are driven by the activation mechanism, are considered in the following algorithm:

Algorithm for navigation control driven by activation.

1. Establish the centers for the attractive and repulsive autowave sources, using destination as an attractive center and obstacles as repulsive centers.

2. From these centers, generate attractive and repulsive autowaves, using a RDCNN for each type. This step is equivalent to step 1 of the algorithm described in Section 20.5.1.

3. For every sampled position of the mobile robot, apply steps 3.1-3.6:

3.1 For every direction north, south, east and west apply a first order motion detection CNN to detect if the corresponding attractive fronts have nonzero intensities at the robot position. Determine the atractive direction $\mathbf{d}_1$ as a sum of the inverted directions corresponding to the attractive fronts with nonzero intensity.

3.2 For every direction north, south, east and west apply a first order motion detection CNN to detect if the corresponding repulsive fronts have nonzero intensities at the robot position. Determine the repulsive direction $\mathbf{d}_2$ as a sum of the directions corresponding to the repulsive fronts with nonzero intensity.

3.3 If the dot product of $\mathbf{d}_1$ and $\mathbf{d}_2$ is positive, compute $\mathbf{d}_2 = \mathbf{0}$. This action is for suppressing the effect of background obstacles.

3.4 If $\mathbf{d}_1 \neq 0$ (necessary condition for establishing a mechanism driven by activation), select the $\mathbf{d}$ vector in the set

$$\{(1,0),(-1,0),(0,1),(0,-1),(1,1),-(1,1),(1,-1),(-1,1)\}$$

being more similar to $\mathbf{d}_1 + \mathbf{d}_2$.

3.5. Push the robot along direction $\mathbf{d}$.

3.6. If the objective is not reached, return to step 3.

4. End

In Figs. 20.5 to 20.7 three trajectories followed by a small mobile robot are showed over a 70×70 mesh. These trajectories were determined using the above proposed navigation algorithm.

In the left bottom frame of Fig. 20.5, the obstacles (black rectangles) are shown, as well as the goal to be reached (white rectangle). In the left top frame the attractive fronts are shown, meanwhile the repulsive fronts are shown in the right top frame. Also, at the right bottom frame a sample of the detected south fronts is shown. Combining the information given by the north, south, east and west fronts, the robot is able to decide the trajectory to follow. The left bottom frame shows the trajectory of the robot. The final coordinate where the robot arrives is the center of the attractive source. This center is placed a little bit above of the white rectangle, in which the initial conditions to generate the attractive autowave source were defined. The mobile robot is able to move between the obstacles without hitting them, because the repulsive autowaves put it far away from obstacles.

In Fig. 20.6 the frames are interpreted similarly as in Fig. 20.5, but the robot begins its trajectory from a different place. Again, the mobile robot has a trend to put far away from obstacles, meanwhile it approaches to the objective.

In Fig. 20.7 a new disposition of obstacles is presented. The mobile robot tends to go around obstacles, such as the trajectory used to reach the goal is minimized, meanwhile the hitting with obstacles is avoided.

20.6 Conclusions

The importance of the models for producing autowaves is justified by their capability for explaining oscillatory phenomena, which might be present in the analysis of models in different scientific disciplines. They can be used also in the implementation of algorithms for the navegation with autowaves and the synchronization of mobile robots.

For the model proposed in this chapter, the cells are self-organized to produce a synchronized and periodic behavior in the time. The reaction functions $f(U,V)$ and $g(U,V)$ are differentiable, and the model does not use saturated values of U and V in the diffusion terms [10],[11]. It can be computationally advantageous, and provides an ordinarily continuous reaction-diffusion model. In Section 20.3 a revision was made of the linear theory for reaction-diffusion models, and it was used to establish a mathematical criterion for determining when a small perturbation grows periodically in the time. This criterion was applied to the reaction-diffusion model proposed in this chapter. The simulations show that using appropriated initial conditions the model can produce autowave sources and isolated fronts. The random initial conditions produce a large number of autowave sources, confirming the emergent and self-organized nature of the model.

Two RDCNN are used to produce attractive and repulsive autowaves. However, several problems arise, such as standstill, close contact with obstacles, dominant effect of a type of wave over other, deficient detection of fronts, and asynchronous reception of the two types of fronts. In this chapter, a detailed procedure accounting such problems has been described. Computer simulations show the effectivity of the proposed procedure for three configurations of fix obstacles. Future research involves to verify the performance of the proposed algorithm for a labyrinth and for moving obstacles.

Acknowledgements

The authors thank the funding by The Center for Research and Advanced Studies of the National Polytechnic Institute (CINVESTAV-IPN), the National Council for Science and Technology (CONACYT) and of the Aguascalientes Autonomous University (UAA), Mexico.

References

1. J. D. Murray, Mathematical Biology, 2 ed. Berlin, Germany: Springer-Verlag, 1993.
2. V. I. Krinsky, "Autowaves: Results, Problems, Outlooks", in *Self-Organization: Autowaves and Structures Far from Equilibrium*. Berlin, Germany: Springer-Verlag, 1984, pp. 9-18.
3. Turing, A.M. "The Chemical Basis of Morphogenesis", *Proc. Roy. Soc. Lond. B* **327** (1952) 37-72.
4. A. Perez-Munuzuri, V. Perez-Munuzuri, M. Gomez-Gesteira, L. O. Chua, V. Perez-Villar "Spatiotemporal Structures in Discretely-Coupled Arrays of Nonlinear Circuits: a Review", *International Journal of Bifurcation and Chaos*, **5**, No. 1, pp. 17-50, 1995.
5. L. O. Chua, M. Hasler, G. S. Moschytz and J. Neirynck, "Autonomous Cellular Neural Networks: A Unified Paradigm for Pattern Formation and Active Wave Propagation", *IEEE Trans. Circuits Syst. I*, **42**, No. 10, pp. 559-577, Oct. 1995.
6. L. Goras and L. O. Chua, "Turing Patterns on CNNs-Part II: Equations and Behaviors", *IEEE Trans. Circuits Syst. I*, **42**, No. 10, pp. 612-626, Oct. 1995.
7. L. Pivka, "Autowaves and Spatio-Temporal Chaos in CNNs-Part I:A Tutorial", *IEEE Trans. Circuits Syst. I*, **42**, No. 10, pp. 638-649, Oct. 1995.
8. L. Pivka, "Autowaves and Spatio-Temporal Chaos in CNNs-Part II:A Tutorial", *IEEE Trans. Circuits Syst. I*, **42**, No. 10, pp. 650-664, Oct. 1995.
9. Cross, M. and Hohenberg, P. "Pattern Formation Outside Equilibrium", *Rev. Mod. Phys.* **65**, pp. 851-1112, 1993.
10. Arena P, Baglio S, Fortuna L, Manganaro G. "Self-Organization in a Two-layer CNN", *IEEE Transactions on Circuits and Systems-I* **45**, No. 2, pp. 157-162, 1998.
11. Adamatzky A, Arena P, Basile A, Carmona-Galan R, Costello BDL, Fortuna L, Frasca M, Rodriguez-Vazquez A. "Reaction Difusion Navigation Robot Control: from Chemical to VLSI Analogic Processors", *IEEE Transactions on Circuits and Systems-I*, **51**, No. 5, pp. 926-938, 2004.
12. Arena P, Fortuna L, Banciforte M "Reaction Difusion CNN Algorithms to Generate and Control Artificial Locomotion", *IEEE Transactions on Circuits and Systems-I*, **46**, No. 2, pp. 253-260, 1999.

Fig. 20.1. Source of autowaves

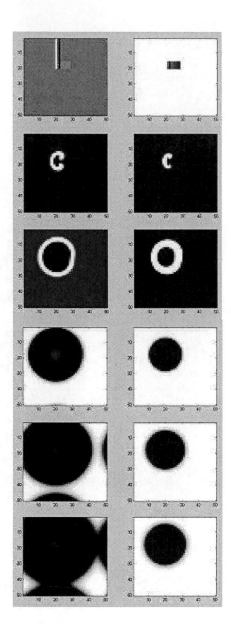

Fig. 20.2. Source of isolated fronts

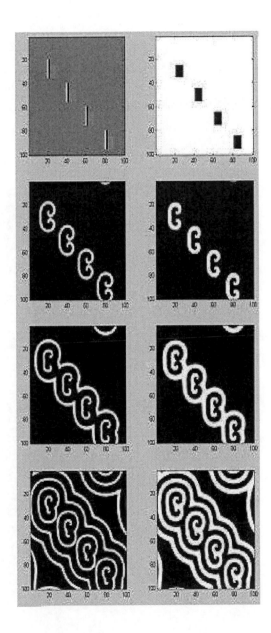

Fig. 20.3. Diagonal array of autowaves sources

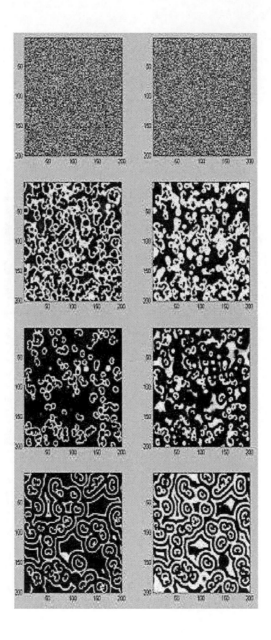

Fig. 20.4. Random autowaves sources

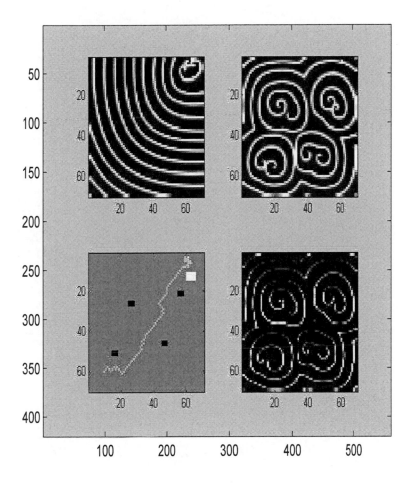

Fig. 20.5. First trajectory followed by mobile robot

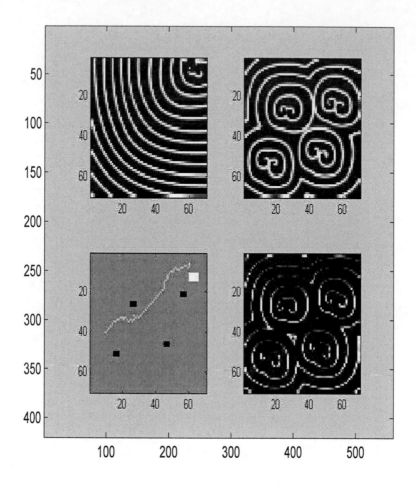

Fig. 20.6. Second trajectory followed by mobile robot

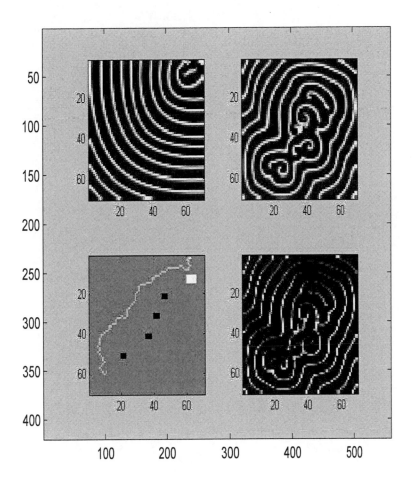

Fig. 20.7. Third trajectory followed by mobile robot

Part IV

Computer Science and Computer Engineering

Optimization of a Type-2 Fuzzy Controller for Output Regulation of a Servomechanism with Backlash via Genetic Algorithms

Nohe R. Cazarez-Castro[1], Luis T. Aguilar[2], and Oscar Castillo[3]

[1] Facultad de Ciencias Químicas e Ingeniería. UABC. Calzada Universidad No. 14418. Tijuana B.C., México.
[2] CITEDI-IPN, Tijuana B.C., México.
[3] División de Estudios de Posgrado e Investigación. Instituto Tecnológico de Tijuana, Tijuana B.C., México.

Summary. A Genetic - Type-2 Fuzzy Logic Controller (FLC) is proposed to achieve the output regulation of a servomechanism with backlash. The problem is the design of a Type-2 FLC, which will be optimized by a Genetic Algorithm (GA) to obtain the closed-loop system in which the load of the driver is regulated to a desired position. The provided servomotor position is the only measurement available for feedback. Simulations results illustrate the effectiveness of the optimized closed-loop system.

Keywords: Type-2 Fuzzy Control, Backlash, Genetic Algorithm

21.1 Introduction

A major problem in control engineering is a robust feedback design that asymptotically stabilizes a nominal plant while also attenuating the influence of parameter variations and external disturbances. In the last decade this problem was heavily studied and considerable research efforts have resulted in the development of systematic design methodology for nonlinear systems.

In the present paper, the output regulation problem is studied for an electrical actuator consisting of a motor part driven by DC motor and a reducer part (load) operating under uncertainty conditions in the presence of nonlinear backlash effects. The objective is to drive the load to a desired position while providing the boundedness of the system motion and attenuating external disturbances. Due to practical requirements [1], the motor angular position is assumed to be the only information available for feedback.

Type-2 Fuzzy Logic Systems (FLS) allow us to deal with uncertainty, but this uncertainty must to be modeled in the form of Type-2 MFs, which can carry a new problem in the designing of FLCs. In [2], authors show that making a uniform modification to the MF's parameters to a certain limit, the closed-loop system keep some properties like stability, but change in some others, like performance.

GAs [3], are derivative free optimization methods that have been used in a wide range of issues. Particulary in [4], GAs are presented as class of optimization methods for FLSs.

Problem statement. The study is motivated by the backlash affecting each joint of PEGASUS robot manipulator (see Fig. 21.1) installed in the Robotics & Control Laboratory of CITEDI-IPN, where the backlash problem occurs due to the chains and gears transmission elements. It should be pointed out that the measurements are provided from the motor side only, while the links, attached into the load of the motor, must be positioned at the desired point.

Fig. 21.1. PEGASUS robot manipulator of Robotics & Control Laboratory of CITEDI-IPN.

Our approach is to design a Type-2 FLS to be used as Type-2 FLC in the closed-loop system. The Type-2 FLC will be optimized by a GA by searching for the best performance of the Type-2 FLC.

Contributions. The main contributions of the paper are a)we present a hybrid approach, including a GA [3] in the desatgn of the Type-2 FLC, looking for an optimization method that allow us to deal with the management of uncertainty in MFs, keeping desired properties, without loss of performance; b)we solve the output regulation problem of a servomechanism with backlash; c)the system contains a great deal of uncertainty, both by the presence of backlash, as by the only source of measurement with which account. The presence of uncertainty allows us to argue and probe (by simulations) the opportunity to use Type-2 FLC in such applications. A difficult optimization problem is raised by the GA.

Organization. The paper is organized as follows. The dynamic model is in Section 21.2. Section 21.3 describes the problem statement. Type-2 Fuzzy Sets and Systems are presented in Section 21.4. GAs are described in Section 21.5. The Genetic Optimization of a Type-2 FLC is presented in Section 21.6. Simulations results are presented in Section 21.7. Finally, conclusions are given in Section 21.8.

21.2 Dynamic Model

The dynamic model of the angular position $q_i(t)$ of the DC motor and $q_o(t)$ the angular position of the load are given according to

$$J_0 N^{-1} \ddot{q}_0 + f_0 N^{-1} \dot{q}_0 = T + w_0$$
$$J_i \ddot{q}_i + f_i \dot{q}_i + T = \tau_m + w_i, \qquad (21.1)$$

hereafter, J_0, f_0, $\ddot{q}_0$ and $\dot{q}_0$ are, respectively, the inertia of the load and the reducer, the viscous output friction, the output acceleration, and the output velocity. The inertia of the motor, the viscous motor friction, the motor acceleration, and the motor velocity denoted by J_i, f_i, $\ddot{q}_i$ and $\dot{q}_i$, respectively. The input torque τ_m serves as a control action, and T stands for the transmitted torque. The external disturbances $w_i(t)$, $w_0(t)$ have been introduced into the driver equation (21.1) to account for destabilizing model discrepancies due to hard-to-model nonlinear phenomena, such as friction and backlash.

The transmitted torque T through a backlash with an amplitude j is typically modeled by a dead-zone characteristic [5]:

$$T(\Delta q) = \begin{cases} 0 & |\Delta q| \leq j \\ K\Delta q - Kj\mathrm{sgn}(\Delta q) & \text{otherwise} \end{cases} \qquad (21.2)$$

with

$$\Delta q = q_i - N q_0, \qquad (21.3)$$

where K is the stiffness, and N is the reducer ratio. Such a model is depicted in Fig. 21.2. Provided the servomotor position $q_i(t)$ is the only available measurement on the system, the above model (21.1)-(21.3) appears to be non-minimum-phase because along with the origin the unforced system possesses a multivalued set of equilibria (q_i, q_0) with $q_i = 0$ and $q_0 \in [-j, j]$.

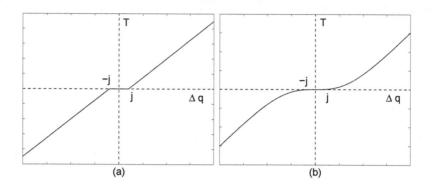

Fig. 21.2. a) The dead-zone model of backlash; b) The monotonic approximation of the dead-zone model.

To avoid dealing with a non-minimum-phase system, we replace the backlash model (21.2) with its monotonic approximation:

$$T = K\Delta q - K\eta(\Delta q) \tag{21.4}$$

where

$$\eta = -2j\frac{1 - \exp\left\{-\frac{\Delta q}{j}\right\}}{1 + \exp\left\{-\frac{\Delta q}{j}\right\}}. \tag{21.5}$$

Coupled to the drive system (21.1) subject to motor position measurements, it is subsequently shown to continue a minimum phase approximation of the underlying servomotor, operating under uncertainties $w_i(t)$, $w_0(t)$ to be attenuated. As a matter of fact, these uncertainties involve discrepancies between the physical backlash model (21.2) and its approximation (21.4) and (21.5).

21.3 Problem statement

To formally state the problem, let us introduce the state deviation vector $x = [x_1, x_2, x_3, x_4]^T$ with

$$x_1 = q_0 - q_d$$

$$x_2 = \dot{q}_0$$

$$x_3 = q_i - Nq_d$$

$$x_4 = \dot{q}_i$$

where x_1 is the load position error, x_2 is the load velocity, x_3 is the motor position deviation from its nominal value, and x_4 is the motor velocity. The nominal motor position Nq_d has been pre-specified in such a way to guarantee that $\Delta q = \Delta x$, where

$$\Delta x = x_3 - Nx_1.$$

Then, system (21.1)-(21.5), represented in terms of the deviation vector x, takes the form

$$
\begin{aligned}
\dot{x}_1 &= x_2 \\
\dot{x}_2 &= J_0^{-1}[KNx_3 - KN^2x_1 - f_0x_2 + KN\eta(\Delta q) + w_o] \\
\dot{x}_3 &= x_4 \\
\dot{x}_4 &= J_i^{-1}[\tau_m + KNx_1 - Kx_3 - f_ix_4 + K\eta(\Delta q) + w_i].
\end{aligned}
\tag{21.6}
$$

The zero dynamics

$$
\begin{aligned}
\dot{x}_1 &= x_2 \\
\dot{x}_2 &= J_0^{-1}[-KN^2x_1 - f_0x_2 + KN\eta(-Nx_1)]
\end{aligned}
\tag{21.7}
$$

of the undisturbed version of system (6) with respect to the output

$$y = x_3 \tag{21.8}$$

is formally obtained by specifying the control law that maintains the output identically zero.

The objective of the Type-2 FLC output regulation of the nonlinear driver system (21.1) with backlash (21.4) and (21.5), is thus to design a FLC so as to obtain the closed-loop system in which all these trajectories are bounded and the output $q_0(t)$ asymptotically decays to a desired position q_d as $t \to \infty$ while also attenuating the influence of the external disturbances $w_i(t)$ and $w_0(t)$.

21.4 Type-2 Fuzzy Sets and Systems

21.4.1 Type-2 Fuzzy Sets

A type-2 fuzzy set, denoted $\tilde{A}$ is characterized by a type-2 membership function $\mu_{\tilde{A}}(x, u)$ [6], where $x \in X$ and $u \in J_x \subseteq [0, 1]$, i.e,

$$\tilde{A} = \{((x, u), \mu_{\tilde{A}}(x, u)) | \forall x \in X, \forall u \in J_x \subseteq [0, 1]\} \tag{21.9}$$

in which $0 \leq \mu_{\tilde{A}}(x,u) \leq 1$. $\tilde{A}$ can alse be expressed as follows [6]:

$$\tilde{A} = \int_{x \in X} \int_{u \in J} \mu_{\tilde{A}}(x,u)/(x,u) \tag{21.10}$$

where $J_x \subseteq [0,1]$ and $\int\int$ denotes union over all admissible x and u [6].

J_x is called primary membership of x, where $J_x \subseteq [0,1]$ for $\forall x \in X$ [6]. The uncertainty in the primary memberships of a type-2 fuzzy set $\tilde{A}$, consists of a bounded region that is called the *footprint of uncertainty* (FOU) [6]. It is the union of all primary memberships [6].

21.4.2 Type-2 Fuzzy Logic Controller (FLC)

As in the Type-1 case, a Type-2 Fuzzy Inference System (FIS) is used as FLC. A type-2 FLC scheme is depicted in Fig. 21.3, it contains five components which are fuzzifier, rule base, fuzzy inference engine, type-reducer and defuzzifier. In our case both, the inputs and outputs will be represented by interval type-2 fuzzy sets like proposed in [7]-[8]. We use interval type-2 fuzzy sets as they are simple to use [6] and they distribute the uncertainty evenly among all admissible primary memberships [9]. Furthermore, the general type-2 FLC is computationally intensive and the computation simplifies a lot when using interval type-2 FLC [10](using interval type-2 fuzzy sets).

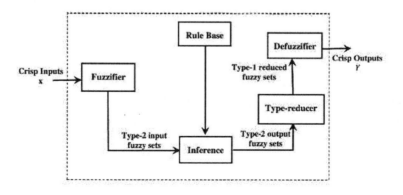

Fig. 21.3. Type-2 Fuzzy Logic Controller.

The type-2 FLC works like a type-1 FLC, of course in this case the fuzzifier take crisp inputs and this inputs are fuzzified into, in general, input type-2 fuzzy sets, which then activate the inference engine and the rule base to produce output type-2 fuzzy sets. These output type-2 fuzzy sets are then processed by the type-reducer which combines the output sets and then perform a centroid calculation, which leads to type-1 fuzzy sets called the type-reduced set [9]. The defuzzifier can then defuzzify the type-reduced type-1 fuzzy outputs to produce crisp outputs.

The uncertainty is handled by the antecedents and consequents interval type-2 fuzzy sets which include FOUs to accommodate the linguistic and numerical uncertainties associated with changing unstructured environments.

The fuzzy rule base generally is in the form of (21.11), which is a extended form for the Mamdani [11] type of Fuzzy Inference:

$$\text{IF } y_1 \text{ is } \tilde{A}_1^l \text{ AND } y_2 \text{ is } \tilde{A}_2^l \text{ THEN } y_3 \text{ is } \tilde{B}^l. \tag{21.11}$$

We select triangular interval type-2 membership functions like is proposed in [7]-[8] for each input (error and change of error) and output (control) variables.

These input and output variables are combined in fuzzy rules in the form of (21.11), selecting the seven fuzzy rules shown in Table 21.1.

Table 21.1. Fuzzy IF-THEN rules

No.	error	change of error	control
1	n	n	p
2	n	p	z
3	n	z	p
4	p	p	n
5	p	n	z
6	p	z	n
7	z	z	z

21.5 Genetic Algorithms

GAs are derivative-free optimizations methods based on the concepts of natural selection and evolutionary process [3]. They were first proposed and investigated in [12]. Their popularity can be attributed to their freedom from dependence on functional derivatives and other characteristics reported in [3].

The main idea of a GA is to maintain a population solution of a problem that evolves over a time through a process of competition and controlled variation. Each individual in the population represents a candidate solution to the specific problem and has associated a fitness to determine which individuals are used to form (by sexual reproduction and mutation) new ones in the process of competition.

The sexual reproduction of GAs consists basically in a Selection Process [12], where a set of individuals are selected to be passed through a Crossover Operation [12], which consist in to take a pair of individuals and interchanging its gens from one (or more) random selected cross point to the end of the chromosome. Mutation [12] consist in to change one or more randomly selected gens of the chromosome in some of the selected individuals.

The objective function [12] of a GA is a the value (fitness) that the method must maximize or minimize.

21.6 Genetic Optimization of the Type-2 FLC

In this paper we use a GA to optimize the parameters of the MFs of a Type-2 FLC, considering Triangular Interval MFs to each one of our three variables: *error* and *change of error* as input variables and *control* as output variable; each one of our variables is granulated in three MFs: *negative, zero* and *positive*; following the parameterization proposed in [8]-[7] we need six parameters for each Triangular Interval Type-2 MF, that is, we need to encode a total of 54 parameters for each individual (FLC) of our population, to make this encoding we design a chromosome structure of 54 consecutive real gens, where are represented the six parameters of each one of the three MFs of each one of the three variables of each Type-2 FLC (individual) of our population, see Fig. 21.4.

Fig. 21.4. Genotype for the Genetic - Type-2 Fuzzy Logic Systems optimization approach.

The optimization problem is a minimization, and with the objective function:

$$\text{fitness}_i = min(mean|error|) \tag{21.12}$$

we express that we want to minimize the mean error in our solutions space. The set of the GA parameters are shown in Table 21.2.

To solve the Type-2 FLC Output Regulation problem, we implement a two-input one-output fuzzy system, selecting triangular Interval Type-2 MFs like is proposed in [8]-[7] for each variable. These input and output variables are combined in fuzzy rules in the form of (21.11), selecting the seven fuzzy rules shown in Table 21.1.

21.7 Simulations Results

To perform simulations we use the dynamical model (21.1) of the experimental testbed installed in the Robotics & Control Laboratory of CITEDI-IPN (see Fig. 21.1), which involves a DC motor linked to a mechanical load through an imperfect contact gear train [5]. The parameters of the dynamical model (21.1) are in Table 21.3, while $N = 3$, $j = 0.2$ [rad], and $K = 5$ [N-m/rad]. These paremeters are taken from the experimental testbed.

The GA was implemented using the Genetic Algorithm and Direct Search Toolbox, each individual (Type-2 FLC) was implemented in the Interval Type-2

Table 21.2. Parameters of the genetic algorithm

Parameter	Value
Representation	real
Population size	10
Selection method	Roulette [3]
Cross method	two points
Rate of cross	0.8
Mutation method	Gaussian
Rate of mutation	0.1
Elitism [3]	2
Generations	100

Table 21.3. Nominal parameters.

Description	Notation	Value	Units
Motor inertia	J_i	2.8×10^{-6}	Kg-m^2
Load inertia	J_o	1.07	Kg-m^2
Motor viscous friction	f_i	7.6×10^{-7}	N-m-s/rad
Load viscous friction	f_o	1.73	N-m-s/rad

Fuzzy Logic Toolbox reported in [8]-[7], and each individual of the population was tested in a closed-loop system modeled in Simulink, in that model we consider the angular motor position as the only information available for feedback. In the simulations, the load was required to move from the initial static position $q_0(0) = 0$ [rad] to the desired position $q_d = \pi/2$ [rad]. In order to illustrate the size of the attraction domain, the initial load position was chosen reasonably far from the desired position. The GA was executed in a PC Computer with Intel Pentium processor of 2.4 GHz and 512 Mb of RAM.

The GA was executed in about 200 hours, producing the results shown in Table 21.4, which includes the settling time of each individual. As can be seen, a best fitness do not mean a best settling time, this because the objective function (21.12) is not designed to achieve this performance measurement.

From Table 21.4, the individual number six is the best solution of the GA, because it has the best fitness. Table 21.5 shows the chromosome of individual number six whose phenotype [3] that is depicted in Fig. 21.5. Fig. 21.6 shows its surface of control and Fig. 21.7 shows the response the system for the individual number six.

The fitness function (21.12) is not designed to guarantee ideal performance, by this reason individual number six in Fig. 21.7 results to achieve a settling criterion for ±1% in 15.2611 sec, but the error on this time is not zero, that is, maybe we can find another individual with less fitness but with an error of zero at a settling time less than 50 sec, in our case, for example, individual number eight has worse fitness

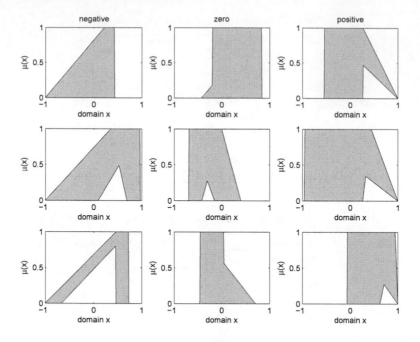

Fig. 21.5. Phenotype visualization of individual number six.

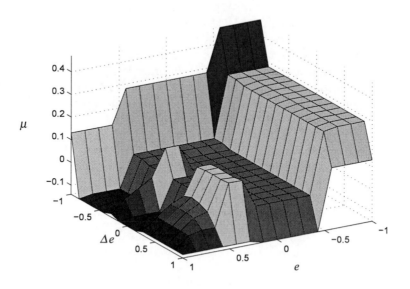

Fig. 21.6. Surface of control for individual number three.

Table 21.4. Results given by the Genetic Algorithm

Individual	Fitness	Settling Time
1	0.3032	17.1036
2	0.8898	24.1193
3	0.8593	15.3227
4	0.3046	17.0761
5	0.3809	16.0508
6	0.3004	15.2611
7	0.3308	16.4957
8	2.2201	22.9921
9	0.6113	20.8435
10	2.3717	35.2812

Table 21.5. Data of the individual number six

Variable	μ	a1	b1	c1	a2	b2	c2
	negative	-1.0000	0.2176	0.2176	0.3012	0.4293	0.4293
error	zero	-0.1982	-0.1982	-0.1912	-0.4260	0.8378	0.8378
	positive	0.2669	0.2669	1.0000	-0.5479	-0.5479	1.0000
	negative	-1.0000	0.3504	0.6897	0.0926	0.9659	0.9659
change of error	zero	-0.6963	-0.6963	-0.1490	-0.4170	0.0000	0.4005
	positive	0.2611	0.4405	1.0000	-0.9561	-0.9561	1.0000
	negative	-1.0000	0.4586	0.4586	-0.6673	0.7247	0.7247
control	zero	-0.4571	-0.4571	0.7067	0.0446	0.0446	0.0446
	positive	0.6220	0.9432	1.0000	-0.0605	-0.0605	1.0000

that individual number six, but the response (see Fig. 21.8) of the system under this solution can be considered better that the one given by individual number six.

21.8 Conclusions

The main goal of this work was to design a Genetic - Type-2 FLC architecture for the optimization of Type-2 FLC for the output regulation of a servomechanism with backlash. The GA provide us an automatic method to design and optimize FIS having a proposed set of fuzzy rules and a proposed set of MFs, which is a heavy task due the nonlinearity of the proposed problem.

The GA was implemented and performed in a satisfactory fashion, giving us a whole family of solutions to the problem at hand. Some of the resulting solutions give us better performance that other ones, allowing us to decide which one is the better to each case, that is, the solutions reported in this paper are evidently different, confirming that FIS can be obtained from human expertise, and in this case, each solution can be a representation of the expertise of different experts.

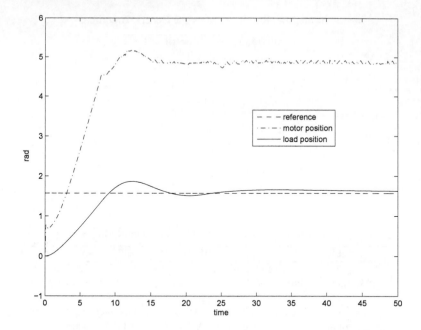

Fig. 21.7. Simulation result for individual number six.

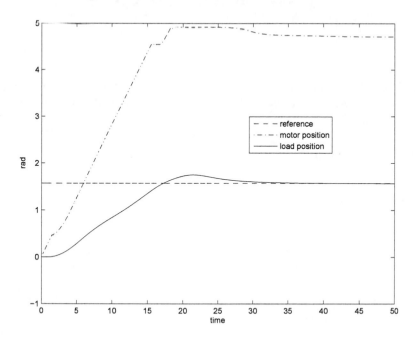

Fig. 21.8. Simulation result for individual number eight.

To the author's knowledge, the Type-2 FLC has been tested by first time in the proposed problems, demonstrating their ability to deal with the uncertainty induced by the backlash and the only source of measurement.

The combination GA and Type-2 FIS in a single architecture results to be a good method to the proposed problem, but the time necessary to run each execution of the GA is to long.

References

1. Lagerberg, A., Egardt, B.: Estimation of backlash with application to automotive powertrains. Proc. of the 42th Conf. on Decision and Control (1999) 135–147
2. Castillo, O., Aguilar, L.T., Cazarez, N., Cardenas, S.: Systematic design of a stable type-2 fuzzy logic controller. Journal of Applied Soft Computing **8**(3) (June 2008) 1274–1279
3. Castillo, O., Melin, P.: Hybrid Intelligent Systems for Pattern Recognition Using Soft Computing: An Evolutionary Approach for Neural Networks and Fuzzy Systems. Springer-Verlag, Berlin (2005)
4. Cordón, O., Herrera, F., Hoffman, F., Magdalena, L.: Genetic fuzzy systems: Evolutionary tuning and learning of fuzzy knowledge base. World Scientific, Singapore (2001)
5. Aguilar, L.T., Orlov, Y., Cadiou, J.C., Merzouki, R.: Nonlinear H_∞-output regulation of a nonminimum phase servomechanism with backlash. Journal of Dynamic Systems, Measurement, and Control **129**(4) (2007) 544–549
6. Mendel, J., John, R.: Type-2 fuzzy sets made simple. IEEE Transactions on Fuzzy Systems **10**(2) (April 2002) 117 – 127
7. Castro, J.R., Castillo, O., Melin, P.: An interval type-2 fuzzy logic toolbox for control applications. In: Proc. FUZZ-IEEE 2007. IEEE, London (July 2007) 61 – 67
8. Castro, J.R., Castillo, O., Martinez, L.G.: Interval type-2 fuzzy logic toolbox. Journal of Engineering Letters **15**(1) (August 2007) 89–98 online version.
9. Mendel, J.: Uncertain Rule-Based Fuzzy Logic Systems: Introduction and New Directions. Prentice Hall, Upper Saddle River, NJ (2001)
10. Liang, Q., Mendel, J.: Interval type-2 fuzzy logic systems: theory and design. IEEE. Transactions Fuzzy Systems **6**(5) (October 2000) 535 550
11. Mamdani, E.H., Assilian, S.: An experiment in linguistic synthesis with a fuzzy logic controller. Int. J. Hum.-Comput. Stud. **51**(2) (1999) 135–147
12. Holland, J.M.: Adaptation in Natural and Artificial Systems. University of Michigan Press, Ann Arbor, MI (1975)

Integrating vision-based motion planning and defeasible decision making for differential-wheeled robots

A case of study with the Khepera 2 robot

Edgardo Ferretti[1], Roberto Kiessling[2], Alejandro Silnik[2], Ricardo Petrino[2], and Marcelo Errecalde[1]

[1] Laboratorio de Investigación y Desarrollo en Inteligencia Computacional
[2] Laboratorio de Electrónica, Investigación y Servicios
Universidad Nacional de San Luis. San Luis, Argentina.
{ ferretti,rkiessling,aasilnik,rpetrino,merreca
}@unsl.edu.ar

Summary. This chapter presents a general approach to combine autonomous navigation with high-level reasoning; a key issue in intelligent mobile robotics. This work successfully integrates vision-based motion planning and defeasible decision making for differential-wheeled robots. As a case of study the *Khepera* 2 robot is used. The problem faced comprises a scenario where the robot has to perform removal tasks of different size objects. This problem can be conceptually divided into two stages: (a) deciding the objects transportation order, and (b) planning and controlling the robot motion. The decision making task is carried out using Defeasible Logic Programming, while motion planning is performed using a smooth shortest path control generated from visual information.

22.1 Introduction

Today's research on mobile robotics has produced a large number of control techniques in real environments [1, 2, 3, 4], that do not typically engage in abstract reasoning processes. On the other hand, research on knowledge representation and reasoning has resulted in a number of interesting theories to reason about actions and plans [5, 6, 7]. These theories are mostly stated at a very abstract level, and do not consider the oddities and uncertainties that arise from operating in a real physical environment.

In a former work [8], was presented *Khe*-DeLP, a framework to develop multi-robot applications with *Khepera* 2 robots [9]. *Khe*-DeLP facilitates the implementation of cognitive robotic tools and provides Defeasible Logic Programming (DeLP) [10] to describe the robots' deliberative behavior.

In this work, we follow some of the ideas exposed in [11] about the convenience of integrating high-level reasoning facilities with robot autonomous navigation. We share the approach of seeing the planning and control module as a black box which: (a) receives from the high-level component goals to be achieved, (b) generates plans to achieve those goals, and (c) sends an acknowledgment to the high-level component to inform failures or if everything finished as planned.

In [8, 12] results of the use of *Khe*-DeLP in simulated scenarios are provided. In this chapter, we describe some extensions to these works by incorporating to *Khe*-DeLP a vision-based motion planning and control module, in order to perform experiments on a real *Khepera* 2 robot.

The chapter is organized as follows. In the first place, Sect. 22.2 characterizes the experimental setting used in our study. Secondly, the *Artificial Vision* module is described in Sect. 22.3. This module performs the first step of the robot's functional cycle and provides the input data to the *Motion Planning* module described in Sect. 22.4. Thirdly, Sect. 22.5 explains how is performed the set up of the robot's knowledge base, that provides to the robot the symbolic representation of its environment. Section 22.6 gives a brief overview of the *Khe*-DeLP framework and the formalism it uses for knowledge representation and reasoning. Besides, the *Cognitive* module of the system is specified. This module is dedicated to decide which is the more appropriate object to be transported next; which in turn becomes the target waypoint of the robot *Motion Control* module introduced in Sect. 22.7. Then, Sect. 22.8 presents a description of the first iteration performed by the robot's functional cycle and discusses the results. Finally, Sect. 22.9 puts forward related works and conclusions.

22.2 Experimental Setting

The working environment (arena) consists of a 120cm × 150cm flat wooden surface with no texture. Over this surface, high reflectivity expanded polystyrene objects are disposed. A JAI MCV-50 zenithal camera captures a frame of the arena and feeds visual information to a Matrox Corona frame grabber on a PC. The camera is mounted on a pipe structure with cylindrical coordinate system adjustment capability, that along with interchangeable lenses allows to modify the scale and viewpoint for different experiments.

The objects' size in the image defines the following classification criteria: objects bigger than the *Khepera* 2 gripper are considered obstacles, while those that can be manipulated by the robot are classified according to its size as *small*, *medium* and *big*. These objects are scattered through the working area and they have to be transported by the robot to a defined storage area (*store*) in the arena. As obstacles, simple shape objects are used, which can be reduced after image processing to polygonal obstacles. Figure 22.1 shows the particular initial setting. To perform the objects removal tasks several independent modules interact as follows:

- Robot, objects and obstacles positions are obtained from the camera image by the vision module.
- Navigable paths between robot, objects and storage area are calculated by the motion planning module.

- Objects picking order is defined by the cognitive module.
- Robot control to successfully perform the specified task is carried out by the control module.

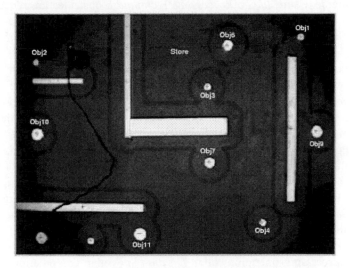

Fig. 22.1. Experimental environment

The vision and motion planning modules algorithms were written in C language using Matrox MIL libraries [13] for image processing and run on a PC. The control module was also written in C using the K-Team cross-compiler and it executes completely on the robot. The *Khe*-DeLP framework runs on a *Ciao* Prolog interpreter [14].

The autonomy of the robot is limited and it cannot measure the state of its battery, thus, it cannot perform a globally optimized task.[3] Because of this drawback, a *greedy* strategy is used to select the next object. To perform the reasoning, the robot will use its perceptual information about the environment (Sect. 22.5) and its preferences (Sect. 22.6).

22.3 Artificial Vision

In the last 20 years [15], Vision for Mobile Robot Navigation has been widely used to face indoor and outdoor navigation problems. Indeed, machine vision techniques have been proven to be useful to find obstacle positions when the environment of the robot is partially known or dynamic.

---

[3]This matter has been considered by K-Team, the manufacturer of the *Khepera* 2 robots. This is why the *Khepera* 3 robots include this feature.

This section describes the *Artificial Vision* module used in our experiments for image processing. This module can be conceptually conceived as a black box, whose input is an image of the environment (where the robot perform its activities) and the output is a useful description of the obstacles to be avoided and the objects to be gathered. Using *Image Processing Techniques*, obstacles are recognized and described by the coordinates of their vertices. This model of representation is closely related to the path planning strategy chosen, in our case, the *Visibility Graph Method* [16] which is described in Sect. 22.4.

Objects are represented by their *area* and the *coordinates* of their *centers*. In fact, objects are also considered as obstacles, because the robot has to avoid them when following a certain path in its way to the object to be transported. As shown in Fig. 22.2, the results from this vision processing stage are used by (a) the *cognitive module* and (b) the *motion planning module*.

For the *cognitive module* it computes the position and the objects' size. The way the cognitive module uses this information is described in Sect. 22.5. For the *motion planning module*, the vision module returns an array containing a description of the obstacles using *cartesian coordinates* for the vertices. The visibility graph method for motion planning, represents the robot as a point in an appropriate space, the configuration space, and maps the obstacles in this space [16]. This implies that obstacles have to be grown up in proportion to the robot's size; this is performed by an image processing operation called *Dilation* [17] or in computational geometric terms, Minkowski Sums [18].

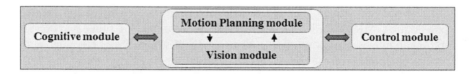

Fig. 22.2. Functional description of the system

Obstacle representations by their vertices coordinates are obtained through the following image processing steps [19]:

1. Image capture of the robot's working environment. This gives information about the position and size of the robot and the elements in the arena, but in image format (Fig. 3(a)).
2. Filtering, binarization and segmentation of obstacles and robot. This gives a binary image with objects and obstacles segmented from the background. These images can be stored for later processing (Fig. 3(b)).
3. Computing the Minkowski sum operation (dilation) over the obstacles grow them and obtains the configuration space, in which the obstacles are larger and the robot is represented as a point.
4. Vertices detection. Objects in the image are represented by their vertices coordinates. The algorithm used in this work is described in [19]. First, object

boundaries are computed (Fig. 3(c)). Later, vertices detection is performed computing a boundary following algorithm that detects the vertices coordinates (Fig. 3(d)).

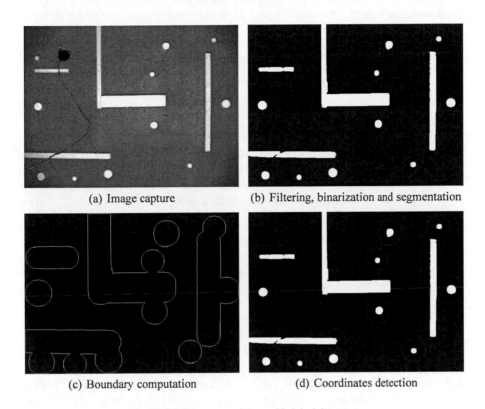

(a) Image capture

(b) Filtering, binarization and segmentation

(c) Boundary computation

(d) Coordinates detection

Fig. 22.3. Main steps of the artificial vision stage

22.4 Robot Motion Planning

Planning a path for a mobile robot means to find an obstacle-free continuous trajectory leading from its initial position (*pstart*) to its target position (*pgoal*). In this section we consider a case where the robot is a point and where the scene in which the robot travels is a two-dimensional plane. Robot motion planning for a mobile robot in a two-dimensional plane can be solved using *Visibility Graphs* and *Voronoi Diagrams*. In our work the visibility graph method and Dijkstra's algorithm for the shortest path have been chosen [20].

The visibility graph is a non-directed graph G whose nodes are the initial and goal positions and all the obstacles vertices (S). The links of G are all the straight line segments connecting two nodes that do not intersect the interior of the obstacle region. In Sect. 22.3, obstacles were described by their vertices, now the convenience of that description is evident. In this stage, obstacles are proportionally augmented to the robot's shape and size so that the robot can be treated as a point and the problem is easier to be solved. The shortest path algorithm from *pstart* to *pgoal* is described by the following steps [18]:

1. Compute the Visibility Graph:

$$Gvis = \text{VisibilityGraph}(S \cup \{pstart, pgoal\})$$

2. Assign to each arc (p, w) in *Gvis* a weight, which is the *Euclidean length* of the segment $\overline{pw}$; where p and w are any point of $(S \cup \{pstart, pgoal\})$.
3. Use Dijkstra's algorithm to compute a shortest path between *pstart* and *pgoal* in *Gvis*. The visibility graph, includes segments that connect any pair of points which are mutually visible, among all this segments, a shortest path can be found using Dijkstra's algorithm. A shortest path between two points among a set of polygonal obstacles with a total of n edges, can be computed in time $O(n^2 \log n)$.

This algorithm was written using C language and develops principles of computational geometry to save time and avoid rounding errors. The motion planning module takes as inputs a set of obstacles, described by their vertices, and two points (*start* and *pgoal*) and returns as output the shortest path among obstacles between these two points. This shortest path is described by segments. The length of the path is also computed.

The output of the motion planning module serves as input to the cognitive and robot control modules (see Fig. 22.2). The cognitive module requires real navigable paths between the robot and each object, and between the store and each object. These distances are also calculated by the motion planning algorithm because it computes the shortest path, and therefore the navigable path between any two selected places or points on the scenario. The calculated navigable paths are written on a file following the format: *object label, type, area, X coordinate, Y coordinate, navigable path object-to-robot* and *navigable path object-to-store*. Table 22.1 shows the above-mentioned data corresponding to the scenario depicted in Fig. 22.1. Figure 22.4 shows the navigable paths calculated by the system drawn on the arena picture.

22.5 Knowledge Base Set Up

This section describes the operational steps performed to build the robot's knowledge base (KB) from the data presented in Table 22.1. In the first place, each object obj_i is represented as a fact $cyl(obj_i)$ (facts (1)-(9) of Fig. 22.5). Although there are eleven objects in the arena (see Fig. 22.1), objects $obj5$ and $obj8$ are not reachable by the robot until object $obj11$ be removed (see Fig. 22.4). Therefore, the motion planning module do not include them in the input provided to the cognitive module.

Moreover, the robot and store are implicitly represented by the literals *khep* and *store*, respectively.

Each preference criterion provided to the robot has a comparison literal that represents it. In our experimental study three criteria were selected, but of course more (or less) criteria can be used. In order of higher to lower preference, these criteria are: *proximity to the robot, proximity to the store* and the *boxes' size*. In turn, they are represented by the following comparison literals: *nearer_robot, nearer_store* and *smaller*. These comparison literals are defined by the strict rules (37)-(41) of Fig. 6(a). The semantics of these rules will be explained in Sect. 22.6. The two former comparison literals represent binary criteria, this is, an object is near or far from the robot / store. On the other hand, the third comparison literal represent the boxes' size; a criterion that in this case has more than two possible values.

Arbitrarily those objects being more than 150 pixels away from the robot / store are considered far. For each binary criterion ($comp\_lit_{pc}$), it is obtained the set of objects ($\Upsilon_{comp\_lit_{pc}}$) that make it true; *e.g.*, the *nearer_robot* criterion set only contains *obj2*, while for the *nearer_store* criterion it holds that $\Upsilon_{nearer\_store} = \{obj3, obj6\}$. Then, for each $x \in \Upsilon_{comp\_lit_{pc}}$ the facts that help to state the status of these literals are asserted in KB. For instance, as can be observed in Fig. 22.5 facts (19)-(27) were asserted for the literal *nearer_robot* and facts (28)-(36) were included for the literal *nearer_store*.

For the remaining criteria, the non binary ones, facts will be asserted in the knowledge base in a convenient way considering the values of the different alternatives with respect to the particular criterion. For example, facts (10)-(18) were incorporated to the KB for the literal *smaller*, as shown in Fig. 22.5. Finally, if there is only one object *obj* in the environment KB will only contain the literal *unique(obj)*. Similarly, if there are no objects in the environment KB will contain the literal *no_objects*.

Table 22.1. Environment description for the cognitive module (measures are given in pixels)

Object	Type	Area	X	Y	Path-Robot	Path-Store	Object	Type	Area	X	Y	Path-Robot	Path-Store
khep	robot		118	87		232	obj7	cyl	462	387	295	396	288
store	store		350	76	232	0	obj9	cyl	586	603	233	772	479
obj1	cyl	205	568	45	765	233	obj10	cyl	562	41	238	229	681
obj2	cyl	176	38	95	80	804	obj11	cyl	594	248	439	421	495
obj3	cyl	227	382	144	653	75	obj12	obs	10992	290	189	200	127
obj4	cyl	222	493	416	535	382	obj13	obs	5351	550	228	454	251
obj6	cyl	469	423	62	706	74	obj14	obs	1339	83	132	56	272
							obj15	obs	4571	133	384	297	377

22.6 *Khe*-DeLP and DeLP Overview

Prolog is a programming language that has already been used to develop applications in the field of cognitive robotics [21, 22]. Following this trend, the *Khe*-DeLP

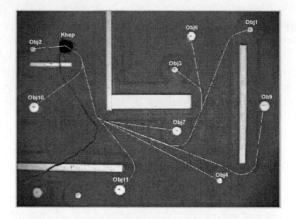

(a) Robot to object

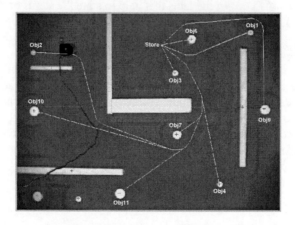

(b) Store to object

Fig. 22.4. Navigable paths calculated by the system shown over the working environment

framework has also been developed in Prolog. Below, it is presented a brief overview of this framework that uses Defeasible Logic Programming (DeLP) for reasoning and knowledge representation.

22.6.1 DeLP

DeLP is an argumentative formalism suitable for reasoning in real and dynamic environments, *i.e.*, scenarios where the information that the agent (a robot in our case) has about its environment, is usually incomplete or contradictory.

$cyl(obj1)$ (1)	$cyl(obj10)$ (8)	$medium(obj7)$ (15)	$far(obj4,khep)$ (22)	$far(obj2,store)$ (29)
$cyl(obj2)$ (2)	$cyl(obj11)$ (9)	$big(obj9)$ (16)	$far(obj6,khep)$ (23)	$far(obj4,store)$ (30)
$cyl(obj3)$ (3)	$small(obj1)$ (10)	$big(obj10)$ (17)	$far(obj7,khep)$ (24)	$far(obj7,store)$ (31)
$cyl(obj4)$ (4)	$small(obj2)$ (11)	$big(obj11)$ (18)	$far(obj9,khep)$ (25)	$far(obj9,store)$ (32)
$cyl(obj6)$ (5)	$small(obj3)$ (12)	$far(obj1,khep)$ (19)	$far(obj10,khep)$ (26)	$far(obj10,store)$ (33)
$cyl(obj7)$ (6)	$small(obj4)$ (13)	$near(obj2,khep)$ (20)	$far(obj11,khep)$ (27)	$far(obj11,store)$ (34)
$cyl(obj9)$ (7)	$medium(obj6)$ (14)	$far(obj3,khep)$ (21)	$far(obj1,store)$ (28)	$near(obj3,store)$ (35)
				$near(obj6,store)$ (36)

Fig. 22.5. Symbolic representation of the environment (context Φ), built by the cognitive module from the raw numerical data presented in Table 22.1

A DeLP-program $\mathscr{P}$ is denoted (Π, Δ), where $\Pi = \Pi_f \cup \Pi_r$, distinguishing the subsets Π_f of facts, Π_r of strict rules and the subset Δ of defeasible rules. *Facts* are ground literals representing atomic information or the negation of atomic information. In our application example facts represent the robot's perception; for example, Fig. 22.5 shows the robot's perception corresponding to the experimental environment presented in Fig. 22.1.

Strict Rules are denoted $L_0 \leftarrow L_1, \ldots, L_n$ and represent firm information, whereas *Defeasible Rules* are denoted $L_0 \prec L_1, \ldots, L_n$ and represent tentative information. In both cases, the *head* L_0 is a literal and the *body* $\{L_i\}_{i>0}$ is a set of literals. As usual in *Logic Programming*, variables are denoted with an initial uppercase letter. Observe that strong negation is allowed in the head of program rules, and hence may be used to represent contradictory knowledge.

In the robot's knowledge base (see Fig. 6(a)) there are six strict rules ((37)-(42)). Rules (37)-(41) represent physical properties of the objects; in particular, rules (37) and (38) state when an object X is nearer to the robot / store than an object Y, while rules (39)-(41) determine when an object X is smaller than Y. On the other hand, the strict rule (42) states that when there is only one box B in the environment, it must be chosen. Besides, there are nine defeasible rules ((43)-(51)) representing tentative

$nearer\_robot(X,Y) \leftarrow near(X,khep), far(Y,khep)$	(37)	
$nearer\_store(X,Y) \leftarrow near(X,store), far(Y,store)$	(38)	
$smaller(X,Y) \leftarrow small(X), medium(Y)$	(39)	
$smaller(X,Y) \leftarrow small(X), big(Y)$	(40)	
$smaller(X,Y) \leftarrow medium(X), big(Y)$	(41)	$nearer\_store(Z,X) > smaller(X,Y)$ (52)
$choose(B) \leftarrow unique(B)$	(42)	$nearer\_robot(Z,X) > nearer\_store(X,Y)$ (53)
$choose(B) \prec better(B,OB)$	(43)	$nearer\_robot(Z,X) > smaller(X,Y)$ (54)
$\sim choose(B) \prec better(OB,B)$	(44)	$nearer\_robot(Z,Y) > smaller(X,Z)$ (55)
$\sim choose(B) \prec better(B,OB), better(BB,B)$	(45)	$nearer\_store(Z,X) > smaller(Y,Z)$ (56)
$better(B,OB) \prec nearer\_robot(B,OB)$	(46)	$nearer\_robot(Z,Y) > nearer\_store(X,Z)$ (57)
$better(B,OB) \prec nearer\_store(B,OB)$	(47)	$smaller(Z,X) > smaller(X,Y)$ (58)
$better(B,OB) \prec smaller(B,OB)$	(48)	
$\sim better(B,OB) \prec nearer\_robot(OB,B)$	(49)	
$\sim better(B,OB) \prec nearer\_store(OB,B)$	(50)	
$\sim better(B,OB) \prec smaller(OB,B)$	(51)	

    (a) $\mathscr{P}_k$: Robot's knowledge base   (b) L-order over c-lits($\mathscr{P}_k$)

Fig. 22.6. Defeasible rules, strict rules and priorities for our experimental setting

information about which cylinder to choose (preferences over objects). For example, rule (43) provides a reason to choose B if it is a better alternative than OB with respect to a particular criterion (rules (46)-(48)). On the other hand, rule (44) supports the opposite. Moreover, rule (45) provides a reason for not choosing alternative B, because despite B being better than OB, there is another alternative BB that is in turn better than B. Finally, rules (49)-(51) are the counterparts of rules (46)-(48).

In DeLP, contradictory literals could be defeasibly derived; for example, from $\mathscr{P}_k$ (see Fig. 6(a)) both *choose(obj2)* and $\sim$*choose(obj2)* are defeasibly derived, but as we will show in Sect. 22.8, only one of them will be accepted as *warranted*. Furthermore, to deal with contradictory and dynamic information, *arguments* for conflicting pieces of information are built and then compared to decide which one *prevails*. The prevailing argument provides a warrant for the information that it supports. A literal L is *warranted* from (Π, Δ) if a non-defeated argument $\mathscr{A}$ supporting L exists. To put it briefly, an *argument* for a literal L, denoted $\langle \mathscr{A}, L \rangle$, is a minimal set of defeasible rules $\mathscr{A} \subseteq \Delta$, such that $\mathscr{A} \cup \Pi$ is non-contradictory and there is a derivation for L from $\mathscr{A} \cup \Pi$. To establish if $\langle \mathscr{A}, L \rangle$ is a non-defeated argument, *argument rebuttals* or *counter-arguments* that could be *defeaters* for $\langle \mathscr{A}, L \rangle$ are considered, *i.e.*, counter-arguments that by some criterion are preferred to $\langle \mathscr{A}, L \rangle$. Since counter-arguments are arguments, defeaters for them may exist, and defeaters for these defeaters, and so on. Thus, a sequence of arguments called *argumentation line* is constructed, where each argument defeats its predecessor in the line (for a detailed explanation of this dialectical process see [10]).

The prevailing argument provides a warrant for the information it supports. A literal L is *warranted* from (Π, Δ) if a non-defeated argument $\langle \mathscr{A}, L \rangle$ exists. In DeLP the comparison criterion among arguments can be established in a modular way. Thus, an appropriate domain-dependent criterion can be used. The criterion that we will use in our approach is defined next.

Definition 1 (L-order). *Let $\mathscr{P}$ be a DeLP program and* c-lits($\mathscr{P}$) *be the set of literals that belong to $\mathscr{P}$. An L-order over $\mathscr{P}$ is a strict partial order over the elements of* c-lits($\mathscr{P}$).

As above-mentioned in Sect. 22.5, the autonomy of the robot is limited and it cannot measure the state of its battery. Because of this drawback, a *greedy* strategy is used to select the next box. In our setting, the robot will *prefer its nearest boxes, then the boxes nearest to the store, and finally the smallest ones*. These preferences will be explicitly established using a preference order among the literals *smaller*, *nearer_store* and *nearer_robot*. As will be shown below, an L-order must be provided as a set of facts within the program. These facts are written as $X > Y$, stating that a literal X is preferred to a literal Y, and they will be used to decide when an argument is better than another. In particular, the L-order defined in Fig. 6(b) represents the robot's preferences over objects, *i.e.*, it shows the L-order defined over the DeLP-program $\mathscr{P}_k$ of Fig. 6(a). Based on a given L-order, the following argument comparison criterion can be defined.

Definition 2. *Let* $\mathscr{P} = (\Pi, \Delta)$ *be a DeLP program and let* ">" *be an L-order over* $\mathscr{P}$. *Given two conflicting argument structures* $\langle \mathscr{A}_1, h_1 \rangle$ *and* $\langle \mathscr{A}_2, h_2 \rangle$, *the argument* $\langle \mathscr{A}_1, h_1 \rangle$ *will be preferred over* $\langle \mathscr{A}_2, h_2 \rangle$ *iff:*

1. *there are two literals* l_1 *and* l_2 *such that* $l_1 \in^* \mathscr{A}_1$, $l_2 \in^* \mathscr{A}_2$, $l_1 > l_2$, *and*
2. *there are no literals* l_1' *and* l_2' *such that* $l_1' \in^* \mathscr{A}_1$, $l_2' \in^* \mathscr{A}_2$, *and* $l_2' > l_1'$.

where $l \in^* \mathscr{A}$ *iff there exists a defeasible rule* $(l_0 \prec l_1, l_2, \ldots, l_n)$ *in* $\mathscr{A}$ *and* $l = l_i$ *for some* i $(1 \leq i \leq n)$.

22.6.2 *Khe*-DeLP

This is a layered framework that provides facilities for programming high-level robots' behaviors using DeLP. In *Khe*-DeLP, lower-level layers are hardware-oriented and allow interaction with *Khepera* robots. In contrast, upper layers are dedicated to cognitive robotics implementations. To provide Defeasible Logic Programming, *Khe*-DeLP uses a DeLP-Server [23]. A DeLP-Server is an implementation of a DeLP interpreter that provides a reasoning service for a group of agents. This service runs as a stand-alone program that can interact with the robots using the TCP/IP protocol. We will only describe that features of DeLP-Server that will be needed in the rest of the chapter; however, DeLP-Server has other features that can be very useful for multi-robot programming.

A query for a DeLP-Server is a pair $[\Phi, Q]$, where Q is the literal that DeLP will try to warrant and Φ is the *context* for that query. The context Φ of a query can be any DeLP-program, in our experimental application, Φ will be the set of facts that represent the perception of the current state of the environment (see Fig. 22.5). For instance, if a robot *khep* has to decide which object to choose, and in its environment there are two cylinders: *obj2* near the robot and *obj1* far from it, then it may send the query $[\{near(obj2, khep), far(obj1, khep)\}, choose(X)]$. Since the environment changes, the robot will send queries and the context will correspond to the current state of the robot's perception of its environment. The answer returned for a query $[\Phi, Q]$ will depend on the DeLP-program loaded in the DeLP-Server and the context Φ. Thus, the server would respond YES, when Q is warranted; NO, when the complement of Q is warranted; and UNDECIDED, if neither Q nor its complement are warranted.

Our methodology consist of having the robot's knowledge (Fig. 6(a)) represented as a DeLP-program loaded in one or many DeLP-Server's, and to send the current state of the environment as contextual information in every query. The *Khe*-DeLP *cognitive layer* provides a predicate `answer` (Φ, Q, Ans) that sends to a DeLP-Server a query $[\Phi, Q]$ and obtains the answer.

Programming the Functional Cycle

Figure 22.7 shows the implementation in Prolog of the funcional cycle depicted in Fig. 22.2. The predicate `behave/0` implements a simple control cycle of the kind "sense-deliberate-act," of course, more sophisticated methods can be implemented. In the first place, the robot builds its knowledge base following the stages described in Sects. 22.3, 22.4 and 22.5 using the `sense/1` predicate that returns in CC the world representation as a list of facts, as those shown in Fig. 22.5. The finalization detection is done by checking that the list of facts returned in CC (by the `sense/1` predicate) is not a list containing only one element; the literal *no_objects*. If there is still any object in the environment, the list of facts CC is received by `decide_action/2` and used as context to query the DeLP-Server using `answer(CC,choose(X),Ans)`. In this case, the robot queries the DeLP-Server with the predicate `choose(X)`.

```
:-use_module(khe_delp)
:-use_module(vision_planning).
behave:-
        sense(CC),
        CC \= [no_objects],
        decide_action(CC,Act),
        perform(Act),
        behave.
decide_action(CC, Act):-
        answer(CC,choose(X),Ans)
    (   Ans = yes, Act = goto(X)
    ;   Ans = no, Act = wait
    ;   Ans = undecided, Act = goto(X)   ).
```

Fig. 22.7. Sense-deliberate-act robot control cycle

22.7 Robot Motion Control

The motion control module is in charge of actually driving the robot along the path. This task is performed based on the robot position and orientation gathered by the vision module, and the optimum path selected by the cognitive layer. This is a simple and not computationally heavy algorithm that can be performed inside the robot in spite of its limited capabilities. Odomethry and speed control methodologies are used in a closed control loop to generate the wheels target (or reference) speeds.

The robot can sense its wheels speeds by reading two encoders coupled to its axes. By using (22.1) and (22.2), the robot position can be calculated at any moment.

$$x(t) = x(t_0) + \int_{t_0}^{t} v_x(t)dt \tag{22.1}$$

$$y(t) = y(t_0) + \int_{t_0}^{t} v_y(t)dt \tag{22.2}$$

These equations imply high computing capabilities not usually found in robots' microcontrollers. Considering that in the instant under study dt wheels speeds are constant, the robot describes an arc as shown in Fig. 22.8, along with the relations between the arcs described by the left and right wheels, S_l and S_r respectively.

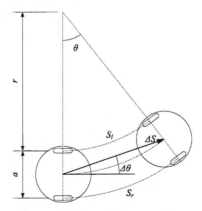

Fig. 22.8. Displacement in dt

$$S_l = r\theta \tag{22.3}$$

$$S_r = (r+a)\theta \tag{22.4}$$

$$S_c = (r+\frac{a}{2})\theta \tag{22.5}$$

The arc described by the center of the robot (its reference for positioning) in this differential time is the average between the arcs described by the wheels (values that can be easily acquired from the encoders). The change in the robot pose can be calculated as:

$$\Delta\theta = \frac{S_r - S_l}{a} \tag{22.6}$$

$$\Delta U = \frac{S_r + S_l}{2} \tag{22.7}$$

The robot current position and orientation can be calculated from the last position and orientation known and the changes observed in the wheels coupled encoders, so the integral formulas (22.1) and (22.2) become:

$$\theta_i = \theta_{i-1} + \Delta\theta_i \tag{22.8}$$

$$x_i = x_{i-1} + \Delta U_i \cos\theta_i \tag{22.9}$$

$$y_i = y_{i-1} + \Delta U_i \sin\theta_i \tag{22.10}$$

These equations are more suitable to the robot calculus capabilities, allowing it to update information of its own position and orientation independently of the vision module.

Figure 22.9 depicts the robot location and orientation and the next waypoint location in the path provided by the motion planning module. Disorientation ($\Delta\alpha$) is defined as the angular difference between the direction the robot aims and the segment defined by the position of the robot and position of the next waypoint.

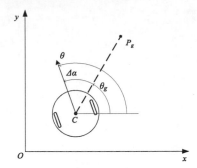

Fig. 22.9. Orientation error

From this disorientation $\Delta \alpha$, the robot calculates the linear and angular velocities it needs to get closer to the next waypoint by using the control laws described by (22.11), (22.12) and Fig. 22.10, where V_M and ω_M are the maximum linear and angular speeds for non-sliding displacement of the robot, experimentally found for each surface.

$$V_l = \begin{cases} (1 - \frac{|\Delta \alpha|}{90})V_M & \text{if } |\Delta \alpha| < 90 \\ 0 & \text{if } |\Delta \alpha| \geq 90 \end{cases} \qquad (22.11)$$

$$\omega = \frac{\Delta \alpha}{180} \omega_M \qquad (22.12)$$

As can be seen in Fig. 22.10, the linear velocity reaches its maximum value when the robot is aiming at the next waypoint, and decreases to zero as the absolute value of the disorientation angle increases to 90 degrees (Fig. 10(a)). Greater disorientation values also produces zero linear velocity. On the other hand, when orientation error is small, no angular velocity is applied to the robot. If the disorientation increases, the angular velocity applied increases in the direction that makes this error decrease (Fig. 10(b)). So, the greater the misorientation, the greater the angular velocity and the lower the linear velocity should be. In cases of great disorientation, no linear speed will be applied and the robot will make an in-place rotation.

When the disorientation is small, maximum linear velocity are achieved with no angular velocity, making the robot approach quickly to the waypoint. In differential drive robots, calculating wheels speeds from linear and angular velocities, is straightforward using the following velocities relations:

$$V_r = V + \frac{a}{2}\omega \qquad (22.13)$$

$$V_l = V - \frac{a}{2}\omega \qquad (22.14)$$

$$V = \frac{V_r + V_l}{2} \qquad (22.15)$$

$$\omega = \frac{V_r - V_l}{a} \qquad (22.16)$$

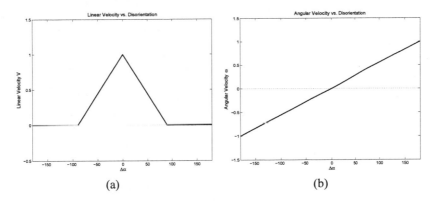

Fig. 22.10. Linear and angular velocities versus disorientation

These calculated velocities, are applied to each wheel and thus modify the robot position and orientation making move it toward to the next waypoint. The described process repeats in a loop, calculating robot position and orientation, orientation error and wheels speeds that tend to minimize that error, in a "Closed Control Loop". This should not repeat indefinitely, as in a given moment the robot will get to the waypoint. If the robot surpasses the waypoint, the orientation error will become instantly $180°$, making the robot stop and rotate in place. This odd behavior is avoided by switching waypoints.

When the robot distance to the next waypoint gets smaller than a predefined "switching threshold", this waypoint is discarded and the algorithm switches to the next one in the path as the next goal, recalculating speeds and producing a smooth reorientation of the robot. This procedure repeats until there are no more waypoints in the path (defined by the motion planning module). Setting of the switching threshold implies a compromise between accuracy and smoothness in the path execution, as shown in Fig. 22.11.

The switching threshold is directly related to the size of the growing mask used to define the navigation prohibited area of the image, and the best results are obtained for the empirical value of $\frac{a}{2}$. A greater value for the switching threshold leads to a late turn and an overpass as shown in Fig. 11(a) and oppositely, a bigger threshold value produces an early turn (Fig. 11(b)). Under these situations, the robot can collide with obstacles close to the path. The switching threshold is also dependant of the values chosen for the maximum linear and angular speeds.

When the waypoint switching is performed, the relative position of the new waypoint respect to the last one will determinate the new wheels speeds. If the new waypoint position is not aligned with the robot, there should be a noticeable change in speeds. This situation does not occur in practice because the vision module used a circular growing mask to perform obstacles dilation (Fig. 3(c)). Next, that module used a vertices description of the objects using a minimum vertices distance

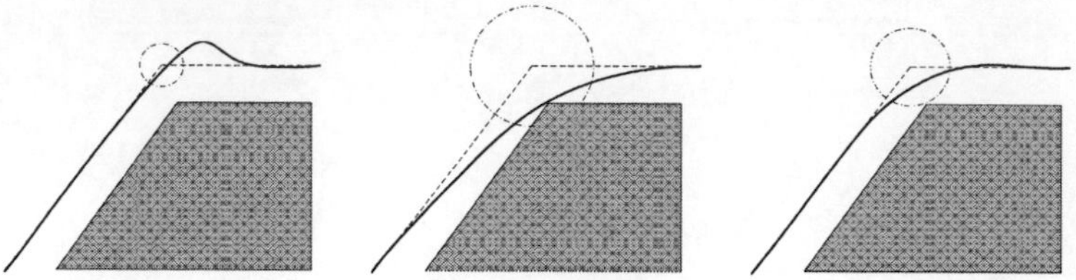

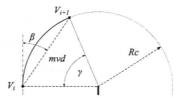

(a) Too small   (b) Too large   (c) Optimum switching

Fig. 22.11. Waypoints switching thresholds

separation between waypoints. This leads to the waypoint configuration in the path depicted in Fig. 22.12.

Fig. 22.12. Relative position of successive waypoints

It can be demonstrated [24] that the inter-waypoint orientation change is a function of the growing mask radius and the minimum vertices distance, and keeping these parameters small insures small orientation changes at waypoint switching.

Table 22.2 compares the paths generated using round and square shaped dilation masks by the motion planning module. Although an increment in the number of waypoints can be noted, there are reductions in the total path length, in the accumulated rotations and in the inter-waypoint distances. Thus the circular dilation mask used is particularly suited to the motion control method implemented.

Table 22.2. Paths comparison using Round / Square masks based on 28 synthetic images

	Square mask	Round mask	Round / Square relations
Total waypoints number	204	283	38.72%
Accumulated Rotations	6565°	4301°	−34.48%
Accumulated displacements	20710 pixels	16129 pixels	−22.11%
Inter-waypoint mean distance	117.67 pixels	63.25 pixels	−46.24%

22.8 Experimentation

This section discuss each of the predicates involved in the robot's functional cycle presented in Fig. 22.7. As this is a recursive Prolog program, only its first execution will be explained, but the whole picking order of the objects will be provided.

The robot functional cycle starts by calling the `sense(CC)` predicate. This predicate interacts with the vision and motion planning modules which perform the tasks described in Sects. 22.3 and 22.4. The output of the motion planning module is the input of a subprocess of the `sense` predicate that builds the context Φ (perceptions), as explained in Sect. 22.5. When this subprocess ends, it returns in CC a list with the facts shown in Fig. 22.5 and then `decide_action(CC,Act)` is executed. In turn, this predicate uses the `answer(CC,choose(X),Ans)` primitive provided by the cognitive module of *Khe*-DeLP, to query a DeLP-Server which has been previously loaded with the program and the preferences of Fig. 22.6.

Given the scenario depicted in Fig. 22.1, for each possible object to be transported a dialectical tree is generated alternating reasons in favor and against the selection of the object. In this way, given the query `choose(X)`, the answer returned by the DeLP-Server is `Ans=yes` and $X=obj2$.[4] This means that the first object to be transported is $obj2$. Figure 22.13(a) shows the dialectical tree calculated to decide that $obj2$ is the best alternative, and a description of the arguments is presented in Fig. 22.13(b).

As can be observed from Fig. 22.13, the argument structure $\langle \mathscr{A}_1, choose(obj2) \rangle$ is the first argument generated to support the selection of $obj2$, but it is defeated by arguments $\langle \mathscr{A}_2, {\sim}choose(obj2) \rangle$, $\langle \mathscr{A}_4, {\sim}choose(obj2) \rangle$ and $\langle \mathscr{A}_6, {\sim}choose(obj2) \rangle$ which are based on comparison literals with higher priority (Fig. 6(b)). But in turn these arguments are also defeated by arguments $\langle \mathscr{A}_3, choose(obj2) \rangle$ and $\langle \mathscr{A}_5, choose(obj2) \rangle$, thus reinstating $\langle \mathscr{A}_1, choose(obj2) \rangle$.

Next, following the functional cycle, the predicate `perform(goto(obj2))` is executed. As indicated in Sect. 22.4 all the shortest paths (store-to-object; robot-to-object) were calculated by the motion planner (see Fig. 22.4), and hence, the control module generates and executes a smooth trajectory motion to the selected object ($obj2$) and then carries it to the *store*.

This cycle is repeated until all the objects in the environment have been transported to the store. In each execution step, it is needed to recalculate the navigable paths to each objects because: (a) the map of obstacles changes as consequence of the removal of objects, and (b) new objects can be reachable because of the removal of the last selected object. Indeed, in our example, after removing object $obj11$, object $obj5$ is made reachable and the same occurs after removing $obj5$, with object $obj8$. As can be observed in Table 22.1, objects $obj5$ and $obj8$ are not initially present.

The transportation order determined by the cognitive module will be: $obj2$, $obj3$, $obj6$, $obj1$, $obj4$, $obj7$, $obj9$, $obj10$, $obj11$, $obj5$ and $obj8$. It is important to notice that this order is given by the preferences of Fig. 6(b), and changing them

---

[4]Due to space constraints we only show the dialectical tree of the only object elegible in this particular initial situation.

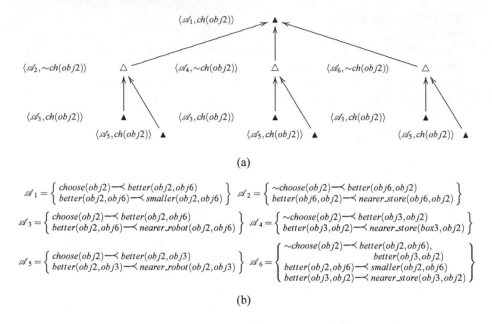

(a)

$$\mathscr{A}_1 = \left\{ \begin{array}{l} choose(obj2) \multimap better(obj2,obj6) \\ better(obj2,obj6) \multimap smaller(obj2,obj6) \end{array} \right\} \quad \mathscr{A}_2 = \left\{ \begin{array}{l} \sim choose(obj2) \multimap better(obj6,obj2) \\ better(obj6,obj2) \multimap nearer\_store(obj6,obj2) \end{array} \right\}$$

$$\mathscr{A}_3 = \left\{ \begin{array}{l} choose(obj2) \multimap better(obj2,obj6) \\ better(obj2,obj6) \multimap nearer\_robot(obj2,obj6) \end{array} \right\} \quad \mathscr{A}_4 = \left\{ \begin{array}{l} \sim choose(obj2) \multimap better(obj3,obj2) \\ better(obj3,obj2) \multimap nearer\_store(box3,obj2) \end{array} \right\}$$

$$\mathscr{A}_5 = \left\{ \begin{array}{l} choose(obj2) \multimap better(obj2,obj3) \\ better(obj2,obj3) \multimap nearer\_robot(obj2,obj3) \end{array} \right\} \quad \mathscr{A}_6 = \left\{ \begin{array}{l} \sim choose(obj2) \multimap better(obj2,obj6), \\ \qquad\qquad better(obj3,obj2) \\ better(obj2,obj6) \multimap smaller(obj2,obj6) \\ better(obj3,obj2) \multimap nearer\_store(obj3,obj2) \end{array} \right\}$$

(b)

Fig. 22.13. Arguments and dialectical tree

will produce an alternative transportation order. It is important to note that this is a flexible way of representing the preferences of the robot, since modifying, adding or removing facts to the L-order, allows to easily change the robot's decision policy with minor variations in its knowledge base.

Given the same experimental environment, let us suppose that the preferences provided to the robot are modified, giving higher priority to the proximity to the store than the proximity to the robot. Besides, let us maintain the preference order among the objects' size and the proximity to the store / robot. Considering this new preference order, facts (53) and (57) do not belong to the L-order presented in Fig. 6(b) and in their place, facts $nearer\_store(Z,X) > nearer\_robot(X,Y)$ and $nearer\_store(Z,Y) > nearer\_robot(X,Z)$ are included. This means that with these new preferences, the cylinder $obj3$ would be in an analogous situation to $obj2$ in the previous setting, since it is as nearest as $obj6$ from the store but it is smaller, therefore $obj3$ is the best object to be transported first. In Fig.22.14 it is depicted the path followed by the robot to transport cylinder $obj3$ to the store.

22.9 Related Work and Conclusions

After a proliferation period of a wide variety of reactive robotic architectures [25, 26], it was clear the necessity of introducing high-level deliberative processes in the decision making of autonomous robots. Regardless of the fact that deliberation has many advantages for decision making of an agent, it has the disadvantage of

Fig. 22.14. A sequence of image frames showing the *Khepera* 2 performing a removal task

being slow compared to generating actions in a reactive fashion. Consequently, several hybrid architectures [27, 28, 29] that combine the advantages of reactive and goal-directed aspects were proposed, having as their main difference the way they incorporate the deliberative component.

Arkin [27] was among the first to advocate the use of both deliberative and reactive control systems within the autonomous robot architecture, incorporating a traditional planner that could reason over flexible and modular reactive control system. Gat [29] proposed a three-level hybrid system [30] (Atlantis) incorporating a Lisp-based deliberator, a sequencer that handled failures of the reactive system, and a reactive controller. Estlin *et al.* [28] presented a two-layered architecture where the top decision layer contains techniques for autonomy creating a plan of robot commands, and the bottom functional layer provides standard robot capabilities that interface to system hardware. The main attention focus of the above-mentioned works has been the definition of the architecture related components necessary to achieve a successful behavior of the robots in real-life complex problems. Aspects like the description of the reactive component, the support of planning capabilities and the interaction of both components in an adequate planning-execution system for the robots, are considered in detail.

Our work successfully integrates vision-based motion planning and control with high-level reasoning facilities based on argumentation. Although the *Khepera* 2 robot is used as a case of study of a differential-wheeled robot, the approach presented in this work is general enough to be applied to other differential-wheeled robots. The vision stage is independent of the robot used for experimentation since the vision module only uses the robot's shape and size to calculate the navigable paths, and the control laws described by (22.3)-(22.5) are specially defined to be used with differential-wheeled robot. Moreover, the engine used by *Khe*-DeLP

(DeLP-Server) is a stand-alone application, therefore following this approach it could be easily integrated in a particular robot architecture (in our case in [31] it was integrated to the *Khe*-DeLP framework [8]).

It is important to note that our aim is not to propose a high-level planner to produce a plan describing the optimal transportation order of the objects, instead, as above-mentioned, we aim to integrate a one-shot decision technique with a vision-based navigation system. However, following the ideas exposed in [6], the proposal presented in [32] (which formalizes the work presented in [12]) could be used to design a high-level planner.

Our approach is closely related to the one adopted in [11] which incorporates a BDI deliberative component to decide which goal to achieve next. We share the approach of having a two-layer system, the upper layer makes the decisions while the lower layer plans and controls the robot motion. Nonetheless, our work has same differences with the proposal of [11] in that we do not use a BDI deliberator as high-level reasoning layer, instead we use a non-monotonic reasoning module based on a defeasible argumentation system.

At present, the cognitive module only provides Defeasible Logic Programming but we plan to incorporate other capabilities like learning and high-level planning. Other important issue we are considering, is to extend this work to a more general environment, where more than one robot perform removal tasks. On one hand, from the high-level specification of the problem, this will imply to reason about other robots' decisions and how they are considered in the argumentation process, as well as their impact in the robots coordination. On the other hand, from the robots' control view this requires to perform real time planning with mobile objects.

Acknowledgment

This work is partially supported by, Universidad Nacional de San Luis and Agencia Nacional de Promoción Científica y Tecnológica (ANPCyT).

References

1. Parasuraman, S., Ganapathy, V., Shirinzadeh, B.: Behaviour based mobile robot navigation technique for real world environments using fuzzy logic system. In: Proceedings of the IEEE International Conference on Systems, Man & Cybernetics. (2004) 3359–3364
2. Suraj, Z., Peters, J.F., Grochowalski, P.: A controller design for the khepera robot: A rough set approach. Fundamenta Informaticae **67**(1-3) (2005) 219–231
3. Pilat, M.L., Oppacher, F.: Evolution of khepera robotic controllers with hierarchical genetic programming techniques. In: Evolvable Machines. Springer (2005) 43–71
4. Germán L. Osella Masa, Hernán Vinuesa, L.L.: Modular creation of neuronal networks for autonomous robot control. Revista Iberoamericana de Inteligencia Artificial **35** (2007) 43–53

5. Artale, A., Franconi, E.: A temporal description logic for reasoning about actions and plans. Journal of Artificial Intelligence Research **9** (1998) 463–506
6. Simari, G.R., García, A.J., Capobianco, M.: Actions, planning and defeasible reasoning. In: 10th International Workshop on Non-Monotonic Reasoning (NMR). (2004) 377–384
7. Chang, L., Lin, F., Shi, Z.: A dynamic description logic for representation and reasoning about actions. In: Knowledge Science, Engineering and Management (KSEM). Volume 4798 of LNAI., Springer (2007) 115–127
8. Ferretti, E., Errecalde, M., García, A., Simari, G.: KheDeLP: A framework to support defeasible logic programming for the khepera robots. In: International Symposium on Robotics and Automation (ISRA). (2006) 98–103
9. K-Team: Khepera 2. http://www.k-team.com (2002)
10. García, A.J., Simari, G.R.: Defeasible logic programming: An argumentative approach. Theory and Practice of Logic Programming **4**(1) (2004) 95–138
11. Parsons, S., Pettersson, O., Saffiotti, A., Wooldridge, M.: Robots with the Best of Intentions. In: Artificial Intelligence Today: Recent Trends and Developments. Springer (1999)
12. Ferretti, E., Errecalde, M., García, A., Simari, G.: An application of defeasible logic programming to decision making in a robotic environment. In Baral, C., Brewka, G., Schlipf, J.S., eds.: Proceedings of the 9th International Conference on Logic Programming and Nonmonotonic Reasoning (LPNMR). Volume 4483 of LNAI., Springer (2007) 297–302
13. : Matrox imaging library (2007) Release 8.0.
14. Bueno, F., Cabeza, D., Carro, M., Hermenegildo, M., López-García, P., Puebla, G.: The ciao prolog system. reference manual. Technical Report CLIP3/97.1, School of Computer Science, Technical University of Madrid (UPM) (August 1997)
15. DeSouza, G.N., Kak, A.C.: Vision for mobile robot navigation: A survey. IEEE Transactions on Pattern Analysis and Machine Intelligence (2) (February 2002) 237–267
16. Latombe, J.C.: Robot Motion Planning. Kluwer Academic Publishers (1991)
17. Jain, R., Kasturi, R., Schunck, B.: Machine Vision. McGraw-Hill (1995) page 50.
18. Berg, M.D., Kreveld, M.V., Overmars, M., Schwarzkopf, O.: Computational Geometry: Algorithms and Applications. Springer Verlag, Berling Heidelberg (1997)
19. Petrino, R., Nasisi, O., Gellón, H., Guarnes, M.: Reconocimiento del ambiente de trabajo con visión artificial. In: XVIII Congreso Argentino de Control Automático, Buenos Aires, Argentina (2002)
20. Petrino, R., Grosso, A., Guarnes, M., Gellón, H., Nasisi, O.: Algoritmo para la determinación del camino más corto entre obstáculos en 2d. In: IX RPIC, Santa Fe, Argentina (2001)
21. García, A.J., Simari, G.I., Delladio, T.: Designing an agent system for controlling a robotic soccer team. In: X Computer Science Argentine Conceference. (2004) 1646–1656
22. Levesque, H.J., Pagnucco, M.: Legolog: Inexpensive experiments in cognitive robotics. In: 2nd International Cognitive Robotics Workshop. (2000) 104–109
23. García, A.J., Rotstein, N.D., Tucat, M., Simari, G.R.: An argumentative reasoning service for deliberative agents. In: 2nd KSEM. Volume 4798 of LNCS., Springer (2007) 128–139
24. Silnik, A., Kiessling, R., Petrino, R., Gellón, H.: Una estrategia para el suavizado de la trayectoria de un robot móvil diferencial. In: XII RPIC, Río Gallegos, Argentina (2007)
25. Arkin, R.C.: Behaviour-Based Robotics. The MIT Press (1998)
26. Nolfi, S., Floreano, D.: Evolutionary Robotics. The MIT Press (2000)
27. Arkin, R.: Integrating behavioral, perceptual and world knowledge in reactive navigation. Robotics and Autonomous Systems **6** (1990)

28. Estlin, T., Volpe, R., Nesnas, I., Muts, D., Fisher, F., Engelhardt, B., Chien, S.: Decision-making in a robotic architecture for autonomy. In: Interl. Symp., on AI, Robotics and Automation for Space. (2001)

29. Gat, E.: On three-layer architectures. In: Artificial Intelligence and Mobile Robots. (1998)

30. Gat, E.: Integrating planning and reacting in a heterogeneous asynchronous architecture for mobile robots. SIGART Bulletin **2** (1991) 70–74

31. Ferretti, E., Errecalde, M.L., García, A.J., Simari, G.R.:   Khepera robots with argumentative reasoning. In: Proceedings of the 4th International AMIRE Symposium. (2007) 199–206

32. Ferretti, E., Errecalde, M., García, A., Simari, G.:   Decision rules and arguments in defeasible decision making.   In Besnard, P., Doutre, S., Hunter, A., eds.: Proceedings of 2nd International Conference on Computational Models of Argument (COMMA). Volume 172 of Frontiers in Artificial Intelligence and Applications., IOS Press (2008) 171–182

23

Moodle UML model and its Security Vulnerabilities

Juan Carlos Galán Hernández and Miguel Ángel León Chávez

Computer Science Faculty, Benemérita Universidad Autónoma de Puebla,
14 Sur y Av. San Claudio, CP 72570, Puebla, México.
jcgalanh@gmail.com, mleon@cs.buap.mx

Summary. Moodle (Object-Oriented Dynamic Learning Environment) is an open source software e-learning platform that provides educators tools to create a course web site. Moodle has been developed under the General Public License and many of its components were developed without a specific design documentation including its security services. This paper presents an object oriented model of Moodle using the Unified Model Language and an analysis of its security services as well as solutions to its security vulnerabilities.

23.1 Introduction

Moodle (Modular Object-Oriented Dynamic Learning Environment) [1] is an open source software e-learning platform (also known as a Course Management System (CMS), or Learning Management Systems (LMS), or Virtual Learning Environment (VLE)) that provides educators tools to create a course web site. Moodle is used all over the world (193 countries) by more of 400,000 registered users. This free e-learning platform as well as others proprietary platforms are used in our University to support five undergraduate degree programs.

Even thought Moodle has a long list of developers and its web page provides developer information, roadmap, coding guide, and concurrent versioning system guide to access its source code, it does not provide a formal model for future developments, even more the web side receive reports of security issues and the security team [2] will work with reporters to fix problems and publicize patches to Moodle users as quickly as possible. In previous work [3] a model of Moodle has been presented using the Unified Modeling Language (UML), this paper is focused on the Moodle security services, therefore it presents and discusses a UML model of the Moodle security services, and it presents some security attacks in order to discover its vulnerabilities, finally some solutions are proposed.

The rest of this paper is organized as follows: section 2 presents a resumed UML model of Moodle which is composed of the following models: analysis, design and components, special attention is dedicated to its security services. Section 3 briefly describes the security attacks to our Moodle web server in order to discover its vulnerabilities. Section 4 proposes the solutions for these vulnerabilities, and finally section V presents some conclusions and future research work.

23.2 Moodle UML Model

Moodle is developed under the General Public License (GPL) as an open source project write in PHP and makes use of a database for data persistency. As in many cases with the open source methodology, Moodle offers public documentation to its users and developers [4]. However, the documentation for developers is not enough, increasing the learning curve of its development. Even more, Moodle also contains several modules from third party contributions that, without a common vision with the leading developer team, do not share the same code structure because while some libraries are procedure oriented some others are object oriented. This makes the architecture of Moodle confusing.

For a security services analysis it is required that the system architecture and its services are clear and well documented, something that Moodle does not have. This section presents the Moodle UML model developed for this work; it covers the basic architecture of Moodle and some of its services. This UML model is represented into three models: analysis, design, and components.

23.2.1 Analysis Model

The analysis model consists of two segments: use cases diagram and several rules defined for a preliminary class diagram. The use cases diagram represents both the services that Moodle offers to its actors (users with different roles) and the services that are needed for internal function such as session authentication. The use cases diagram takes into account the following five roles which are modeled as actors:

- Student: The lowest role in the hierarchy. Users with this role can review public course content, search for courses, and enroll on courses. When a student is enrolled on a course, he/she can review his grades and make use of the different activities inside the course such as downloading files inside the directory of the course.
- No Editing Teacher: This role is intended to work as an administrator of a given course. It has permissions for checking the history reports of student activities over the course and grades. Also, this role can send messages to as many users as needed at once.
- Teacher: A teacher can add or remove course activities, modify grade criteria, create course backups, and upload files into the course. When a course is created, it will have the global configuration, which can be modified later by this role.
- Course Creator: A course creator can add or remove courses from the Moodle system.
- Administrator: This role is the super user. It has tools for creating reports of activities that took place over the site. Create new user accounts, and change

the global configuration of the site as a whole. It also can add and remove new modules either service modules or interface modules, and delete user accounts.

All actors inherit the permissions of the roles below in the hierarchy, for example the Teacher actor will also have the permissions of the No Editing Teacher, and so on. Beside the roles of Moodle, the analysis model takes into account one more actor called Anonymous, which is not taken into the inherit chain of the hierarchy. This actor is needed for the use cases diagram because Moodle allows unauthenticated user to navigate on the main page, login page and courses marked as public by the course owner. Therefore the use cases diagram has a sixth actor:

- Anonymous: This actor represents an unauthenticated user, which only has enough permission for searching public courses, request account creation, and request the authentication service. When an Anonymous actor is authenticated it will become the actor of the role associated with the account of the user.
  The Anonymous actor is extremely important in the model because an attacker will start as an anonymous actor and will try to escalate in the actor hierarchy. The permissions for the different actors are stored by groups of services and managed independently as well as the data for such services. In this model, such services where grouped and modeled as actors too, as follows:
- Course Manager: Set of services oriented to retrieve course data and data storage.
- Session Manager: This actor represents the session subsystem that it is implemented in the web server.
- User Manager: Set of services oriented to retrieve user data and data storage.
- Module Manager: Set of services oriented to retrieve module data and data storage.
- Report Storage: Set of services oriented to retrieve non history data of actions taken by the users of the system.

Figure 23.1 presents the use cases diagram for the Anonymous actor, this figure shows both the services that Moodle offers to this actor and the services that are needed for internal function such as session authentication. These last services are represented as included use cases. In this figure, the use case named Session Retrieval return to the calling use case the session data from the Session Manager. This use case is mostly used by the Authentication use case.

Also, the use case named Authentication will return a Boolean result when another use case request if the invoking actor has permissions for using a given service. In web based systems, the authentication service is the target of most security attacks. The Authentication use case is the login service of the Moodle site. The Session Authentication use case checks upon any user request if it holds an ID session number, if it is valid and if the session is still active. The Add to History is the use case of the non-repudiation service, as it will be discussed latter.

The last segment of the analysis model is to define a set of rules for creating class diagrams. Overall, PHP scripts can be written on object oriented programming, procedure oriented programming or a mix of both. The scripts inside of the many packages of Moodle were written on all three styles making hard to create class diagrams unless some rules are defined, as the following ones:

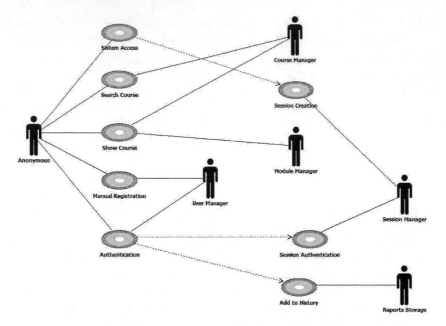

Fig. 23.1. Use case diagram for the anonymous actor

- All scripts have a main method.
- PHP files are classes.
- Classes inside PHP files are marked as needed for the PHP file class.
- All directories are packages.
- All classes and methods are public unless a modifier is presented.
- The include_once instruction represents a dependency.

23.2.2 Design model

The design model is composed of several diagrams such as the refined class, interaction, sequence, and state. Because analyzing the security services of Moodle is the main goal of this paper, this section presents both the most important classes for implementing the Moodle security services and an interaction diagram.

Moodle optionally implements the following security services: authentication, access control, and non-repudiation. These services are not mandatory during the installation process of a Moodle server. These services are defined by ISO [5] as follows: The authentication service verifies the supposed identity of a user or a system. The access control service protects the system resources against non-authorized users. The non-repudiation service prevents an entity from denying previous commitments or actions.

Figure 23.2 shows the class diagram with the classes involved in the security services, and they are described below.

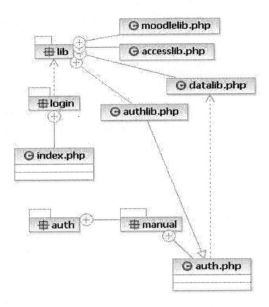

Fig. 23.2. Fragment of the class diagram of the Moodle security services

The class acceslib.php inside the lib package has the methods needed to save and retrieve the permissions of each role. Those permissions are stored in several tables inside the Moodle database.

The class datalib.php defines three methods for the non repudiation service: add_to_log, get_logs_usercourse, and get_logs_userday. With these methods any module can implement non-repudiation service trough history tags saving trough add_to_log the required information to identify such actions. The other two methods are used for report generation and displaying into the user interface.

The class moodlib.php defines methods for multiple services in Moodle, including the security services.

The most complex service is the authentication service, it consists of two main actions: user authentication, also called user login, and session authentication.

User authentication is implemented trough several modules, each module retrieves user data, i.e., user name and password, from different sources such as cookies and database. But at the end, the module will resolve the user authentication by matching the user name and password retrieved with the data stored in the database.

All authentication modules are descendant of the auth_plugin_database class that is inside the authlib.php class.

The configuration data of the Moodle web site is stored in the class config.php stored in the root package.

On the other hand, the interaction diagram describes how objects, components, and actors interact with each other when a use case is executed. The interaction diagram, shown in figure 23.3, helps to spot when an attack can occur at a given request. The diagram shows how the authentication use case is consolidated in the class accesslib.php

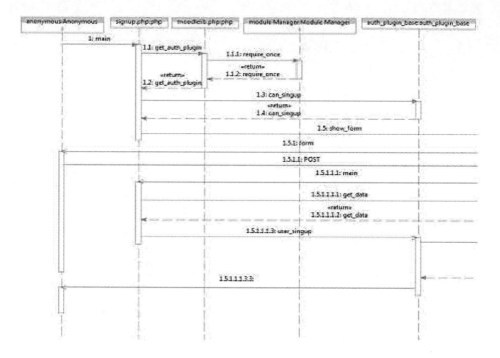

Fig. 23.3. Interaction diagram for the Use Case "Register". It shows the interaction between several objects in a top down sequence when an actor request the registration service

23.2.3 Components model

As in all web based systems, some of the security services are not directly implemented but they rely in the web server configuration. In order to identify such services it is needed to identify the components of a Moodle implementation. The most important components are shown in the figure 23.4.

The web server offers the service of web pages to the remote users, and it uses the services of the security server. This server provides the following services: non-repudiation, server authentication, integrity, and confidentiality. They are implemented by means of the Socket Secure Layer (SSL) protocol when the web server requests them. Such request is realized when the URL of the web page stored inside the web server starts with HTTPS instead of HTTP.

Other two components are the PHP pre-processor and the Moodle package. PHP is an interpreted language and its implementation needs that the source code can be accessed by the pre-processor.

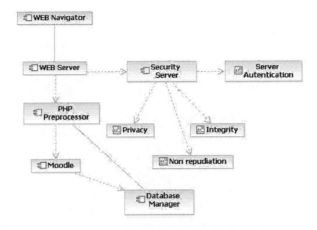

Fig. 23.4. Components diagram of Moodle

Also, Moodle stores the data inside a database trough a defined connection. A recommended configuration for Moodle [1] is as follows:

- Web Server: Apache.
- Security Server: OpenSSL.
- Database: MySQL.
- PHP pre-processor: Official PHP pre-processor.

23.3 Security Attacks to Moodle

With the growing demand of Internet service there have been appearing multiple Internet Service Providers along with Hot Spots (open access points) available to general population. Several of this hot spots are unsecured or with weak security. Under these conditions, Moodle is vulnerable to combined techniques of network monitoring [6] and web based attacks [7]. Such attacks can be divided in two groups, session attacks and design attacks.

Two session attacks are effective against Moodle: session hijacking and session fixation.

1. Session hijacking. This is part of the eavesdropping attacks, where an attacker will listen in the communication between client and server trying to find some information inside the payload, in this case the HTTP request, that can be used to impersonate the user and taking control of his/her session. The figure 23.5 shows a diagram of a Session Hijacking attack.

   Moodle manages its session's trough two values that identify an active session: MoodleSession and MoodleSessionTest. Those values are stored in the cookie that is sent on each HTTP request inside the header of the message. In order to impersonate a target user, an attacker must obtain such values.

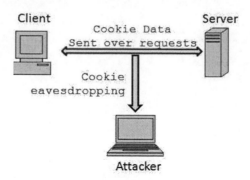

Fig. 23.5. Session hijacking attack

Obtaining a full HTTP request data with the cookie included is easy because Moodle only uses SSL tunnels on the login service and a few administrative services. Because of that, most HTTP request are done on plain text that can be intercepted and easily decoded. After obtaining the cookie, the attacker can use this data on its own HTTP request, taking control of the target user's session.

2. Session Fixation. This attack also targets the session data of a user. However, this attack is classified as an active attack [8] or an interception attack [9]. Instead of eavesdrop the communication between a target user and the server, the attacker intercepts the HTTP request of the target user. Each time an anonymous user access Moodle, a MoodleSession and a MoodleSessionTest are granted. Due to this behavior, an attacker can get such values as an anonymous user and then intercept a request of a target user that is not yet authenticated. On such interception, the attacker replaces the user's MoodleSession and MoodleSessionTest values with the ones obtained previously. If the target user is authenticated, the session is granted with the user's permissions allowing the attacker to have the same permissions because he/she already has the MoodleSession and MoodleSessionTest values that identifies the fixated session. Figure 23.6 shows a diagram of a Session Fixation attack.

Figure 23.7 shows a request to Moodle's server with the cookie. Session hijacking was done by copying the cookie to a web navigator. Username prediction was done with this data as well as by decoding the MOODLEID_ field. In the figure this field has the value %25E2%25C8%2513E%25BD which is decoded into admin, the default username for the administrator of Moodle. The request was captured with Wireshark [6], a tool that allows monitoring network transmissions data.

Regarding the design attacks, Moodle is vulnerable to password prediction and username prediction.

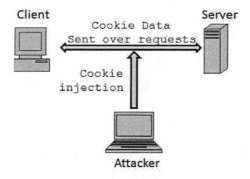

Fig. 23.6. Session fixation attack

```
0220  64 6c 65 2f 63 6f 75 72   73 65 2f 76 69 65 77 2e    dle/cour se/view.
0230  70 68 70 3f 69 64 3d 32   0d 0a 43 6f 6f 6b 69 65    php?id=2 ..Cookie
0240  3a 20 4d 6f 6f 64 6c 65   53 65 73 73 69 6f 6e 3d    : Moodle Session=
0250  64 62 32 32 35 63 65 38   35 32 32 62 31 32 61 31    db225ce8 522b12a1
0260  63 31 61 39 32 32 61 34   31 63 62 37 36 35 64 31    c1a922a4 1cb765d1
0270  3b 20 4d 6f 6f 64 6c 65   53 65 73 73 69 6f 6e 54    ; Moodle SessionT
0280  65 73 74 3d 65 6a 33 33   49 72 46 64 43 63 3b 20    est=ej33 IrFdCc;
0290  4d 4f 4f 44 4c 45 49 44   5f 3d 25 32 35 45 32 25    MOODLEID _=%25E2%
02a0  32 35 43 38 25 32 35 31   33 45 25 32 35 42 44 0d    25C8%251 3E%25BD.
02b0  0a 0d 0a                                             ...
```

Fig. 23.7. Fragment of a request done to a Moodle site captured with Wireshark

1. Password Prediction. This attack was done by sending multiple requests to the Moodle server with the cookie field empty.
   Because of a design flaw in Moodle, the login failures count is reset to zero when inside the request the cookie field is with no values or no cookie at all. This flaw allows an attacker to perform a brute force attack for password prediction.

2. Username Prediction. This may be done by two methods. Intercepting a cookie and by brute force.
   With the cookie intercepted, the field MOODLEID_ was decoded with URL decoding and RC4 decoding. The private key for RC4 is hard coded inside the file moodlelib.php with the fixed value nfgjeingjk (Moodle v. 1.8.6).
   The brute force method is done like in password prediction. However, instead of sending several passwords, several usernames are sent with a random password. The response from Moodle will take longer with a valid username than with an invalid one and this was used to differentiate between them in the attacks realized.

Figure 23.8 shows a comparative graphic of the average response time to login requests done to Moodle with valid and invalid usernames. This graph represents five rounds of 1000 consecutive login request each. The known username data (light gray) shows the average of each round with a valid username "admin" paired with a wrong password "foo", and the invalid username data (dark gray) shows the average of the rounds with the unknown username "admi" and the password "foo". It can be noted how in the last four rounds a response to a request with a valid username takes longer.

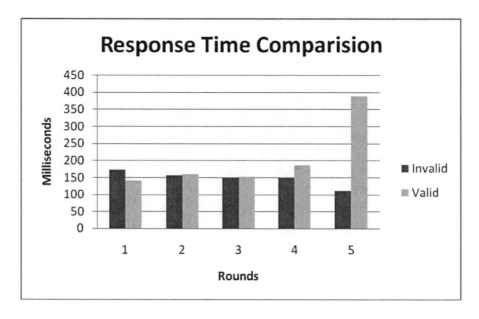

Fig. 23.8. Invalid username against valid username. Five rounds of the average of 1000 login request with a valid username "admin" against an invalid username "admi" both with wrong password

23.4 Solutions to Moodle Vulnerabilities

As it was shown in previous section, Moodle is vulnerable to the following attacks: session fixation, session hijacking, prediction of usernames, and prediction of passwords by brute force.

Such vulnerabilities can be avoided by modifying certain portions of the Moodle code and adding new functions. These modifications are described next.

1. To use SSL over all site. Moodle already has an option for using SSL over certain critical actions. However such method cannot prevent session fixation, session hijacking, and username prediction. In order to avoid such attacks the entire

site must create SSL connections with its clients. This can be done by adding a PHP scripts that changes the content of the object that holds the environment configuration named CFG. Inside CFG there are the following four variables that are SSL related:

a) Themewww. This variable holds the location of resources for building the graphical interface as a full URL string. The script has to change the HTTP protocol for HTTPS (SSL request).

b) Wwwroot. Moodle uses this variable to know the URL assigned to it for quick navigation. The script has to change the HTTP protocol for HTTPS.

c) Loginhttps. The value of this flag is retrieved from the database and when it is on the login page, it is encrypted trough SSL. The script turns it on, even if the main configurations say otherwise.

d) Httpstheme. When the loginhttps is turned on, the original source code changes the URL protocol from HTTP to HTTPS. This script also changes this value to HTTPS, overriding the original loginhttps value.

The script also has to change the value of the global flag HTTPSPAGERE-QUIRED to true. This flag is part of the Moodle's default configuration. Such fixes were implemented in a script called buap_security that is invoked when the main configuration script config.php is called on every user request. The security server configuration page was also changed to reflect the new option of ciphering the entire site by modifying the source code of the security.php located inside the admin package.

2. ID Session regeneration. The SSL protocol cannot protect the site from session fixation when the user requests a connection for the first time if that petition is done by HTTP protocol without SSL. This was fixed by generation of a new ID Session when the user is authenticated by login/password matching. PHP lets to change the ID by placing the instruction session_regenerate_id after switching privileges, at the line 2684, from the source code moodlelib.php stored inside the lib package.

3. Login with CAPTCHA. The authentication service, implemented as a login page, can be subject of automatic brute force attack. The login page can be protected by using CAPTCHA [10]. The login page was added with the official CAPTCHA implementation known as reCaptcha [11]. This makes the authentication service stronger against brute force attacks automated with software tools. Moodleib.php has to be modified in order to check the new data associated with the reCaptcha library as described in [11].

4. Correct permissions loading. As the author suggest in [7], usernames stored inside Moodle can be predicted because the permissions are loaded before username and password checking making responses to take longer when a request with a valid username is sent. This was fixed by just changing the

order of the actions taken by the authenticate_user_login method coded in the moodlelib.php file.

Figure 23.9 shows a comparative graphic of response times to request done to Moodle with valid and invalid usernames as in Fig. 8. But this time the permissions load order is changed.

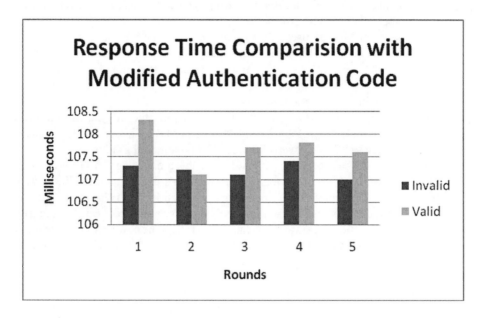

Fig. 23.9. Invalid username against valid username with permission loading corrected

5. Username obfuscation. As part of its authentication implementation, Moodle stores the username inside the cookie of the HTTP protocol. This field is ciphered with the RC4 algorithm using a private key defined in the source code. This is highly insecure because even with SSL an attacker can capture the user cookie of older sessions and decode it with the information of the source code.

In order to avoid such decoding it was implemented a new configuration file that lets the administrator of Moodle to change the obfuscation private key and even to choose the algorithm used because instead of using the implementation of Moodle of RC4 the new configuration file implements the mcrypt library [12] of PHP. It is possible to change also the algorithm used for storing passwords. Currently, Moodle uses MD5 for storing passwords inside its database, but there are works at [13] that shows how a MD5 hash can be broken. The new configuration file also offers the possibility of selecting a new hash algorithm available in the hash library [12] of PHP. However, when changing this parameter, the administrator of Moodle must be aware that every password hash previously stored with the old algorithm, even the password of the administrator, will become invalid.

23.5 Conclusions

When implementing electronic learning courses (e-learning), Moodle offers enough services and reliability as a web based solution. However, the documentation available for Moodle developers is poor and makes the training time of new developers longer and also provokes that new outside developing to not share same conventions such OOP. Also, the security services of Moodle are not well documented and in some cases vulnerable to certain attacks.

This work has presented an UML model of Moodle security services for reducing the learning curve of the Moodle architecture. Also, this work has shown that a Moodle's server is vulnerable to the following attacks: session fixation, session hijacking, prediction of usernames, and prediction of passwords by brute force.

This work has presented the non intrusive modifications of the source code of Moodle that fix its vulnerabilities and can be implemented in the last two versions of Moodle 1.8.5 and 1.9.1.

On the other hand, Moodle is an e-learning platform based on constructivist and social constructionist approach to education, therefore the security requirements of the courses is our future research work.

References

1. Rice, W. H. IV: Moodle E-Learning Course Development, PACKT Publishing, pp. 6, 31–72 (2006).
2. Moodle Modules Forum, http://moodle.org/mod/forum/view.php?id=7128
3. Galán, J. C., León, M. A.: Modelo UML de Moodle, La Computacin en Puebla en el siglo XXI, Vol. IV, pp. 67–75, Puebla, Mxico (2007).
4. Moodle Documentation Site, http://docs.moodle.org/
5. ISO 7498-2 Information processing systems - Open Systems Interconnection - Basic Reference Model - Part 2: Security Architecture, International Organization for Standardization, (1989).
6. Spivey, M. D.: Practical Hacking Techniques and Countermeasures, AuerBach Publications, pp. 193–260, (2007).
7. Stuttard, D., Pinto, M.: The Web Application Hacker's Handbook: Discovering and Exploiting Security Flaws, Wiley Publishing Inc., (2007).
8. Eric M.: Network Security a Beginner's Guide, 2 Ed., McGraw-Hill/Osborne, pp. 19–24, (2003).
9. Mcclure, S., Scambray, J., Kurtz, G.: Hacking Exposed: Network Security Secrets And Solutions, 3 Ed., McGraw-Hill/Osborne, pp. 340, (2001).
10. Von Ahn, L., Blum M., Hopper, N. J., Langford, J.: CAPTCHA: Using Hard AI Problems for Security, Conference on the Theory and Applications of Cryptographic Techniques, EUROCRYPT, pp. 294–311, (2003).
11. Recaptcha Project of Carnegie Mellon University, http://www.recapcha.net

12. Snyder, C., Southwell, M.: Pro PHP Security, Apress, pp.66-68, 80-85, (2005)
13. Wang, X., Feng, D., Lai, X., Yu, H.: Collisions for hash functions MD4, MD5, HAVAL-128 and RIPENMD, In Cryptology ePrint Archive, http://eprint.iacr.org, Report 2004/199. (2004).

Towards a Reconfigurable Platform to Implement Security Architectures of Wireless Communications Standards Based on the AES-CCM Algorithm

Ignacio Algredo-Badillo[1], Claudia Feregrino-Uribe[2], René Cumplido[2], and Miguel Morales-Sandoval[3]

[1] University of Istmo, Campus Tehuantepec,
   Computer Engineering
   Ciudad Universitaria S/N, Sto. Domingo Tehuantepec, Oax., Mexico. 70760.
   `algredobadillo@sandunga.unistmo.edu.mx`
[2] National Institute for Astrophysics, Optics and Electronics,
   Sta. Ma. Tonantzintla, Pue., Mexico,
   `cferegrino@inaoep.mx`, `rcumplido@inaoep.mx`,
[3] Polytechnic University of Victoria
   Information Technology Department
   Calzada Luis Caballero No. 1200
   Cd. Victoria, Tamps., Mexico. 87070
   `mmoraless@upv.edu.mx`

Summary. Over decades the design of communication devices has been mainly focused on hardware, where upgrading a design meant to completely abandon it and start a new design. For modern communication networks, the ideal device is the one capable of operating in different networks employing basically the same hardware resources. Recently, software radios offer the capability of being multiband, multimode, software-intensive radios. A key element in software radios is security, this is because these devices are able to access different wireless networks and employ the atmosphere as transmission medium, thus may be vulnerable to malicious attacks when data are transmitted. Several security architectures have been standardized for different networks, such as IEEE 802.11i-2004 for WLANs (Wireless Local Area Networks) and IEEE 802.16e-2005 for WMANs (Wireless Metropolitan Area Networks), operating on the MAC (Medium Access Control) sublayer. This chapter shows how by means of the software radio paradigm, two hardware implementations of these standard security architectures can be implemented in a single and flexible platform.

24.1 Introduction

In digital communications, mobility is a desirable feature that has motivated the development and the growth of wireless systems. Wireless networks use different set of rules or protocols for governing communication among diverse devices, and each network can support different applications that in turn can use different protocols.

Ideally, a device should operate in diverse applications on different wireless networks. This desired feature motivated the development of the software radio devices, which are radios able to operate with different hardware/software configurations in order to support the functions needed to access different communication networks. Software radios have evolved in response to advances in the technology. Initially, only basic radio architectures were available, these radios then evolved to software capable radios, then into software programmable radios, and finally to software-defined radios (SDR). Future developments will be focused on cognitive radios that will provide new capabilities in order to support more complex functions, such as: awareness, adaptability and the ability to learn, see Fig. 24.1 [1].

Fig. 24.1. Evolution of the software radios.

Radio processing platforms usually combine the use of highly efficient Application Specific Integrated Circuits (ASICs) and very flexible programmable General Purpose Processors (GPPs) or Digital Signal Processors (DSPs). ASICs are well suited for performing dedicated functions like A/D converters and are extensively used to build digital receivers that require custom hardware structures to operate in real time. GPPs and DSPs are preferred for performing tasks that carry out some sort of data analysis or decision making which depend on a particular operation mode or application.

After the received signal is digitalized by the A/D converter, all required operations in modern radio systems are executed by digital hardware. This digital signal processing hardware has been traditionally hardwired to achieve the highest possible performance. However, this customization has limited the development of processing platforms with enough flexibility to operate on networks that use different standards.

The development of software radios is currently focused on the lower layers of the OSI model, which are normally implemented in hardware [2]. A key element in software radios is security, this is because these devices are able to access different wireless networks and employ the atmosphere as transmission medium, thus may be vulnerable to malicious attacks when data are transmitted. A number of solutions have been developed to protect these networks. These solutions are based on a variety of techniques and algorithms, examples are: firewalls, cryptography, antivirus, intrusion detectors, secure routing, and security policy management. One of the most important security mechanisms are based on cryptographic algorithms that are used on different operation modes. Security based in cryptography has become a key element in recent standards of wireless communication networks.

Systems must be able to deal with the extra computations required by the use of cryptographic algorithms, especially in demanding applications that transmit large amounts of data. Cryptographic algorithms are characterized by a large number of complex operations, that may result in system processing bottlenecks. It is important to highlight that hardware cryptographic implementations have shown to have better performance than software only implementations. Hardware architectures are able to better exploit the parallelism of cryptographic algorithms and also, when compared against general purpose processors, are better suited to perform operations directly on bits or values represented by different word sizes.

Reconfigurable logic devices, in particular FPGAs, are becoming an increasingly important part of software radio platforms. Current FPGA devices offer a number of features that make them suitable to efficiently perform a large number of functions and tasks required to implement digital communications systems. Among these features we can find dedicated hardware multipliers, flexible memory structures, plenty of I/O pins, gigabit serial transceivers and other user-configurable system interfaces, and high density of logic slices. All these features combined with the availability of new and powerful design tools allow engineers to deal with very complex design is a relatively short time. In addition, it is now also possible to acquire highly efficient intellectual property (IP) cores from specialized vendors. This plethora of cores that are ready to drop into the FPGA, allows building complex application-specific architectures. Like ASICs, all functions implemented in FPGAs can potentially be executed in parallel. This ability of performing a large number of operations in parallel and the availability of distributing memory along the signal path results in highly efficient architectures. Also, unlike GPPs and DSPs, FPGAs offer the possibility of assigning to each variable exactly the number of bits required. All the advantages offered by configurable logic devices come at the expense of higher power consumption and in some cases higher costs. Thus, for the foreseeable future it is not possible to think that they can fully replace ASICs, GPPs and DSPs. Instead, the combined use of all these type of devices seems to be the best option to implement complex system like those required in software radio applications.

The main idea of the software radio platforms is to provide support for different applications. These platforms must be able to dynamically reconfigure themselves in order to perform the tasks associated to different access technologies and standards. There are several ways to provide flexible platforms [3], being the use of reconfigurable techniques one of the most promising. Different works have explored these ideas for different applications. For example, in [4], a cognitive radio is presented, and in [5], a DSP/FPGA reconfigurable architecture is described. Few works have focused on evaluating implementations of security algorithms on reconfigurable architectures [6]. In this chapter, the aim is to examine two hardware architectures that implement the security architectures of two wireless communication standards as the first step towards developing a single processing reconfigurable platform for security. This platform is developed for software-radio applications operating in the MAC sublayer, which executes different configurations, and in this case, for two protocols of two different networks (WMAN and WLAN). To design this platform, characteristics such as hardware resources, throughput, efficiency and reconfigurable modules, are evaluated.

In this work, two hardware architectures that implement the security architectures of two of the most important wireless communications standards are evaluated. The aim is to evaluate the feasibility of implementing a single hardware module that will be able to provide security services for both standards. The selected standards are the IEEE 802.11i-2004 and IEEE 802.16e-2005.

24.2 Security Protocols

Security protocols for the widely-used wireless communication networks propose the use cryptographic solutions based on diverse cryptographic algorithms. Security is defined in the MAC sublayer, enabling communication networks to provide the security services of privacy, authentication and confidentiality, based on cryptographic algorithms that use several iterative mathematic operations. These algorithms protect data transmissions at the expense of high computational costs that may cause bottlenecks in the data transmissions, thus to cope with the high speeds of future data transmissions, such as in the wireless networks [7] with application to transmit high-quality TV, movies in DVD, and great amount of digital files using personal computers, architectures with high throughput are required, at least performing at 1 Gbps. The IEEE 802.11i-2004 standard is designed to provide enhanced security in the MAC sublayer for 802.11 networks. In this standard, the security architecture is based on AES-CCMP (AES-CCM Protocol) that in turn is based on the Advanced Encryption Standard (AES) in CCM operation mode to provide robust security features for data transfers.

The Mobile WirelessMAN standard defined by the IEEE 802.16e-2005 amendment to the 802.16-2004, includes better support for Quality of Service and the use of Scalable OFDMA, and is sometimes called 'Mobile WiMAX'. This standard, as well as IEEE 802.11i-2004 standard, proposes to implement the CCM operation mode of the AES algorithm. It has specified security mechanisms to provide better security services, although it is required to execute a great number of operations, several iterations, and multiple processes. In this work, proposed hardware architectures are based on the AES-CCM, using parallelization and modular specialization, and reducing critical path without increasing the execution latency.

IEEE802i-2004 Standard.

The IEEE 802.11i-2004 or 802.11i is an amendment to the IEEE 802.11 standard that specifies mechanisms for wireless networks. It supersedes the Wired Equivalent Privacy (WEP) security specification which have inherent weaknesses [8] [9] [10]. As an intermediate solution to WEP insecurities, the Wi-Fi Alliance proposed the Wi-Fi Protected Access (WPA). IEEE 802.11i-2004 makes use of the AES block cipher, whereas WEP and WPA use the RC4 stream cipher [11].

The components of 802.11i architecture are: 802.1X for authentication [12] (implicating the use of EAP -Extensible Authentication Protocol- and an authentication server), Robust Security Networks (RSN) for keeping track of associations, and

AES-based CCMP (explained in detail later) to provide data confidentiality, integrity, and origin authentication. Another important element of the authentication process is a four-way handshake, a pairwise key management protocol. The IEEE 802.11i amendment introduced the concept of a Robust Security Network (RSN), defined as a wireless security network that only allows the creation of Robust Security Network Associations (RSNA). 802.11i defines two classes of security algorithms: RSNA, for establishing a logical connection between communicating IEEE 802.11 entities established through the IEE 802.11i key management scheme (4-way handshake)[13] and Pre-RSNA. Pre-RSNA security consist of Wired Equivalent Privacy (WEP) and 802.11 entity authentication. RSNA provides two data confidentiality protocols, called the Temporal Key Integrity Protocol (TKIP) and the Counter Mode with Cipher Block Chaining Message Authentication Code (CBC-MAC) Protocol (CCMP). To avoid a range of security problems, the primary recommendation in [14] is the use of CCMP for data confidentiality whenever possible because WEP and TKIP have inherent weaknesses. CCMP uses the CCM (Counter with CBC-MAC) operation mode [15] of the Advanced Encryption Standard (AES) algorithm.

CCMP uses the Advanced Encryption Standard (AES) algorithm, see Fig. 24.2 (a). Unlike in TKIP, key management and message integrity is handled by a single component built around AES using a 128-bit key and a 128-bit block. AES-CCMP operates on the Medium Access Protocol Data Unit (MPDU) which comprises five sections: 1) MAC header, 2) CCMP header, 3) Data unit, 4) Message Integrity Code (MIC) and 5) Frame Ckeck Sequence (FCS), where only the Data Unit and the MIC are encrypted. In general, the security architecture based on AES-CCMP ciphers data input (plaintext MPDU), using AES-CCM algorithm, and resulting the data output Cipher MPDU. AES-CCMP disassembles each packet in KeyID, packet number (PN) and plaintext MPDU. PN is a 48-bit number stored across 6 octets. The PN never repeats for the same temporal key (TK) since reuse of a PN with the same temporal key voids all security guarantees, so, a fresh TK is required for every ciphering session. MPDU is expanded in several fields, such as payload DataP, Address 2 (A2), a priority octet, and the MAC Header. With these fields, a CCMP Header is constructed as well as a Nonce value for the CCM algorithm (unique for each frame protected by a given TK and PN) and the additional authentication data (AAD). CCMP combines the MPDU Address 2 and priority field, and the PN to create the Nonce value. Then, it feeds the TK, the constructed Nonce, certain header information, and the data unit, to the CCM originator. The CCM origitator returns this encrypted data, and a MIC, which is combined with the unencripted CCMP and MAC headers, and sequence check for transmission.

The payload, TK, Nonce value and ADD are input to the AES-CCM. It outputs the cipher data and MIC that are used together with the CCMP and MAC headers to build the Cipher MPDU. AES-CCM is the main cryptographic algorithm, which executes two related processes: generation-encryption and decryption-verification. CCMP uses CCM with the following parameters: M = 8 - indicating that the MIC is 8 octets and L = 2 - indicating that the Length field is 2 octets. According to an analysis performed to IEEE 802.11i by [14], CCM provides a level of confidentiality and authenticity comparable to other authenticated encryption modes.

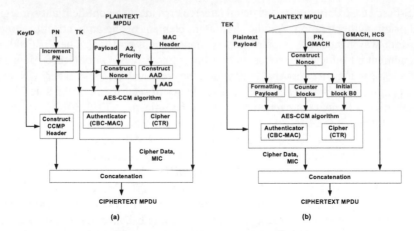

Fig. 24.2. Security architecture based on the AES-CCM protocol for IEEE 802.11i networks and related processes for ciphering in the IEEE 802.16e-2005 standard.

For the purposes of this work, which is focused on the reconfiguration of a transmission platform, the generation-encryption process is considered to design the architecture. CBC-MAC process is applied to the payload DataP, the data associated AAD, and the nonce to generate a MIC (Message Integrity Code) whereas CTR mode is applied to the MIC and the payload DataP to obtain the ciphertext (Cipher MPDU).

IEEE802.16e-2005 Standard.

The IEEE 802.16e-2005 is the latest standard on broadband wireless metropolitan area networks. It aims at maintaining clients connected to a MAN while moving around and supports enhanced security, mobility management, and improved support for fast handovers while addressing many design and security flaws in original baseline IEEE 802.16-2004 standard [16]. IEEE 802.16e-2005 is also called WiMAX, meaning Worldwide Interoperability for Microwave Access, provides wireless transmission using a variety of transmission modes, from point-to-multipoint links to portable and fully mobile internet access. The WiMAX Fourum has more than 500 members including operators, component and equipment manufacturers, and many others in the communication ecosystem [17] and was formed to certify and promote the compatibility and interoperability of the standard and describes WiMAX as 'a standards-based technology enabling the delivery of last mile wireless broadband access as an alternative to cable and DSL' [18]. To date, commercial WiMAX coverage service has been achieved over 19.8 km. This year, it is expected to see Mobile WiMAX expanding to 20 MHz channel bandwidths with peak download rates exceeding 144 Mbps per sector. In 2010, the WiMAX industry is expected to transition to its next release. Features will include higher channel and VoIP capacity, higher user and data rates and will offer mobility for up to 500 km/hr with a link layer latency less than 10 ms [17].

IEEE 802.16e-2005 security scheme incorporates two component protocols: Privacy and Key Management (PKM) responsible for providing the secure distribution of keying material and the encapsulation [19] protocol for securing packet data. In the general operation of the encapsulation protocol, ciphering is applied to the MAC PDU payload for privacy service, whereas the PKM allows for authentication. In the encapsulation, data are protected by ciphering the information or plaintext payload, and by providing a value for the message integrity. Ciphering payload requires that two values shall be appended: packet number (PN) and message authentication code (MIC), and AES-CCM algorithm shall be applied to the plaintext payload, see Fig. 24.2 (b). For applying AES-CCM algorithm, other related main functions should be executed, formatting data input such as plaintext payload, counter blocks, initial block, nonce value, packet number (PN), and generic MAC header (GHMAC). These functions are described in the security scheme of the standard. In Section 3, hardware architectures using the AESCCM algorithm are proposed, which combine parallelized structures with low hardware resource requirements.

AES-CCM algorithm.

Traditionally, two different cryptographic algorithms are used to provide confidentiality and authentication, but AES-CCM algorithm provides these two security services with the same algorithm, using the AES block cipher and the same key. CCM uses the CTR (Counter) mode and CBC-MAC (Cipher Block Chaining - Message Authentication Code) [20]. The confidentiality is provided by the AES algorithm in CTR mode, requiring a value that ensures uniqueness. The authentication is performed by the AES algorithm in CBC-MAC mode and provides additional capabilities.

AES-CBC-MAC and AES-CTR use AES block cipher as the main module, working in cascade, where the first one generates an intermediate value MIC T for the second one, which then generates the cipherdata and the final value MIC U. On one hand, AES-CBC-MAC, which is an integrity method, works sequentially and it cannot be parallelized, ensuring that every ciphered block depends on every preceding part of the plaintext, where ciphering two identical blocks results in different cipher blocks. AES-CBC-MAC is used when there is an exact number of blocks and hence requires padding. By the other hand, when ciphering two identical input blocks, AES-CTR mode produces different cipher blocks, which is based on a nonce value rather than starting it from a fixed value. This mode provides authentication by adding extra capabilities. Some properties of AES-CTR is that ciphering can be done in parallel, and the message is not required to break into an exact number of blocks [21].

As it was mentioned, AES-CBC-MAC and AES-CTR modes use the AES block cipher as their main module, which can process data blocks of 128 bits using a 128-bit key [22]. AES executes an initial round followed by ten rounds. These last ten rounds have four transformations: 1) byte-to-byte substitution (SubByte),

2) rotation of rows (ShiftRow), 3) mixing of columns (MixColumn), and 4) addition of round key (AddRoundKey). Other operation is key expansion, which computes a key schedule or a 128-bits key in each round.

24.3 Hardware Implementation of the Security Architectures

One of the most simple design techniques for hardware implementations is based on modular designs. The next subsections show the modular designs and implementations of the security architectures for the IEEE-802.11i-2004 standard (AESCCMP) and for the IEEE 802.16e-2005 (AESCCM6), they focus on high throughput. This is reached by making an analysis to reduce critical path by developing specialized modules, proposing compact control units, identifying parallelization of the data buses and modules, and balancing paths formed by the combinational and sequential elements. For evaluation purposes and for developing the reconfigurable platform, these architectures are implemented in FPGA devices. The design of the architectures is written in VHDL and simulated using FPGA Advantage 6.3.

The methodology followed for the hardware implementation of the security architectures consisted on firstly modeling the security architectures in software in order to create the test data for validating the corresponding hardware modules. After that, a straightforward initial hardware architecture was designed and implemented with the aim of providing baseline architectures to evaluate improvement strategies. This also allowed identifying processing bottlenecks, critical paths and potential for hardware reutilization. Some techniques, such as loop unrolling, pipelining, and the use of embedded hardware resources [23], allowed reducing the critical path. The focus was on achieving higher throughput and to require lower hardware resources, which results in high througput/area ratio. After simulations, potential modules that could be parallelized were identified as well as a strategy for balancing the processing workload among the more complex modules. The complete functionality of the architecture was tested using the test vectors available in the standards FIPS 190-7 (AES), NIST SP800-38C (AES-CCM), IEEE 802.11i-2004 (AES-CCM Protocol) and IEEE 802.16e-2005 (AESCCM6).

24.3.1 AESCCM Hardware Architecture.

The general operation of the AESCCM module 24.3 is divided into these two sub-modules, where the first sub-module is required to calculate the authentication field value T, and the second sub-module computes the cipher_MPDU and the value MIC U. The plaintext message input is divided into 128-bit data blocks BX, and they are used for processing in the two sub-modules, whereas counter blocks are used only for AESCTR sub-module. These sub-modules have a main component: AES_Cipher [24], which computes AES algorithm with its rounds, transformations and key calculation.

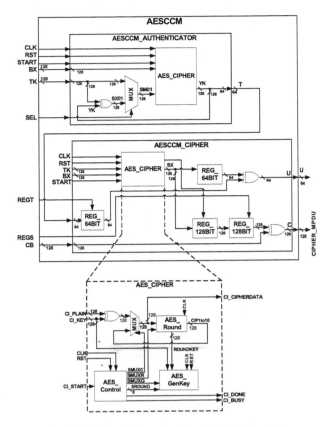

Fig. 24.3. Block diagram of the AESCCM module.

24.3.2 AESCCMP Hardware Architecture.

For the IEEE 802.11i-2004 standard, it is proposed the AESCCMP hardware architecture. The AES-CCMP hardware architecture is illustrated in Fig. 24.4. From Fig. 24.2, Increment PN and Construct CCMP Header blocks are considered to be executed in an upper layer. The AESCCMP hardware architecture is constituted by specialized modules to format data (Format N&Q, Format AAD, Format Payload, and Format CB), to compute AES-CCM algorithm (AESCCM) and main control (Control CCMP). Each module to format data has its particular control submodule. The main control module is based on Finite State Machines (FSMs). AESCCM module executes AES-CBC-MAC and AES-CTR submodules in parallel, which compute AES-CBC-MAC and AESCTR algorithm, respectively. The general operation consists on processing two sources of data, parsed in 128-bit data blocks, and the same 128-bit key block through AESCCM module. The first source generates data blocks from three different modules (PAY N&Q, PAY AAD, and PAY PAY) to compute the MIC value in the AES-CBC-MAC submodule, whereas the second source takes data blocks from the same module (Format CB) to compute cipher data in the AES-CTR submodule. After processing all data blocks, AESCCM generates the cipherdata Cipher MPDU and the U value.

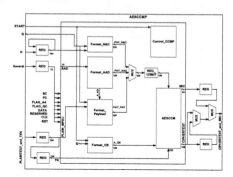

Fig. 24.4. Block diagram of the AESCCMP.

24.3.3 AESCCM6 Hardware Architecture.

The security architecture of the IEEE 802.16e-2005 standard is based on the AES-CCM algorithm as illustrated in Fig. 24.5. For this work, the hardware implementation of this security architecture will be called AESCCM6. The architecture is constituted by specialized modules to format data (Modifying GMACH, Construct Nonce, Format Payload, Format B0, and Format CB), to compute AES-CCM algorithm (AESCCM module), more details are in Section 2. The dataflow is managed by the main control. Format Payload executes a complex process due to the variable length L of the plaintext payload, so, this module has a particular control submodule. Similar to AESCCMP hardware architecture, main control is based on an FSM, generating flag and control signals to the dataflow, whereas AESCCM module computes AES-CBC-MAC and AES-CTR processes in parallel form.

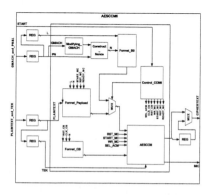

Fig. 24.5. Block diagram of the AESCCM6.

The general operation consists on processing two sources of data, parsed in 128-bit data blocks, and the same 128-bit key block through AESCCM module. The first data source is taken from two different data blocks (Format Payload and Format B0) to compute the MIC value in the AES-CBCMAC submodule, whereas the second data source is taken from the module Format CB to compute ciphertext in the AES-CTR submodule. After processing all data blocks, AESCCM generates the Ciphertext and MIC value. Computation of the AES algorithm in the AESCCM blocks is executed by an iterative and compact module, which reports high performance based on several studies [24].

24.4 Implementations

The design and development of the proposed reconfigurable platform is based on trade-off analysis in order to take advantage of parallelism in order to achieve high performance with moderate power consumption. This also allows to reduce hardware resources and to increase the flexibility while maintaining the performance. To reach these objectives, several reconfigurable schemes should be evaluated (see Fig. 24.6): full reconfiguration to reconfigure the application totally, see Fig. 24.6 (a) and partial reconfiguration for a partial reconfiguration of the application, see Fig. 24.6 (b) and Fig. 24.6 (c). This last type requires a precise analysis of the different functions to be mapped on the same platform in order to find the common components or reconfigurable components. These two reconfigurable schemes have a drawback due to the fact that the device operation is stopped while the reconfiguration is performed, resulting on a timing overhead. This drawback can be solved by implementing other application on the device before switching from one to another.

The synthesis results of the AESCCMP and AESCCM6 hardware architectures are presented in this section. For the purpose of validation and comparison, these architectures were synthesized, mapped, placed and routed for three different FPGA technologies: Virtex-5, Virtex-4 and Spartan-3. The synthesized architectures were simulated and verified considering real-time operation condition by using the design conformance test data, which are provided by the IEEE 802.11i-2004 and IEEE 802.16e-2005 standards.

The AESCCM6 architecture processes 65 data blocks (64 for the message, and 1 for the initial data block), but the initial data block is considered overhead, so only 64 data blocks have effective bits, i. e., (64) (128 bits) = 8192 bits = Plain_data_block_size. These data blocks for authentication (initial block, and message) and ciphering (CBs and message) are processed in parallel, so, it is necessary to process 65 data blocks, requiring (65) (10 clock cycles) + 10 clock cycles + 3 clock cycles = 663 clock cycles. In the same way, the AESCCMP architecture processes 67 data blocks (64 for the message, one for the initial data block and two for the AAD) in 683 clock cycles. The value of the Plain_data_block_size is (66) (128 bits) = 8448 bits. The throughput of these iterative architectures is given by (24.1).

$$Throughput = \frac{(Plain\ data\ block\ size)\ (Clock\ period)}{Clock\ cycles} \qquad (24.1)$$

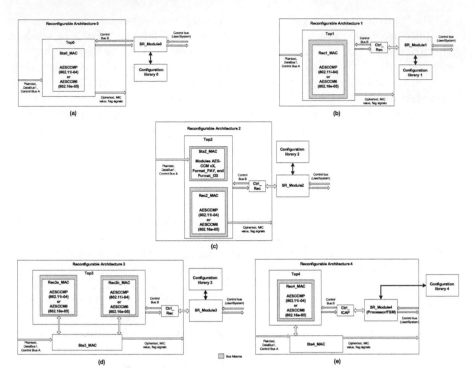

Fig. 24.6. Reconfigurable schemes for the platform.

Table 24.1 shows implementation results of these hardware architectures in three different FPGA devices, Virtex5, Virtex4 and Spartan3, where all but one implementations support more than 1 Gbps. The designs reported in this work were implemented in FPGA devices for evaluation and validation purposes. These architectures can be also considered in the design of application-specific hardware devices. The implementation efficiency (Gbps/slices) is a measurement of this type of cryptographic hardware implementations and it is defined as the ratio between the reached throughput and the number of slices that each implementation consumes [25].

In this Section, FPGA implementation costs and performance evaluation are discussed for the design and development of a security software-radio platform with reconfigurable architecture. To evaluate implementation costs, some characteristics such as utilized resources, period, clock frequency, and latency have been considered, see Table 24.1. For performance evaluation, characteristics such as throughput and efficiency have been considered. These studies and performance measurements of the AESCCMP and AESCCM6 implementations are used to design the security software-radio platform with reconfigurable architecture.

Firstly, considering only the AESCCMP or AESCCM6 hardware architecture, different device families (Virtex versus Spartan) will yield different implementation

Table 24.1. Implementation results of the AESCCMP and AESCCM6 architectures for three technologies

Parameter	CCMP	CCM6	CCMP	CCM6	CCMP	CCM6
Device	Xc5v1x85-1		Xc4vlx60-10		Xc3s4000-4	
Period (ns)	7.064	9.241	7.952	9.091	11.562	15.028
Clock (MHz)	141.56	108.21	125.75	110.00	66.54	66.54
IOBs	269	309	269	309	269	309
Slice-LUT pairs	2920	2761	-	-	-	-
Slices	-	-	2294	1977	2012	1691
BRAMs	10	10	20	20	20	20
Power Cons. (mW)	1425	1396	668	997	576	944
Throughput (Gbps)	1.782	1.321	1.583	1.342	1.089	0.812
Efficiency (Gbps/slice x10-3)	0.610	0.478	0.690	0.679	0.541	0.480
Efficiency (Gbps/MHz)	12.590	12.208	12.590	12.208	12.590	12.208

cost and performance, and newer technologies such as Virtex-4 and Virtex-5 present shorter periods or higher operation clock frequencies. When comparing AESCCMP against AESCCM6, it can be noted that the first one uses slightly more hardware resources for the hardware platform. This is due to the use of specific modules, which execute different formatting of data and specifications. This difference in LUTs and slices is not very significant, considering that a predefined part of the FPGA will be selected for the reconfiguration. The two architectures for two different networks can be supported in the same reconfigurable platform. All implementations use ten BRAMs, situation that enables a consistency in the architectures. These BRAMs are used to implement S-boxes, which are required by the AES cipher. An important detail is the disparity on the use of the IOBs. These pins should be distributed, considering the reconfiguration of the device. The designs of these architectures for the reconfiguration should select tasks to be executed by the input/output data, evaluating reconfigurable/nonreconfigurable modules, which require connections to the exterior. Except for Virtex-4, the AESCCMP implementation reports better performance than compared with AESCCM6 implementation. The minimized area resources of AESCCMP and AESCCM6 do not decrease the system performance, which reach throughput superior to 1 Gbps. These security architectures of the standards are the elements that execute more operations at high computational cost. According to equation (24.1), *Plain_data_block_size* and *Clock_cycles* have fixed values (8448 bits and 683 clock cycles for AESCCMP, and 8192 bits and 663 clock cycles for AESCCM6), but *Clock_period* is defined by the implementation results, which produces different throughput and efficiency for the implementations in the diverse technologies. If this value is fixed, both AESCCMP and AESCCM6 architectures will report the same throughput on the reconfigurable platform, where hardware resources are just selected for reconfiguration or configuration, where a similar efficiency can be obtained. Comparing against related works, see Table 24.2, it is important to highlight that these implementations report high throughput and efficiency, characteristics that can be affected by implementing on a reconfigurable platform. There are many works reporting AES-CCM operations, but few works present complete security architectures executing operations based on AES-CCM. The period of each implementation should be analyzed to improve the performance of the platform, which is affected when reconfigurable architectures are mapped and reconfigurable modules are reused. For AESCCM6 architecture, few works have been reported, implemented on FPGAs [26]. For AESCCMP architecture, related works report different AES-CCMP implementations on FPGA [27]-[30], the proposed AESCCMP implementation reports the highest throughput and efficiency, allocating less area than [29] and [30], with higher operation frequency.

Table 24.2. Related works of 802.11i-2004 and 802.16e-2005 hardware architectures.

Work/ Standard	Slices	BRAM	Clock (MHz)	Throughput (Gbps)
[27]-802.11i	523	-	63.70	0.127
[28]-802.11i	3750	-	50.00	0.243
[29]-802.11i	3474	15	80.30	0.275
[30]-802.11i	5605	-	50.00	0.258
[26]-802.16e	-	-	93.00	-
[26]-802.16e	-	-	197.00	-

[31] presents an analysis, mentioning that CBC-MAC and CTR are the more computationally demanding modules. They propose a minimalist design of AES-CCMP for energy saving. [32] presents a commercial programmable processor with several security functions, including AES-CCMP and reporting a throughput of 1 Mbps at 166 MHz with a maximum-transmission current consumption of 350 mA at 3.3 V. [33] is a commercial processor for different applications and has several security functions as well as AES-CCMP. It consumes 475 mA in the typical transmit mode at 3.3 V, for computations of power consumption. [34] presents a commercial processor with several features for transmission/reception data, considering modulation and security. Elliptic Technology Inc., offers two processors: [35] has the functionality of the standard, supporting AES-CCM and DES-CBC, whereas [36] has a single functionality based on AES-CCM. These two documents do not report implementation results. [37] is a commercial module on ASIC, reporting two architectures implemented on the same technologies. As AESCCM is the main module, there are several special cores for IEEE 802.11i-2004 (such as [38] and [6]) and IEEE 802.16e-2005 security schemes (such as [39] and [6]).

24.5 Conclusions

Software radio is a hot research topic with focus on designing and developing hardware elements that present capabilities of high flexibility and performance. In these radios, security is a key characteristic to protect the transmissions of data in the wireless networks. The implementation costs and performance evaluation of the proposed security architectures from two different networks, such as WMAN and WLAN, enable to design and propose a reconfigurable platform, which supports these implementations, using similar hardware resource and reporting high efficiency. Due to the latency of ten clock cycles in both implementations, the same throughput is reached when that same clock frequency is applied to the reconfigurable platform. Custom hardware architectures are able to better exploit the parallelism of cryptographic algorithms and also are well suited to perform operations directly on bits or values represented by different word sizes. The results obtained in this work indicate that it is feasible to implement a single hardware module to provide security services for two of the most important wireless communications standards.

References

1. Fette, B. A.: Cognitive Radio Technology: Using TPM in Embedded Systems. Newnes, ISBN 0750679522, Ch. 4, 119–133 (2006)
2. Center for Software Defined Radio, Software Defined Radio: Terms, Trends and Perspectives, White Paper. Available: www.csdr.ck. (2007)
3. Shah A., An Introduction to Software Radio, available at: http://www.vanu.com/resources/intro/SWRprimer.pdf
4. Jrg Lotze, Suhaib A Fahmy, Juanjo Noguera *et al.* (2008) An FPGA-based Cognitive Radio Framework, 138-143. In Irish Signals and Systems Conference (ISSC).
5. J. P. Delahaye, G. Gogniar, C. Roland, P. Bomel, Software Radio and Dynamic Reconfiguration on a DSPFPGA Platform, Journal of Telecommunications, pages 152-159, N58, 5-6/2004.
6. Algredo-Badillo I., Feregrino-Uribe C., Cumplido R., Morales-Sandoval M., FPGA Implementation and Performance Evaluation of AES-CCM Cores for Wireless Networks, 2008 International Conference on ReConFigurable Computing and FPGAs (ReConFig'08), pp. 421-426, ISBN: 978-1-4244-3748-1, 3-5 December, 2008.
7. ICT-Centre, Multi Gigabit Millimeter Wave Wireless, Innovative ICT transforming Australian industries, Available: www.ict.csiro.au/index.php. (2009)
8. W. A. Arbaugh, N. Shankar, and J. Wang. Your 802.11 Network has no Clothes. In Proceedings of the First IEEE International Conference on Wireless LANs and Home Networks, pages 131-144, December, 2001.
9. N. Cam-Winget, R. Housley, D. Wagner, and J. Walker. Security flaws in 802.11 data link protocols. Special Issue: Wireless networking security, Communications of the ACM, Volume 46, Issue 5, pages 35-39, May, 2003.
10. J. S. Park and D. Dicoi. WLAN security: current and future. IEEE Internet Computing, Volume 7, No. 5, pages 60-65. September/October , 2003.

11. LAN/MAN Standards Committee, Part 11: Wireless LAN Medium Access Control (MAC) and Physical Layer (PHY) Specifications, IEEE Std 802.11i-2004, IEEE Computer Society, July (2004).

12. IEEE Standard 802.1X-2001. IEEE Standard for Local and Metropolitan Area Networks. June, 2001.

13. Xinyu Xing, Elhadi Shakshuki, Darcy Benoit, Tarek Sheltami, Security Analysis and Authentication Improvement for IEEE 802.11i Specification, Global Telecommunications Conference, 2008. IEEE GLOBECOM 2008, Pp:1 - 5. Nov-Dec. 2008.

14. Changhua He, John C. Mitchell, Security Analysis and Improvements for IEEE 802.11i. The 12th Annual Network and Distributed System Security Symposium (NDSS'05), pages 90-110. Feb. 2005.

15. D. Whiting, R. Housley, and N. Ferguson. Counter with CBC-MAC (CCM). RFC 3610, September, 2003.

16. Sheraz Naseer, Muhammad Younus and Attiq Ahmed, Ninth ACIS International Conference on Software Engineering, Artificial Intelligence, Networking, and Parallel/Distributed Computing, pp. 344-349, 2008.

17. WiMAX Insight, white paper, p. 14, www.wimaxforum.org. Retrieved September 24th, 2009.

18. WiMAX Forum Overview, www.wimaxforum.org. Retrieved September 24th, 2009.

19. LAN/MAN Standards Committee of the IEEE Computer Society and the IEEE Microwave Theory and Techniques Society, "Part 16: Air Interface for Fixed and Mobile BroadbandWireless Access Systems", IEEE Std 802.16e-2005, IEEE Standard for Local and Metropolitan Area Networks,February (2006).

20. M. Dworkin,: NIST Special Publication 800-38C. Recommendation for Block Cipher Operation modes: The CCM Mode for Authentication and Confidentiality, National Instituteof Standards and Technology (NIST), May (2004).

21. Morris Dworkin, NIST Special Publication 800-38C, Recommendation for Block Cipher Modes of Operation: The CCM Mode for Authentication and Confidentiality, available at: http://csrc.nist.gov/publications/nistpubs/800-38C/SP800-38C_updated-July20_2007.pdf.

22. FIPS-197. Announcing the Advanced Encryption Standard (AES). Federal Information Processing Standards Publication, November 2001.

23. R. Chaves, G. K. Kuzmanov, S. Vassiliadis, and L. A. Sousa, Reconfigurable Memory Based AES Co-processor, International Parallel and Distributed Processing Symposium 2006 (IPDPS 2006), IEEE Computer, 446–455, 2006.

24. Algredo-Badillo I., Feregrino-Uribe C., Cumplido-Parra R.,Design and Implementation of an FPGA-Based 1.452-Gbps Non-pipelined AES Architecture, ICCSA 2006, Lecture Notes in Computer Science 3982, 446–455, Springer-Verlag, (2006).

25. N. Sklavos, G. Selimis and O. Koufopavlou, FPGA Implementation Cost and Performance Evaluation of IEEE 802.11 Protocol Encryption Security Schemes, Journal of Physics: Conference Series 10 (2005), 361–364, Second Conference on Microelectronics, Microsystems and Nanotechnology, (2005).

26. Jetstream Media Technologies, JetCCM-6: 802.16e WiMAX AES-CCM Core, Datasheet, (2006). Available: www.security-cores.com.

27. A. Aziz, A. Samiah, and N. Ikram, A Secure Framework for Robust Secure Wireless Network (RSN) using AESCCMP, 4th International Bhurban Conference on Applied Sciences and Technology, June (2005).

28. J. H. Shim, T. W. Kwon, D. W. Kim, J. H. Suk, Y. H. Choi, and J. R. Choi, Compatible Design of CCMP and OCB AES Cipher Using Separated Encryptor and Decryptor for IEEE 802.11i, Proceedings of the International Symposium on Circuits and Systems, 2004. ISCAS'04, pp. III- 645-8 vol. 3, ISBN: 0-7803-8251-X, (2004).

29. N. Smyth, M. McLoone, and J. V. McCanny, WLAN Security Processor, IEEE Transactions on Circuits and Systems I: Fundamental Theory and Applications, Vol. 53, Issue 7, 1506–1520, ISSN: 1057-7122, (2006).
30. D. Bae, G. Kim, J. Kim, S. Park, and O. Song, An Efficient Design of CCMP for Robust Security Network, ICISC 2005, Lecture Notes in Computer Science 3935, 352–361, Springer-Berlin, (2006).
31. M. Razvi Doomun, K.M. Sunjiv Soyjaudah, Resource Saving AES-CCMP Design with Hybrid Counter Mode Block Chaining - MAC, IJCSNS International Journal of Computer Science and Network Security, VOL.8 No.10, October 2008.
32. Lantronix , Inc., WiFi Embedded DeviceSserver - Network Processor Module: MatchPort b/g Pro Wireless Device Server, Datasheet, 2008. Available at: www.lantronix.com.
33. Quatech, Inc., Airbone Embedded Radio Modules (802.11b/g): WLRG-RA-DP100 series, Datasheet, 2007. Available at: www.quatech.com
34. ASK Fujitsu Microelectronics Europe, MB86K21 Baseband SoC: The Mobile WiMAX 802.16e-2005 SoC, Factsheet, 2009. Available at: http://emea.fujitsu.com/microelectronics
35. Elliptic Technologies Inc, LLP-02 Product Brief: PDU Processor for 802.16/WiMAX, Datasheet, 2009. Available at: www.elliptictech.com
36. Elliptic Technologies Inc, LLP-03 Product Brief: PDU Processor for WiMAX Mobile Profile, Datasheet, 2009. Available at: www.elliptictech.com
37. IPCores Inc., CCM6 IEEE 802.16e (WiMAX) AES Core, Datasheet, 2008. Available at: www.ipcores.com/wimax_802.16e_aes_ccm_core.htm
38. Helion Technology Limited, AES-CCM Core Family for Xilinx FPGA, Datasheet, 2008. Available at: www.heliontech.com.
39. Elliptic Technologies Inc., CLP-28 AES Core for 802.16/WiMax, Datasheet, 2009. Available at: www.elliptictech.com.

25

A Method to Generate Automatically a Triangulated Irregular Network from Contour Maps

Luis Gerardo de la Fraga

Cinvestav, Computer Science Department
Av. Instituto Politécnico Nacional 2508. 07360 Mexico City, México.
fraga@cs.cinvestav.mx

Summary. In this chapter, a method to build efficiently a mesh of triangles from a contour map of a terrain is presented. This triangular mesh is a three-dimensional reconstruction of the terrain that can be used to visualize it interactively. The input data to the proposed method are 3D points of polygonal lines that represent contour maps with the elevation information included; such data is available at INEGI in the case of Mexico. The main idea of the proposed method is to include intermediate points among contour lines to avoid the problem of flat triangles, thus a good reconstruction with a surface that follow the terrain trend can be achieved. The intermediate points are selected from the skeleton –which is an approximation of the Medial Axis Transform– of the contour lines, and their heights are calculated automatically. An algorithm to sampling efficiently every contour line, in order to calculate the skeleton, is also given. The method is tested with simulations and real data to shown its effectiveness.

Keywords: Visualization, computational geometry, terrain reconstruction, crust and skeleton, triangulated irregular network.

25.1 Introduction

Even today most of the world's databases of terrain elevation from traditional mapping agencies (i.e., INEGI [1]) are stored in the form of contour overlays. These maps have the advantage that they were developed using human understanding of the observed landforms, using photogrammetry techniques. But their disadvantage consist in the fact that they can not be easily converted to a useful three-dimensional format. Due to the high cost of direct terrain elevation capture methods, cartographic maps are preferred [2]. Therefore, methods that obtain a 3D reconstruction of a terrain from cartographic contour maps have to be developed.

The problem of terrain reconstruction can be seen as a special case of the general problem of three-dimensional reconstruction from cross bi-dimensional sections. Reconstruction in this field means an interpolation problem of n contours each one with a specific elevation. The characteristic feature of these contours is they are not intersected and they can be nested.

In [3] was proposed a method to perform 3D reconstructions of terrains. Their method uses an image of the contour map as the input, and apply several image processing techniques to calculate the skeleton. Their method also uses all the skeleton points. The disadvantages of that method are that it does not use vectorial data as the input, and it uses all skeleton points, increasing the total number of points in the final reconstruction.

In this work I propose a new efficient method to obtain a triangulated irregular network (TIN), that represents also a three-dimensional reconstruction of the terrain, from vectorial data that are offered by INEGI. The method samples efficiently every contour curve of the map, in order to use the minimum set of points of the skeleton that avoid the flat triangles into the mesh. Flat triangles, triangles with the same height in its three vertices, pose a problem because they do not represent the trend of the terrain surface and therefore they produce a rough visualization. This article is an extended form of paper [4]. Other alternative to TINs to represent digital terrain models are grids [5].

In the next section, the proposed method together with some results to illustrate it will be presented. Finally, in section 25.3 conclusions of this work are drawn.

25.2 Terrain reconstruction

In a very general form, in order to obtain a terrain reconstruction from vectorial data, the following tasks must be performed:

1. Re-sampling the contour lines. New points are interpolated among the given ones, all these points are necessary to be able to calculate automatically the skeleton.
2. Calculate the skeleton points that avoid flat triangles. The skeleton is an approximation of the Medial Axis Transform among the contour lines.
3. Calculate the Delaunay triangulation with all the points, the resampled ones and the skeleton points that eliminate the flat triangles; and finally,
4. Visualize the reconstruction

Now, every of these four tasks will be explained with detail.

25.2.1 Sampling the contour lines

The skeleton is an approximation of the Medial Axis Transform [6]. From objects defined by a set of points that samples its boundary, the skeleton can be approximated using the Voronoi diagram [7, 6].

Amenta et al [8] define the Local Feature Size, LFS(p), for a point p on a smooth curve to be the distance from p to a closest point on the medial axis of the curve. A curve is r-sampled, for some $0 < r < 1$, if, for every point p on the curve, there is a sample point within $r * $ LFS(p) of p. They prove two theorems for the globally defined crust of an r-sampled curve: First, when $r < 0.40$, every edge between adjacent samples is included in the crust. Second, when $r < 0.252$, no extra edges are included.

Gold and Snoeyink [9] prove that when a smooth curve is r-sampled at $r = 0.42$, it is an enough sampling density to build the crust from the set of resampled points in a way that exists only one edge between each pair of adjacent points.

Therefore, in order to re-sampling the contour lines we need to calculate first the LFS of each point p on the contours. In [3], the $LFS(p)$ was calculated with image processing techniques using a distance transform of images of the contour lines. In [4] it is proposed to sample all the curves at a same sampling length of 0.42 of the shortest distance among all curves. Finding the closest pair of points between all the curves could become a really high consuming time task if it is not designed carefully: the closest pair between two polygonal curves takes an $O(n^2)$ effort, and checking all the pairs of curves will take also an $O(n^2)$ effort, thus this brute force solution has a $O(n^4)$ complexity.

Taking a single value of r to resample all the contour presents two additional problems: a large number of resampled points will be generated, this is because no all curves needs to be sampled at that minimum value of r; and the other problem is that the calculated skeleton from high density sampled curves will have a very large detail, with lot of branches [7] that are not necessary because all the added details are below the real map resolution.

Now, in the next subsection 25.2.2 will be explained an new procedure, with an $O(n^2)$ complexity instead of an $O(n^4)$ one, to calculate the r_i value for each contour curve i. This procedure will avoid the three described problems: high execution time, large number of points to represent the contours, and very large details into the skeleton.

Once those r_i values are calculated, resampling the contour lines is performed in a very simple way: Let d the Euclidean distance between to contour points; new points are introducing by interpolating new points using Catmull-Rom splines when $\sqrt{2}d > r_i$, where r_i is the sampling size for the curve i. The Catmull-Rom spline is easily calculated using (25.1).

$$Q(t) = [x(t), y(t)] = \frac{1}{2}[t^3, t^2, t, 1] \begin{bmatrix} -1 & 3 & -3 & 1 \\ 2 & -5 & 4 & -1 \\ -1 & 0 & 1 & 0 \\ 0 & 2 & 0 & 0 \end{bmatrix} \begin{bmatrix} p_{i-1} \\ p_i \\ p_{i+1} \\ p_{i+2} \end{bmatrix} \quad (25.1)$$

Where p_{i-1}, p_i, p_{i+1}, p_{i+2} are points along the curve, (note however that the curve drawn actually only passes through points p_i and p_{i+1})

25.2.2 All closest distances among curves

With an example will be visualized in a better way the problem of finding the sampling distance for each contour. In Fig. 25.1 are shown 15 level curves composed by 14 closed polylines and one open polyline, it last open curve is labeled with the number '15' in Fig. 25.2.

Fig. 25.1. Contours taken from INEGI's topographic map E1403 inside a box between coordinates $(641246, 2110916)$ and $(654470, 2121303)$

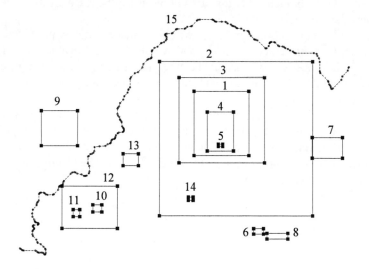

Fig. 25.2. Boxes calculated from contours in Fig. 25.1

To reduce the complexity from $O(n^4)$ (comparing all the point of all the curves) to $O(n^2)$, each closed polyline is reduced to a box and then all boxes are sorted according which contains the others. Thus the problem is simplified by taken into account this reduction of curves to boxes and the sorting of them.

To find the sampling distance for each contour, the following steps must be performed:

1. Calculate the boxes
2. Sort all the boxes
3. Calculate the minimum distance among all boxes
4. Calculate the real distances, using the contour data points, among the curves
5. Calculate the sampling distance for each contour

Step (1) is very simple: for each contour, this is, for each closed polyline, the minimum and maximum x and y coordinates are calculated.

To perform the step (2), to sort the calculated boxes, the area of each box is calculated and then boxes are sorting according to their areas, from the biggest to the smallest. Now, starting from the smallest box, it is checked which of the others boxes contains the previous one. At the end a tree of boxes is generated; by example, from the boxes shown in Fig. 25.2, the generated tree is in Fig. 25.3. In this figure 25.2, labels correspond to the boxes in Fig. 25.2, and the node '0' could be interpreted as a virtual box that contains the whole map. All the not closed polylines can not be nested in other boxes and therefore they are children of the root node. This step (2) has a complexity of $O(n)$ if all boxes are nested or complexity $O(n^2)$ if all boxes are disjoint. An algorithm to perform the method until this step (2) is shown in the following pseudocode:

Require: The list of boxes, B_i, $1 \leq i \leq n$.
Ensure: A list of relations child-parent boxes

```
    Calculate_areas(B);
    C = Sort(B);                    ▷ C is the list of boxes sorted according to their areas
    for  1 ≤ i ≤ n  do
        a(i) = −1                                           ▷ a is an auxiliary array
    end for
    for  (k = n; k >= 1; k − −)  do
        m = k;
        j = m − 1;
        while  j >= 1  do
            if [ then Box m still is not processed] a(m) < 0
                r = is_included(C, m, j);
                if  r == 1  then
                    child = m;     parent = j;
                    print "child:parent";
                    a(m) = 1                                ▷ Box m was processed
                    m = j;
                end if
```

```
        else
            m − −;
        end if
        j − −;
    end while
    if  m == k  then
        print "m:0";
    end if
end for
```

Function is_included(C, m, j) returns true if box C_m is included within box C_j, this is when C_m.xmin $> C_j$.xmin AND C_m.ymin $> C_j$.ymin AND C_m.xmax $< C_j$.xmax AND C_m.ymax $< C_j$.ymax is true.

The other steps (3), (4) and (5) will be explained with detail in the following subsections.

Minima distances among boxes

Once boxes are sorted and a tree like the shown in Fig. 25.3 is obtained, the following tasks are performed:

- For the tree's root node, distances among all children boxes and the possible open polylines are calculated. For every box or polyline the nearest box, or polyline, is calculated.
- For any other parent node, different that the root node, which contains children nodes, distances among boxes are not calculated because this could origin a really bad approximation. Only it is indicated that the real distances among the parent and children, and among all children, must be calculated instead.
- If a parent node has only one child, the distance between them is not calculated, because each box is the nearest to the other.

The result of apply this step (3) is another tree with nodes grouped according to the minima distances among them, as can be seen in Fig. 25.4. Those minima distances are only an approximation to the real ones because they were calculated using boxes.

To calculate the nearest box to each box, the following algorithm could be used:

Require: The list of boxes, B_i, $1 \leq i \leq n$.
Ensure: The nearest box o every box

```
for  1 ≤ i ≤ n  do
    for  i + 1 ≤ y ≤ n  do
        D(i, j) = D(j, i) = distance_boxes(B_i, B_j)
    end for
end for
for  1 ≤ i ≤ n  do
    m(i) = 0
end for
for  1 ≤ i ≤ n  do
```

if $m(i) > 0$ **then**
   continue
end if
min $= 1$
for $2 \leq j \leq n$ **do**
   if $j == i$ **then**
      continue
   end if
   if $D(i,j) < D(i,\text{min})$ **then**
      min $= j$;
   end if
end for
if min $> i$ **then**
   $m(\text{min}) = i$
end if
end for

To calculate the distance between boxes, that distance is the value returned by distance_boxes(B_i, B_j) procedure, two situations must be considered: if the projections of both boxes on x and y axes do not intersect, the distance between them will be the Euclidean distance of their nearest corners (as is shown in Fig. 25.5(a)); if only one of their projections intersects, the distance between them will be the distance between the boxes' edges that do not intersect (as is illustrated in Fig. 25.5(b)). There is an addition situation: if both boxes intersects, then their distance will be equal to zero.

A box is defined by coordinates of its left-bottom corner (xmin,ymin) and its right-top corner (xmax, ymax). One implementation of distance_boxes(B_i, B_j) could be as follow:

Require: Two boxes, represented each one by their coordinates (xmin,ymix), (xmax, ymax).

Ensure: The distance between the two input boxes
$d = 0$;
if $B1$.ymax $< B2$.ymin **then**
   $d = d_y = B1$.ymin $- B2$.ymax;
   if $B1$.xmin $> B2$.xmax **then**
      $d_x = B1$.xmin $- B2$.xmax; $d = \sqrt{d_x^2 + d_y^2}$;
   else if $B1$.xmax $< B2$.xmin **then**
      $d_x = B2$.xmin $- B1$.xmax; $d = \sqrt{d_x^2 + d_y^2}$;
   end if
else if $B1$.ymin $> B2$.ymax **then**
   $d = d_y = B1$.ymin $- B2$.ymax;
   if $B1$.xmin $> B2$.xmax **then**
      $d_x = B1$.xmin $- B2$.xmax; $d = \sqrt{d_x^2 + d_y^2}$;
   else if $B1$.xmax $< B2$.xmin **then**

$$d_x = B2.\text{xmin} - B1.\text{xmax}; \ d = \sqrt{d_x^2 + d_y^2};$$

   end if
else
   if $B1.\text{xmin} > B2.\text{xmax}$ **then**
      $d = B1.\text{xmin} - B2.\text{xmax};$
   else if $B1.\text{xmax} < B2.\text{xmin}$ **then**
      $d = B2.\text{xmin} - B1.\text{xmax};$
   end if
end if
return d;

To calculate the minima distance among boxes and polylines, the distance between each polyline vertex and the box must be checked. This task can be performed using a function distance_box_point(B,x,y), that gets as arguments a box B and the coordinates of a point (x,y); the behavior of this function is similar to distance_boxes(B_i, B_j) function: distance can be the Euclidean distance from one box corner to the point, or distance from one box's edge to the given point.

Real distances among curves

Now, the real distances among the level curves, associated with graphs built according the boxes distances, are calculated. For example, for each edge in graphs shown in Fig. 25.4, the minimum distance among all the vertices of each contour, corresponding to each node, are calculated. This task has a complexity $O(n^2)$. To calculate the minimum distance between two curves, also the distances among edges and vertices should be checked, but it is not necessary to perform because curves are defined by a large number of points. By this way, for each edge in graphs shown in Fig. 25.4, the real distance between curves corresponding to the two vertices of that edge is obtained.

To calculate the distance at which contour i must be sampled, the minimum distance of all the associated edges that arrive to node i must be calculated. The sampling distance r_i, for node (or contour) i, is a half of that calculated distance. For example, for the minima distance graph shown in Fig. 25.4, for node (or contour) 2, the calculated distances are $d(2,3) = 519$ meters, $d(2,7) = 951$ m, $d(2,14) = 258$ m (all distances are in meters); the minimum distances is $d(2,14)$, thus $r_2 = 258/2 = 129$. The sampling distance for node 5 is $d(5,4)/2$ because node 5 has only the edge 5-4. For node number 4, sampling distance is $\min[d(4,5), d(4,1)]/2$. And so on for all the rest of nodes.

The resampling task, as it was explained in the end of Sec. 25.2.1, is performed by introducing new points between two given points of the curve using Catmull-Rom splines.

One observation here, by taking a half of the minimum distance, instead of 0.42 of that minimum distance, to resampling the curves has no consequences: the crust is given by the resampled points, and it is necessary only to add the skeleton points that eliminate flat triangles. But some of the contours do not present the condition of

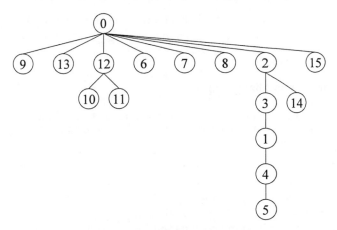

Fig. 25.3. Tree of the sorted boxes

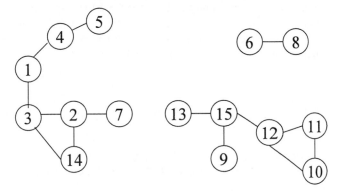

Fig. 25.4. Graph of the minima distance among boxes and polyline, corresponding to example in Figs. 25.2 and 25.1

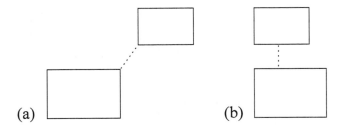

(a)   (b)

Fig. 25.5. Examples to calculate the distance between two boxes. The calculated distance is equal to the length of the dotted line.

smoothness, this produce that the skeleton cross the crust edges. The calculation of crust and skeleton and the correction applied to avoid the described problem will be described in the next subsection.

25.2.3 Points to avoid flat triangles

I am going to explain how the geometrical forms of the skeleton and crust are calculated and how at the same time, the skeleton's edges which avoid the flat triangles are selected.

A fast algorithm to calculate the skeleton and crust were proposed in [9]; here will be described how this algorithm works. It is necessary to calculate before the Voronoi diagram of the set of resampled points. For this task, Qhull program [10] was used.

For each edge (A,C) on the contour and its corresponding Voronoi edge (B,D), as it is shown in Fig. 25.6(1), the test *InsideCircle* is computed. If the test is positive (it is greater than zero), meaning that D lie outside of the circle drawn through (A, B, C), then edge (A,C) is part if the crust. And if the test fails, the Voronoi edge (B,D) is included in the skeleton. Therefore, the crust are the edges corresponding to the points in the contour lines and the skeleton is built with the Voronoi edges. The InsideCircle(A,B,C,D) is calculated as the value of the determinant of 3×3 matrix:

$$\begin{bmatrix} (x_A - x_D)^2 + (y_A - y_D)^2 & (x_A - x_D) & (y_A - y_D) \\ (x_B - x_D)^2 + (y_B - y_D)^2 & (x_B - x_D) & (y_B - y_D) \\ (x_C - x_D)^2 + (y_C - y_D)^2 & (x_C - x_D) & (y_C - y_D) \end{bmatrix}.$$

Besides the situation described by Fig. 25.6(1), there are other three possibilities that must be checked: in Fig. 25.6(2) points A,B,C are clockwise and D is inside the circle, in Fig. 25.6(3) points A,D,C are counterclockwise and B is inside the circle, and in 25.6(4) points A,D,C are clockwise and B is inside the circle.

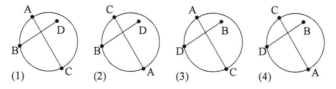

Fig. 25.6. Possible cases on points positions for the InsideCircle(A,B,C,D) test.

To check if three points, say A,B,C, are counterclockwise oriented, the double of the triangle area formed by points A,B,C is calculated; it is equal to calculate the determinant of matrix:

$$\begin{bmatrix} x_A - x_B & x_C - x_B \\ y_A - y_B & y_C - y_B \end{bmatrix}$$

If the value of the determinant is negative, then triangle's vertices are oriented counterclockwise. If its value is positive then they are oriented clockwise.

Now, to check the four cases shown in Fig. 25.6, note that in Fig. 25.6(1), triangle ABC has an area bigger than area of triangle CDA. By using this characteristic, one can orient the given points in order to apply *InsideCircle* test, such as the following subroutine does:

Require: Four points A, B, C y D. AC is the Delaunay edge, BD is the Voronoi edge.
Ensure: The value of *InsideCircle* test.

```
 1: a₁ =areCounterclockwise(A,B,C);
 2: a₂ =areCounterclockwise(A,D,C);
 3: r = 0;
 4: if |a₁| > |a₂| then
 5:                                              ▷ Area |a₁| is greater than |a₂|
 6:     if |a₁|/|a₂| > 30 then
 7:         return 0;
 8:     end if
 9:     if a₁ < 0 then
10:         r =inCircle(A,B,C,D);                ▷ Case (1) in Fig. 25.6
11:     else
12:         r =inCircle(C,B,A,D);                ▷ Case (2) in Fig. 25.6
13:     end if
14: else
15:                                              ▷ Area |a₂| is greater or equal than area |a₁|
16:     if |a₁| == 0 then
17:         return 0;
18:     end if
19:     if |a₂|/|a₁| > 30 then
20:         return 0;
21:     end if
22:     if a₂ < 0 then
23:         r =inCircle(A,D,C,B);                ▷ Case (3) in Fig. 25.6
24:     else
25:         r =inCircle(C,D,A,B);                ▷ Case (4) in Fig. 25.6
26:     end if
27: end if
28: return r;
```

In lines 6 and 19 we can see the correction that is performed to *InsideCircle* test. Those lines correct the situation shown in Fig. 25.7: this figure is part of the curve corresponding to box 4 in Figs. 25.2 and 25.1, the curve is not smooth producing *InsideCircle* test is valid and that the skeleton cross the curve. The correction is that *InsideCircle* test is considered invalid if the ratio of the triangles' area is greater than 30. This value was set by the practical experience and here is not given a mathematical proof for using that value.

By definition, a flat triangle has its three vertices with the same height. Therefore, if the corresponding heights of the two points over the contour, corresponding to the

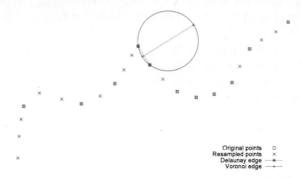

Fig. 25.7. The situation when a skeleton edge cross a crust edge. This situation is because the contour is not smooth and its correction is explained on the text

Voronoi edge that was selected to be part of the skeleton, are equal, then that Voronoi edge is selected and all the other skeleton edges are discharged.

In Fig. 25.8(c) three contour lines and parts of the skeleton that was selected until this step of the proposed method are shown.

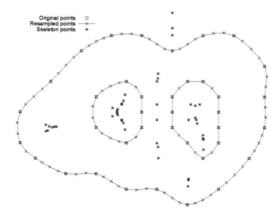

Fig. 25.8. An example of a map with three contour lines. Resampled and skeleton points are also shown.

25.2.4 Graphs classification

As a result from the previous step, the selection of skeleton's edges that to eliminate the flat triangles, these edges form connected graphs. It is possible to distinguish four cases of graphs:

1. Graphs inside summits, as the two graphs marked with '1' in Fig. 25.8(c).
2. Graphs between two level curves with the same height, as the one marked with '2' in Fig. 25.8(c).
3. Re-entrant graphs inside a contour line as the graph marked with '3' in Fig. 25.8(c).
4. Re-entrant graphs outside a contour line, it is marked with '4' in Fig. 25.8(c).

The problem now is how to recognize every graph and how to assign automatically the heights to every vertex in the graph. This is an easy task for graphs type '2': they come from two different contour lines, and their vertices edges must be the height

$$h = h_1 - \frac{a}{2} \qquad (25.2)$$

where h_1 is the height of the associated contour lines of the graph and a is the height between two consecutive contour lines (by definitions a is the same in the whole map).

Graphs of type '1' and '3' are similar: the assigned heights to their vertices must be higher than the associated contour line (from which they were selected). Thus, the height of the vertices for these graphs are calculated as:

$$h = h_1 + \frac{a r_i}{2 r_{MAX}} \qquad (25.3)$$

where r_i is the distance of vertex i on the graph to the associated contour line, and r_{MAX} is the maximum of all these distances. The expression (25.3) to calculated the heights was proposed in [11]. A good approximation of r distances can be calculated at the same time when skeleton edges were selected: r is the distance to each vertex to any of the associated vertices on contour lines.

Finally, the height for vertices of graphs type '4' are lower that the height of the associated contour line, and it is calculated as:

$$h = h_1 - \frac{a r_i}{2 r_{MAX}} \qquad (25.4)$$

How to recognize automatically each graph type? Type '2' is different of the others and can be recognized when it is built. Graphs '1' and '3' are inside the associated contour line, and graphs '4' are outside of the associated contour line. To distinguish graphs type '1' and '3' from type '4', an algorithm to test whether any point of the graph lies within the polygon represented by the contour line [12, p. 239–245] is used.

25.2.5 Visualization

In Fig. 25.9 a simulated example is used to compare the 3D reconstruction obtained from the original raw data and the reconstruction obtained with the proposed method. Figs. 25.9(a) and 25.9(c) shown the triangulations with both sets of data; and in Figs. 25.9(b) and 25.9(d) shown visualizations of both reconstructions. We can see clearly that the reconstruction obtained with the proposed method is much better that the one obtained with the raw data.

Also, the reconstruction obtained with the real data in Fig. 25.1, is shown in Fig. 25.10.

Visualization in Figs. 25.9(b) and 25.9(d) and 25.10(c) were made in OpenGL by setting the smooth option to paint the triangles. A single color was used per each contour. Colors corresponding to skeleton vertices were approximated lineally from the colors in their two nearest contours.

25.3 Conclusions

A method to obtain a digital representation, in form of a triangulated irregular network, of a terrain was presented. The input to the presented method is vectorial data in form of polylines that represent contour lines. An efficient algorithm, with $O(n^2)$ complexity, to calculate automatically the sampling size for each contour is proposed. Resampling contour lines is needed in order to calculate the skeleton, and some points of the skeleton, the ones that eliminate the flat triangles, are selected. The method automatically also compute the heights of the vertices selected from the skeleton. Flat triangles is a problem because they do not represent the trend of the terrain surface and therefore they produce a rough visualization. Without the flat triangles we obtain a good 3D visualization of the reconstructed terrain.

The method uses several algorithms of Computational Geometry applied to vectorial data of contour lines maps with elevation information that can be obtained from traditional mapping agencies, i.e. from INEGI in the case of Mexico.

References

1. :   Instituto Nacional de Estadstica, Geografa e Informtica (INEGI) http://www.inegi.gob.mx.
2. Thibault, D., Gold, C.M.: Terrain reconstruction from contours by skeleton construction. GeoInformatica **4**(4) (2000) 349–373
3. Rivas, A., de la Fraga, L.: Terrain reconstruction from contour maps. In: Proceedings of the 14th International Congress on Computing, IEEE Press (2005) 36–39
4. de la Fraga, L.: Smooth three-dimensional reconstruction from contour maps. In: Proceedings of the 2008 5th International Conference on Electrical Engineering, Computing Science and Automatic Control (CCE 2008), IEEE Press (2008) 364–367
5. El-Sheimy, N., Valeo, C., Habib, A.: Terrain digital modeling: acquisition, manipulation, and applications. Artech House, Inc (2005)

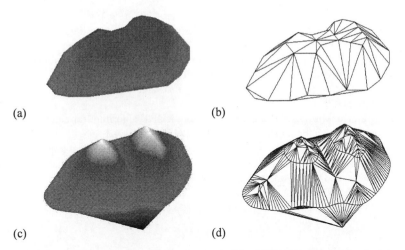

(a)   (b)

(c)   (d)

Fig. 25.9. Examples of reconstructions of maps in Fig. 25.8(a-b) without the skeleton points that eliminate the flat triangles, and in (c-d) using points that eliminate flat triangles.

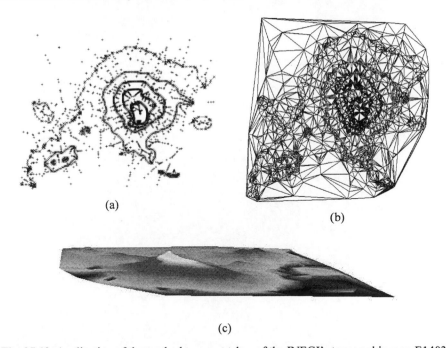

(a)

(b)

(c)

Fig. 25.10. Application of the method to a part taken of the INEGI's topographic map E1403, in (a) resampled and skeleton points, (b) Delaunay triangulation with all the points in (a), and (c) a visualization of the 3D reconstruction.

6. Castleman, K.: Digital Image Processing. Prentice-Hall (1996)
7. Attali, D., Montanvert, A.: Computing and simplifying 2d and 3d continuos skeletons. Computer Vision and Image Understanding **67**(3) (1997) 261–273
8. Amenta, N., Bern, M., Eppstein, D.: The crust and the β-skeleton: combinatorial curve reconstruction. Graphical Models and Image Processing **60**(2) (1998) 125–135
9. Gold, C.M., Snoeyink, J.: A one-step crust and skeleton extraction algorithm. Algorithmica (2001) 144–163
10. Barber, C., Dobkin, D., Huhdanpaa, H.: The quickhull algorithm for convex hull. ACM Transactions on Mathematical Software **22**(4) (1996) 469–483
11. Gold, C., Dakowicz, M.: Visualizing terrain models from contours – plausible ridge, valley and slope estimation. In: Proceedings of the International Workshop on Visualization and Animation of Landscape, Kumming, China (2002)
12. O'Rourke, J.: Computational Geometry in C. 2nd edn. Cambridge University Press (1998)

Image classification with colour and texture using the Complex Wavelet Transform

Maria E. Barilla[1] and Michael Spann[2]

[1] Universidad Autónoma del Estado de México Toluca, Cerro de Coatepec S/N, Toluca, México
barillam@fi.uaemex.mx
[2] The University of Birmingham, Birmingham B15 2TT, United Kingdom
M.Spann@bham.ac.uk

Summary. In this paper we present an experimental analysis of colour-based texture image classification in order to evaluate whether colour and texture information should be used jointly or separately. Various colour spaces are used for colour information extraction. The complex wavelet transform is used to extract texture information. Results show that colour and texture information should be treated separately. As well, they evidence that colour is highly important when present in textured images.

26.1 Introduction

The analysis of colour-based texture images poses the question of how to incorporate colour and texture information; together or separately.

Maenpaa et al. [19] argue that colour and texture give more accuracy if they are treated separately. This is based on the results of classification experiments where they tested the accuracy obtained by grey-level texture features and colour-based texture features. They conclude that a parallel procedure for colour and texture obtained separately is better than a combination of them. This means performing a process with one feature and improving the accuracy with information from the other one. Moreover, they explain that colour texture features are best compared to grey level texture features only under static illumination conditions, which is not useful for the analysis of natural images. These experiments were done over Gabor and local binary pattern operator [22] as texture features.

A set of classification experiments performed by Drimbarean and Whelan [7] shows that colour improves the performance of colour texture classification. When using Discrete Cosine Transform (DCT), Gabor filters and Co-occurrence matrices from each colour plane of the RGB colour space, DCT produces the highest classification accuracy.

Also, they ran experiments using DCT over every colour plane of the HSI, XYZ, YIQ and $L^*a^*b^*$ colour spaces comparing them with the results from using DCT over the intensity, luminance or lightness of the same colour spaces in a combination with the chrominance components. $L^*a^*b^*$ colour space showed a better performance, but not considerably better than the other colour spaces.

On the contrary, Singh et al. [27] argues that the chosen colour space for image analysis determines the results. This conclusion is based on colour image texture analysis using colour moments, correlogram features and a combination of those features. The colour spaces tested were $L^*a^*b^*$, $L^*u^*v^*$, $I1I2I3$, RGB, $sRGB$, XYZ, YCC, YIQ and a non-linear. Mahalanobish distance was the chosen for the experiments. Singh et. al. [27] also obtained that for different kinds of images, the distance was different, $L^*a^*b^*$ and "non-linear" colour spaces showing the highest distances. Paradoxically, $L^*a^*b^*$ colour space happens to be the best colour space for a classification experiment.

The apparent lack of consensus for colour-based texture feature extraction suggests testing the performance of feature vectors based on colour and texture taken together against taken separately.

This paper is organized as follows. In section 26.2 we describe the feature vectors extraction process. Section 26.2 has four subsections: listing the colour spaces used (subsection 26.2.1), describing the Complex Wavelet Transform and its advantages (subsection 26.2.2), the extraction of feature vectors set 1 (subsection 26.2.3) and the extraction of feature vectors set 2 (subsection 26.2.4). We detail the classification process in section 26.3. The results follow in section 26.4. Finally, in section 26.5 we present conclusions and suggest some ideas for future work.

26.2 Feature Vectors

The first feature vector is created by obtaining textural information from every colour plane. The second feature vector is created by obtaining textural information from the luminance/lightness/intensity colour plane and adding the colour information from the rest of the colour bands. In order to do this, we propose grouping colour spaces as described in section 26.2.1. Due to the many advantages offered by Dual-tree $\mathbb{C}WT$, as described in section 26.2.2, it is used to obtain textural information.

26.2.1 Colour

Table 26.1 (a) contains the colour spaces that are normally used in colour-based image classification. Every colour plane gives information about a certain quality of colour. Note that this table contains two primary colour spaces: RGB and XYZ. The colour spaces in Table 26.1 (b) can be separated by luminance/lightness/intensity and chrominance.

Colour space	C_1	C_2	C_3
RGB	R	G	B
XYZ	X	Y	Z
Yxy	Y	x	y
YUV	Y	U	V
YIQ	Y	I	Q
AC_1C_2	A	C_1	C_2
$L^*u^*v^*$	L^*	u^*	v^*
$L^*a^*b^*$	L^*	a^*	b^*
$I_1I_2I_3$	I_1	I_2	I_3
Yuv	Y	u	v
YQ_1Q_2	Y	Q_1	Q_2

Colour space	$L/Y/I$	C_2C_3
Yxy	Y	xy
YUV	Y	UV
YIQ	Y	IQ
AC_1C_2	A	C_1C_2
$L^*u^*v^*$	L^*	u^*v^*
$L^*a^*b^*$	L^*	a^*b^*
$I_1I_2I_3$	I_1	I_2I_3
Yuv	Y	uv
YQ_1Q_2	Y	Q_1Q_2

Table 26.1. (a) Left - Colour spaces and their colour attribute planes.
(b) Right - Separable colour spaces in $L/Y/I$ and *Chrominance* information.

26.2.2 Texture obtained by the Complex Wavelet Transform

The complex wavelet transform[3] ($\mathbb{C}WT$) has been developed in the last decade to overcome some of the drawbacks that the DWT presents [16, 12, 15]. Two main attributes of the $\mathbb{C}WT$ are particularly attractive: near shift invariance and good directional selectivity. In order to achieve these advantages, the $\mathbb{C}WT$ introduces mild signal redundancy.

The structure of $\mathbb{C}WT$ is similar to the DWT's except that $\mathbb{C}WT$ filters have complex coefficients and generate complex output samples [12].

To achieve perfect reconstruction while keeping the properties of shift invariance and directional selectivity, another structure of the $\mathbb{C}WT$ has been suggested: dual-tree complex wavelet transform[4] (dual-tree $\mathbb{C}WT$) [11, 13, 14, 24]. As its name suggests, a dual-tree structure of the wavelet transform is used instead of the well known single tree structure.

The salient features of dual-tree $\mathbb{C}WT$ along with their advantages are described as follows.

Near shift invariance When downsampling is executed, there is an aliasing effect. This causes non-redundant wavelet transform methods based on FIR filters to lack shift invariance [2]. Consequently, if the overlap between aliased passbands is reduced, shift dependence is also reduced [13]. Aliasing is reduced by using two trees, as dual-tree $\mathbb{C}WT$ does.

Directional selectivity The concept of directionality has been considered an important attribute even before the advent of complex wavelets. The Gabor transform is most frequently discussed in the literature. It gives a large range of directions, but they are extremely redundant and so computationally expensive [20].

---

[3]The symbol $\mathbb{C}$ is used to make a distinction between the Complex Wavelet Transform ($\mathbb{C}WT$) and the Continuous Wavelet Transform (CWT) [24].

[4]Dual-Tree Complex Wavelet Transform is abbreviated as DT-CWT by Kinsbury [15] and as dual-tree $\mathbb{C}WT$ by Selesnick [24]. In this work, Selesnick's abbreviation is adopted.

There also exist steerable wavelet transform [26], curvelet transform [1, 6] and contourlet transform [5]. Complex wavelets provide a reasonable amount of directions to separate subbands for positive and negative orientations: 15°, 45°, 75°, -15°, -45°, -75°.

Moderate Redundancy Redundancy is inevitable if one wants to reduce aliasing and, therefore, be shift invariant . The redundancy introduced by complex wavelets is $2^m : 1$ for m dimension [2, 13], i.e. for 1-D, redundancy is $2 : 1$ and for 2-D redundancy is $4 : 1$.

An improvement to the dual-tree $\mathbb{C}WT$ is the Q-shift dual-tree $\mathbb{C}WT$, which utilizes a different set of filters for level 1 than for the next levels. Orthonormal perfect reconstruction filters are chosen so that the reconstruction filters are just the time reverse of the equivalent analysis filters. There are a number of choices of possible filter combinations. Kingsbury [15] suggested a few. The Q-shift transform retains the near shift invariance and directionality properties of the original while also improving the sampling structure. Complex wavelet transform has been used for texture modelling by a number of different researchers [3, 9, 10, 18, 23, 25], but has not been so widely used yet.

26.2.3 Set 1 of feature vectors

This set of feature vectors is obtained by getting textural information from every colour plane of the list of colour spaces from Table 26.1(a) at every scale s. Then the set of feature vectors $\mathbf{f}_1^{(s)}$ is given by

$$\mathbf{f}_1^{(s)} = \{|z|^{(s,b)}\}, \forall b \tag{26.1}$$

where $\{b = 1, ..., 6\}$ indicates the number of subbands produced by the wavelet transform at scale s and $|z|$ represents the modulus of the complex wavelet coefficients for every subband b at every scale s.

26.2.4 Set 2 feature vectors

This set of feature vectors is obtained by getting textural information only from the $L/Y/I$ plane and adding the colour information. This feature vectors are extracted from the list of colour spaces from Table 26.1(b). Note that Table 26.1(b) does not contain colour spaces RGB nor XYZ. Then the set of feature vectors $\mathbf{f}_2^{(s)}$ is given by

$$\mathbf{f}_2^{(s)} = \{|z|^{(s,b)}, C_2, C_3\}, \forall b \tag{26.2}$$

where $\{b = 1, ..., 6\}$ indicates the number of subbands produced by the wavelet transform at scale s and $|z|$ represents the modulus of the complex wavelet coefficients for every subband b at every scale s. C_2 and C_3 are the chromatic information of every colour space.

26.2.5 Feature vectors comparison test

In order to compare our feature vectors, we will make use of the McNemar's test. One-to-one matching techniques as McNemar's test are often used by researchers to increase the precision of a comparison [8]. McNemar's test is useful for comparing frequencies in matched samples. It is applicable to situations in which the same subjects are observed on two occasions. In particular, McNemar's test is an excellent tool when comparing two classifier when the error generated for each one is an independent event [4, 28]. In our method we will compare the classification of different feature vectors with the same classification method. As different texture images are classified, every classification error is an independent event, therefore McNemar's test is good to be used.

McNemar's test is applied as follows. We first divide the image data S, into a training set R and a test set T. The two classifiers A and B are then trained over R and tested over T. For every $x \varepsilon T$, the frequency of classification is recorded according to the following table:

Number of samples misclassified by A but not by B	Number of samples misclassified by A and B
Number of samples misclassified by neither A nor B	Number of samples misclassified by B but not by A

Table 26.2. Frequencies of misclassification in McNemar's test for classifiers A and B.

We will use the notation

where $a + b + c + d$ is the total of the samples in the set T.

Since interest lies in any difference between a and d, frequencies b and c of the table are of little interest. Frequency b refers to matched pairs both of which misclassified the same sample and frequency c refers to pairs both of which did not misclassified the same sample. The comparison is thus confined to the frequencies a and d. Under the hypothesis that a and d do not differ as regards the sample, a and d are expected to be equal, with estimated value $(a + d)/2$.

The McNemar's formula can be defined as in equation 26.3

$$X^2 = \frac{(|a - d| - 1)^2}{a + d} \tag{26.3}$$

McNemar's formula is for testing an association in a 2×2 table under the null hypothesis for which there is no difference between the error rate of two classifiers, which means that $a = d$. The test statistics has a chi-square distribution with one *degree of freedom* (d.f.).

An specific threshold for X^2 has to be chosen in order to reject the null hypothesis. A common value for X^2 threshold is 3.841459, which has an associated probability value of 0.05. Therefore if $X^2 > 3.841459$, we may reject the null hypothesis in favor of the hypothesis that the performance of classifiers A and B are different [28].

26.3 Classification method

The classification method we use was proposed by Southam [28]. In particular, we have implemented his method for real world scene images test 1 called RW_Test1. RW_Test1 is implemented using colour-based images albeit the processing is grey-level. Using this method allow us to compare how the results are affected by including colour information as well as how colour must be treated; jointly or separately. Methods tested by Southam are briefly described in table 26.4 and their success rate results are shown in table 26.5.

Method RW_Test1 is implemented as follows. First of all, a new database of images is created. The new database is obtained from the Outex_NS_00000 test suite of the Outex database [21] by interposing every image with its ground truth image and hand cropping out rectangular shaped regions in such a way that the sample area is maximised. The image database is labelled Outex_NS_ALT_00000 [5]. The same class types are used (sky, tree, bush, grass, road and building) and the number of samples are shown below giving a total of 91 images (table 26.3).

Class Name	Sky	Tree	Bush	Grass	Road	Building
No. Samples	14	17	15	20	16	9

Table 26.3. The classes and number of samples in the new Outex_NS_ALT_00000 natural scene database. This table has been extracted from Southam's PhD thesis [28] on page number 101.

The images in the new alternative image database are then used in 91 leave-one-out-cross-validation experiments. PCA is applied to the feature vectors and a knn classifier and Euclidean distance measure is used to test to find the optimal k. It is worth to point out again that these experiments were made over grey-scale images.

The results obtained for the benchmark methods compared are as shown in table 26.5.

We have repeated the same experiments over the two feature vectors $\mathbf{f}_1^{(s)}$ and $\mathbf{f}_2^{(s)}$ over the colour spaces listed in table 26.1 (a) and table 26.1 (b) respectively. We have denoted these new experiments as RW_Test1_$\mathbf{f}_1$ and RW_Test1_$\mathbf{f}_2$.

---

[5] The Outex_NS_ALT_00000 can be found at http://fi.uaemex.mx/barillam/outex/index.html

Fig. 26.1. Sample sky, tree, bush, grass, road, and building class data from the new Outex NS ALT 00000 database.

Methods	Description
oc-sieve	Morphological scale-space operation with one opening filter (o-sieve) and a closing filter (c-sieve) [28].
2D-sieve	Morphological scale-space operation with an M-filter using opening and closing in one operation [28].
1D-sieve	Similar to 2D-sieve but it also includes the orientation at which the filter is applied.
cooc5	Co-occurrence Matrices with $d = [1, 2, 5, 13, 30]$.
DTCWT	Dual-Tree $\mathbb{C}WT$
LM	Convolution with Leung and Malik's filter-bank [17].
GLA	Granulometric method with line structuring elements of length 1, 2, 5, 13 and 30 at angles 0°, $\pm 35^o$, $\pm 45^o$ and $\pm 90^o$.
MR8	Convolution with Varma and Zisserman's filter bank [29]. It uses only the 8 maximum filter respondes and takes mean, standard deviation and skewness of the filtered images as feature vectors.
GDV	Granulometric method which uses a combination of vertical line and disk shaped structuring elements.
cooc1	Co-occurrence Matrices with $d = 1$.
GDS	Granulometric method which uses disk and square structuring elements.
MR8All	Convolution with Varma and Zisserman's filter bank [29] which uses mean, standard deviation and skewness of the filtered images as feature vectors.
LBP	Local Binary Pattern
EdgeLOG	Edge density features obtained by a Laplacian of Gaussian filter.
EdgeCAN	Edge density features obtained by a Canny edge detector.
EdgeSOB	Edge density features obtained by a Sobel edge detector.

Table 26.4. Description of the texture feature extraction methods tested by Southam [28].

26.4 Results

Table 26.5 shows the results obtained by Southam [28]. As it can be observed, morphological scale-space operations have the highest scores for the mean success rate. Dual-tree $\mathbb{C}WT$ is the fifth highest score in the comparison only after co-occurrence matrices. Edge detector methods are the less successful methods not even achieving fifty percent of success.

Comparing directly dual-tree $\mathbb{C}WT$ with oc-sieve, we can observe that dual-tree $\mathbb{C}WT$ can classify grass and buildings better than oc-sieve. However, oc-sieve classifies trees, bushes and roads better than dual-tree $\mathbb{C}WT$. Both methods can perfectly classify the sky.

The question now is how including colour might improve the success rate. This is explain as follows.

	Sky	Tree	Bush	Grass	Road	Building	mean
oc-sieve	1.000	0.933	0.700	0.737	0.714	0.500	0.764
2D-sieve	1.000	1.000	0.500	0.632	0.714	0.500	0.724
1D-sieve	1.000	0.933	0.400	0.842	0.714	0.375	0.711
cooc5	1.000	0.867	0.400	0.632	0.714	0.625	0.706
DTCWT	1.000	0.800	0.300	0.842	0.643	0.625	0.702
LM	0.857	0.933	0.300	0.737	0.643	0.375	0.641
GLA	1.000	1.000	0.400	0.632	0.429	0.375	0.639
MR8	1.000	0.933	0.400	0.737	0.429	0.250	0.625
GDV	0.929	0.933	0.200	0.684	0.500	0.500	0.624
cooc1	0.929	0.867	0.300	0.421	0.643	0.500	0.610
GDS	0.929	1.000	0.300	0.421	0.357	0.500	0.584
MR8All	0.857	0.933	0.200	0.632	0.500	0.250	0.562
LBP	0.786	0.600	0.500	0.526	0.571	0.125	0.518
EdgeLOG	0.929	0.200	0.100	0.368	0.071	0.250	0.320
EdgeCAN	0.500	0.133	0.300	0.368	0.357	0.250	0.318
EdgeSOB	0.286	0.467	0.100	0.421	0.143	0.000	0.236

Table 26.5. The mean success rate and success rate for each class for RW_Test_1. This table has been extracted from Southam's PhD thesis [28] on page number 104. Oc-sieve method shows the highest success rate.

26.4.1 RW_Test1_f_1

The results for these experiments of RW_Test1_f_1 are shown in Table 26.6.

The best performance of all was from AC_1C_2. $L^*u^*v^*$ and $L^*a^*b^*$ are the colour spaces have been the most used in research so far, but only achieved the fourth best and seventh best.

The worst performers were XYZ and Yxy, which occupy the last two places in the success rate table. These two colour spaces misclassified buildings. Even though RGB colour space is not in the last places on the table, it showed problems in classifying bushes and building.

YQQ performed better than expected, even better in fact than $L^*u^*v^*$ and $L^*a^*b^*$. The YQQ colour space has the advantage that it uses less computational resources and would therefore be a good choice where limited computing power is a consideration.

The McNemar's test results in table 26.8 show that the the two top performers and the worst performer of colour spaces are considerably different from each other. The success rate results are consistent with McNemar's test. The poor performance of XYZ is also reveal on McNemar's test.

The only colour space on RW_Test1_f_1 that improve the classification with respect to Southam's experiments was AC_1C_2. The rest of the colour spaces in these experiments do not seem to make any improvement in the grey scale results. However, the separation of the colour information is tested in the next subsection.

	Sky	Tree	Bush	Grass	Road	Building	mean
AC_1C_2	1.000	0.882	0.667	0.950	0.938	0.444	0.814
YQQ	1.000	0.882	0.267	0.650	0.875	0.333	0.668
$L^*u^*v^*$	1.000	0.824	0.233	0.900	0.750	0.333	0.657
Yuv	1.000	0.706	0.467	0.750	0.625	0.222	0.628
$L^*a^*b^*$	1.000	0.882	0.133	0.700	0.812	0.222	0.625
RGB	1.000	0.647	0.200	0.750	0.750	0.222	0.595
YUV	1.000	0.529	0.200	0.850	0.750	0.222	0.592
YIQ	1.000	0.471	0.133	0.700	0.750	0.333	0.565
$I_1I_2I_3$	1.000	0.412	0.133	0.550	0.812	0.222	0.522
XYZ	0.929	0.706	0.267	0.400	0.562	0.000	0.477
Yxy	0.786	0.412	0.467	0.500	0.188	0.000	0.392

Table 26.6. The mean success rate and success rate for each class for RW_Test1_f_1.

	Sky	Tree	Bush	Grass	Road	Building	mean
$L^*u^*v^*$	1.000	0.824	0.667	0.850	0.875	0.667	0.814
$L^*a^*b^*$	1.000	0.765	0.667	0.950	0.875	0.333	0.765
YUV	1.000	0.824	0.533	0.850	0.938	0.444	0.765
$I_1I_2I_3$	1.000	0.824	0.533	0.850	0.875	0.444	0.754
Yuv	1.000	0.882	0.467	0.900	0.938	0.222	0.735
AC_1C_2	0.429	0.941	0.733	0.950	1.000	0.333	0.731
YQQ	1.000	0.765	0.200	0.750	1.000	0.444	0.693
YIQ	1.000	0.706	0.400	0.800	0.938	0.111	0.659
Yxy	0.786	0.353	0.467	0.500	0.250	0.000	0.393

Table 26.7. The mean success rate and success rate for each class for RW_Test1_f_2.

	AC_1C_2	$I_1I_2I_3$	$L^*a^*b^*$	$L^*u^*v^*$	RGB	XYZ	Yuv	YIQ	YQQ	YUV	Yxy
AC_1C_2	0.00	24.50	12.57	10.89	15.39	24.64	10.70	19.20	9.80	15.39	31.04
$I_1I_2I_3$	-	0.00	5.26	9.80	6.40	0.36	7.12	2.00	8.17	5.33	3.90
$L^*a^*b^*$	-	-	0.00	0.82	0.60	7.00	0.00	2.13	0.60	0.53	12.10
$L^*u^*v^*$	-	-	-	0.00	2.25	8.76	0.43	6.25	0.00	3.60	15.24
RGB	-	-	-	-	0.00	4.48	0.69	1.60	1.80	0.00	11.65
XYZ	-	-	-	-	-	0.00	7.00	1.58	12.57	3.90	2.46
Yuv	-	-	-	-	-	-	0.00	2.88	0.43	0.53	17.29
YIQ	-	-	-	-	-	-	-	0.00	4.17	2.00	6.82
YQQ	-	-	-	-	-	-	-	-	0.00	1.80	16.03
YUV	-	-	-	-	-	-	-	-	-	0.00	10.31
Yxy	-	-	-	-	-	-	-	-	-	-	0.00

Table 26.8. McNemar's p-values for RW_Test1_f_1. Values $p > 3.84$ mean that null hypothesis can be rejected at $\alpha = 0.05$. Values $p > 2.71$ mean that null hypothesis can be rejected at $\alpha = 0.01$.

26.4.2 RW_Test1_f_2

The results for these experiments are shown in Table 26.7.

The highest success rate of all was from $L^*u^*v^*$. $L^*a^*b^*$ is the second best.

The worst performer was again the Yxy which are occupied the last place in the success rate table. McNemar's test also shows this great difference on performance between Yxy and the rest of the methods (Table 26.9).

The most interesting thing is that eight colour spaces improve the success rate obtained by Southam's experiments on $DTCWT$ on grey-scale images of 0.702 on Table 26.5.

These results on feature vectors $\mathbf{f_2}$ support what Maenpaa et al. [19] have said; colour and texture give more accuracy if they are treated separately.

	AC_1C_2	$I_1I_2I_3$	$L^*a^*b^*$	$L^*u^*v^*$	Yuv	YIQ	YQQ	YUV	Yxy
AC_1C_2	0.00	1.64	0.80	0.36	2.00	8.90	6.00	1.32	31.04
$I_1I_2I_3$	-	0.00	0.29	0.53	0.00	2.33	2.25	0.20	27.92
$L^*a^*b^*$	-	-	0.00	0.07	0.29	4.77	3.56	0.07	29.88
$L^*u^*v^*$	-	-	-	0.00	0.53	5.00	3.52	0.25	30.86
Yuv	-	-	-	-	0.00	2.88	2.00	0.14	26.56
YIQ	-	-	-	-	-	0.00	0.06	4.00	16.90
YQQ	-	-	-	-	-	-	0.00	3.27	16.95
YUV	-	-	-	-	-	-	-	0.00	26.27
Yxy	-	-	-	-	-	-	-	-	0.00

Table 26.9. McNemar's p-values for RW_Test1_f_2. Values $p > 3.84$ mean that null hypothesis can be rejected at $\alpha = 0.05$. Values $p > 2.71$ mean that null hypothesis can be rejected at $\alpha = 0.01$.

26.5 Conclusions and future work

We can conclude that colour contributes important information when classifying colour-based texture images. However, colour and texture must be treated separately for this to happen. Even though there is a considerable amount of colour spaces, we can conclude that it is better to use $L^*u^*v^*$ and $L^*a^*b^*$. Dual-tree $\mathbb{C}WT$ plays an important roll in the classification success due to its properties to characterize texture so we recommend it for other image processing tasks like image segmentation. Hence colour-based texture image segmentation by dual-tree $\mathbb{C}WT$ and $L^*u^*v^*$ and $L^*a^*b^*$ in combination seem to have potential for future experiments.

References

1. E. J. Candes and D. L. Donoho. Curvelets, multiresolution representation, and scaling laws. In *SPIE Wavelet Applications in Signal and Image Processing VIII*, volume 4119, 2000.
2. P. de Rivaz. *Complex Wavlet Based Image Analysis and Synthesis*. PhD thesis, University of Cambridge, 2000.
3. P. de Rivaz and N. Kingsbury. Complex wavelet features for fast texture image retrieval. In *Proceedings of IEEE Conference on Image Processing*, pages 25–28, Kobe Japan, October 1999.
4. T.G. Dietterich. Approximate statistical tests for comparing supervised classification learning algorithms. *Neural Computation*, 10(7):1895–1923, September 1998.
5. M. N. Do and M. Vetterli. Contourlets: a directional multiresolution image representation. In *IEEE International Conference on Image Processing*, volume 1, pages 357–360, 2002.
6. M.N. Do and M. Vetterli. Pyramidal directional filter banks and curvelets. In *IEEE International Conference on Image Processing*, volume 3, pages 158–161, Thessaloniki, Greece, October 2001.
7. A. Drimbarean and P. F. Whelan. Experiments in colour texture analysis. *Pattern Recognition Letters*, 22(10):1161–1167, 2001.
8. B. Everitt. *The analysis of contingency tables*. Chapman and Hall, London, 1992.
9. S. Hatipoglu, S. K. Mitra, and N. G. Kingsbury. Image texture description using complex wavelet transform. In *International Conference on Image Processing*, volume 2, pages 530–533, Vancouver, BC, Canada, September 2000.
10. P. R. Hill, D. R. Bull, and C. N. Canagarajah. Rotationally invariant texture features using the dual-tree complex wavelet transform. In *International Conference on Image Processing*, volume 3, pages 901–904, Vancouver, BC, Canada, September 2000.
11. N. G. Kingsbury. The dual-tree complex wavelet transform: a new technique for shift invariance and directional filters. In *IEEE Digital Signal Processing Workshop*, number 86, Bryce Canyon UT, USA, August 1998.
12. N. G. Kingsbury. Image processing with complex wavelets. Phil. Trans. Royal Society London A, February 24-25, 1999. On a Discussion Meeting on "Wavelets: the key to intermittent information?".
13. N. G. Kingsbury. Shift invariant properties of the dual-tree complex wavelet transform. In *International Conference on Acoustics, Speech and Signal Processing ICASSP99*, Phoenix, AZ, March 16-19, 1999. Paper SPTM 3.6.
14. N. G. Kingsbury. A dual-tree complex wavelet transform with improved orthogonality and symmetry properties. In *IEEE International Conference on Image Processing*, volume 2, pages 375–378, Vancouver, BC, Canada, September 2000.
15. N. G. Kingsbury. Complex wavelets for shift invariant analysis and filtering of signals. *Journal of Applied and Computational Harmonic Analysis*, 10(3):234–253, May 2001.
16. N. G. Kingsbury and J. F. A. Magarey. Wavelet transforms in image processing. In *First European Conference on Signal Analysis and Prediction*, pages 23–34, Prague, Czech Republic, June 24-27, 1997. (Invited paper).
17. T. Leung and J. Malik. Representing and recognizing the visual appearance of materials using three-dimensional texton. In *International Conference of Computer Vision*, volume 43, pages 29–44, 1999.
18. E. Lo, M. Pickering, M. Frater, and J. Arnold. Scale and rotation invariant texture features from the dual-tree complex wavelet transform. In *ICIP*, pages 227–230, 2004.
19. T. Maenpaa and M. Pietikainen. Classification with color and texture: jointly or separately? *Pattern Recognition*, 37(8):1629–1640, August 2004.

20. B. S. Manjunath and W. Y. Ma. Texture features for browsing and retrieval of image data. *IEEE Trans. Pattern Anal. Mach. Intell.*, 18(8):837–842, 1996.

21. University of Oulu. University of oulu texture database. URL, http://www.outex.oulu.fi/.

22. T. Ojala, M. Pietikainen, and T. Maenpaa. Multiresolution gray scale and rotation invariant texture analysis with local binary patterns. *IEEE Trans. Pattern Anal. Mach. Intell.*, 24(7):971–987, 2002.

23. J. Portilla and E. P. Simoncelli. Texture modelling and synthesis using joint statistics of complex wavelet coefficients. In *IEEE Workshop on Statistical and Computational Theories of Vision*, 1999.

24. I.W. Selesnick, R.G. Baraniuk, and N.C. Kingsbury. The dual-tree complex wavelet transform. *IEEE Signal Processing Magazine*, 22(6):123 – 151, 2005.

25. E. Simoncelli and J. Portilla. Texture characterization via joint statistics of wavelet coefficient magnitudes. In *Fifth IEEE Int'l Conf on Image Proc*, volume I, Chicago, 4-7 1998. IEEE Computer Society.

26. E. P. Simoncelli, W. T. Freeman, E. H. Adelson, and D. J. Heeger. Shiftable multi-scale transforms. *IEEE Transactions on Information Theory, Special Issue on Wavelets*, 38(2):587–607, 1992. MIT Media Laboratory Vision and Modeling Technical Report 161.

27. M. Singh, M. Markou, and S. Singh. Colour image texture analysis: Dependence on colour spaces. In *ICPR (1)*, pages 672–675, 2002.

28. P. Southam. *Texture Analysis with the Sieve*. PhD thesis, School of Computing Science, University of East Anglia, June 2006.

29. M. Varma and A. Zisserman. Statistical approaches to material classification. In *Proceedings of the Indian Conference on Computer Vision, Graphics and Image Processing*, page 167–172, 2002.

A Software Performance Comparison of Blind Signature Schemes

Lourdes López-García, Luis Martínez-Ramos and Francisco Rodríguez-Henríquez

Computer Science Department, CINVESTAV-IPN, Av. Instituto Politécnico Nacional 2508 Col. San Pedro Zacatenco, C.P. 07300, México, D.F., México

Summary. A blind digital signature is an important cryptographic primitive that has been utilized as the main building block of several security protocols, such as secure e-voting and e-cash systems. In this chapter, we report a software performance comparison of ten blind signature schemes that have been proposed since 1983 until 2008, spanning a total of 25 years of active research in this topic. This chapter begins with a brief introduction to the basic concepts of blind digital signatures, followed by an algorithmic description of the ten schemes studied here along with the corresponding main arithmetic building blocks utilized by them. We coded all the selected schemes using a C library and ran experiments on an Intel Core Duo processor working at 2.0 GHz.

27.1 Introduction

In 1976, Diffie and Hellman introduced in [12] the concept of public key cryptography. Public key crypto-schemes are characterized by the fact that a pair of public and private keys is assigned to each user in the system. In modern cryptography, public key crypto-schemes are extensively used for generating digital signatures. The signer's private key, that must be known only by its owner, is required by the signer entity in order to produce a unique and unforgeable digital signature of a given document, whereas the signer's public key has to be known by the verifier entity in order to decide whether the document's signature is valid or not. Roughly speaking, a digital signature should exhibit the following three properties,

1. *Integrity*: It implies that the received document is a genuine identical copy of the one that was sent.
2. *Identity*: It ensures that the received document was created by a given entity.
3. *Non-repudiation*: Neither the sender nor the receiver, can deny having sent or having received a document.

In practice, a digital signature scheme is composed of three different algorithms: *key generation*, *signature* and *signature verification*. The key generation algorithm permits to generate the signer's private and public key pair. The signature algorithm takes as input the signer's private key and the message to produce the message signature as the output. The signature's verification is performed by a third entity, which executes the signature verification algorithm taking as input the tuple (signer's public key, signature, message) and given as an output either true or false. One of the most important building blocks of a digital signature, and cryptography in general, is the *hash function* that we described in the following.

Formally, a hash function H is a computationally efficient function that maps fixed binary chains of arbitrary length $\{0,1\}^*$ to bit sequences of fixed length m, i.e., $H : \{0,1\}^* \longrightarrow \{0,1\}^m$. Let B be an arbitrary binary chain. Then, we say that $h = H(B)$ is the hash value or digest of B. In practice, modern hash functions are specifically designed for having a short bit-length hash value (usually from 128 bits up to 512 bits). This characteristic is especially attractive for the application of hash functions in digital signature schemes. Therefore, rather than attempting to sign the whole message (which by definition has arbitrary length), it becomes more practical (and also more secure) to sign the digest of the document. Hash functions should exhibit the following properties:

- Given B, it should be easy to compute $H(B)$. However, given h, it should be computationally intractable to find a value B such that $H(B) = h$.
- For any given message B, it is computationally infeasible to find a different message C, such that $H(B) = H(C)$.
- Finding a pair (B,C) such that $H(B) = H(C)$, should be computationally infeasible.

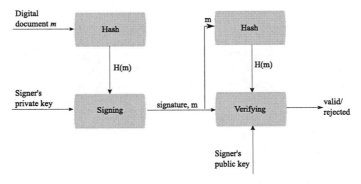

Fig. 27.1. Basic signature-verification procedure

Fig. 27.1 shows the typical process followed in order to sign/verify a digital document. First, the sender must sign the hash value of the digital document using his/her private key. Thereafter both, the document and its signed hash value are sent to the receiver. The receiver can then verify the document's signature by using the sender's public key. Only if the received document is identical to the one that was sent and if the correct public key (the one corresponding to the private key that was used for signing) is utilized, the signature will pass the verification process. However, digital signatures do not guarantee by themselves that a given public key belongs to a given user and consequently they are not suitable to deal with the following four issues,

1. *Secure key authentication.* It is crucial to avoid attacks like man-in-the-middle, replay attacks and identity usurpation attacks among others.
2. *Key revocation.* In the case that A's private key has been compromised by the opponent, then A has no option but to generate a new pair of keys while his/her old keys must not be used anymore (an action known as key revocation). However, it remains as an open problem how to announce to all A's correspondents that A's keys have just been revoked.
3. *Non-repudiation.* One of the main goals of a digital signature is to offer the non-repudiation security service, under the assumption that if A keeps his/her private key in secret, then nobody else can generate a digital signature but himself/herself. However, A could deny his/her alleged digital signature by arguing that the signature does not correspond to his/her secret key.
4. *Policy application.* The only concerted way to enforce security policies among a large community of users is by mean of an external infrastructure of authority entities.

As a consequence of the above concerns, it has been necessary to create an infrastructure able to cover the aforementioned security gaps. That infrastructure is known as Public Key Infrastructure (PKI). PKI integrates digital certificates, public key cryptography and Certificate Authorities (CA) into a single security architecture, where the CAs are the entities in charge of generating digital certificates for the users of the system. A digital certificate endorses the relation among the user and her/his public key. Figure 27.2 shows a simplified version of the messages that must be exchanged among the entities in order to securely performed the authentication secure service using digital signatures.

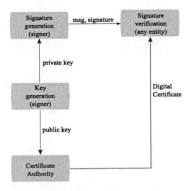

Fig. 27.2. A Digital signature scheme with Digital Certificates (Simplified Version)

The first practical digital signature scheme, RSA, was proposed by Rivest, Shamir and Adleman in 1978 [34]. At present date, RSA is the most widely deployed digital signature method. Shortly after it, the ElGamal's digital signature was proposed by ElGamal in 1985 [15]. This scheme was standardized by the National Institute of Standards and Technology as DSA in 1994 [13]. Across the years, many other digital signatures schemes have been proposed [24]. One notorious example is elliptic curve cryptography that was independently proposed by neal Koblitz [27] and Victor Miller [29] in 1985. Within the framework of elliptic curves cryptography, the most noteworthy standard is the Elliptic Curve Digital Signature Algorithm (ECDSA) [21]. Due to the high difficulty of computing the discrete logarithm problem in elliptic curves over finite fields, one can obtain with ECDSA the same security provided by other existing public-key cryptosystems such as RSA or DSA, but using shorter key lengths (up to ten times smaller).

In 1994, Menezes, Okamoto & Vanstone [30] and Frey & Rück [14] introduced the Weil and Tate pairings in cryptography as a tool to attack the discrete logarithm problem on some classes of elliptic curves defined over finite fields. A few years later, Sakai, Oghishi & Kasahara [36], and Joux [23] discovered constructive cryptographic properties of pairings. After these works, pairing-based cryptography has emerged as a major public key cryptographic system with many novel pairing-based protocols proposed every year. The first digital signature using bilinear pairing was the short signature protocol proposed by Boneh et al. in 2001 [1]. Moreover, bilinear pairings paved the way for practical implementations of Identity-Based Cryptography (IBC) that has two peculiarities: a user can select any of his/her public information as public key; and the certificate authorities are replaced by Private Key Generators (PKG). This ingenious idea had been theoretically proposed by Shamir, back in 1984 [35], and solved in practice by Boneh and Franklin in 2001 [2]. Most digital signature schemes based on IBC use the bilinear pairing function to verify the signature. In 2003, Cha et al. [5] proposed one of the first digital signatures using IBC and bilinear pairings. Figure 27.3 shows the messages among the entities within the identity-based digital signature scenario.

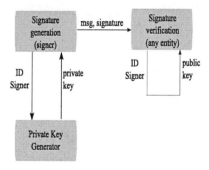

Fig. 27.3. Identity-based digital signature schemes

Security notions of digital signatures

We say that a digital signature scheme is secure if an attacker commonly called the *Adversary*, cannot perform a signature forgery, i.e., the adversary is unable to generate one or more valid signatures for any given message without the knowledge of the signer's private key. The notion of *unforgeability* was introduced by Goldwasser et al. [16], where the authors defined three type of attacks. The first one is called *known message attack*. In this attack the adversary strives to forge a signature under the restriction that he/she only knows the signer's public key and a fixed list of valid pairs (message, signature). In the second attack, named *chosen message attack*, the adversary can query a special entity, called *oracle*, which can give the hash values and digital signatures corresponding to messages chosen by the adversary. The adversary's aim is to create a valid signature for a message whose signature has not been previously queried, using the information obtained from all the answers given by the special entity. The third attack is called *adaptive chosen message attack*, which is similar to the previous attack, except that the message is chosen by taking into account the special entity's responses to all previously chosen messages. The adaptive message attack is the most powerful attack against digital signature schemes.

As it was mentioned before, digital signatures are used to authenticate a user who signs a digital document. However, for some applications it is convenient and sometimes even required that the signer has no knowledge of the message that is signing. This kind of signatures are called *blind signatures* and they are useful for providing anonymity and unlinkability in e-voting and e-cash systems. This chapter is devoted to analyze and give a performance comparison of ten blind signature schemes that have been reported since David Chaum first proposal of this concept in 1981 [6].

The rest of this chapter is organized as follows. Section 27.2 explains the main concepts around blind digital signatures. Then, in Section 27.3 we present a short description of the mathematical problems used by the blind signature schemes as their security guarantees. Then, in Section 27.4, the algorithmic description of ten blind signature schemes is presented. Comparative efficiency analysis and a general performance comparison of all the schemes are presented in Section 27.5. Finally, concluding remarks are given in Section 27.6.

27.2 Digital blind signatures

In 1981, Chaum [6] introduced the concept of *blind signatures*. A blind signature is a special class of digital signature where the message is *blinded* before it gets signed. The blinding process is typically achieved by combining the original message with randomly selected "blind" factors. This blinding process prevents the signer to know the message content. The blinded message so obtained is then transmitted to the signer entity that proceeds to sign it by means of a standard public key cryptoscheme. Then, the requester entity receives the blind signature and removes the blind factor(s) in such a way that the resulting signature can be verified by a third party.

Blind signatures have been shown to be useful for providing anonymity and unlinkability in e-voting and e-cash systems. In the case of e-voting systems, blind signatures help to keep in secret the vote's value when the electoral entity authenticates the voter. For e-cash systems, blind signatures prevent to link an electronic coin with the user that pretends to spend it. This is so, because the blind signature primitive prevents the signer from linking the blinded message signed by it to a posteriori unblinded version that can be published and it can be called upon for verification. In the following, we give a formal definition of blind signatures that has been directly adapted from [3].

The procedure to produce a blind signature can be mathematically described as follows: Let x, y be the signer's private and public keys, respectively. Let us consider two message domains M and Z, a *blind factor'* domain A; a function $z = f_x(m)$ where f_x takes as input $m \in M$, and produces $z \in Z$; two functions $\phi_{y,\alpha} : M \rightarrow M$, and $\phi_{y,\alpha}^{-1} : Z \rightarrow Z$, where $\alpha \in A$, and $\phi_{y,\alpha}^{-1}(f_x(\phi_{y,\alpha}(m))) = f_x(m)$; a function $m = f_y(z)$ where f_y takes as input a value $z \in Z$, and produces an output $m \in M$.

First, the *signer* generates a pair (x, y) and publishes y. Then, the *user* chooses a message m, generates a blind factor $\alpha \in A$ at random, and calculates $m' = \phi_{y,\alpha}(m)$. Then, the signer receives m' and returns to the user its signature $z' = f_x(m')$. The user computes $z = \phi_{y,\alpha}^{-1}(z')$. Finally, z can be verified by a third party, computing $m = f_y(z)$. In practice, a blind signature scheme should include the following five primitives:

1. **Key Generation** *(signer)*: signer's private/public key pair generation.
2. **Blinding** *(user)*: the user blinds a message using the function ϕ.
3. **Signature** *(signer)*: the signer entity signs the blind message utilizing the function f_x.
4. **Unblinding** *(user)*: the user unblinds the blind signature using the function ϕ^{-1}.
5. **Signature verification** *(third party)*: A third party can verify the signature, utilizing f_y.

Security notions of blind digital signatures

The security of a blind signature is defined by its degree of blindness and its degree of unforgeability. Blindness is a requirement where firstly, the contain of the message must be hidden from the signer, and secondly, the user anonymity must be satisfied even after the pair (m, s) has been published (*unlinkability*). In short, the signer should not be able to learn "anything" about the message being signed [3].

The forgery attack for blind signatures is defined after the signer has produced the blind signature. That is, the requester tries to obtain one or more signatures from the blind signature generated by the signer. Hence, in the context of blind signatures, is the requester who becomes the adversary. Pointcheval et al. defined in [32, 33] the *one-more forgery attack*, which tries to cover the scenery aforementioned. The idea is to obtain ℓ blind signatures after the same number or interactions with the special entity and with this information the adversary can obtain $\ell + 1$ valid signatures. The goal is that the adversary requests signatures (to the special entity) of blind messages that contain somehow useful information for producing one more signature at the end of the attack.

Digital blind signatures schemes

The vast majority of blind digital signature schemes that have been proposed up-today are modifications of traditional digital signatures. The first practical proposal of this kind of schemes was presented by Chaum in [7], where he used an ingenious modification of the RSA algorithm. Soon after that, he also presented another proposal in [8], which was based in the DSA algorithm. Since then, many other blind signatures schemes have been reported in the literature. Among them, we will mention here the following eight schemes.

In [10], Camenisch et al. proposed two blind signature methods. One of them was based on a modification of the digital signature algorithm DSA, whereas the second one was defined after a modification of the Nyberg-Rueppel digital signature scheme. Cao and Liu [9] used a modification of Zhu's digital signature utilizing the strong RSA assumption as the underlying mathematical problem. Then, Jena et al. [22] published a scheme based on elliptic curve cryptography (CCE). Boldyreva [4] devised an elegant scheme based on bilinear pairings. Zhang & Kim [39], Cui [11], and Gao et al. [17], proposed different schemes based on identity-based cryptography and bilinear pairings.

In the following sections, we analyze and classify the aforementioned schemes according to the specific group utilized by them, namely, multiplicative groups over the integers (DSA, RSA), additive groups over the points in an elliptic curve (ECC) or additive groups mapped to multiplicative groups defined over finite fields (bilinear pairings). We also present a blind signature scheme that uses elliptic curve cryptography, which is a direct adaptation of one of the DSA Camenisch blind signature schemes. Finally, we give a general feature and performance comparison of the ten blind signature schemes analyzed in this work. Specifically, we compare the number of blind factors, number of messages exchanged among the entities, signature length in bits, security assumption or complexity algorithm and time computational cost employed by each scheme.

We hope that this analysis will be useful for interested readers trying to decide which scheme is more appropriate for a given scenario.

27.3 Mathematical Background and Definitions

In the following, we give a short definition of the hard computational problems in which the security guarantees of the blind signature schemes to be studied in this paper are based.

Integer Factorization Problem (IFP)

Given a positive integer $n > 2$, the integer factorization problem consists on finding a set of prime numbers $p_i, i = 1, 2, \ldots, k$, such that $n = p_1^{e_1} p_2^{e_2} \cdots p_k^{e_k}$ holds, where each p_i is an arbitrary prime number and $e \geq 0$. Solving the integer factorization problem for arbitrary large composite numbers is considered a hard computational problem.

Strong RSA Assumption

Given a RSA module N and a random number $c \in \mathbb{Z}_N^*$, it is computationally infeasible to find $a, b \in \mathbb{Z}_N^*$ such that $a^b \equiv c \bmod N$.

Discrete Logarithm Problem (DLP)

Given p an integer number, a generator $g \in \mathbb{Z}_p^*$ and an arbitrary element $a \in \mathbb{Z}_p^*$, finding the unique number i, $0 \leq i < p - 1$ such that $a \equiv g^i \bmod p$ is a difficult computational problem.

Elliptic Curve Discrete Logarithm Problem (ECDLP)

Given an elliptic curve E defined over a finite field $\mathbb{F}_{p^m}$, where p is a prime number and m is a positive integer, a point $P \in E(\mathbb{F}_{p^m})$ of order n, and a point $Q \in \langle P \rangle$, finding the integer $k \in [0, n-1]$ such that $Q = kP$ is considered a difficult computational problem.

Bilinear Pairing

Let E be a supersingular elliptic curve defined over the finite field $\mathbb{F}_{p^m}$. Define $N = \#E(\mathbb{F}_{p^m})$ as the number of rational points of E, and l the largest prime factor of N. Let $\mathbb{G}_1 = E(\mathbb{F}_{p^m})[l]$ denote the set of points in E of order l, and let k be the smallest positive integer such that $n|(p^{km} - 1)$ then, the multiplicative group $\mathbb{F}_{p^{km}}^*$ of the extension field $\mathbb{F}_{p^m}$ has a unique subgroup $\mathbb{G}_2$ of order l. The integer k is called the *embedding degree*.

A bilinear pairing on $(\mathbb{G}_1, \mathbb{G}_2)$ is a mapping $\hat{e} : \mathbb{G}_1 \times \mathbb{G}_1 \longrightarrow \mathbb{G}_2$ satisfying the following properties:

- Bilinearity: $\forall R, S, T \in \mathbb{G}_1$
  $\hat{e}(S+R, T) = \hat{e}(S,T) \cdot \hat{e}(R,T)$
  $\hat{e}(S, R+T) = \hat{e}(S,R) \cdot \hat{e}(S,T)$
- Non-degeneracy: $\hat{e}(P,P) \neq 1$
- Computability: $\hat{e}$ can be efficiently computed.

The Diffie-Hellman Problems

Let $\mathbb{G}_1$ and $\mathbb{G}_2$ be an additive and multiplicative groups of order n, respectively. Let P be a prime order point in $\mathbb{G}_1$, and let $\hat{e}$ be a bilinear map from $\mathbb{G}_1$ to $\mathbb{G}_2$.

- *The Computational Diffie-Hellman Problem (CDHP)* Given (P, aP, bP) where $a, b \in [1, n-1]$, compute abP.
- *The Decisional Diffie-Hellman Problem (DDHP)* Given (P, aP, bP, cP) where $a, b, c \in [1, n-1]$, deciding whether $c \equiv ab \bmod n$. It is noticed that the DDHP in $\mathbb{G}_1$ can be solved in polynomial time by verifying $\hat{e}(aP, bP) = \hat{e}(P, cP)$.
- *The Bilinear Diffie-Hellman Problem (BDHP)* Given the tuple (P, aP, bP, cP), compute $e(P, P)^{abc}$.
- *The Modified Generalized Bilinear Inversion (MGBI)* Given $h \in \mathbb{G}_2$, find a point $S \in \mathbb{G}_1$ such that $\hat{e}(P, S) = h$.
- *The q-strong Diffie-Hellman Problem (QSDHP)* Given a $(q+2)$−tuple $(P, Q, \alpha Q, \alpha^2 Q, ..., \alpha^q Q)$ as input, find a pair $(c, (c+\alpha)^{-1})$ with $c \in \mathbb{Z}_n^*$.

We know that if the ECDLP can be efficiently solved, then so can the CDHP. In other words, the DHP is not harder than the DLP. Moreover, it is generally assumed (it has been proved for some scenarios) that the ECDLP is not harder than the CDHP. Further, if the CDHP is easy, then the BDHP is easy. Nothing else is known about the hardness of the BDHP. Finally, we remark that there exists solid evidence that the CDHP is strictly harder than the QSDHP [25, 26].

27.4 Blind signature schemes

In this Section we give algorithmic descriptions of ten schemes that have been proposed across the years in the topic of blind signatures.

27.4.1 Chaum blind signature scheme

The blind signature scheme proposed by Chaum was the first blind signature scheme reported in the literature. This scheme is based on the RSA signature scheme. Algorithm 4 shows the steps to produce a blind signature. We note that in order to blind a message m, this scheme uses a blind factor b. Moreover, just two messages are exchanged between the signer and the user; the signature is an integer s modulus n. In order to achieve a security level equivalent to 128 bits, we decided to use a modulus n with a bitlength of 3072-bits. This blind signature scheme is secure against a one-more forgery attack and achieves unconditional blindness [7].

27.4.2 Cao and Liu blind signature scheme

Cao and Liu proposed a scheme based on the strong RSA assumption [9]. This scheme, shown in algorithm 5, requires an exchange of two messages between the user and the signer. One of the main differences with respect to the Chaum scheme described in the previous subsection is the length of the signature which in this method consists of two values, namely, e and y.

According to the authors, this blind signature scheme is secure against chosen-message attacks and achieves the unlinkability requirement [9].

Algorithm 4 Chaum's Blind Signature Scheme [7]

Require: bit-length k, a public exponent e, where e is a small prime number, the message m to be signed, the hash function H.

Key Generation (signer)

Randomly find two $k/2$-bit prime numbers p and q.

$n = pq$;

$\phi(n) = (p-1)(q-1)$, with $gcd(e, \phi(n)) = 1$;

Find d such that $d \equiv e^{-1} \pmod{\phi(n)}$

Private key: (d, n), Public key: (e, n)

Blinding (user)

$h = H(m)$;

Randomly select $b \in \mathbb{Z}_n^*$.

$h' = hb^e \bmod n$;

Signature (signer)

$s' = (h')^d = (hb^e)^d \bmod n$

Unblinding (user)

$s = s'b^{-1} = (h^d)bb^{-1} \bmod n$

A valid signature s of the message m is obtained

Signature Verification (third party)

Check whether $s^e \bmod n = H(m)$, holds.

Algorithm 5 Cao & Liu Blind Signature Scheme [9]

Require: The message m to be signed, a special hash function $H_1(\cdot) : \{0,1\}^* \longrightarrow \{0,1\}^{768}$, a hash function H.

Key Generation (signer)

Randomly find two $k/2$-bit prime numbers p and q, with $|p| = |q| = 1536$ bits, where $p = 2p' + 1, q = 2q' + 1$, and where p' and q' are two prime numbers.

$n = pq$;

Randomly find two quadratic residue generators $X, g \in \mathbb{Z}_n$;

Private key : (p, q), Public key :(n, X, g)

Blinding (user)

$h = H(m)$;

Randomly find a, d, e of 769 bit-length.

$e' = ade$;

$\bar{y} = (Xg^{H_1(h||e||X)})^d \bmod n$.

send $(e', \bar{y})$ to sender.

Signature (signer)

Given: $\phi(n) = (p-1) \cdot (q-1); d' = (1/e') \bmod \phi(n)$;

Compute $\hat{y} \equiv (\bar{y})^{d'} \bmod n$;

The resulting blind signature is $\hat{y}$.

Unblinding (user)

Obtain y by computing $y = \hat{y}^a \bmod n$.

The pair (e, y) is the signature of the message m.

Signature Verification (third party)

Check whether $X = y^e g^{-H_1(h||e||X)} \bmod n$, holds.

27.4.3 Camenisch et al. blind signature schemes

Camenisch et al. defined a blind signature scheme where the signer's view v of the message and the message-signature pair $(m, sig(m))$ are statistically independent [10]. Camenisch et al. actually proposed two schemes that are shown in algorithms 6 and 7, respectively. The first one is derived from a variation of the DSA [13], and the last one is based on the digital signature proposed by Nyberg-Rueppel [31].

Algorithm 6 A DSA-based Blind Signature Scheme [10]

Require: The message m to be signed, a prime p and a prime factor q of $(p-1)$, an element $g \in Z_p^*$ of order q, a hash function H.

Key Generation (signer)

Randomly choose $x \in \mathbb{Z}_q^*$. Compute $y \equiv g^x \bmod p$;

Private key: x; Public key: y.

Randomly choose $k \in \mathbb{Z}_q^*$; $\hat{R} \equiv g^k \bmod p$.

Send $\hat{R}$ to User.

Blinding (user)

$h = H(m)$;

Randomly choose $\alpha, \beta \in \mathbb{Z}_q^*$.

$R = \hat{R}^\alpha g^\beta \bmod p$;

$\hat{h} = \alpha h \hat{R} R^{-1} \bmod q$.

Send $(\hat{R}, \hat{h})$ to Signer.

Signature (signer)

$\hat{s} = k\hat{h} + \hat{R}x \bmod q$.

Unblinding (user)

$s = \hat{s} R \hat{R}^{-1} + \beta h \bmod q$;

$r = R \bmod q$.

The pair (r, s) is the signature of the message m.

Signature Verification (third party)

Check whether $r = (g^s y^{-r})^{h^{-1}} \bmod q$, holds.

We note that in these two schemes, the session key k is generated by the signer. This implies one additional message among the entities. On the other hand, the user employs two blind factors and they are not removed from the signature during the execution of the unblindness primitive, due to the fact that the session key k is used to produce r and s, and the blind factors are utilized to hide these values. This way, the signer cannot link the pair (r, s) with some k previously stored.

Unfortunately, the authors failed to present security proofs of their two proposals. This fact probably motivated that several authors have tried to find weaknesses in these schemes although, until now, without success. In particular, several authors claimed that the first blind digital signature proposal of Camenisch et al. did not fulfill the unlinkability requirement. In 1995, Harn [18] claimed that it was possible to link the pair (user, signature), however, in that same year, Horster [20] proved the incorrectness of Harn's arguments. A similar case happened when Lee et al. in

---

Algorithm 7 A Nyberg-Rueppel-based Blind Signature Scheme [10]

Require: The message m to be signed, a prime p, a prime factor q of $p-1$, an element $g \in Z_p^*$ of order q, and a hash function H.

Key Generation (signer)

Randomly choose $x \in \mathbb{Z}_q^*$; $y \equiv g^x \bmod p$;

Private key: x, Public key: y

Randomly choose $k \in \mathbb{Z}_q^*$; $\hat{r} \equiv g^k \bmod p$.

Send $\hat{r}$ to User.

Blinding (user)

Randomly find $\alpha, \beta \in \mathbb{Z}_q^*$

$h = H(m)$

$r = hg^\alpha \hat{r}^\beta \bmod p$

$\hat{h} = r\beta^{-1} \bmod q$.

Signature (signer)

$\hat{s} = \hat{h}x + k \bmod q$;

Unblinding (user)

$s = \hat{s}\beta + \alpha \bmod q$;

The pair (r, s) is the signature of the message m.

Signature Verification (third party)

Check whether $H(m) = (g^{-s}y^r)r \bmod p$, holds.

2003 [28] assured that the first blind signature scheme presented in this subsection did not exhibit the unlinkability requirement and proposed two blind signatures schemes that they claimed were able to repair the problems detected. However, the conclusions of this work were refuted by Wu and Wang[38], who find out mistakes in the accurateness requirement of Lee's proposals. Finally, T. Wu demonstrated in 2006 [37] that the first blind signature of Camenisch did enjoy the property of unlinkability. Hence, we are lean to say that the two proposals presented in this subsection fulfill the blindness, unlinkability and unforgeability properties.

27.4.4 Jena et al. blind signature scheme

Jena et al. proposed in [22] a scheme based on the elliptic curve discrete logarithm problem (ECDLP). This scheme, shown in algorithm 8, is similar to the first Camenisch's blind signature scheme, when translated to elliptic curve arithmetic. It is noticed that this blind signature scheme uses two blind factors to produce the elliptic curve point R. The modular operations are mainly multiplications and one inversion in the verification primitive. Authors defined the signature as the pair (r, s), and R should be only known by the user. However, in order to verify the signature, the verifier should take the point R because trying to compute the scalar multiplication $H(m)R$ only with the information of r is not possible. The authors showed that their scheme fulfills the accuracy requirement, nevertheless, the security proof for the rest of requirements are not considered.

Algorithm 8 Jena et al. Blind Signature Scheme [22]

Require: The message m to be signed, an elliptic curve E, a base point P of prime order n, the hash function H.

Key Generation (signer)

Randomly choose $d \in \mathbb{Z}_n^*$; $Q = dP$;

Private key: d, Public key: Q.

Randomly choose $k \in \mathbb{Z}_n^*$.

$\hat{R} = kP = (\hat{x}_r, \hat{y}_r)$; $\hat{r} = \hat{x}_r \bmod n$;

Blinding (user)

$h = H(m)$;

$a, b \in [1, n-1]$;

$R = a\hat{R} + bP = (x_r, y_r)$; $r = x_r \bmod n$;

$\hat{h} = ah\hat{r}r^{-1} \bmod n$;

Signature (signer): $\hat{s} = d\hat{r} + k\hat{h} \bmod n$;

Unblinding (user): $s = \hat{s}\hat{r}^{-1}r + bh \bmod n$;

The pair (r, s) is the signature of the message m.

Signature Verification (third party)

Check whether $P = s^{-1}(rQ + H(m)R)$, holds.

27.4.5 An ECDSA-based Blind Signature Scheme

The DSA-based blind signature scheme proposed by Camenisch, can be easily translated to the addition group of the points that belong to an elliptic curve defined over finite fields. This elliptic curve version of the second Camenisch scheme is shown in algorithm 9. This Camenisch ECC version has exactly the same blind factors and number of messages between the entities that the original scheme. The main difference with the original version is that the signature consists in two values: a point R and an integer s. Having the point R as a part of the signature, it is essential to verify the signature, since the algorithm requires both, the point R and its coordinate x, namely $r = x_r \bmod n$. The security proof of this scheme has not been presented yet.

27.4.6 Boldyreva's Blind Signature Scheme

Boldyreva [4] proposed in 2003 a blind signature scheme based on bilinear pairings. Her method, shown in algorithm 10, uses the bilinear property of pairings to solve the decisional Diffie-Hellman problem (DDHP) as the main operation within signature's verification. Notice that, scalar multiplication is the most dominant cryptographic operation during the first three primitives, whereas bilinear pairing computations are only required in signature verification. Thus, the blind signature of a message m is a point S, which is generated with just two messages among the entities. This scheme requires a special hash function H_1 which maps arbitrary strings to an elliptic curve point, and it is strictly necessary to use it, otherwise the scheme would not be able to offer unforgeability. Boldyreva presented the security proof of her scheme and concluded that her proposal is secure against one-more forgery attack.

Algorithm 9 A Nyberg-Rueppel-based Blind Signature Scheme over points

---

Require: The message m to be signed, an elliptic curve E, a base point P of prime order n, a hash function H.

Key Generation (signer)

Randomly choose $d \in \mathbb{Z}_n^*$; $Q = dP$;

Private key: d, Public key: Q

Randomly choose $k \in \mathbb{Z}_n^*$; $\hat{R} = kP$.

Send $\hat{R}$ to User.

Blinding (user)

Randomly choose $\alpha, \beta \in \mathbb{Z}_n^*$;

$h = H(m)$;

$(x_R, y_R) = R = (hP + \alpha P + \beta \hat{R})$;

$r = x_R \bmod \ $;

$\hat{h} = r\beta + \alpha \bmod n$.

Signature (signer)

$\hat{s} = \hat{h}d + k \bmod n$;

Unblinding (user)

$s = \hat{s}\beta + \alpha \bmod n$;

The pair (R, s) is the signature of the message m.

Signature Verification (third party)

$r = x_R \bmod n$

Compute $(x_T, y_T) = T = -sP + rQ + R$,

Check whether $H(m) = x_T \bmod n$ holds.

Algorithm 10 Boldyreva's Blind Signature Scheme [4]

Require: The message m to be signed, a supersingular elliptic curve E, a based point P of order n, a special hash function $H_1 : \{0,1\}^* \longrightarrow \mathbb{G}_1$.

Key Generation (signer)

Find d such that $d \in [2, n-1]$; $Q = dP$;

Private key : d, Public key :Q

Blinding (user)

$M = H_1(m)$.

Randomly find $b \in \mathbb{Z}_n^*$;

$\hat{M} = bM$.

Signature (signer)

$\hat{S} = d\hat{M}$.

Unblinding (user)

$S = b^{-1}\hat{S}$;

S is the signature of the message m.

Signature Verification (third party)

Check whether $\hat{e}(Q, H_1(m)) = \hat{e}(P, S)$, holds.

27.4.7 Zhang and Kim Blind Signature Scheme

Zhang and Kim [39] proposed an identity based blind signature scheme using bilinear pairings. The identity-based public key cryptosystem was proposed by Shamir [35] to simplify the public key exchange without the need of a trusted authority. Bilinear pairings are the major building block required to construct new ID-based cryptographic schemes. The main parameters of this scheme are $\mathbb{G}_1, n, P, P_{pub}, H, H_1$, where $\mathbb{G}_1$ is an additive group generated by the base point P of prime order n. The schemes makes use of two hash functions, namely, $H : \{0,1\}^* \rightarrow \mathbb{Z}/n$ and $H_1 : \{0,1\}^* \rightarrow \mathbb{G}_1$. Additionally, $P_{pub} = sP$ and $s \in [2, n-1]$ are the public and private key of the Public Key Generator (PKG) entity, respectively. As shown in algorithm 11, this method does not require the computation of a field inversion operation. In contrast with the scheme proposed by Boldyreva, Zhang and Kim require a pairing computation along with two special hash functions to blind the message m. Zhang and Kim included a security proof of their scheme and concluded that their scheme is secure against one-more forgery attack.

Algorithm 11 Zhang & Kim blind signature scheme [39]

Require: The message m, an elliptic curve E, a based point P of prime order n, two hash functions H, H_1. Additionally, the PKG public and private keys, $P_{pub} = sP$ and $s \in [2, n-1]$, respectively.

Key Generation (signer)

ID is a Signer's identity

Private key: $S_{ID} = sQ_{ID}$, Public key: $Q_{ID} = H_1(ID)$.

Randomly choose $r \in \mathbb{Z}_n^*$; $R = rP$

Send R to User.

Blinding (user)

Randomly choose $a, b \in \mathbb{Z}_n^*$.

$t = \hat{e}((bQ_{ID} + R + aP), P_{pub})$;

$c = H(m, t) + b \bmod n$.

Signature (signer)

$S = cS_{ID} + rP_{pub}$.

Unblinding (user)

$S' = S + aP_{pub}$;

$c' = c - b$.

The pair (S', c') is the signature of the message m.

Signature Verification (third party)

Check whether $c' = H(m, \hat{e}(S', P)\hat{e}(Q_{ID}, P_{pub})^{-c'})$, holds.

27.4.8 Gao et al. Blind Signature Scheme

Gao et al. [17] developed an ID-based blind signature based on the Bilinear Diffie-Hellman Inversion (BDHI) assumption, previously defined in Section 27.3. The system's parameters are $\mathbb{G}_1, \mathbb{G}_2, n, P, P_{pub}, H_1, H_2$, where $\mathbb{G}_1$ is an additive group generated by the base point P of order n, the hash functions used by the scheme are defined as $H_1, H_2 : \{0,1\}^* \rightarrow \mathbb{G}_1$. Additionally, $P_{pub} = sP$ is a public key and s the private key of the Public Key Generator (PKG). The algorithm 12 uses eight pairing operations. The signature consists of three points. The main disadvantage of this scheme is the computational cost of calculating four bilinear pairing in order to perform the unblindness primitive. The verifier needs to compute four bilinear pairing functions. Hence, we can state that this scheme has the most costly verification primitive. The proposed ID-blind signature meets the blindness requirement, and it is secure against one-more forgery attack.

Algorithm 12 Gao's Blind Signature Scheme [17]

Require: The message m to be signed, an elliptic curve E, a based point P of prime order n, hash functions H_1, H_2.

KeyGen (signer)

ID is a Signer's identity

Private key: $S_{ID} = sQ_{ID}$, Public key: $Q_{ID} = H_1(ID)$.

Blinding (user)

Randomly choose $r_1 \in \mathbb{Z}_n$; $P'_m = r_1 H_2(m)$;

Signature (signer)

Randomly choose $x_{ID} \in \mathbb{Z}_n$;

$A' = x_{ID} P'_m$; $B' = x_{ID}^{-1} S_{ID}$, $C' = x_{ID} P$.

Unblinding (user)

Verify $\hat{e}(A', P) = \hat{e}(P'_m, C')$ and,

$\hat{e}(Q_{ID}, P_{pub}) = \hat{e}(B', C')$

Randomly choose $r_2 \in \mathbb{Z}_n$;

$A = r_2 r_1^{-1} A'$, $B = r_2^{-1} B'$, $C = r_2 C'$;

The triple (A, B, C) is the signature of the message m.

Signature Verification (third party)

Check whether $\hat{e}(A, P) = \hat{e}(H_2(m), C)$, $\hat{e}(Q_{ID}, P_{pub}) = \hat{e}(B, C)$, hold.

27.4.9 Cui et al. Blind Signature Scheme

Cui et al. [11] developed an ID-based blind signature. The system's parameters are $\mathbb{G}_1, \mathbb{G}_2, n, P, P_{pub}, H_1, H_2$, where $\mathbb{G}_1$ is an additive group generated by the base point P of order n; $\mathbb{G}_2$ is an multiplicative group of order n; $P_{pub} = sP$ is a public key and s is the private key of the Public Key Generator (PKG); two hash function are used by the scheme defined as $H_1 : \{0,1\}^* \rightarrow \mathbb{Z}_n^*$, and $H_2 : \{0,1\}^* \times \mathbb{G}_2 \rightarrow \mathbb{Z}_n^*$. In order to perform the unblindness primitive, Algorithm 13 calculates two pairing operations. The three messages transmitted by the entitties are elements in $\mathbb{G}_2$. The main disadvantage of this scheme is that the signature is composed by one element in $\mathbb{G}_2$ and one point in $\mathbb{G}_1$, which implies that this scheme is the one that requires more storage and bandwidth of all the schemes presented before. The proposed ID-blind signature meets the blindness requirement, and it is secure against one-more forgery attack.

Algorithm 13 Cui's Blind Signature Scheme [11]

Require: The message m to be signed, an elliptic curve E, a based point P of prime order n, two hash functions H_1 and H_2.

Key Generation (signer)

ID is a Signer's identity

Private key: $S_{ID} = (s + H_1(ID))^{-1}P$, Public key: $Q_{ID} = H_1(ID)P + P_{pub}$;

Randomly choose $x \in \mathbb{Z}_n^*$;

Compute $r = \hat{e}(P, Q_{ID})^x$.

Blinding (user)

Randomly choose $a, b \in \mathbb{Z}_n^*$

Calculate $r' = r^a \hat{e}(P,P)^b$ and $h' = a^{-1}(H_2(m, r') + b)$.

Signature (signer)

Compute $V = xP + h'S_{ID}$.

Unblinding (user)

Calculate $V' = aV$.

The triple (r', V') is the signature of the message m.

Signature Verification (third party)

Check whether $r' \hat{e}(P,P)^\beta = \hat{e}(V', Q_{ID})$, where $\beta = H_2(m, r')$

27.5 Performance comparison of blind signature schemes

In this Section we give a comparative efficiency analysis as well as a general performance comparison of all the ten blind digital signature schemes studied in the previous Section. We will analyze the computational costs of the schemes when trying to obtain AES 128-bit equivalent security.

27.5.1 General characteristics

As it has been mentioned, a blind signature scheme consists of five primitives, namely: Key Generation, Blinding, Signature, Unblinding and Signature Verification. Each one of those primitives has special characteristics. For example, in the blinding primitive, the user needs to choose one or more blind factors, that will be used to blind the message. For example, Cao et al. scheme requires two blind factors, whereas Chaum uses only one. Table 27.1 shows the security assumptions, number of blind factors and indicates the computational cost of the blind signature primitives for each one of the ten schemes. The number of messages between entities and bit-length of the signatures generated in the schemes are listed in table 27.2, where we can see that the Boldyreva scheme requires only one value for the signature with a length of 807 bits. In contrast, Cui's scheme requires a signature with a length of 5648 bits.

All the schemes where implemented considering a cryptographic strength of approximately 128 security bits. In Table 27.3, we show the minimal specifications that the schemes need to meet in order to offer that security level. Notice that, the schemes defined over integers have a larger bit-length than the ones that are defined over elliptic curves or the ones that make use of bilinear pairings. In order to compare

Table 27.1. General comparison among schemes

Scheme	Underlying hard problem	Security proof	Most costly primitive
Chaum 1982	IFP	Yes	Signing
Cao and Liu 2007	Strong RSA	Yes	Signing
Camenisch et al. 1994 (Modification DSA)	DLP	No	Blinding
Camenisch et al. 1994 (Nyberg-Rueppel)	DLP	No	Blinding
Our scheme 2008	ECDLP	No	Blinding
Jena et al. 2008	ECDLP	No	Blinding
Boldyreva 2003	DDHP	Yes	Verifying
Zhang and Kim 2002	BDHP	Yes	Verifying
Gao et al. 2007	BDHI	Yes	Unblinding
Cui et al. 2007	q-Strong DHP	Yes	Verifying

Table 27.2. General comparison among schemes

Scheme	Message	Blind factors	Signature	Length
Chaum	2	1	s	3072 bits
Cao and Liu	2	2	(e,y)	(513,3072) bits
Camenisch et al. 1	3	2	(r,s)	(256,256) bits
Camenisch et al. 2	3	2	(r,s)	(3072,256) bits
ECDLP scheme	3	2	(R,s)	((283,1),283) bits
Jena et al.	3	2	(R,r,s)	((283,1),283) bits
Boldyreva	2	1	(S)	(509,1) trits ($\sim$ 807 bits)
Zhang and Kim	3	2	(S',c')	((509,1),509) trits ($\sim$ 1614 bits)
Gao et al.	2	1	(A,B,C)	((509,1),(509,1),(509,1)) trits ($\sim$ 2421 bits)
Cui et al.	3	2	(r',V')	(3054,(509,1)) trits ($\sim$ 5648 bits)

the performance of the ten schemes analyzed in this work, in the rest of this Section we classified them according to the abelian group where the schemes operate. The software library was coded in the C language and all the timings were measured on an Intel Core Duo running at 2.0 GHz.

Table 27.3. 128-bit Security Strength

Scheme	Parameter Length
Chaum	3072-bit modulus
Cao and Liu	3072-bit modulus
Camenisch et al. 1	$|p| = 3072, |q| = 256$
Camenisch et al. 2	$|p| = 3072, |q| = 256$
ECDLP scheme	$m = 283$ bits
Jena et al.	$m = 283$ bits
Boldyreva	$m = 509$ trits, $k = 6$
Zhang and Kim	$m = 509$ trits, $k = 6$
Gao et al.	$m = 509$ trits, $k = 6$
Cui et al.	$m = 509$ trits, $k = 6$

27.5.2 Blind Signature Schemes defined over $\mathbb{Z}_n$ Multiplicative Groups

Table 27.4 compares the performance of the two blind signature schemes that are defined over the ring of integers modulo $n = pq$. Chaum and Cao-Liu use integer numbers in the ring $\mathbb{Z}_n$, where n is the product of the two prime numbers p and q. On the other hand, Table 27.5 shows the two schemes proposed by Camenisch that were built as extensions of DSA.

Although all schemes include at least one unblinding primitive with negligible cost, Chaum's scheme has an advantage, which is that this scheme only uses one major cryptographic operation per primitive. Hence, one can say that Chaum's scheme is more efficient than the Cao-Liu scheme.

Camenisch calculates one modular exponentiation in the signing primitive, however, this operation is computed before the user blinds the message. The reason why the signer has to compute this operation at the beginning of the procedure is because he/she generates a session key which must be hidden to the user, as a consequence the exponentiation before the blinding primitive is required.

Table 27.4. Schemes based on module $n = 3072$-bit RSA

	Chaum		Cao-Liu	
	Exp.	Time μs	Exp.	Time μs
Blinding	1	4554.38	2	9108.76
Signing	1	61288.63	1	61288.63
Unblinding	–	–	1	4554.38
Verifying	1	4554.38	2	9108.76
Total	3	70397.39	6	84060.53

Table 27.5. Schemes based on module ($p = 3072, q = 256$)-bit DSA

	Camenisch 1		Camenisch 2	
	Exp.	Time μs	Exp.	Time μs
Blinding	2	37892.2	2	37892.2
Signing	1	18946.1	1	18946.1
Unblinding	–	–	–	–
Verifying	3	56838.3	2	37892.2
Total	6	113676.6	5	94730.5

27.5.3 Blind Signatures Schemes Defined over Elliptic Curve Points

In order to obtain more efficiency when computing elliptic curve's arithmetic, we used two useful techniques: fixed point scalar multiplication algorithms and Shamir's trick for simultaneous scalar multiplication. The first technique takes advantage of the precomputation of multiples of the point P in order to accelerate the scalar multiplication computation. This technique is possible only if the point is known in advance. The second one allows to compute multiple scalar multiplications with more efficiency [19].

Table 27.6 compares two blind signature schemes based on elliptic curves cryptography, using the NIST recommended K-283 Koblitz elliptic curve. As it can be seen, Jena et al. scheme computes the same number of operations than the Camenisch scheme adaptation for ECC. However, the running time in Camenisch is smaller than Jane's scheme. The reason for that is because, we used fixed point and Shamir's trick in the primitives of blindness and verification. In our proposal, we have three scalar multiplications of which two are calculated by fixed point, while in Jena et al., we need to compute three scalar multiplications with three different points of which two of them are unknown in advance. For comparison purposes, we also present in Table 27.7 the implementation for the NIST recommended binary elliptic curve B-283, where the arithmetic is more inefficient than Koblitz curves. We note that for these two schemes, the most expensive operations are the blinding and signature verification primitives, whereas the signing primitive is the most efficient.

Table 27.6. Schemes with NIST recommended K-283 Koblitz Elliptic Curve

	Our Proposal		Jena et al.	
	Scalar mult.	Time μs	Scalar mult.	Time μs
Blindness	3	3124.42	2	2282.01
Signing	1	842.41	1	842.41
Unblindness	–	–	–	–
Verification	2	2282.01	3	4318.8
Total	6	6248.84	6	7443.22

Table 27.7. Schemes with NIST recommended B-283 Binary Elliptic Curve

	Our Proposal		Jena et al.	
	Scalar mult.	Time μs	Scalar mult.	Time μs
Blindness	3	6262.63	2	5411.81
Signing	1	850.82	1	850.82
Unblindness	–	–	–	–
Verification	2	5411.81	3	6434.62
Total	6	12525.26	6	12697.25

27.5.4 Blind Signature Schemes that utilizes Bilinear Pairings

The blind digital schemes of Gao et al., Zhang-Kim, and Cui et al. are identity-based cryptographic schemes. Therefore, they need the Private Key Generator (PKG) authority, which is responsible of generating private/public key pairs for all the users. On the other hand, Boldyreva's scheme requires a Certification Authority (CA). In order to establish the difference between these schemes, we decided to present the results in two separate Tables. First, in Table 27.8, we present the performance obtained by Boldyreva's scheme. Table 27.9 presents the other two schemes analyzed in this Section. It is noticed that the Boldyreva's scheme requires less scalar multiplications and bilinear pairing computations than the rest of the schemes. Furthermore, this scheme has a great advantage: it produces the shortest signature. Among the identity-based blind signature proposals, Gao's scheme has the most inefficient signature algorithm, requiring the computation of eight bilinear pairings, which contrasts with the Zhang and Kim scheme that computes only three pairings.

Table 27.8. Pairing-based schemes with $m = 509$-trits

Boldyreva blind signature

Primitive	Operations	Time μs
Blinding	1 scalar multiplication	3975.1
	1 special hash function	629.3
Signing	1 scalar multiplication	3975.1
Unblinding	1 scalar multiplication	3975.1
Verifying	2 bilinear pairings	19772.0
	1 special hash function	629.3
Total	–	32955.9

Table 27.9. Identity-based blind signature schemes using bilinear pairings with $m = 509$-trits

	Zhang-Kim		Gao et al.		Cui et al.	
	Oper.	Time μs	Oper.	Time μs	Oper.	Time μs
Blinding	2 s. mult.	7950.2	1 s. mult.	3975.1	1 hash	629.3
	1 pairing	9886.0	1 hash	629.3	1 pairing	9886.0
Signing	3 s. mult.	11925.3	3 s. mult.	11925.3	2 s. mult.	7950.2
					1 pairing	9886.0
Unblinding	1 s. mult.	3975.1	4 pairings	39544.0	1 s. mult.	3975.1
			3 s. mult.	11925.3		
Verifying	2 pairings	19772.0	4 pairings	39544.0	1 hash	629.3
			1 hash	629.3	2 pairings	19772.0
Total	–	53508.6	–	108172.3	–	52727.9

27.6 Conclusion

In this chapter, we analyzed ten blind digital signature schemes, and we classify them according to the kind of group where they are operating, namely, multiplicative groups over the integers (DSA, RSA), additive groups over the points in an elliptic curve (ECC) or additive groups mapped to multiplicative groups defined over finite fields (bilinear pairings). We also presented a blind signature scheme that uses elliptic curve cryptography, which is a direct adaptation of one of the DSA Camenisch's blind signature schemes.

The main aim of this chapter is to discuss the fundamental concepts of blind digital signatures. We also present a general feature and performance comparison of the ten selected blind signature schemes. Our comparative analysis studies the number of blind factors, number of messages exchanged among the entities, signature length in bits or trits, hardness of the underlying mathematical problem, whther the scheme was or was not presented with a security proof, timing computational cost employed by each scheme and the bit-length security offered by all schemes. In fact, it is possible to implement a blind signature scheme efficiently, using only one blind factor and one single value for the signature, as in the case of Chaum's and Boldyreva's schemes. However, we consider that the Boldyreva scheme that is defined over elliptic curve points has the advantage of producing a smaller signature than Chaum's signature, while achieving the same degree of security.

The reader should be reminded that the actual timings reported in this Chapter can be greatly improved if the C library takes advantages's of Intel Streaming SIMD Extensions (SSE) making use of the 128-registers. Another promising avenue for obtaining better results is to parallelize the cryptographic computations taking advantage of Intel's multicore architectures.

References

1. Boneh, D., Lynn, B., Shacham, H.: Short Signatures from the Weil Pairing. ASIACRYPT, Vol. 2248, 514-532 (2001)
2. Boneh, D. and Franklin M.: Identity-Based Encryption from the Weil Pairing. CRYPTO, series Lecture Notes in Computer Science, 213-229 (2001)
3. Bleumer, G.: Encyclopedia of Cyrptography and Security. Springer, 37–41, (2005)
4. Boldyreva, A.: Threshold Signatures, Multisignatures and Blind Signatures Based on the Gap-Diffie-Hellman-Group Signature Scheme. PKC '03: Proceedings of the 6th International Workshop on Theory and Practice in Public Key Cryptography, 31–46, (2003)
5. Cha, J., Cheon, J.: An Identity-Based Signature from Gap Diffie-Hellman Groups. PKC '03: Proceedings of the 6th International Workshop on Theory and Practice in Public Key Cryptography. Springer-Verlag, 18–30, (2003)
6. Chaum, D.: Blind Signatures for Untraceable Payments. Advances in Cryptology Proceedings of CRYPTO'82. Springer-Verlag, 199–203 (1983)
7. Chaum, D.: Security without identification: transaction systems to make big brother obsolete. Commun. ACM, Vol. 28, 10, 1030–1044, (1985)

8. Chaum, D.: Blinding for Unanticipated Signatures. Advances in Cryptology - EUROCRYPT '87, LNCS, (1987)
9. Cao, Z., Liu, L.: A Strong RSA Signature Scheme and its Application. Software Engineering, Artificial Intelligence, Networking, and Parallel Distributed Computing. Vol. 01, 111–115, (2007)
10. Camenisch, J., Piveteau J., Stadler, M.: Blind Signatures Based on the Discrete Logarithm Problem. In Advances in Cryptology - EUROCRYPT '94, LNCS, Vol. 950, 428–432, (1994)
11. Cui, W., Yang, X., Yixian, Y., Xinxin, N.: A New Blind Signature and Threshold Blind Signature Scheme from Pairings. Proceedings of the 2007 International Conference on Computational Intelligence and Security Workshops CISW '07, 699–702, (2007)
12. Diffie W. and Hellman M.: New Directions In Cryptography. IEEE Transactions on Information Theory, 22(6): 644–654, November 1976
13. NIST.: Digital Signature Standard (DSS) Federal Information Processing Standards Publications (FIPS PUBS), (1994)
14. G. Frey and H.-G. Rück. A remark concerning m-divisibility and the discrete logarithm in the divisor class group of curves. *Mathematics of Computation*, 62(206): 865–874, April 1994.
15. ElGamal, T.: A Public Key Cryptosystem and a Signature Scheme Based on Discrete Logarithms, IEEE Transactions on Information Theory, Vol. 31, 4, 469–472, (1985)
16. Goldwasser, S., Micali, S., Rivest, R.: A Digital Signature Scheme Secure Against Adaptive Chosen-Message Attacks. SIAM J. Comput., Vol. 17, 2, 281–308, (1988)
17. Gao, W., Wang, X., Wang, G., Li, F.: One-Round ID-Based Blind Signature Scheme without ROS Assumption. Cryptology ePrint Archive, Report 2007/007, (2007)
18. Harn, L.: Cryptanalysis of the Blind Signatures Based on the Discrete Logarithm Problem. Electronics Letters, 31(14):1163, (1995)
19. Hankerson D., Menezes A. and Vanstone S.: Guide to Elliptic Curve Cryptography. Springer, 2004
20. Horster, P., Michels, M., Petersen, H.: Comment: Cryptanalysis of the Blind Signatures Based on the Discrete Logarithm Problem. Electronics Letters, 31(21):1827, (1995)
21. IEEE standards documents. IEEE P1363: Standard Specifications for Public Key Cryptography. Draft Version D18. IEEE, November 2004. http://grouper.ieee.org/groups/1363/.
22. Jena, D., Kumar, S., Acharya, B., Kumar-Jena S.: A Novel ECDLP-Based Blind Signature Scheme. National Conference on Information Security NCISIC'08, 37–40, (2008)
23. Joux A. A one round protocol for tripartite Diffie-Hellman. In W. Bosma, editor, *Algorithmic Number Theory – ANTS IV*, number 1838 in Lecture Notes in Computer Science, pages 385–394. Springer, 2000
24. Koblitz, N. and Menezes A.: A Survey of Public-Key Cryptosystems. SIAM Review, Vol. 46, 4, 599–634, (2004)
25. Koblitz, N. and Menezes A.: The brave new world of bodacious assumptions in cryptography Notices of the AMS, 57 (2010), 357-365
26. Koblitz, N. and Menezes A.: Intractable problems in cryptography Proceedings of $\mathbb{F}_q9$, to appear.
27. Koblitz N.. Elliptic curve cryptosystems. *Mathematics of Computation*, 48(177):203–209, January 1987.
28. Lee, C-C., Yang, W-P., Hwang, M-S.: Untraceable Blind Signature Schemes Based on Discrete Logarithm problem. Fundam. Inform., 55(3-4):307–320, (2003)

29. Miller V. S. Use of elliptic curves in cryptography. In Hugh C. Williams, editor, *Advances in Cryptology - CRYPTO '85*, volume 218 of *Lecture Notes in Computer Science*, pages 417–426. Springer, 1985.

30. Menezes A., Okamoto T., and Vanstone S.A. Reducing elliptic curves logarithms to logarithms in a finite field. *IEEE Transactions on Information Theory*, 39(5):1639–1646, September 1993.

31. Nyberg, K., Rueppel R.: A New Signature Scheme Based on the DSA Giving Message Recovery. ACM Conference on Computer and Communications Security, 58–61, (1993)

32. Pointcheval, D.: Strengthened Security for Blind Signatures. EUROCRYPT'98, 391–405, (1998).

33. Pointcheval, D., Stern, J.: Provably secure blind signature schemes. Advances in Cryptology-ASIACRYPT'96, Vol. 1163, 252–265, (1996).

34. Rivest, R., Shamir, A., Adleman, L.: A Method for Obtaining Digital Signatures and Public-Key Cryptosystems. Communications of the ACM, Vol. 21, 2, 120–126, (1978)

35. Adi Shamir, A.: Identity-Based Cryptosystems and Signature Schemes. Advances in cryptology, CRYPTO'84, 47–53, (1984)

36. Sakai R., Oghishi K., and Kasahara M. Cryptosystems based on pairing. In *2000 Symposium on Cryptography and Information Security (SCIS2000), Okinawa, Japan*, pages 26–28, January 2000.

37. Wu, L-C.: Analysis of Traceability Attack on Camenisch et al.'s Blind Signature Schemes. In Ferng-Ching Lin, Der-Tsai Lee, Bao-Shuh Lin, Shiuhpyng Shieh, and Sushil Jajodia, editors, ASIACCS, ACM, page 366, (2006)

38. Wu, T., Wang, J-R.: Comment: A new Blind Signature Based on the Discrete Logarithm Problem for Untraceability. Appl. Math. Comput., 170(2):999–1005 (2005)

39. Zhang, F., Kim, K.: ID-Based Blind Signature and Ring Signature from Pairings. ASIACRYPT '02, 533–547, (2002)

A Compression Algorithm for Mining Frequent Itemsets

Raudel Hernández León[1,2], Airel Pérez Suárez[1,2], and Claudia Feregrino-Uribe[1]

[1] National Institute for Astrophysics, Optics and Electronics, INAOE, MEXICO
  {raudel,airel,cferegrino}@ccc.inaoep.mx
[2] Advanced Technologies Application Center, CENATAV, CUBA
  {rhernandez,asuarez}@cenatav.co.cu

Summary. In this chapter, we propose a new algorithm for mining frequent itemsets. This algorithm is named *AMFI* (*Algorithm for Mining Frequent Itemsets*), it compresses the data while maintains the necessary semantics for the frequent itemsets mining problem and, for this task, it is more efficient than other algorithms that use traditional compression algorithms. The *AMFI* efficiency is based on a compressed vertical binary representation of the data and on a very fast support count. *AMFI* introduces a novel way to use equivalence classes of itemsets by performing a breadth first search through them and by storing the class prefix support in compressed arrays. We compared our proposal with an implementation that uses the *PackBits* algorithm to compress the data.

Keywords: data mining, frequent patterns, compression algorithms

28.1 Introduction

Mining association rules in transaction datasets has been demonstrated to be useful and technically feasible in several application areas, particularly in retail sales [1, 2, 3, 4], document datasets applications [5], and also in intrusion detection [6]. Association rule mining is usually divided in two steps. The first one consists of finding all itemsets appearing on the dataset (or having a support) above a certain threshold, these itemsets are called *frequent itemsets* (FI). The second one consists on extracting association rules from the FI found in the first step. The FI mining task is very difficult because of its exponential complexity, for that reason the work developed in this chapter will focus on the first step.

The management and storage of large datasets have always been a problem to solve in data mining, particularly in FI mining where the representation of data is decisive to compute the itemset supports.

Conceptually, a dataset is a two-dimensional matrix where the rows represent the transactions and the columns represent the items. This matrix can be implemented in the following four different formats [7]:

- Horizontal item-list (HIL): The dataset is represented as a set of rows (transactions) where each row stores an ordered list of items.
- Horizontal item-vector (HIV): This is similar to HIL, except that each row stores a bit-vector of 1's and 0's to represent the presence or absence of each item.
- Vertical Tid-list (VTL): The dataset is represented as a set of columns (items) where each column stores an ordered list of transactions *ids* in which the item is contained. Note that the VTL format needs exactly the same space as the HIL format.
- Vertical tid-vector (VTV): This is similar to VTL, except that each column stores a bit-vector of 1's and 0's to represent the presence or absence of the item in each transaction. Note that the VTV format needs exactly the same space as the HIV format.

Many algorithms have been proposed using vertical binary representations (*VTV*) in order to improve the process of obtaining FI [7, 8, 9, 10]. The fact that the presence or absence of an item in a transaction can be stored in a bit and that thousands of transactions can be present in a single matrix suggests the use of compression algorithms on the data. To find an efficient algorithm that compresses the data while maintaining the necessary semantic for the FI mining problem is the goal of this work.

We propose an algorithm based on a breadth first search [1] (BFS) through equivalence classes [11] combined with a compressed vertical binary representation of the dataset. This compressed representation, in conjunction with the equivalence class processing, produces a very fast support count and it produces a less expensive representation, specially in large sparse datasets.

In this chapter formal definitions are given and some compression algorithms including *PackBits* method are described following with the explanation of *AMFI* and the pseudo code of the algorithm. Experimental results are discussed and finally the conclusions drawn from this work are presented.

28.2 Preliminaries

Let $I = \{i_1, i_2, \ldots, i_n\}$ be a set of items. Let D be a set of transactions, where each transaction T is a set of items, so that $T \subseteq I$. An itemset X is a subset of I. The support of an itemset X is the number of transactions in D containing to X. If the support of an itemset is greater than or equal to a given support threshold (*minSup*), the itemset is called a *frequent itemset* (FI). The size of an itemset is defined as its cardinality; an itemset containing k items is called a k-itemset.

For example, in Table 28.1, if we have a support threshold equal to three, the FI obtained are: {coke}, {diaper}, {beer} and {diaper, beer}.

---

[1] In graph theory, BFS is a search algorithm that begins at the root node and explores all the neighboring nodes. Then for each of those nearest nodes, it explores their unexplored neighbor nodes, and so on, until it finds the goal.

Table 28.1. Transactional datasets

id	tems
1	coke, milk
2	bread, diaper, beer
3	coke, diaper, beer
4	pan, diaper, beer
5	coke, milk, diaper

In [11], the authors proposed partitioning the itemset space into equivalence classes. These equivalence classes are defined by the equivalence relation "*k-itemsets sharing its first $k-1$ items belong to the same equivalence class EC_k*", therefore all elements of EC_k have size k (see Fig.28.1).

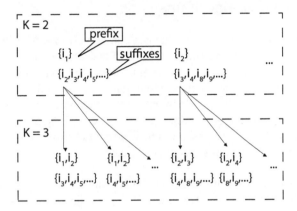

Fig. 28.1. Equivalence classes

Each equivalence class of level $k-1$ generates several equivalence classes at level k.

Most of the algorithms for finding FI are based on the *Apriori* algorithm [2]. To achieve an efficient frequent patterns mining, an anti-monotonic property of frequent itemsets, called the *Apriori* heuristic, was formulated [2]. The basic intuition of this property is that any subset of a frequent itemset must be frequent. *Apriori* is a BFS algorithm, with an *HIL* organization, that iteratively generates two kinds of sets: C_k and L_k. The set L_k contains the frequent k-itemsets. Meanwhile, C_k is the set of candidate k-itemsets, representing a superset of L_k. This process continues until a null set L_k is generated.

The set L_k is obtained by scanning the dataset and determining the support for each candidate k-itemset in C_k. The set C_k is generated from L_{k-1} following the next procedure.

$$C_k = \{c \mid \text{Join}(c, L_{k-1}) \wedge \text{Prune}(c, L_{k-1})\} \tag{28.1}$$

where:

$$\text{Join}(\{i_1, i_2, \ldots, i_{k-1}, i_k\}, L_{k-1}) \equiv$$
$$\langle \{i_1, \ldots, i_{k-1}\} \in L_{k-1} \wedge \{i_1, \ldots, i_k\} \in L_{k-1} \rangle, \qquad (28.2)$$

$$\text{Prune}(c, L_{k-1}) \equiv$$
$$\langle \forall s[(s \subset c \wedge |s| = k-1) \rightarrow s \in L_{k-1}] \rangle, \qquad (28.3)$$

The main problem about the computation of FI is the support counting, that is, computing the number of times that an itemset appears in the dataset. To choose a compression algorithm suitable for data compacting and later on to compute the FI is not an easy task.

28.3 Compression Algorithms

The compression of a transactional dataset can be performed horizontally or vertically. Taking into account the characteristics of the problem, horizontal compaction can be a throwaway due to transactions being defined as sets of items: the sets do not have repeated elements, for which no redundancy is present for a compression algorithm be able to work properly. When data are vertically represented, a transactional identifier list or vector can be obtained per every item. Then, to compute the support for an itemset under this representation, it is required to intersect a transactional identifier list from transactions associated to each item.

Recently there have been in the literature some works about using compression for frequent itemsets. One of them is [12] that proposes to use a dynamic clustering method to compress the frequent itemsets approximately. *Expression similarity* and *support similarity* are defined according to the requirements of the frequent itemset compression. Authors mention that with their method user's do not need to specify the number of frequent itemsets clusters explicitly and user's expectation of compression ratio is incorporated. They claim that their method is feasible and the compression quality is good. Another work is [13] that proposes an algorithm for vertical association rule mining that compresses a vertical dataset using bit vectors. Authors claim that their algorithm needs only a small amount of memory compared to other compression techniques that had been used by many association rule mining algorithms. Yet another scheme that uses compression to find the most interesting frequent itemsets is [14] that uses the principle of *the best set of frequent itemsets is that set that compresses the database best*. Rather than compressing the set of frequent items, they compress the database.

Since the last mid-century, many compression algorithms have been developed [15, 16, 17, 18, 19]. In all lossless compression implementations, there is a trade-off between computational resources and the compression ratio. Often, in both statistical and dictionary-based methods, the best compression ratios are obtained at expenses of long execution times and high memory requirements. Statistical compressors are characterized by consuming higher resources than dictionary based when they are implemented in both software and hardware, however they can achieve compression ratios near to the source entropy. The most demanding task in this kind of algorithms is the implementation of the model to get the statistics of the symbols and to assign the bit string. Perhaps, the most representative statistical method is the proposed by Huffman [15] in 1952. In this algorithm a tree is built according to the frequency of the symbols. All symbols are placed at the leaves of the tree. The Huffman method achieves compression by replacing every symbol by a variable bit string. The bit string assigned to every symbol is determined by visiting every internal node from the root up to the leaf corresponding to the symbol. Initially the bit string is the null string. For every internal node visited, one bit is concatenated to the bit string, 1 or 0, depending on the current visited node whether it is a right or left child of its father. Symbols at longer branches will be assigned larger bit strings.

In the dictionary-based methods, the most time-consuming task is searching for strings in a dictionary, which usually has hundreds of locations. Dictionary-based algorithms are considered simpler to implement than statistical ones but the compression ratio that can be achieved is lower. Another kind of compression algorithms, *ad-hoc*, that were developed in early days of data compression are Run Length Encoding-like (RLE) algorithms [20]. RLE takes advantage of the presence of consecutive identical single symbols often found in data streams. It replaces long runs of repeated symbols with a special token and the length of the run. This method is particularly useful for small alphabets and provides better compression ratios when symbols are correlated with their predecessors.

Selecting a compression method among the existent ones is non-trivial. While one method may be faster, other may achieve better compression ratio and yet another may require less computational resources. Furthermore, due to the nature of mining frequent itemsets, using these algorithms for the transactional identifier list compression, the semantics required for the intersection are lost, bringing as a consequence the necessity of decompressing before intersecting.

After a careful analysis of existing compression algorithms, we concluded that RLE type of algorithms are more suitable for compressing our data. In [21] several variants of RLE algorithm are described, however, [22] describes a variant that in our opinion, can adjust better to the type of data managed here and it may compress with higher compression rates besides allowing intersecting without requiring decompression.

28.3.1 PackBits Algorithm

PackBits algorithm is a fast and simple compression scheme for run-length encoding of data. A *PackBits* data stream consists of packets of one byte of header followed by data. The header is a signed byte; the data can be signed or unsigned.

In the following table, let n be the value of the header byte as a signed integer.

Note that interpreting 0 as positive or negative makes no difference in the output. Runs of two bytes adjacent to non-runs are typically written as literal data. It should also be noticed that it is impossible, from the *PackBits* data, to determine the end of the data stream; i.e., one must know a priori the size of the uncompressed data before reading a *PackBits* data stream to know where it ends.

Table 28.2. Data stream of PackBits

Header byte	Data following the header byte
0 to 127	$(1+n)$ literal bytes of data
-1 to -127	One byte of data, repeated $(1-n)$ times in the decompressed output
-128	No operation (skip and treat next byte as a header byte)

28.4 Characteristic of *AMFI* Algorithm

A new algorithm for FI mining is proposed in this section. The efficiency of this algorithm is based on a compressed vertical binary representation of the data and on a very fast support count.

28.4.1 Storing the Transactions

Let us call *filtered transaction* to the itemset that is obtained by removing no-frequent items from a transaction. The size of the filtered transactions is obviously smaller than the size of the dataset. Based on the anti-monotonic property of FI [2], all FI of a dataset can be computed even if only filtered transactions are available.

The set of filtered transactions can be represented as an m x n matrix, where m is the number of transactions and n is the number of frequent items. We can denote the presence or absence of an item in each transaction by a binary value (1 if it is present, 0 otherwise).

If the maximum number of transactions is not greater than the *CPU* word size w (32 or 64 bits), the dataset can be stored as a simple set of integers. However, a dataset is normally much greater than the *CPU* word size. For this reason, we propose to use an array of integers to store the presence or not of each frequent item along the transactions. It will be explained later on how to extend these integer arrays to a frequent itemset.

Let M be the binary representation of a dataset, with n items and m transactions. Retrieving from M the columns associated to frequent items, we can represent each item j as an integer array I_j where each integer has size w, as follows:

$$I_j = \{W_{1,j}, W_{2,j}, \ldots, W_{q,j}\}, q = \lceil m/w \rceil \tag{28.4}$$

where each integer of the array can be defined as:

$$W_{k,j} = \sum_{r=1}^{w} 2^{w-r} * M_{((k-1)*w+r),j} \tag{28.5}$$

being $M_{i,j}$ the bit value of item j in transaction i, in case of $i > m$ then $M_{i,j} = 0$. This representation of FI allows a fast counting of the itemset support in large datasets.

28.4.2 Reordering of Frequent 1-itemsets

As other authors, in *AMFI*, we have used the heuristic of reordering the frequent 1-itemsets in increasing support order. This will cause a reduction of candidate sets in the next level. This heuristic was first used in MaxMiner [23], and has been used in other methods since then [8, 24, 25, 26, 27, 28, 29]. In the case of our algorithm, reordering frequent 1-itemsets contributes to a faster convergence, as well as saving memory.

28.4.3 AMFI Algorithm

AMFI is a BFS algorithm through equivalence classes with a compressed vertical binary representation. This algorithm iteratively generates a list EC_k. The elements of this list represent the equivalence classes of size k and have the format:

$$\langle \text{Prefix}_{k-1}, \text{IA}_{\text{Prefix}_{k-1}}, \text{Suffixes}_{\text{Prefix}_{k-1}} \rangle, \tag{28.6}$$

where Prefix_{k-1} is the $(k-1)$-itemset that is common to all the itemsets of the equivalence class, $\text{Suffixes}_{\text{Prefix}_{k-1}}$ is the set of all items j which extend to Prefix_{k-1}, where j is lexicographically greater than any item in the prefix, and $\text{IA}_{\text{Prefix}_{k-1}}$ is an array of non null integers that is built with the intersection (using *AND* operation) of the arrays I_j, where j belongs to Prefix_{k-1}. The *IA* arrays store the accumulated supports of the prefix of each equivalence class EC_k. If k is larger then the number of elements of *IA* is lesser because the *AND* operation generates null integers, and null integers are not stored because they do not have influence on the support. The procedure for obtaining *IA* is as follows: Let i and j be two frequent items,

$$IA_{\{i\} \cup \{j\}} =$$
$$\{(W_{k,i} \,\&\, W_{k,j}, k) \,|\, (W_{k,i} \,\&\, W_{k,j}) \neq 0, k \in [1,q]\}, \tag{28.7}$$

similarly, let the frequent itemset X and the frequent item j

$$IA_{X \cup \{j\}} = \{(b \,\&\, W_{k,j}, k) \,|$$
$$(b,k) \in IA_X, (b \,\&\, W_{k,j}) \neq 0, k \in [1,q]\}, \tag{28.8}$$

This representation not only reduces the required memory space to store the integer arrays but also eliminates the *Join* step described in (28.2).

In order to compute the support of an itemset X with an integer-array IA_X, the following expression is used:

$$\text{Support}(IA_X) = \sum_{(b,k) \in IA_X} \text{BitCount}(b) \tag{28.9}$$

where $\text{BitCount}(b)$ is a function that calculates the Hamming Weight of b. The *IA* cardinality is reduced with the increment of the itemsets size due to the downward

---

Algorithm 14 AMFI

Input: Dataset in binary representation
Output: Frequent itemsets
 1: Answer $= \emptyset$
 2: $L = \{$frequent 1-itemsets$\}$
 3: **for all** $i \in L$ **do**
 4:     ECGenAndCount($\langle \{i\}, I_i, \text{Suffixes}_{\{i\}} \rangle, EC_2$)
 5:     $k = 3$
 6:     **while** $EC_{k-1} \neq \emptyset$ **do**
 7:         **for all** $ec \in EC_{k-1}$ **do**
 8:             ECGenAndCount(ec, EC_k)
 9:         **end for**
10:         Answer $=$ Answer $\cup EC_k$
11:         $k = k + 1$
12:     **end while**
13: **end for**
14: **return** Answer

---

closure property. It allows for improvement of the processes (28.8) and (28.9). The *AMFI* algorithm pseudo code is shown in Algorithm 14.

In line 2 of Algorithm 14, the frequent 1-itemsets are calculated and ordered in increasing support order. In line 4, the equivalence classes of each frequent 1-itemset are built. From line 6 to 12, each equivalence class of size 2 is processed using ECGenAndCount function. The ECGenAndCount function takes an equivalence class of length $k - 1$ as argument and generates a set of equivalence classes of length k (see Algorithm 15).

In line 2 of Algorithm 15, all the items i that form the suffix of the input equivalence class (EC_{k-1}) are crossed. In line 3 the prefixes Prefix$'$ of the equivalence classes of level k are built by adding each suffix i to the prefix of EC_{k-1}. In line 4, the IA array associated to each Prefix$'$ is calculated by means of AND operation between the IA of EC_{k-1} and the I_i associated to the item i (28.8). From lines 6 to 13, the suffix items j of EC_{k-1}, lexicographically greater than i, are crossed and the support of the sets Prefix$' \cup j$ is calculated.

28.4.4 Memory Considerations

As mentioned, a data set can be represented in four ways, horizontally (HIL and HIV) and vertically (VTL and VTV). Making a decision between VTL and VTV representations is a non trivial task. Burdick, Calimlim and Gehrke in [8] analyzed these two vertical formats when they proposed the MAFIA algorithm. They showed, experimentally, that the memory efficacy of these representations depends on the density of the dataset. Particularly, on 32 bit machines, the VTL format is guaranteed to be a more expensive representation in terms of space if the support of an item (or itemset) is greater than $1/32$ or about 3% of the number of transactions. In the VTL

Algorithm 15 ECGenAndCount

Input: An equivalence class in $\langle \text{Prefix}, \text{IA}_{\text{Prefix}}, \text{Suffixes}_{\text{Prefix}} \rangle$ format
Output: The equivalence classes set generated

1: Answer $= \emptyset$
2: **for all** $i \in \text{Suffixes}_{\text{Prefix}}$ **do**
3:      $\text{Prefix}' = \text{Prefix} \cup \{i\}$
4:      $\text{IA}_{\text{Prefix}'} = \text{IA}_{\text{Prefix} \cup \{i\}}$
5:      $\text{Suffixes}'_{\text{Prefix}'} = \emptyset$
6:      **for all** $(i' \in \text{Suffixes}_{\text{Prefix}})$ and $(i' > i)$ **do**
7:          **if** $\text{Support}(\text{IA}_{\text{Prefix}' \cup \{i'\}})$ **then**
8:              $\text{Suffixes}'_{\text{Prefix}'} = \text{Suffixes}'_{\text{Prefix}' \cup \{i'\}}$
9:          **end if**
10:     **end for**
11:     **if** $\text{Suffixes}'_{\text{Prefix}'} \neq \emptyset$ **then**
12:         $\text{Answer} = \text{Answer} \cup \{ \langle \text{Prefix}', \text{IA}_{\text{Prefix}'}, \text{Suffixes}'_{\text{Prefix}'} \rangle \}$
13:     **end if**
14: **end for**
15: **return** Answer

---

representation, we need an entire word to represent the presence of an item versus the single bit of the VTV approach.

In the compressed *IA* array of the CA algorithm, a pair of integers for each one of the simple (uncompressed) VTV format is required. As the CA algorithm representation includes pairs of words only for non null integers, the memory overhead of this representation is higher than the simple VTV if the support of an item (or itemset) is greater than $1/2$. Furthermore, considering a dataset of m transactions on 32 bit machines and an item (or itemset) with a support *sup*, a simple VTV requires $m/8$ bytes of memory while the CA algorithm, in its worst case, when the item (or itemset) transactions are sparsely distributed, requires $8 * \min(m * sup, m/32)$ bytes.

So, we can infer that if the support of an item (or itemset) is less than $1/64$ or about 1.5% the CA algorithm has a representation less expensive. Therefore, we can conclude that the compressed integer array representation is better than the simple VTV representations in sparse datasets with *minSup* less than 1.5% or, in the dense datasets, less than 50%.

28.5 Experimental Results

Several experiments were carried out where our proposed algorithm, *AMFI*, was compared against a version that compresses the data using PackBits. Time consumption and memory requirements were considered as measurements of efficiency.

Experiments were developed with two newsitem datasets and two synthetic datasets (Table 28.3).

Table 28.3. Summary of the main datasets characteristics

	Transactions	Items Count	Avg. Length
El Pais	550	14489	173.1
TDT	8169	55532	133.5
Kosarak	990002	41935	8.1
Webdocs	1704140	5266562	175.98

Some of these datasets are sparse, such as *El Pais*, and some, very sparse, such as *Webdocs*. Newsitems datasets were lemmatized using the *Treetagger* program [30], and the *stopwords* were eliminated.

The *Kosarak* dataset was provided by Ferenc Bodon to *FIMI* repository [31] and contains (anonymized) click-stream data of a Hungarian on-line news portal. The *Webdocs* dataset was built from a spidered collection of web html documents and was donated to *FIMI* repository by Claudio Lucchese *et al. TDT* dataset contains news (newsitems) data collected daily from six news sources in American English, over a period of six months (January - June, 1998). The *El Pais* dataset contains 550 news, published at El Pais (Spain) newspaper in June in 1999.

Our tests were performed on a PC with an Intel Core 2 Duo at 1.86 GHz CPU and 1 GB DDR2 RAM. The operating system was Windows XP SP2. We considered CPU+IO time (in seconds) at execution time for all algorithms included in this paper.

In figures 28.2, 28.3, 28.4 and 28.5, a comparison of memory consumption by level is shown, meaning that the comparison is done for frequent 1-itemsets, frequent 2-itemsets and so on until frequent 6-itemset. We plot the values until level 6 in order to not overload the graphics, but the performance is the same.

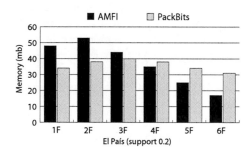

Fig. 28.2. Memory consumption (El Pais dataset)

As it can be seen from the figures, the *AMFI* algorithm requires less memory than the variant of *PackBits* as the size of the FI increases. In level 1, *AMFI* consumes more memory since it stores all the bytes while *PackBits* compresses the bytes with equal value (as the datasets are very sparse, see Table 28.3, many bytes are equal to 0).

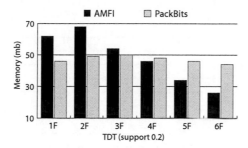

Fig. 28.3. Memory consumption (TDT dataset)

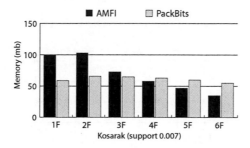

Fig. 28.4. Memory consumption (Kosarak dataset)

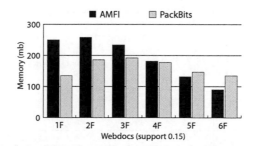

Fig. 28.5. Memory consumption (Webdocs dataset)

As the levels increase, *PackBits* requires always to compress a constant amount of bytes, while *AMFI* will store only the bytes that are different from 0, which diminish fast due to the intersection operations.

In figures 28.6, 28.7, 28.8 and 28.9 a comparison of execution time with different supports is shown. As it can be seen, *AMFI* algorithm not only requires less memory but also is more efficient. This is mainly due to the number of blocks different from 0 decreases with increasing levels (the size of the itemsets) and intersecting only these blocks is faster than iterating two compressed byte flows that intersect all the blocks.

To show how decreasing the number of nonzero blocks in *AMFI* algorithm, we took the two largest datasets and we computed the average value of the number of elements of the IA array in the equivalence classes from EC_2 to EC_9 (Table 28.4). This average value indicates the average number of intersections needed to compute

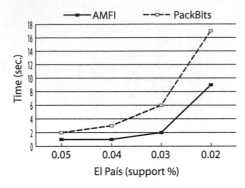

Fig. 28.6. Time consumption (El Pais dataset)

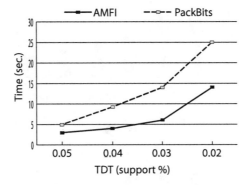

Fig. 28.7. Time consumption (TDT dataset)

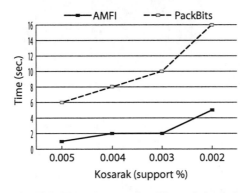

Fig. 28.8. Time consumption (Kosarak dataset)

the support of an FI in each equivalence class. It is important to highlight that the number of intersections is large only in the processing of EC_2 and, the average of zero blocks generated in EC_k is equal to the difference between the average of intersections needed for processing EC_{k+1} and EC_k respectively.

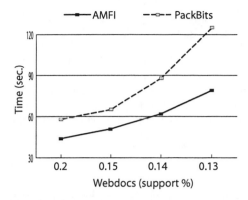

Fig. 28.9. Time consumption (Webdocs dataset)

The values shown in table 28.4 corroborate the results obtained in the conducted experiments.

Table 28.4. Average value of the number of elements of IA from EC_2 to EC_9

Datasets	EC_2	EC_3	EC_4	EC_5	EC_6	EC_7	EC_8	EC_9
Kosarak	30938	3703	2981	2420	2114	2029	2000	1980
Webdocs	53255	4208	3927	3221	2415	2212	2183	1813

28.6 Conclusions

In this chapter we have presented a compressed vertical binary approach for mining FI. In order to reach a fast support computing, our algorithm uses a BFS through equivalence classes storing the class prefix support in compressed arrays. The experimental results showed that *AMFI* algorithm achieves better performance than *PackBits* as much in consumption of memory as in run time. It can be concluded that although existing compression methods are good, they are not always suitable for certain problems due to when compressing the required semantic is lost.

References

1. Agrawal, R., Imielinski, T., Swami, A.: Mining association rules between sets of items in large databases. In Proceedings of the ACM SIGMOD International Conference on Management of Data, Washington, D.C. (1993) 207–216
2. Agrawal, R., Srikant, R.: Fast algorithms for mining association rules. In Proceedings of the 20*th* International Conference on Very Large Data Bases, VLDB'94, Santiago de Chile, Chile (1994) 487–499

3. Savasere, A., Omiecinski, E., Navathe, S.: An efficient algorithm for mining association rules in large databases. In Technical Report GIT-CC-95-04, Institute of Technology, Atlanta, USA (1995)

4. Brin, S., Motwani, R., Ullman, J.D., Tsur, S.: Dynamic itemset counting and implication rules for market basket data. In Proceedings of the ACM SIGMOD International Conference on Management of Data, Tucson, Arizona, USA (1997)

5. Feldman, R., Dagan, I.: Kdt-knowledge discovery in texts. In Proceedings of the First International Conference on Knowledge Discovery (KDD) (1995) 112–117

6. Silvestri, C., Orlando, S.: Approximate mining of frequent patterns on streams. Intell. Data Anal., Vol. 11, No. 1 (2007) 49–73

7. Shenoy, P., Haritsa, J., Sudarshan, S., Bhalotia, G., Bawa, M., Shah, D.: Turbo-charging vertical mining of large databases. In Proceedings of the ACM SIGMOD International Conference on Management of Data, Dallas, USA (2000)

8. Burdick, D., Calimlim, M., Gehrke, J.: Mafia: A maximal frequent itemset algorithm for transactional databases. In Proceedings of the International Conference on Data Engineering (ICDE), Heidelberg, Germany (2001)

9. Grahne, G., Zhu, J.: Fast algorithms for frequent itemset mining using fp-trees. IEEE Transactions on Knowledge and Data Engineering, Vol. 17, No. 10 (2005) 1347–1362

10. Han, J., Pei, J., Yin, Y.: Mining frequent patterns without candidate generation. In Proceedings ACM-SIGMOD International Conference on Management of Data, New York, NY, USA (2000)

11. Zaki, M.J., Parthasarathy, S., Ogihara, M., Li, W.: New algorithms for fast discovery of association rules. In Proceedings of the 3rd International Conference on KDD and Data Mining, EU (1997)

12. Yan, H., Sang, Y.: Approximate frequent itemsets compression using dynamic clustering method. IEEE Conference on Cybernetics and Intelligent Systems (2008) 1061–1066

13. Mafruz Zaman Ashrafi, D.T., Smith, K.: An efficient compression technique for frequent itemset generation in association rule mining. Advances in Knowledge Discovery and Data Mining, Lecture Notes in Computer Science, Springer Berlin / Heidelberg, Vol. 3518 (2005) 125–135

14. Arno Siebes, Jilles Vreeken, M.v.l.: Item sets that compress. Proceedings of the SDM'06 (2006) 393–404

15. Huffman, D.: A method for the construction of minimum redundancy codes. In Proceedings of the IRE 40(9) (1952) 1098–1101

16. Ziv, J., Lempel, A.: A universal algorithm for sequential data compression. IEEE Transactions on Information Theory IT-23(3) (1977) 337–343

17. Ziv, J., Lempel, A.: Compression of individual sequences via variable-rate coding. IEEE Transactions on Information Theory IT-24(5) (1978) 530–536

18. Fiala, E.R., Greene, D.H.: Data compression with finite windows. Communications of the ACM 32(4) (1989) 490–505

19. Phillips, D.: Lzw data compression. The Computer Application Journal Circuit Cellar Inc., 27 (1992) 36–48

20. W., G.S.: Run-length encodings. IEEE Transactions on Information Theory, 12 (1966) 399–401

21. Salomon, D.: Data compression: The complete reference. 3rd Edition, Published by Springer. ISBN 0-387-40697-2. LCCN QA76.9 D33S25, 899 pages (2004)

22. http://www.fileformat.info/format/tiff/corion-packbits.htm

23. Bayardo, R.J.: Efficiently mining long patterns from databases. In ACM SIGMOD Conf. on Management of Data (1998) 85–93

24. Agrawal, R., Aggarwal, C., Prasad, V.: Depth first generation of long patterns. In 7th Int'l Conference on Knowledge Discovery and Data Mining (2000) 108–118
25. Gouda, K., Zaki, M.J.: Genmax: An efficient algorithm for mining maximal frequent itemsets. Data Mining and Knowledge Discovery, 11 (2005) 1–20
26. Zaki, M.J., Hsiao, C.J.: Charm: An efficient algorithm for closed itemset mining. In 2nd SIAM International Conference on Data Mining (2002) 457–473
27. Calders, T., Dexters, N., Goethals, B.: Mining frequent itemsets in a stream. Proceedings of the IEEE International Conference on Data Mining (2007) 83–92
28. Calders, T., Dexters, N., Goethals, B.: Mining frequent items in a stream using flexible windows. Intelligent Data Analysis, Vol. 12, No. 3 (2008)
29. Kalpana, B., Nadarajan, R.: Incorporating heuristics for efficient search space pruning in frequent itemset mining strategies. CURRENT SCIENCE 94 (2008) 97–101
30. Schmid, H.: Probabilistic part-of-speech tagging using decision trees. In: International Conference on New Methods in Language Processing, (Software in: www.ims.uni-stuttgart.de/ftp/pub/corpora/tree-tagger1.ps.gz), Manchester, UK (1994)
31. FIMI-Frequent Itemset Mining Implementations Repository, (Software developed by Ferec Bodon, URL: http://fimi.cs.helsinki.fi/src)